WATCH THE NEW
MIS VIDEOS ONLINE

Want to see how the information systems concepts you're learning apply to the real world? Use the password card in this textbook to check out relevant MIS video clips online via **www.course.com/mis/stair**. Go to *Fundamentals of Information Systems, Third Edition*, then click on the video clip that corresponds to the chapter you're studying and discover more today.

If your book does not contain a password card, please ask your professor to contact their Thomson Course Technology representative.

Clip	Chapter
Go Inside Krispy Kreme	Chapter 1: An Introduction to Information Systems in Organizations
Getting Started	Chapter 2: Hardware and Software
Predicting Huge Surf	Chapter 3: Organizing Data and Information
Online Storage	Chapter 4: Telecommunications, the Internet, Intranets, and Extranets
Find the Best Deals Online	Chapter 5: Electronic Commerce and Transaction Processing Systems
Army's Virtual World	Chapter 6: Information and Decision Support Systems
Predicting the Future of AI— Marvin Minsky	Chapter 7: Specialized Information Systems: Artificial Intelligence, Expert Systems, Virtual Reality, and Other Systems
Design Your Own Video Game	Chapter 8: Systems Development
Controversial Digital Bouncers	Chapter 9: Security, Privacy, and Ethical Issues in Information Systems and the Internet

LOOK FOR THESE OTHER POPULAR THOMSON COURSE TECHNOLOGY
MIS TITLES

Database Systems: Design, Implementation, and Management, Sixth Edition

by Peter Rob and Carlos Coronel
ISBN: 0-619-21323-X

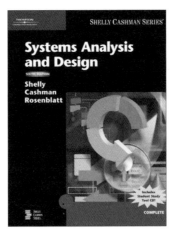

Systems Analysis and Design, Sixth Edition

by Gary B. Shelly, Thomas J. Cashman, and Harry J. Rosenblatt
ISBN: 0-619-25510-2

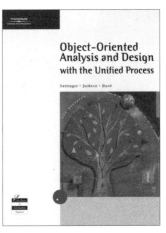

Object-Oriented Analysis and Design with the Unified Process

by John W. Satzinger, Robert B. Jackson, and Stephen D. Burd
ISBN: 0-619-21643-3

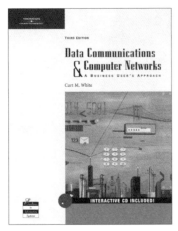

Data Communications and Computer Networks: A Business User's Approach, Third Edition

by Curtis M. White
ISBN: 0-619-16035-7

Information Technology Project Management, Fourth Edition

by Kathy Schwalbe
ISBN: 0-619-21526-7

Principles of Information Security, Second Edition

by Michael Whitman and Herbert Mattord
ISBN: 0-619-21625-5

VIEW OUR ENTIRE COLLECTION OF PRODUCTS ONLINE AT **WWW.COURSE.COM/MIS.**

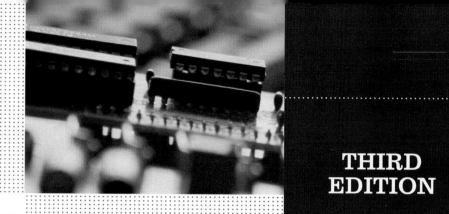

Fundamentals of Information Systems

Third Edition

Ralph M. Stair
Florida State University

George W. Reynolds
The University of Cincinnati

THOMSON

COURSE TECHNOLOGY

Australia · Canada · Mexico · Singapore · Spain · United Kingdom · United States

THOMSON

COURSE TECHNOLOGY

Fundamentals of Information Systems, Third Edition
by Ralph M. Stair and George W. Reynolds

Senior Vice President, Publisher:
Kristen Duerr

Executive Editor:
Mac Mendelsohn

Senior Acquisitions Editor:
Maureen Martin

Senior Product Manager:
Eunice Yeates-Fogle

Development Editor:
Karen Hill

Associate Product Manager:
Mirella Misiaszek

Editorial Assistant:
Jennifer Smith

Production Editor:
Kelly Robinson

Senior Marketing Manager:
Karen Seitz

Text Design:
Jennifer McCall

Cover Designer:
Lisa Rickenbach

Composition House:
Digital Publishing Solutions

Photo Researcher:
Abby Reip

Copy Editor:
Karen Annett

Proofreader:
Sue Forsyth

Indexer:
Rich Carlson

For Lila and Leslie
—Ralph M. Stair

To my grandchildren: Michael, Jacob, Jared, Fievel, and Aubrey Danielle
—George W. Reynolds

CONTENTS

Preface xv

PART 1   Information Systems in Perspective 1

Chapter 1   An Introduction to Information Systems in Organizations 2
Boehringer Ingelheim, GMBH, Germany 3

Information Concepts 4

Data Versus Information 4

The Characteristics of Valuable Information 6

The Value of Information 6

What Is an Information System? 6

Input, Processing, Output, Feedback 7

Manual and Computerized Information Systems 9

Computer-Based Information Systems 9

Business Information Systems 12

Electronic and Mobile Commerce 12

Ethical and Societal Issues: Phishing for Visa Card Customers 14

Transaction Processing Systems and Enterprise Resource Planning 15

Information and Decision Support Systems 16

Specialized Business Information Systems: Artificial Intelligence, Expert
Systems, and Virtual Reality 18

Systems Development 20

Organizations and Information Systems 21

Organizational Culture and Change 22

Technology Diffusion, Infusion, and Acceptance 23

Organizations in a Global Society 23

Competitive Advantage 24

Factors That Lead Firms to Seek Competitive Advantage 24

Strategic Planning for Competitive Advantage 25

Performance-Based Information Systems 27

Productivity 27

Information Systems @ Work: OneBeacon Focuses on Information Systems 29

Return on Investment and the Value of Information Systems 30

Careers in Information Systems 31

Roles, Functions, and Careers in the IS Department 32

Typical IS Titles and Functions 34

Other IS Careers 36

Case One: The Queen Mary 2 and partner 41

Case Two: MyFamily Comforts Its Members 42

PART 2 **Technology 45**

Chapter 2   **Hardware and Software   46**
Porsche AG, Germany   47

Hardware Components   49

Processing and Memory Devices: Power, Speed, and Capacity   49

Processing Characteristics and Functions   50

Memory Characteristics and Functions   50

Multiprocessing   51

Secondary Storage and Input and Output Devices   52

Secondary Storage Access Methods   52

Secondary Storage Devices   53

Input Devices   56

Output Devices   58

Information Systems @ Work: Banks Weigh Move to Improved Check-Clearing Process   59

Computer System Types   61

Handheld Computers   61

Portable Computers   62

Thin Client   62

Desktop Computers   63

Workstations   63

Servers   63

Mainframe Computers   63

Supercomputers   64

Overview of Software   65

Supporting Individual, Group, and Organizational Goals   65

Systems Software   66

Operating Systems   66

Current Operating Systems   69

Workgroup Operating Systems   71

Ethical and Societal Issues: Weighing the Benefits of Open Source   72

Enterprise Operating Systems   73

Operating Systems for Small Computers and Special-Purpose Devices   74

Application Software   75

Types and Functions of Application Software   75

Proprietary Application Software   75

Personal Application Software   76

Workgroup Application Software   82

Enterprise Application Software   82

Application Software for Information, Decision Support, and Specialized Purposes   83

Programming Languages   84

Software Issues and Trends   85

Software Bugs   85

Copyrights and Licenses   85

Global Software Support 85

Case One: PACS at North Bronx Healthcare Network 91

Case Two: UPS Provides Shipping Management Software to Customers 92

Chapter 3   Organizing Data and Information 94
               DDB Worldwide, United States 95

Data Management 97

The Hierarchy of Data 97

Data Entities, Attributes, and Keys 97

The Traditional Approach Versus the Database Approach 99

Data Modeling and the Relational Database Model 102

Data Modeling 102

The Relational Database Model 103

Database Management Systems (DBMS) 106

Overview of Database Types 106

Providing a User View 107

Creating and Modifying the Database 107

Storing and Retrieving Data 109

Manipulating Data and Generating Reports 110

Database Administration 112

Popular Database Management Systems 112

Special-Purpose Database Systems 113

Selecting a Database Management System 113

Using Databases with Other Software 115

Database Applications 115

Linking the Company Database to the Internet 115

**Information Systems @ Work: Web-Based DBMS Empowers Cruise Line
     Personnel 116**

Data Warehouses, Data Marts, and Data Mining 117

Ethical and Societal Issues: The Growing Cost of Data-Related Regulations 120

Business Intelligence 121

Distributed Databases 122

Online Analytical Processing (OLAP) 123

Object-Oriented and Object-Relational Database Management
     Systems 124

Visual, Audio, and Other Database Systems 125

Case One: Brazilian Grocer Gets Personal with Customers 131

Case Two: DBMS Upgrade Faces Employee Opposition 132

Chapter 4   Telecommunications, the Internet, Intranets, and Extranets 134
               The Barilla Group, Italy 135

An Overview of Telecommunications and Networks 137

Telecommunications 138

Networks 142

Use and Functioning of the Internet 148

How the Internet Works 149

Accessing the Internet 150

Ethical and Societal Issues: Entrepreneurs Work to Lessen the Global Divide 151

    Internet Service Providers 152

The World Wide Web 153

    Web Browsers 155

    Search Engines 155

    Web Programming Languages 156

    Developing Web Content 156

    Web Services 156

    Business Uses of the Web 157

Internet and Telecommunications Services 157

    E-Mail and Instant Messaging 157

    Internet Cell Phones and Handheld Computers 159

    Career Information and Job Searching 159

    Web Log (Blog) 159

Information Systems @ Work: Inscene Embassy Changes Marketing Focus with Blogs 160

    Chat Rooms 161

    Internet Phone and Videoconferencing Services 161

    Content Streaming 162

    Shopping on the Web 162

    Web Auctions 163

    Music, Radio, and Video on the Internet 163

    Other Internet Services and Applications 163

Intranets and Extranets 164

Net Issues 166

    Management Issues 166

    Service and Speed Issues 166

    Privacy, Fraud, Security 166

Case One: Eastman Chemical Revamps with Web Services 174

Case Two: Best Western: First to Provide Free Internet 175

Part 2 World Views Case: Australian Film Studio Benefits from Broadband Networks—and Location 177

Part 2 World Views Case: Virtual Learning Environment Provides Instruction Flexibility 179

PART 3 **Business Information Systems 183**

Chapter 5 **Electronic Commerce and Transaction Processing Systems 184**

Sprint, United States; And Sony, Japan 185

An Introduction to Electronic Commerce 186

    The E-Commerce Supply Chain 187

    Business-to-Business (B2B) E-Commerce 188

    Business-to-Consumer (B2C) E-Commerce 189

    Consumer-to-Consumer (C2C) E-Commerce 189

    Mobile Commerce 190

    Global E-Commerce and M-Commerce 191

E-Commerce Applications 191

Ethical and Societal Issues: Canadian Prescription Drug Web Sites 192

Retail and Wholesale 193

Manufacturing 193

Marketing 195

Investment and Finance 196

Auctions 198

E-Commerce, Technology, Infrastructure, and Development 199

Hardware 200

Software 200

Electronic Payment Systems 201

An Overview of Transaction Processing Systems 204

Traditional Transaction Processing Methods and Objectives 205

Transaction Processing Activities 206

Basic TPS Applications 208

TPS Control and Management Issues 209

Business Continuity Planning 209

Transaction Processing System Audit 211

International Issues 211

Enterprise Resource Planning 212

An Overview of Enterprise Resource Planning 212

Advantages and Disadvantages of ERP 214

Information Systems @ Work: ERP Consolidation Opens Doors for College Lender 215

Case One: RealEstate.com: Buying and Selling Homes Online 222

Case Two: Point and Pay in Japan with KDDI 223

Chapter 6   Information and Decision Support Systems 226
Kiku-Masamune Sake Brewing Co., Ltd, Japan 227

Decision Making and Problem Solving 229

Decision Making as a Component of Problem Solving 229

Programmed Versus Nonprogrammed Decisions 230

Optimization, Satisficing, and Heuristic Approaches 231

An Overview of Management Information Systems 232

Management Information Systems in Perspective 232

Inputs to a Management Information System 232

Outputs of a Management Information System 233

Functional Aspects of the MIS 236

Financial Management Information Systems 236

Manufacturing Management Information Systems 238

Marketing Management Information Systems 239

Information Systems @ Work: A "TaylorMade" Information and Decision Support System 240

Human Resource Management Information Systems 241

Other Management Information Systems 242

An Overview of Decision Support Systems 243

Capabilities of a Decision Support System 243

A Comparison of DSS and MIS   245

Components of a Decision Support System   245

The Database   245

The Model Base   246

The Dialogue Manager   247

Group Support Systems   247

Characteristics of a GSS That Enhance Decision Making   247

GSS Software   249

GSS Alternatives   250

Executive Support Systems   252

Executive Support Systems in Perspective   252

Ethical and Societal Issues: CEOs "Called to Action" Regarding Information Security   253

Capabilities of Executive Support Systems   254

Case One: Sensory Systems Provide Better-Tasting Products for Kraft   260

Case Two: Sanoma Magazines Follows Key Performance Indicators to Success   261

Chapter 7   Specialized Information Systems: Artificial Intelligence, Expert Systems, Virtual Reality, and Other Systems   264

Amazon.Com, United States   265

An Overview of Artificial Intelligence   266

Artificial Intelligence in Perspective   267

The Nature of Intelligence   267

The Difference Between Natural and Artificial Intelligence   269

The Major Branches of Artificial Intelligence   270

Ethical and Societal Issues: Biologically Inspired Algorithms Fight Terrorists and Guide Businesses   273

An Overview of Expert Systems   275

Characteristics and Limitations of an Expert System   276

When to Use Expert Systems   277

Components of Expert Systems   277

Expert Systems Development   281

Applications of Expert Systems and Artificial Intelligence   285

Information Systems @ Work: Moving Data—and Freight—Efficiently   286

Virtual Reality   287

Interface Devices   287

Forms of Virtual Reality   289

Virtual Reality Applications   289

Other Specialized Systems   290

Case One: French Burgundy Wines: The Sweet Smell of Success   296

Case Two: Expert System Provides Safety in Nuclear Power Plants   297

Part 3 World Views Case: Kulula.com: The Trials and Tribulations of a South African Online Airline   299

PART 4   Systems Development and Social Issues 303

Chapter 8   Systems Development 304
BMW, Germany 305

An Overview of Systems Development **307**
Participants in Systems Development 307
Information Systems Planning and Aligning Corporate and IS Goals 308

Information Systems @ Work: Fakta Designs Financial Management System to Support Corporate Goals **309**
Importance of IS Planning 310

Systems Development Life Cycles **310**
The Traditional Systems Development Life Cycle 310
Prototyping 311
Rapid Application Development, Agile Development, Joint Application Development, and Other Systems Development Approaches 311
The End-User Systems Development Life Cycle 313
Outsourcing and On Demand Computing 313
Use of Computer-Aided Software Engineering (CASE) Tools 314
Object-Oriented Systems Development 314

Systems Investigation **316**
Initiating Systems Investigation 316
Feasibility Analysis 316
Object-Oriented Systems Investigation 317
The Systems Investigation Report 317

Systems Analysis **317**
Data Collection 318
Data Analysis 319
Requirements Analysis 320
Object-Oriented Systems Analysis 321
The Systems Analysis Report 322

Systems Design **323**
Object-Oriented Design 323
Generating Systems Design Alternatives 324
Evaluating and Selecting a System Design 325
The Design Report 326

Systems Implementation **326**
Acquiring Hardware from an IS Vendor 327
Acquiring Software: Make or Buy? 328
Acquiring Database and Telecommunications Systems 328
User Preparation 329
IS Personnel: Hiring and Training 329
Site Preparation 329
Data Preparation 330
Installation 330
Testing 330
Start-Up 330

User Acceptance   331

Systems Operation and Maintenance   332

Systems Review   332

Ethical and Societal Issues: Finding Trust in Computer Systems   334

Case One: PepsiCo Implements New Procurement System to Minimize Costs   340

Case Two: Segway Stays Light and Nimble with Outsourced Systems   341

Chapter 9   **Security, Privacy, and Ethical Issues in Information Sytems and the Internet   344**
Computer Assisted Passenger Prescreening System, United States   345

Computer Waste and Mistakes   347

Computer Waste   347

Computer-Related Mistakes   347

Ethical and Societal Issues: CAN-SPAM: Deterrent or Accelerant?   348

Preventing Computer-Related Waste and Mistakes   349

Establishing Policies and Procedures   349

Implementing Policies and Procedures   350

Monitoring Policies and Procedures   351

Reviewing Policies and Procedures   351

Computer Crime   351

The Computer as a Tool to Commit Crime   353

Cyberterrorism   353

Identity Theft   353

The Computer as the Object of Crime   354

Illegal Access and Use   354

Data Alteration and Destruction   354

Using Antivirus Programs   357

Information and Equipment Theft   358

Software and Internet Software Piracy   359

Computer-Related Scams   359

International Computer Crime   360

Preventing Computer-Related Crime   361

Crime Prevention by State and Federal Agencies   361

Crime Prevention by Corporations   361

Using Intrusion Detection Software   364

Using Managed Security Service Providers (MSSPs)   364

Internet Laws for Libel and Protection of Decency   365

Preventing Crime on the Internet   366

Privacy Issues   367

Privacy and the Federal Government   367

Information Systems @ Work: UK BioBank Raises Privacy Issues   368

Privacy at Work   369

E-Mail Privacy   370

Privacy and the Internet   370

Fairness in Information Use   371

Federal Privacy Laws and Regulations   371

State Privacy Laws and Regulations   371

Corporate Privacy Policies   372

Individual Efforts to Protect Privacy   373

Ethical Issues in Information Systems   373

The AITP Code of Ethics   374

The ACM Code of Professional Conduct   374

Case One: Working to Reduce the Number of Software Vulnerabilities   379

Case Two: Beware Spyware!   380

Part 4 World Views Case: Brandon Trust Develops MIS Capability for Improved Operations and Services   383

Part 4 World Views Case: Strategic Enterprise Management at International Manufacturing Corporation (IMC)   385

Part 4 World Views Case: Efforts to Build E-Government and an Information Society in Hungary   387

Glossary   389

Index   400

We are proud to publish the third edition of *Fundamentals of Information Systems*. This new edition builds on the success of the previous editions in meeting the need for a concise introductory information systems text. We have listened to feedback from the previous edition's adopters and manuscript reviewers and incorporated many suggestions to refine this new edition. We hope you are pleased with the results.

Like the previous editions, the overall goal of the third edition was to develop an outstanding text that follows the pedagogy and approach of our flagship text, *Principles of Information Systems*, with less detail and content. The approach in the development of *Fundamentals of Information Systems* was to take the best material from *Principles of Information Systems* and condense it into a text containing nine chapters. So, our newest edition of *Principles of Information Systems* was the foundation from which we built this new edition of *Fundamentals of Information Systems*.

We have always advocated that education in information systems is critical for employment in almost any field. Today, information systems are used for business processes from communications to order processing to number crunching and in business functions ranging from marketing to human resources to accounting and finance. Chances are, regardless of your future occupation, you need to understand what information systems can and cannot do and be able to use them to help you accomplish your work. You will be expected to suggest new uses of information systems and participate in the design of solutions to business problems employing information systems. You will be challenged to identify and evaluate IS options. To be successful, you must be able to view information systems from the perspective of business and organizational needs. For your solutions to be accepted, you must identify and address their impact on fellow workers. For these reasons, a course in information systems is essential for students in today's high-tech world.

Fundamentals of Information Systems, Third Edition continues the tradition and approach of the previous editions of this text and *Principles of Information Systems*. Our primary objective is to develop the best IS text and accompanying materials for the first information technology course required of all business students. Using surveys, questionnaires, focus groups, and feedback that we have received from adopters and others who teach in the field, we have been able to develop the highest-quality teaching materials available.

Fundamentals of Information Systems stands proudly at the beginning of the IS curriculum, offering the basic IS concepts that every business student must learn to be successful. This text has been written specifically for the first course in the IS curriculum, and it discusses computer and IS concepts in a business context with a strong managerial emphasis.

APPROACH OF THE TEXT

The overall vision, framework, and pedagogy that made *Principles of Information Systems* so popular have been retained in this text. In particular, this book offers the traditional coverage of computer concepts, but it places the material within the context of business and information systems. Placing IS concepts in a business context has always set the text apart from general computer books and makes it appealing not only to MIS majors but also to students from other courses of study. It approaches information systems from a general management perspective. The text isn't overly technical but rather deals with the role that information systems play in an organization and the general concepts a manager needs to be aware of to be successful. The text stresses IS principles, which are brought together and presented in a way that is both understandable and relevant. In addition, this book offers an overview of the entire IS discipline, as well as solid preparation for further study in advanced IS courses.

IS PRINCIPLES FIRST, WHERE THEY BELONG

Exposing students to fundamental IS principles provides a service to students who do not later return to the discipline for advanced courses. Since most functional areas in business rely on information systems, an understanding of IS principles helps students in other course work. In addition, introducing students to the principles of information systems helps future functional area managers avoid mishaps that often result in unfortunate and sometimes costly consequences. Furthermore, presenting IS principles at the introductory level creates interest among general business students who will later choose information systems as a field of concentration.

AUTHOR TEAM

Ralph Stair and George Reynolds have teamed up again for the third edition. Together, they have more than 60 years of academic and industrial experience. Ralph Stair brings years of writing, teaching, and academic experience to this text. He has written more than 22 books and a large number of articles while at Florida State University. George Reynolds brings a wealth of computer and industrial experience to the project, with more than 30 years of experience working in government, institutional, and commercial IS organizations. He has also authored 14 texts and is an assistant professor at the University of Cincinnati, where he teaches the introductory IS course. The Stair and Reynolds team brings a solid conceptual foundation and practical IS experience to students.

GOALS OF THIS TEXT

Fundamentals of Information Systems has four main goals:

1. To present a core of IS principles with which every business student should be familiar
2. To offer a survey of the IS discipline that will enable all business students to understand the relationship of advanced courses to the curriculum as a whole
3. To present the changing role of the IS professional
4. To show the value of the discipline as an attractive field of specialization

Because *Fundamentals of Information Systems* is written for all business majors, we believe it is important not only to present a realistic perspective of information systems in business but also to provide students with the skills they can use to be effective leaders in their companies.

IS Principles

Fundamentals of Information Systems, Third Edition, provides an essential core of guiding IS principles to use as students face the career challenges ahead. Think of principles as basic truths, rules, or assumptions that remain constant regardless of the situation. As such, they provide strong guidance when facing tough decisions. A set of IS principles is highlighted at the beginning of each chapter. Then these principles are applied to solve real-world problems from the opening vignettes to the end-of-chapter material. The ultimate goal of *Fundamentals of Information Systems* is to develop effective, thinking employees by instilling in them principles to help guide their decision making and actions.

Survey of the IS Discipline

This text not only offers the traditional coverage of computer concepts but also stresses the broad framework to provide students with solid grounding in business uses of technology. In addition to serving general business students, this book offers an overview of the entire IS discipline and solidly prepares future IS professionals for advanced IS courses and their careers in the rapidly changing IS discipline.

Changing Role of the IS Professional

As business and the IS discipline have changed, so too has the role of the IS professional. Once considered a technical specialist, today the IS professional operates as an internal consultant to all functional areas of the organization, being knowledgeable about their needs and competent in bringing the power of information systems to bear throughout the organization. The IS professional views issues through a global perspective that encompasses the entire organization and the broader industry and business environment in which it operates, including the entire interconnected network of suppliers, customers, competitors, regulatory agencies, and other entities—no matter where they are located.

The IS professional assumes the critical responsibility of determining the organization's approach to both overall cost and quality performance and therefore plays an important role in the continued survival of the organization. This new duality in the role of the IS employee—a professional who exercises a specialist's skills with a generalist's perspective—is reflected throughout the book.

IS as a Field for Further Study

Despite the downturn in the economy at the start of the 21st century, especially in technology-related sectors, the outlook for computer and IS managers is bright. In fact, employment of computer and IS managers is expected to grow much faster than the average occupation through the year 2012. Technological advancements are boosting the employment of computer-related workers; in turn, this uptick in hiring will create demand for managers to direct these workers. In addition, job openings will result from the need to replace managers who retire or move into other occupations.

A career in information systems can be exciting, challenging, and rewarding! This text shows the value of the discipline as an appealing field of study and the IS graduate as an integral part of today's organizations.

CHANGES IN THE THIRD EDITION

We have implemented a number of exciting changes based on user feedback and on the way the text can be aligned even more closely with how the IS principles and concepts course is now being taught. A summary of these changes follows:

- *Unifying Theme.* In this edition, we stress the global aspects of information systems as a major theme. As organizations increasingly find themselves competing in a global marketplace, they must recognize the resulting implications on their information systems. Globalization is profoundly changing businesses, markets, and society. With its years of service to the IS discipline, this text retains the traditions and strengths of past successes while helping future managers and decision makers face tomorrow's global challenges.

- *All New World Views Cases.* While the text has always stressed the global factors affecting information systems, these factors are emphasized even more in this edition through the World Views Cases. These cases, written by instructors outside the United States and about real organizations outside the United States, provide the reader with solid insight into the IS issues facing foreign-based or multinational companies.

- *All New Vignettes Emphasize International Aspects.* In addition to the World Views Cases, all of the chapter opening vignettes raise actual issues from foreign-based or multinational companies.

- *Why Learn About Features.* Each chapter has a new "Why Learn About" section at the beginning of the chapter to pique student interest. The section sets the stage for students by briefly describing the importance of the chapter's material to business students—whatever their chosen field.

- *Information Systems @ Work Special-Interest Boxes.* Highlighting current topics and trends in today's headlines, these boxes show how information systems are used in a variety of business career areas.

- *Career Exercises.* New end-of-chapter Career Exercises ask students to research how a topic discussed in the chapter relates to a business area of their choice. Students are encouraged to use the Internet, the college library, or interviews to collect information about business careers.

- *All New Videos and Video Questions.* New video segments are provided for each chapter of the third edition. These segments demonstrate key chapter concepts. Students can actually see IS principles at work in a variety of settings and then answer questions to help them apply what they have learned.

- *Thoroughly Revised End-of-Chapter Material.* The material at the end of each chapter has been thoroughly updated. Summaries linked to the principles, self-assessment questions, review questions, discussion questions, problem-solving exercises, team activities, and Web exercises have been replaced and revised to reflect the theme of the third edition and to give students the opportunity to explore the latest technology in a business setting.

- *All New Cases.* Two new end-of-chapter cases provide a wealth of practical information for students and instructors. Each case explores a chapter concept or problem that a real-world company or organization has faced. The cases can be assigned as individual homework exercises or serve as a basis for class discussion.

CHAPTER CHANGES

Each chapter has been completely updated with the latest topics and examples. Here is a summary of some of the changes.

Chapter 1, An Introduction to Information Systems in Organizations

This chapter is full of new boxes, photos, figures, tables, examples, and more than 60 current references. The new opening vignette focuses on Boehringer Ingelheim, which is among the world's 20 largest pharmaceutical companies, with $7.6 billion in revenues and 32,000 employees in 60 nations. The new "Why Learn About" section motivates students by showing them the importance of information systems to achieve their career goals. The new "Information Systems @ Work" box stresses the use of information systems by OneBeacon Insurance Group to restructure its business with a new $15 million Web-based policy-administration system. The section on "Computer-Based Information Systems" includes many new examples and photos. A new table reveals the many powerful uses of the Internet. The section on "Business Information Systems" includes a new figure that shows how these systems evolved from early transaction processing systems of the 1950s to the advanced decision support and special-purpose systems of today. This section also includes new examples and a new photo of the use of B2B (business-to-business) applications. Mobile commerce (m-commerce) is introduced in this chapter. The new "Ethical and Societal Issues" box includes information on the dangers of *phishing*—the use of e-mail and Web sites as bait to lure consumers into revealing private information. The section on "Virtual Reality" contains new examples and new photos of a head-mounted display and a data glove. New figures show the most common attacks to information systems today and the huge costs of these attacks. The material on organizational change has been enhanced to include sustaining and disruptive change. Christensen, who wrote the business best-sellers *Inventor's Dilemma* and the more recent *Inventor's Solution*, discusses these types of change. The material on competitive advantage has been strengthened with material from Jim Collin's best-selling business book *Good to Great*. Specifically, we introduce the notion of technology acceleration with new examples and a new table. The productivity paradox has been introduced, as recommended by a reviewer. The careers in information systems section has been updated with new positions, such as the chief technology officer (CTO), and several examples and direct quotes from CTOs. The Clinger-Cohen Act, which requires CIOs for certain federal agencies, has also been included, as requested by a reviewer. Of course, there are new questions, exercises, and cases at the end of the chapter.

Chapter 2, Hardware and Software

The new "Why Learn About" and "Information Systems @ Work" features help demonstrate why a business major needs to understand what is presented in this chapter. In addition, the chapter has been revised to have more of a managerial focus on what a non-IS decision maker needs to know about hardware. The latest information on processors, main memory, secondary storage devices, and input/output devices is covered, including topics such as smart phones and tablet PCs. There is added coverage of grid computing and utility computing. Radio-frequency identification (RFID) technology is introduced and practical examples given of its use in industry. The material on computer system types has been modified to conform to the Gartner Group industry standard definitions for the various types of computer systems. The section on personal computer operating systems has been updated to include the latest operating systems and developments. The section on Windows Server has been updated to include a discussion of Windows Server 2003. The section on enterprise operating systems has also been updated and streamlined. The section on consumer appliance operating systems has been changed to operating systems for small computers and special-purpose devices. The sections on Palm OS, Windows Embedded, and Windows Mobile have been completely updated to reflect current developments. The material on proprietary and off-the-shelf software has been redone to address reviewer comments. The material on personal application software has been updated. The material on software suites now includes more information about open-source suites.

Chapter 3, Organizing Data and Information

The material on the traditional approach to database management has been reduced slightly, and the material in the section on the database approach has been updated, including the disadvantages of the database approach. Because hierarchical and network models are no longer widely used, that material has been cut drastically. The section on database management systems has been updated to include a discussion of flat files, single-user, and multiuser databases. The section on manipulating data and generating reports includes new material on query-by-example (QBE) and new examples of SQL. The section on popular database management systems has been updated, and a new section on special-purpose database systems has been added, which contains a number of new examples of real organizations using specialized database systems. A new section on using databases with other software has also been added. Finally, the section on database applications has been updated with new material and new examples, such as Oracle's Warehouse Management software, which can incorporate data from RFID technology. There is also a new section on visual, audio, and other databases near the end of the chapter. As with other chapters, the end-of-chapter material has been completely updated.

Chapter 4, Telecommunications, the Internet, Intranets, and Extranets

The new "Why Learn About" and "Information Systems @ Work" features help demonstrate why a business major needs to understand what is presented in this chapter. The discussion of carriers and services has been modified to discuss the services of local exchange carriers, competitive local exchange carriers, and long distance carriers. The discussion of media types has been divided into two basic categories—guided and wireless. Personal area networks and metropolitan area networks have been added to the discussion of various network types. The emerging wireless protocols including Bluetooth and Wi-Fi are summarized. A discussion of registrars and additional companies seeking accreditation to register domain names from the Internet Corporation for Assigned Names and Numbers (ICANN) has been added. The use of the wireless application protocol (WAP) to connect cell phones and other devices to the Internet is discussed. The section on Java has been expanded to cover a number of Web programming languages and tools including JavaScript, VBScript, ActiveX, and Hypertext Preprocessor. New sections were added on developing Web content and Web services. Discussion of new Internet applications was added, including the use of phones and handheld computers to communicate over the Internet, the use of the Internet to find career information and search for jobs, and the use of Web logs (blogs). The section on management issues has been expanded and updated to cover the lack of central management of the Internet, service and speed issues, and concerns over privacy, fraud, and security.

Chapter 5, Electronic Commerce and Transaction Processing Systems

The new "Why Learn About Electronic Commerce and Transaction Processing Systems?" feature provides motivation for students as they read this chapter. M-commerce is introduced as a new business model, and examples of m-commerce, the technology required, and the anywhere/anytime capability of m-commerce are presented. A brief summary of the status of e-commerce around the world has been added. The use of Web site customer experience technology to analyze the usability of a site is discussed.

A major new section titled "TPS Control and Management Issues" has been added. It covers business continuity planning, the TPS audit, and international issues. This section includes new material on recent legislation associated with controlling the operation and use of these systems, such as the Sarbanes-Oxley Act, the Graham-Leach-Bliley Act, and the Health Insurance Portability and Accountability Act (HIPAA). The list of vendors providing enterprise resource planning (ERP) systems has been updated to reflect recent acquisitions. The section on ERP has been modified to focus more on just what constitutes an ERP and what business processes are involved and affected.

Chapter 6, Information and Decision Support Systems

The new "Information Systems in the Global Economy" opening vignette explores the Kiku-Masamune Sake Brewing Co., Ltd., in Japan. This feature shows how to achieve profits from a state-of-the-art business intelligence system. The section on inputs to a management information system emphasizes supply chain management to a greater extent than before. Data mining is discussed in the section on outputs of a management information system. There are new examples in the section "Financial Management Information Systems," including the 200-year-old New York Stock Exchange, which investigated electronic reports, the Financial Services Authority in England, which is spending about £4 million to identify stock-market abuse and fraud with a new system called Surveillance and Automated Business Reporting Engine (SABRE), and other examples. The manufacturing MIS section also includes new examples, such as European airplane manufacturer Airbus, which is using a manufacturing MIS to monitor and control its suppliers and parts to reduce costs. Procter

& Gamble's, Gillette's, Wal-Mart's, and Target's use of specialized computer chips and tiny radio transmitters is also discussed. The sections on the other functional MISs are also loaded with new examples and references. The "Information Systems @ Work" box reveals how Taylormade used information and decision support systems in making its drivers for golf. There are a number of new examples on the use of a geographic information system (GIS). The section on decision support systems has also been completely updated with new examples and references. The section on group support systems (GSSs) and groupware has been enhanced to include the use of newer wireless systems, including Blackberry and other mobile communications devices and systems. As with all chapters, the end-of-chapter learning materials have been completely updated with new questions, exercises, and cases.

Chapter 7, Specialized Information Systems: Artificial Intelligence, Expert Systems, Virtual Reality, and Other Systems

The title of the chapter has been changed from "Specialized Business Information Systems" to "Specialized Information Systems" to reflect these systems' unique applications in non-profit and military organizations. The chapter begins with a new example of chess master Garry Kasparov competing against an artificial intelligence (AI) software package that runs on a PC called Deep Junior. The section on robotics has many new examples, from NASA to entertainment. The section on vision systems also has new examples, including the use of vision systems to inspect wine bottles in California for fingerprint analysis. The chapter also discusses the use of natural language processing by a hardware company to develop a Web site that allows customers to find what they need. The section on neural networks has examples on improving motor coordination in robots and preventing fraud and terrorism. The chapter also has a new section on other AI applications, which highlights the use of genetic algorithms and intelligent agents and provides numerous examples of these applications. In addition, many new examples of expert systems are included. The section on virtual reality has been updated with new examples such as the following: the use of virtual reality by an automotive company to help design cars and factories; the use of a virtual reality interface device to help a surgeon perform a coronary bypass operation; and the use of computer-generated images in Hollywood movies. The last section on other specialized systems has been completely revised with new examples that range from small microchips planted in the brain to the use of game theory and the development of small networks, called *smart dust*, by the University of California at Berkeley. There is also new material on informatics, including bioinformatics and medical informatics.

Chapter 8, Systems Development

This chapter has been updated with almost 40 new examples and references. The material on project management has been strengthened with new examples and quotes. The new "Information Systems @ Work" feature shows how a Danish information chain uses information systems to support other business activities. The material on agile development includes new examples, such as Sabre Airline Solutions, a $2 billion airline travel company that used extreme programming to eliminate programming errors and shorten program development times. Rapid application development (RAD) tools by IBM and others have been highlighted in a text section. The section on outsourcing has been expanded to include cutting-edge topics such as on-demand or utility computing, along with new material and examples. The material on CASE tools has been streamlined and updated to include new tools, such as VRCASE. As requested by a reviewer, we have deleted the discussion of upper and lower CASE tools. In the material on technical feasibility, we have added a new example, in which technical problems were encountered in developing an automatic tool system for long-haul trucks on the German autobahn. Reusable software is also emphasized, with new examples. Bank of America, for example, reuses previously developed software to deliver new software in 90 days or fewer. The section on acquiring database and network systems has new examples. The Nasdaq stock market, for example, is investing $50 million in a new network system to streamline operations and cut costs. A new section on systems operation has been added that includes information about support, training, and help desks. The section on systems maintenance includes new information on legacy systems. The chapter concludes with new end-of-chapter material and cases.

Chapter 9, Security, Privacy, and Ethical Issues in Information Systems and the Internet

The new "Why Learn About Security, Privacy, and Ethical Issues?" and "Information Systems @ Work" features stress the importance of these topics to the reader's daily life. Many new topics have been added including the following: the Computer Assisted Passenger Pre-screening System, the CAN-SPAM Act, the Sarbanes-Oxley Act, the Health Insurance Portability and Accountability (HIPAA) Act, the 2003 Computer Crime and Security Survey, cyberterrorism, identity theft, virus variants, biometrics, the Children's Internet Protection Act, the Child Online Protection Act, the Gramm-Leach-Bliley Act, software vulnerabilities, spyware, and cyberstalking. Statistics from the FBI/Computer Security Institute on the frequency of computer incidents have been updated to the latest data available.

WHAT WE HAVE RETAINED FROM THE PREVIOIUS EDITIONS

The third edition builds on what has worked well in the previous editions; it retains the focus on IS principles and strives to be the most current text on the market.

- *Overarching Principle.* This book continues to stress a single, all-encompassing theme: The right information, if it is delivered to the right person, in the right fashion, and at the right time, can improve and ensure organizational effectiveness and efficiency.

- *Information Systems Principles.* Information System Principles summarize key concepts that every student should know. This important feature is a convenient summary of key ideas presented at the start of each chapter.

- *Learning Objectives Linked to Principles.* Carefully crafted learning objectives are included with every chapter. The learning objectives are linked to the Information Systems Principles and reflect what a student should be able to accomplish after completing a chapter.

- *Summary Linked to Principles.* Each chapter includes a detailed summary, and each section of the summary is tied to an Information System Principle.

- *Ethical and Societal Issues Special-Interest Boxes.* Each chapter includes an "Ethical and Societal Issues" feature that presents a timely look at the ethical challenges and the societal impact of information systems. Ethics remains a compelling issue for today's business and IS students, and they gain exposure to ethical and societal issues by grappling with the in-depth questions related to the companies that are profiled. New to this edition are "What Would You Do?" scenarios that place students in real-life ethical dilemmas. All boxes relate to the issues discussed in the chapters.

- *Current Examples, Boxes, Cases, and References.* As we have in each edition, we take great pride in presenting the most recent examples, boxes, cases, and references throughout the text. Some of them were developed at the last possible moment, literally weeks before the book went into publication. Information on new hardware and software, the latest operating systems, application service providers, the Internet, electronic commerce, ethical and societal issues, and many other current developments can be found throughout the text. Our adopters have come to expect the best and most recent material. We have done everything we can to meet or exceed these expectations.

- *Self-Assessment Tests*. This popular feature helps students review and test their understanding of key chapter concepts.

STUDENT RESOURCES

MIS Companion CD

We are pleased to include in every textbook a free copy of Thomson Course Technology's MIS Companion CD, which is composed of training lessons in Excel, Access, and MIS concepts. The Companion CD's content is integrated throughout the book. Wherever you see the CD icon in the chapter margins, you know that you can find additional related material on the CD.

Student Online Companion Web Site

We have created an exciting online companion, password protected for students to utilize as they work through the third edition of *Fundamentals of Information Systems*. In the front of this text you will find a key code that provides full access to a robust Web site, located at www.course.com/mis/stair. This Web resource includes the following features:

- **Videos**
  Links to topical video clips relating to every chapter in the book can be found on the student Web site. Questions corresponding to the respective video clips are featured at the end of each chapter in the printed book. These exercises reinforce the concepts taught and provide students with more critical thinking opportunities.

- **PowerPoint Slides**
  Direct access is offered to the book's PowerPoint presentations, which cover the key points from each chapter. These presentations are a useful study tool.

- **Classic Cases**
  A frequent request from adopters is that they wish to have a broader selection of cases from which to choose. To meet this need, an extensive collection of additional cases from previous editions of the text is included here. These cases are the authors' choices of the "best cases" and span a broad range of companies and industries.

- **Links to Useful Web Sites**
  Chapters in *Fundamentals of Information Systems, Third Edition* refer to many interesting Web sites. This resource takes you to links you can follow directly to the home pages of those sites so that you can explore them. There are additional links to Web sites that the authors, Ralph Stair and George Reynolds, think you would be interested in checking out.

- **Hands-On Activities**
  Use the Hands-On Activities to test your comprehension of IS topics and enhance your skills using Microsoft® Office applications and the Internet. Using these links, you can access three critical-thinking exercises per chapter; each activity asks you to work with an Office tool or do some research on the Internet.

- **Test Yourself on IS**
  This tool allows you to access 20 multiple-choice questions for each chapter, test yourself, and then submit your answers. You will immediately find out which questions you got right and which you got wrong. For each question that you answer incorrectly, you are given the correct answer and the page in your text where that

information is covered. Special testing software randomly compiles 20 questions from a database of 50 questions, so you can quiz yourself multiple times on a given chapter and get some new questions each time.

- **Glossary of Key Terms**
  The glossary of key terms from the text is available to search.

- **Online Readings**
  This feature provides you access to a computer database that contains articles relating to hot topics in information systems.

INSTRUCTOR RESOURCES

The teaching tools that accompany this text offer many options for enhancing a course. And, as always, we are committed to providing one of the best teaching resource packages available in this market.

Instructor's Manual

An all-new *Instructor's Manual* provides valuable chapter overviews; highlights key principles and critical concepts; offers sample syllabi, learning objectives, and discussion topics; and features possible essay topics, further readings and cases, and solutions to all of the end-of-chapter questions and problems, as well as suggestions for conducting the team activities. Additional end-of-chapter questions are also included.

Sample Syllabus

A sample syllabus with sample course outlines is provided to make planning your course that much easier.

Solutions

Solutions to all end-of-chapter material are provided in a separate document for your convenience.

Test Bank and Test Generator

ExamView® is a powerful objective-based test generator that enables instructors to create paper-, LAN- or Web-based tests from test banks designed specifically for their Thomson Course Technology text. Instructors can utilize the ultra-efficient QuickTest Wizard to create tests in less than five minutes by taking advantage of Thomson Course Technology's question banks or customizing their own exams from scratch. Page references for all questions are provided so instructors can cross-reference test results with the book.

PowerPoint Presentations

A set of impressive Microsoft PowerPoint slides is available for each chapter. These slides are included to serve as a teaching aid for classroom presentation, to make available to students on the network for chapter review, or to be printed for classroom distribution. Our presentations help students focus on the main topics of each chapter, take better notes, and prepare for examinations. Instructors can also add their own slides for supplemental topics they introduce to the class.

Figure Files

Figure Files allow instructors to create their own presentations using figures taken directly from the text.

DISTANCE LEARNING

Thomson Course Technology, the premiere innovator in management information systems publishing, is proud to present online courses in WebCT and Blackboard.

- *Blackboard and WebCT Level 1 Online Content.* If you use Blackboard or WebCT, the test bank for this textbook is available at no cost in a simple, ready-to-use format. Go to *www.course.com* and search for this textbook to download the test bank.

- *Blackboard and WebCT Level 2 Online Content.* Blackboard 5.0 and 6.0 as well as Level 2 and WebCT Level 2 courses are also available for *Fundamentals of Information Systems, Third Edition.* Level 2 offers course management and access to a Web site that is fully populated with content for this book. For more information on how to bring distance learning to your course, contact your Thomson Course Technology sales representative.

ACKNOWLEDGMENTS

A book of this size and undertaking requires a strong team effort. We would like to thank all of our fellow teammates at Thomson Course Technology and Elm Street Publishing Services for their dedication and hard work. Special thanks to Eunice Yeates-Fogle, our Senior Product Manager. Our appreciation goes out to all the many people who worked behind the scenes to bring this effort to fruition, including Mirella Misiaszek, our Product Manager for supplements. Kelly Robinson, our Production Editor, shepherded the book through the production process, and Beth Paquin drove the Student Online Companion. We would also like to acknowledge and thank Elm Street Publishing Services. Karen Hill, our development editor, deserves special recognition for her tireless effort and help in all stages of this project.

We are grateful to the sales force at Thomson Course Technology, whose efforts make this all possible. You helped to get valuable feedback from current and future adopters. As Thomson Course Technology product users, we know how important you are.

We would especially like to thank Ken Baldauf for his excellent help in writing most of the boxes and cases for this edition. Ken also provided invaluable feedback for many topics discussed in the book.

Ralph Stair would like to thank the Department of Management Information Systems, College of Business Administration, at Florida State University for their support and encouragement. He would also like to thank his family, Lila and Leslie, for their support.

George Reynolds would like to thank the Department of Information Systems, College of Business, at the University of Cincinnati for their support and encouragement. He would also like to thank his family, Ginnie, Tammy, Kim, Kelly, and Kristy, for their patience and support in this major project. He would also like to thank Kristen Duerr and Ralph Stair for asking him to join the writing team back in 1997.

TO OUR PREVIOUS ADOPTERS AND POTENTIAL NEW USERS

We sincerely appreciate our loyal adopters of the previous editions and welcome new users of *Fundamentals of Information Systems, Third Edition.* As in the past, we truly value your needs and feedback. We can only hope the third edition continues to meet your high expectations.

In addition, Ralph Stair would like to thank Mike Jordan, Senior Lecturer, Division of Business Information Management Glasgow Caledonia University for his hospitality and help at the UKAIS conference.

We would especially like to thank reviewers of the third edition and the previous editions.

Reviewers for this Edition

We are indebted to the following individuals for their perceptive feedback on early drafts of this text:

Jane McKay, *Texas Christian University*

Mani Subramani, *Carlson School of Management, University of Minnesota*

Karen Williams, *University of Texas, San Antonio*

Reviewers for Previous Editions

The following people shaped the book you hold in your hands by contributing to the previous editions of *Fundamentals of Information Systems*:

Jill Adams, *Navarro College*

David Anyiwo, *Bowie State University*

Cynthia Barnes, *Lamar University*

Cynthia Drexel, *Western State College*

Brian Kovar, *Kansas State University*

John Melrose, *University of Wisconsin— Eau Claire*

Bertrad P. Mouqin, *University of Mary Hardin-Baylor*

Pamela Neely, *Marist College*

Mahesh S. Raisinghani, *University of Dallas*

Marcos Sivitanides, *Southwest Texas State University*

Anne Marie Smith, *LaSalle University*

Patricia A. Smith, *Temple College*

Herb Snyder, *Fort Lewis College*

OUR COMMITMENT

We are committed to listening to our adopters and readers and to developing creative solutions to meet their needs. The field of information systems continually evolves, and we strongly encourage your participation in helping us provide the freshest, most relevant information possible.

We welcome your input and feedback. If you have any questions or comments regarding *Fundamentals of Information Systems, Third Edition*, please contact us through Thomson Course Technology or your local representative; via e-mail at mis@course.com; via the Internet at *www.course.com*; or address your comments, criticisms, suggestions, and ideas to:

Ralph Stair
George Reynolds
Thomson Course Technology
25 Thomson Place
Boston, MA 02210

PART
· 1 ·

Information Systems in Perspective

Chapter 1 An Introduction to Information Systems in Organizations

CHAPTER
· 1 ·

An Introduction to Information Systems in Organizations

PRINCIPLES	LEARNING OBJECTIVES
- The value of information is directly linked to how it helps decision makers achieve the organization's goals.	- Distinguish data from information and describe the characteristics used to evaluate the quality of data.
- Knowing the potential impact of information systems and having the ability to put this knowledge to work can result in a successful personal career, organizations that reach their goals, and a society with a higher quality of life.	- Identify the basic types of business information systems and discuss who uses them, how they are used, and what kinds of benefits they deliver.
- System users, business managers, and information systems professionals must work together to build a successful information system.	- Identify the major steps of the systems development process and state the goal of each.
- The use of information systems to add value to the organization can also give an organization a competitive advantage.	- Identify the value-added processes in the supply chain and describe the role of information systems within them. - Identify some of the strategies employed to lower costs or improve service. - Define the term *competitive advantage* and discuss how organizations are using information systems to gain such an advantage.
- Information systems personnel are the key to unlocking the potential of any new or modified system.	- Define the types of roles, functions, and careers available in information systems.

INFORMATION SYSTEMS IN THE GLOBAL ECONOMY
BOEHRINGER INGELHEIM, GMBH, GERMANY

Lean and Mean with Information Systems

Boehringer Ingelheim is among the world's 20 largest pharmaceutical companies. A giant company with $7.6 billion in revenues and 32,000 employees in 60 nations, Boehringer has diversified into segments that include manufacturing and marketing pharmaceuticals (such as prescription medicines and consumer healthcare products), products for industrial customers (such as chemicals and biopharmaceuticals), and animal health products.

The sheer size of the company was slowing the flow of information to decision makers in the organization. "I want to be told where I stand and where we are heading," says Holger Huels, chief financial officer, "I like to [be able to] see negative trends and counter them as fast as possible." With each of the company's segments using diverse information systems, it took a significant amount of time to collect and combine all of the financial records. Each month, the accounting department would spend three days collecting and analyzing printed reports to create the company's monthly report.

Top managers decided to totally revamp the company's systems with state-of-the-art information systems from SAP, the world's largest enterprise software company. It took 14 months to roll out the new system, and many employees needed intensive training. In the end, the results were well worth the investment in time and money. The software provided a standard system used across all of Boehringer's business segments and offered convenient Web access to current information. Boehringer is now able to complete monthly reports just two hours after the close of business at the end of each month. The new system has made the accounting department much more productive, allowing staff to run up-to-date reports whenever needed.

Boehringer is committed to providing all employees with the applications and information they need. About one-third of Boehringer's employees work outside the office. To provide its mobile workforce with up-to-the-minute data, the company deployed software from BackWeb Technologies, which allows access to current sales information through a Web portal and a custom Web interface. With the new system, Boehringer's employees can access and change information presented in the portal when they are offline, with updates later when they log on.

By the time Boehringer was finished with its technology makeover, the company had implemented over seven new interconnected information systems and invested millions in hardware, software, databases, telecommunications, and training. But the investment has paid off. Employees can now access up-to-date organization-wide information, wherever they may be, with the click of a mouse. And decision makers can react as nimbly and quickly to changes as many of Boehringer's smaller competitors.

As you read this chapter, consider the following:

- In designing its new information systems, what do you think were Boehringer's most critical goals and considerations?
- How are hardware, software, databases, telecommunications, people, and procedures used in Boehringer's information system to provide valuable data?

Why Learn About Information Systems in Organizations?

Information systems are used in almost every imaginable career area. A management major might be hired to work with computerized employee files and records for a shipping company. A marketing major might work for a large retail store analyzing customer needs with a computer. An accounting major might work for an accounting or consulting firm using a computer to audit other companies' financial records. A real estate major might use the Internet and work within a loose organizational structure with clients, builders, and a legal team located around the world. Regardless of your college major or chosen career, you will find that information systems are indispensable tools to help you achieve your career aspirations. Learning about information systems can help you get your first job, obtain promotions, and advance your career. Why learn about information systems? What is in it for you? Learning about information systems will help you achieve your goals! Let's get started by exploring the basics of information systems.

information system (IS)
A set of interrelated components that collect, manipulate, and disseminate data and information and provide a feedback mechanism to meet an objective.

An **information system (IS)** is a set of interrelated components that collect, manipulate, store, and disseminate data and information and provide a feedback mechanism to meet an objective. The feedback mechanism helps organizations achieve their goals, such as increasing profits or improving customer service.

Information systems are everywhere. A customer at the gas pump waves a keychain tag at a reader that sends the information to a network to verify the customer's profile and credit information. The terminal processes the transaction, prints a receipt, and the customer's credit/check card is automatically billed.

(Source: Courtesy of Texas Instrument Inc. All Rights Reserved.)

Computers and information systems are constantly changing the way organizations conduct business. They are becoming fully integrated into our lives, businesses, and society. They can help organizations carry on daily operations (operational systems). For example, Wal-Mart uses operational systems to pull supplies from distribution centers and, ultimately, suppliers, to stock shelves, and to push out products and services through customer purchases. Computer and information systems also act as command and control systems that monitor processes and help supervisors control them. For example, air traffic control centers use computers and information systems as command and control centers to monitor and direct planes in their airspace.

Computers and information systems will continue to change our society, our businesses, and our lives. In this chapter, we present a framework for understanding computers and information systems and discuss why it is important to study information systems. This understanding will help you unlock the potential of properly applied IS concepts. We begin with information concepts.

INFORMATION CONCEPTS

Information is a central concept throughout this book. The term is used in the title of the book, in this section, and in almost every chapter. To be an effective manager in any area of business, you need to understand that information is one of an organization's most valuable and important resources. This term, however, is often confused with the term *data*.

Data Versus Information

data
The raw facts, such as an employee's name and number of hours worked in a week, inventory part numbers, or sales orders.

Data consists of raw facts, such as an employee's name and number of hours worked in a week, inventory part numbers, or sales orders. As shown in Table 1.1, several types of data can be used to represent these facts. When these facts are organized or arranged in a meaningful manner, they become information. **Information** is a collection of facts organized in such a way that they have additional value beyond the value of the facts themselves. For example, a particular manager might find the knowledge of total monthly sales to be more suited to his or her purpose (i.e., more valuable) than the number of sales for individual sales representatives. Providing information to customers can also help companies increase revenues and profits.

information
A collection of facts organized in such a way that they have additional value beyond the value of the facts themselves.

Data	Represented By
Alphanumeric data	Numbers, letters, and other characters
Image data	Graphic images and pictures
Audio data	Sound, noise, or tones
Video data	Moving images or pictures

Table 1.1

Types of Data

Data represents real-world things. As we have stated, data—simply raw facts—has little value beyond its existence. For example, consider data as pieces of railroad track in a model railroad kit. In this state, each piece of track has little value beyond its inherent value as a single object. However, if some relationship is defined among the pieces of the track, they will gain value. By arranging the pieces of track in a certain way, a railroad layout begins to emerge (Figure 1.1a). Information is much the same. Rules and relationships can be set up to organize data into useful, valuable information.

The type of information created depends on the relationships defined among existing data. For example, the pieces of track could be rearranged to form different layouts (Figure 1.1b). Adding new or different data means relationships can be redefined and new information can be created. For instance, adding new pieces to the track can greatly increase the value—in this case, variety and fun—of the final product. We can now create a more elaborate railroad layout (Figure 1.1c). Likewise, our manager could add specific product data to his or her sales data to create monthly sales information broken down by product line. This information could be used by the manager to determine which product lines are the most popular and profitable.

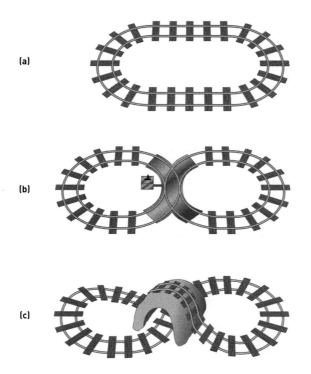

(a)

(b)

(c)

Figure 1.1

Defining and Organizing Relationships Among Data Creates Information

process
A set of logically related tasks performed to achieve a desired outcome.

knowledge
The awareness and understanding of a set of information and the ways it can be used.

Turning data into information is a **process**, or a set of logically related tasks performed to achieve a defined outcome. The process of defining relationships among data to create useful information requires knowledge. **Knowledge** is an awareness and understanding of a set of information and the ways that information can be used to support a specific task or reach a decision. Part of the knowledge needed for building a railroad layout, for instance, is understanding how large an area is available for the layout, how many trains will run on the track, and how fast they will travel. The act of selecting or rejecting facts based on their relevancy to particular tasks is also based on a type of knowledge used in the process of converting data into information. Therefore, information can be considered data made more useful through the application of knowledge. In some cases, data is organized or processed mentally or manually. In other cases, a computer is used. In the earlier example, the manager could have manually calculated the sum of the sales of each representative, or a computer could calculate this sum. What is important is not so much where the data comes from or how it is processed but whether the results are useful and valuable. This transformation process is shown in Figure 1.2.

Figure 1.2

The Process of Transforming Data into Information

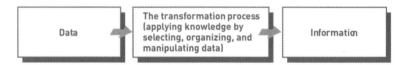

The Characteristics of Valuable Information

To be valuable to managers and decision makers, information should have the characteristics described in Table 1.2. These characteristics also make the information more valuable to an organization. Many organizations and shipping companies, for example, are able to determine the exact location of inventory items and packages in their systems. Recently, the U.S. Army Materiel Command tagged all its cargo and food shipments with radio-frequency identification chips for shipment to the Middle East. Because of the easy electronic retrieval of information from the tags, the time needed to take inventory of the cargo when it arrived was reduced from the usual 2 to 3 days to just 22 minutes.[1]

The Value of Information

The value of information is directly linked to how it helps decision makers achieve their organization's goals. For example, the value of information might be measured in the time required to make a decision or in increased profits to the company. Consider a market forecast that predicts high demand for a new product. If market forecast information is used to develop the new product and the company is able to make an additional profit of $10,000, the value of this information to the company is $10,000 minus the cost of the information. Valuable information can also help managers decide whether to invest in additional information systems and technology. A new computerized ordering system might cost $30,000, but it might generate an additional $50,000 in sales. The *value added* by the new system is the additional revenue from the increased sales of $20,000. Most corporations have cost reduction as a primary goal. Using information systems, some manufacturing companies have been able to slash inventory costs by millions of dollars.

WHAT IS AN INFORMATION SYSTEM?

As mentioned previously, an information system (IS) is a set of interrelated elements or components that collect (input), manipulate (process) and store, and disseminate (output) data and information and provide a feedback mechanism to meet an objective (see Figure 1.3). The feedback mechanism helps organizations achieve their goals, such as increasing profits or improving customer service.

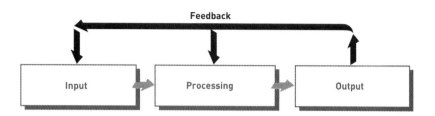

Figure 1.3

The Components of an Information System

Feedback is critical to the successful operation of a system.

Characteristics	Definitions
Accurate	Accurate information is error free. In some cases, inaccurate information is generated because inaccurate data is fed into the transformation process (this is commonly called garbage in, garbage out [GIGO]).
Complete	Complete information contains all the important facts. For example, an investment report that does not include all important costs is not complete.
Economical	Information should also be relatively economical to produce. Decision makers must always balance the value of information with the cost of producing it.
Flexible	Flexible information can be used for a variety of purposes. For example, information on how much inventory is on hand for a particular part can be used by a sales representative in closing a sale, by a production manager to determine whether more inventory is needed, and by a financial executive to determine the total value the company has invested in inventory.
Reliable	Reliable information can be depended on. In many cases, the reliability of the information depends on the reliability of the data collection method. In other instances, reliability depends on the source of the information. A rumor from an unknown source that oil prices might go up may not be reliable.
Relevant	Relevant information is important to the decision maker. Information that lumber prices might drop may not be relevant to a computer chip manufacturer.
Simple	Information should also be simple, not overly complex. Sophisticated and detailed information may not be needed. In fact, too much information can cause information overload, whereby a decision maker has too much information and is unable to determine what is really important.
Timely	Timely information is delivered when it is needed. Knowing last week's weather conditions will not help when trying to decide what coat to wear today.
Verifiable	Information should be verifiable. This means that you can check it to make sure it is correct, perhaps by checking many sources for the same information.
Accessible	Information should be easily accessible by authorized users to be obtained in the right format and at the right time to meet their needs.
Secure	Information should be secure from access by unauthorized users.

Table 1.2

Characteristics of Valuable Information

Input, Processing, Output, and Feedback

Input

In information systems, **input** is the activity of gathering and capturing raw data. In producing paychecks, for example, the number of hours every employee works must be collected before paychecks can be calculated or printed. In a university grading system, individual instructors must submit student grades before a summary of grades for the semester or quarter can be compiled and sent to the students.

input
The activity of gathering and capturing raw data.

Input can be a manual or an automated process. A scanner at a grocery store that reads bar codes and enters the grocery item and price into a computerized cash register is a type of automated input process. Regardless of the input method, accurate input is critical to achieve the desired output.

Processing

processing
The activity of converting or transforming data into useful outputs.

In information systems, **processing** involves converting or transforming data into useful outputs. Processing can involve making calculations, making comparisons and taking alternative actions, and storing data for future use. Processing data into useful information is critical in business settings.

Processing can also be done manually or with computer assistance. In the payroll application, each employee's number of hours worked must be converted into net, or take-home, pay. Other inputs often include employee ID number and department. The required processing first involves multiplying the number of hours worked by the employee's hourly pay rate to get gross pay. If weekly hours worked exceed 40 hours, overtime pay might also be included. Then, deductions—for example, federal and state taxes, and contributions to health and life insurance or savings plans—are subtracted from gross pay to obtain net pay.

After these calculations and comparisons are performed, the results are typically stored. *Storage* involves keeping data and information available for future use, including output, discussed next.

Output

output
The production of useful information, usually in the form of documents and reports.

In information systems, **output** involves producing useful information, usually in the form of documents and reports. Outputs can include paychecks for employees, reports for managers, and information supplied to stockholders, banks, government agencies, and other groups. In some cases, output from one system can become input for another. For example, output from a system that processes sales orders can be used as input to a customer billing system. Often, output from one system can be used as input to control other systems or devices. For instance, the design and manufacture of office furniture is complicated with many variables. The salesperson, customer, and furniture designer can go through several design iterations to meet the customer's needs. Special computer programs and equipment create the original design and allow the designer to rapidly revise it. After the last design mock-up is approved, the computer creates a bill of materials that goes to manufacturing to produce the order.

Output can be produced in a variety of ways. For a computer, printers and display screens are common output devices. Output can also be a manual process involving handwritten reports and documents.

Feedback

feedback
The output that is used to make changes to input or processing activities.

In information systems, **feedback** is output that is used to make changes to input or processing activities. For example, errors or problems might make it necessary to correct input data or change a process. Consider a payroll example. Perhaps the number of hours an employee worked was entered into a computer as 400 instead of 40 hours. Fortunately, most information systems check to ensure that data falls within certain ranges. For number of hours worked, the range might be from 0 to 100 hours because it is unlikely that an employee would work more than 100 hours for any given week. So, the information system would determine that 400 hours is out of range and provide feedback, such as an error report. The feedback is used to check and correct the input on the number of hours worked to 40. If undetected, this error would result in a very high net pay on the printed paycheck! Some blame the August 14, 2003 power blackout in the United States' Northeast on a faulty computer system that wasn't able to provide second-by-second feedback.[2]

Feedback is also important for managers and decision makers. For example, a bedding maker used a computerized feedback system to link its suppliers and plants. The output from an information system might indicate that inventory levels for a few items are getting low—a potential problem. A manager could use this feedback to decide to order more inventory from a supplier. The new inventory orders then become input to the system. In addition to this *reactive* approach, a computer system can also be *proactive*—predicting future events to avoid problems. This concept, often called **forecasting**, can be used to estimate future sales and order more inventory before a shortage occurs.

forecasting
The process of predicting future events to avoid problems.

Manual and Computerized Information Systems

As discussed earlier, an information system can be manual or computerized. For example, some investment analysts manually draw charts and trend lines to assist them in making investment decisions. Tracking data on stock prices (input) over the last few months or years, these analysts develop patterns on graph paper (processing) that help them determine what stock prices are likely to do in the next few days or weeks (output). Some investors have made millions of dollars using manual stock analysis information systems. Of course, today many excellent computerized information systems have been developed to follow stock indexes and markets and to suggest when large blocks of stocks should be purchased or sold (called *program trading*) to take advantage of market discrepancies. It is important to stress, however, that simply computerizing a manual information system does not guarantee improved system performance. If the underlying information system is flawed, the act of computerizing it might only magnify the impact of these flaws.

Program trading systems allow traders to keep up with swift changes in stock prices and make better decisions for their investors.

(Source: © Reuters NewMedia Inc./ CORBIS.)

Computer-Based Information Systems

A **computer-based information system (CBIS)** is a single set of hardware, software, databases, telecommunications, people, and procedures that are configured to collect, manipulate, store, and process data into information. For example, a company's payroll systems, order entry systems, or inventory control systems are examples of a CBIS. The components of a CBIS are illustrated in Figure 1.4. (*Information technology (IT)* is a related term. For our purposes, IT refers to the technology components of hardware, software, databases, and telecommunications.) A business's **technology infrastructure** includes all the hardware, software, databases, telecommunications, people, and procedures that are configured to collect, manipulate, store, and process data into information. The technology infrastructure is a set of shared IS resources that form the foundation of computer-based information systems.

Hardware

Hardware consists of computer equipment used to perform input, processing, and output activities.[3] Input devices include keyboards, automatic scanning devices, equipment that can read magnetic ink characters, and many other devices. Investment firms often use voice response to allow customers to get their balances and other information using ordinary spoken sentences. The Scripps Institution of Oceanography developed a special underwater computer optical input device to allow a diver as deep as 100 feet to control an underwater camera, which was formerly controlled by a computer system and mouse on the surface.[4] Processing devices include the central processing unit and main memory.[5] Processor speed is important in creating video images.[6] Lifelike movie characters such as Gollum in the *Lord of the Rings* shows what is possible with today's fast processors. Mental Images of Germany and Pixar of the United States have used such award-winning image-rendering techniques. The technology is also used to help design cars, such as the sleek shapes of Mercedes-Benz vehicles. Specialized, inexpensive hardware has also been used in schools to help students learn a variety of subjects.[7] Michael Dell, founder of Dell Inc., believes that hardware will increasingly include very small devices that are connected to other hardware devices: "Nanotechnology

computer-based information system (CBIS)
A single set of hardware, software, databases, telecommunications, people, and procedures that are configured to collect, manipulate, store, and process data into information.

technology infrastructure
All the hardware, software, databases, telecommunications, people, and procedures that are configured to collect, manipulate, store, and process data into information.

hardware
The computer equipment used to perform input, processing, and output activities.

Figure 1.4

**The Components of a
Computer-Based
Information System**

and communications will be in everything. All kinds of other devices will attach and link together, centered, I think, with the PC."[8] *Nanotechnology* can involve molecule-sized hardware devices.[9]

Software

software
The computer programs that govern the operation of the computer.

Software is the computer programs that govern the operation of the computer. These programs allow a computer to process payroll, send bills to customers, and provide managers with information to increase profits, reduce costs, and provide better customer service.[10] With software, people can work anytime at any place. On a trip back to the United States from Australia and New Zealand, Steve Ballmer, CEO of Microsoft commented, "I could carry my slides, I could carry my e-mail. I could carry anything I needed to read. I could carry my life with me. It was very powerful."[11] There are two basic types of software: system software, such as Windows XP, which controls basic computer operations such as start-up and printing, and applications software, such as Office 2003, which allows specific tasks to be accomplished, such as word processing or tabulating numbers.

Databases

database
An organized collection of facts and information.

A **database** is an organized collection of facts and information, typically consisting of two or more related data files. An organization's database can contain facts and information on customers, employees, inventory, competitors' sales information, online purchases, and much more. Most managers and executives believe a database is one of the most valuable and important parts of a computer-based information system. Increasingly, organizations are placing important databases on the Internet, discussed next.[12]

Telecommunications, Networks, and the Internet

telecommunications
The electronic transmission of signals for communications; enables organizations to carry out their processes and tasks through effective computer networks.

networks

The connected computers and computer equipment in a building, around the country, or around the world to enable electronic communications.

Telecommunications is the electronic transmission of signals for communications, which enables organizations to carry out their processes and tasks through effective computer networks.[13] Large restaurant chains, for example, can use telecommunications systems and satellites to link hundreds of restaurants to plants and corporate headquarters to speed credit card authorization and report sales and payroll data. **Networks** are used to connect computers and computer equipment in a building, around the country, or around the world to enable electronic communications. Investment firms can use wireless networks to connect thousands of people with their corporate offices. Hotel Commonwealth in Boston uses wireless telecommunications to allow guests to connect to the Internet, get voice messages, and perform other functions without plugging their computers or mobile devices into a wall outlet.[14] Wireless transmission is also allowing drones, like Boeing's Scan Eagle, to monitor power lines, buildings, and other commercial establishments.[15] One company uses a private network to connect offices in the United States, Germany, China, Korea, and other companies. It doesn't use a public network available to everyone, such as the Internet, discussed next.

The **Internet** is the world's largest computer network, actually consisting of thousands of interconnected networks, all freely exchanging information. Research firms, colleges, universities, high schools, and businesses are just a few examples of organizations using the Internet. Table 1.3 lists companies that have used the Internet to their advantage.

Internet
The world's largest computer network, actually consisting of thousands of interconnected networks, all freely exchanging information.

Organization	Objective	Description of Internet Usage
Godiva Chocolatier	Increase sales and profits	The company developed a very profitable Internet site that allows customers to buy and ship chocolates. According to Kim Land, director of Godiva Direct, "This was set up from the beginning to make money." In two years, online sales have soared by more than 70% each year.
Environmental Defense	Alert the public to environmental concerns	The organization, formerly the Environmental Defense Fund, successfully used the Internet to alert people to the practice of catching sharks, removing their fins for soup, and returning them to the ocean to die. The Internet site also helped people fax almost 10,000 letters to members of Congress about the practice. According to Fred Krupp, the executive director of the Environmental Fund, "The Internet is the ultimate expression of 'think global, act local.'"
Buckman Laboratories	Better employee training	The company used the Internet to train employees to sell specialty chemicals to paper companies, instead of bringing them to Memphis for training. According to one executive, "Our retention rate is much higher, and we removed a week [of training] in Memphis, which meant big savings." Using the Internet lowered the hourly cost of training an employee from $1,000 to only $40.
Siemens	Reduce costs	Using the Internet, the company, which builds and services power plants, was able to reduce the cost of entering orders and serving customers. The Internet solution cost about $60,000 compared with a traditional solution that would have cost of $600,000.
Goldman Industrial Group	Save time	The company makes machine tools and was able to slash the time it takes to fill an order from three or four months to about a week using the Internet to help coordinate parts and manufacturing with its suppliers and at its plants.
Partnership America	Make better decisions	The company developed an Internet site for wholesalers of computer equipment and supplies. The wholesalers use the site to make better decisions about the features and prices of various pieces of computer equipment. The system allows wholesalers to connect to Partnership America's site using cell phones. "When many of our customers need information, they're not at their desks," says one company representative.
Altra Energy Technologies	Get energy to companies that need it	The company developed an Internet site to help companies buy oil, gas, and wholesale power over the Internet.

The *World Wide Web (WWW)* or the *Web* is a network of links on the Internet to documents containing text, graphics, video, and sound. Information about the documents and access to them are controlled and provided by tens of thousands of special computers called *Web servers*. The Web is one of many services available over the Internet and provides access to literally millions of documents.

The technology used to create the Internet is now also being applied within companies and organizations to create an **intranet**, which allows people within an organization to exchange information and work on projects. The Virgin Group, for example, uses an intranet to connect its 200 global operating companies and 20,000 employees.[16] According to Ashley Stockwell of the Virgin Group, "One of our key challenges at Virgin is to provide high-quality service to our family of companies. One key tool to help us provide this was the development of an intranet and extranet." An **extranet** is a network based on Web technologies that allows selected outsiders, such as business partners and customers, to access authorized resources on a company's intranet. Companies can move all or most of their business activities to an extranet site for corporate customers. Many people use extranets every day without realizing it—to track shipped goods, order products from their suppliers, or

Table 1.3

Uses of the Internet

intranet
An internal network based on Web technologies that allows people within an organization to exchange information and work on projects.

extranet
A network based on Web technologies that allows selected outsiders, such as business partners and customers, to access authorized resources of the intranet of a company.

access customer assistance from other companies. Log on to the FedEx site to check the status of a package, for example, and you are using an extranet.

People

People are the most important element in most computer-based information systems. Information systems personnel include all the people who manage, run, program, and maintain the system. Large banks can hire hundreds of IS personnel to speed up the development of computer-related projects. Users are any people who use information systems to get results. Users include financial executives, marketing representatives, manufacturing operators, and many others. Certain computer users are also IS personnel.

Procedures

procedures
The strategies, policies, methods, and rules for using a CBIS.

Procedures include the strategies, policies, methods, and rules for using the CBIS. For example, some procedures describe when each program is to be run or executed. Others describe who can have access to facts in the database. Still other procedures describe what is to be done in case a disaster, such as a fire, an earthquake, or a hurricane that renders the CBIS unusable.

Now that we have looked at computer-based information systems in general, we briefly examine the most common types used in business today. These IS types are covered in more detail in Part 3.

BUSINESS INFORMATION SYSTEMS

The most common types of information systems used in business organizations are electronic and mobile commerce systems, transaction processing systems, management information systems, and decision support systems. In addition, some organizations employ special-use systems, such as artificial intelligence systems, expert systems, and virtual reality systems. Together, these systems help employees in organizations accomplish both routine and special tasks—from recording sales, to processing payrolls, to supporting decisions in various departments, to providing alternatives for large-scale projects and opportunities. Figure 1.5 gives a simple overview of the development of important business information systems discussed in this section.

Figure 1.5

The Development of Important Business Information Systems

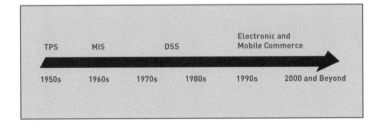

Electronic and Mobile Commerce

e-commerce
Any business transaction executed electronically between parties such as companies (business-to-business, B2B), companies and consumers (business-to-consumer, B2C), consumers and other consumers (consumer-to-consumer, C2C), businesses and the public sector, and consumers and the public sector.

E-commerce involves any business transaction executed electronically between parties such as companies (business-to-business, B2B), companies and consumers (business-to-consumer, B2C), consumers and other consumers (consumer-to-consumer, C2C), businesses and the public sector, and consumers and the public sector. People might assume that e-commerce is reserved mainly for consumers visiting Web sites for online shopping. But Web shopping is only a small part of the e-commerce picture; the major volume of e-commerce—and its fastest-growing segment—is business-to-business (B2B) transactions that make purchasing easier for corporations. This growth is being stimulated by increased Internet access, growing user confidence, better payment systems, and rapidly improving Internet and Web security.

Corporate Express, an office-supply company located in Broomfield, Colorado, uses a sophisticated B2B system to coordinate billions of dollars of office supplies that flow from its suppliers, through its offices, to its customers.[17] Today, more than half of its 75,000 daily orders arrive electronically through B2B on the Internet. E-commerce offers opportunities for small businesses, too, by enabling them to market and sell at a low cost worldwide, thus allowing them to enter the global market right from start-up. **Mobile commerce (m-commerce)** is transactions conducted anywhere, anytime. M-commerce relies on the use of wireless communications to allow managers and corporations to place orders and conduct business using handheld computers, portable phones, laptop computers connected to a network, and other mobile devices.

mobile commerce (m-commerce)
Transactions conducted anywhere, anytime.

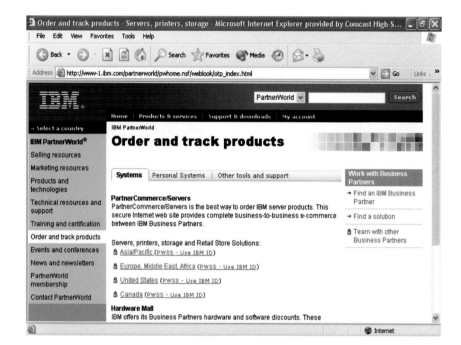

IBM PartnerWorld® is an example of B2B (business-to-business) e-commerce that provides member companies with resources for product marketing, technical support, and training.

Consumers who have tried online shopping appreciate the ease of e-commerce. They can avoid fighting crowds in the malls, shop online at any time from the comfort of their home, and have goods delivered to them directly. As a result, advertisers plan to increase spending by 6.3 percent online versus 4.7 percent in conventional media.[18] In addition, current laws governing online purchases exempt purchasers from paying state sales taxes. However, e-commerce is not without its downside. Consumers continue to have concerns about sending credit card information over the Internet to sites with varying security measures, where high-tech criminals could obtain it. See the "Ethical and Societal Issues" feature, which discusses other potential problems of e-commerce.

Figure 1.6 provides a brief example of how e-commerce can simplify the process for purchasing new office furniture from an office-supply company. Business-to-business e-commerce automates the entire process. Employees go directly to the supplier's Web site, find the item in its catalog, and order what they need at a price set by the employee's company. If approval is required, the approver is notified automatically. As the use of e-commerce systems grows, companies are phasing out their traditional systems. The resulting growth of e-commerce is creating many new business opportunities.

Phishing for Visa Card Customers

A new type of Internet fraud is becoming increasingly prevalent—and costing consumers their money and identity. This latest scam is called *phishing* because it uses e-mail and Web sites as bait to lure consumers into revealing private information.

E-commerce systems rely on the trust of the participants. If they do not trust the technology to provide safe and secure transactions, e-commerce would have no future. Although network research has produced more secure connections between two parties over the Internet, no foolproof systems exist to guarantee that the participants are who they claim to be. Phishing scams exploit this system vulnerability.

A phishing scam was recently launched against Visa card customers and serves as a textbook example of the technique. A mass e-mail was sent to Internet users with an official-looking Visa return address, claiming to have come from Visa International Services. Sending e-mail with a forged return address is a common practice in Internet fraud and is formally referred to as *spoofing*. The e-mail stated that Visa had implemented a new "security system to help you to avoid possible fraud actions" and asked users to click a link to "reactivate your account." The link was printed as *www.visa.com*, but when users clicked the link, it took them to a Web site that resembled the Visa Web site—with an official Visa logo, artwork, and design—but was not owned by Visa. The site asked customers to enter personal information, including their Visa credit card number. The scam artists then had both a customer's account number and name.

The 2003 holiday season saw a 400 percent increase in phishing scams, with 60 unique attacks launched and more than 60 million fraudulent e-mails sent out. It is estimated that 5 to 20 percent of recipients respond to phishing scams. In the Visa scam, the owners of the fraudulent site shut down and disappeared prior to discovery, taking with them an unknown quantity of customer records. The information they stole could be sold in the underground credit card market and used by crooks and thieves to assume the identity of the victims and make illegal purchases.

Phishing scams are increasingly difficult to detect. The fraudulent e-mails and Web sites look identical to original corporate correspondence and Web sites. Web addresses appear legitimate and might even employ secure connections (identified by the closed-lock icon at the bottom of the browser window). Such scams make it difficult for legitimate businesses to communicate electronically with their customers and to conduct business online. "At stake is our very trust that the Internet can be relied upon for safe and secure commerce and communications," says Dave Jevans, chairman of the Anti-Phishing Working Group (*www.antiphishing.org*).

Software tools designed to detect phishing scams typically identify only 50 to 70 percent of all phony systems. The only defense consumers have against such scams is education—and caution. Be leery of any e-mail from a company that asks you to visit a Web page to provide private information. Check with the company at its official Web site to confirm that such requests are legitimate before complying.

Critical Thinking Questions

1. How can people protect themselves from becoming a victim of a phishing scam?
2. What action can people take if they discover that their private information has been stolen?

What Would You Do?

You've received an e-mail from your college's financial aid department that congratulates you on being the recipient of funds from a newly launched grant program. To receive your $2,000 for this semester, you are required to visit the financial aid Web site (*www.financial-aid.yourschool.com*) and submit a brief online application form. After filling out the form, which collects information such as your name, address, phone, date of birth, school ID number, Social Security number, and bank-account number (for automatic deposit), you click the *Submit* button and head out to celebrate your good fortune. After a week, the money has yet to be deposited, and you are getting concerned.

3. What in this scenario suggests that this might be a phishing scam?
4. If you were responsible for information security at your school, what system might you design to assure students that official school correspondence really comes from the school and not from an imposter?

SOURCES: Paul Roberts, "Latest 'Phishing' Scam Targets Visa Customers," *Computerworld*, December 26, 2003, *www.computerworld.com*; "Growth in Internet Fraud to Be Key Concern in 2004," *Electronic Commerce News*, January 5, 2004; the Anti-Phishing group Web site, *www.anti-phishing.org*, accessed January 17, 2004.

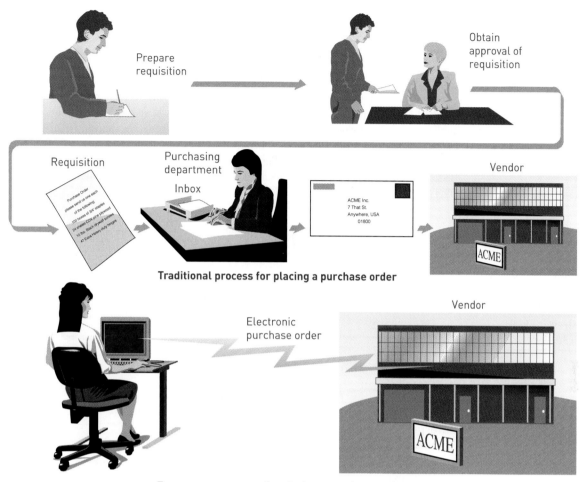

Traditional process for placing a purchase order

E-commerce process for placing a purchase order

In addition to e-commerce, business information systems include the use of telecommunications and the Internet to perform many related tasks. *Electronic procurement (e-procurement)*, for example, involves using information systems and the Internet to acquire parts and supplies using information systems and the Internet. *Electronic business (e-business)* goes beyond e-commerce to include the use of information systems and the Internet to perform all business-related tasks and functions, such as accounting, finance, marketing, manufacturing, and human resource activities. *Electronic management (e-management)* involves the use of information systems and the Internet to manage profit and nonprofit organizations, including governmental agencies, military, religious, and charitable organizations. E-management includes all aspects of staffing and hiring, directing, controlling, and other management tasks.

 Figure 1.6

E-Commerce Greatly Simplifies Purchasing

Transaction Processing Systems and Enterprise Resource Planning

Since the 1950s, computers have been used to perform common business applications. The objective of many of these early systems was to reduce costs by automating many routine, labor-intensive business systems. Today, transaction processing systems and enterprise resource planning are often used to perform common business applications.

Transaction Processing Systems

A **transaction** is any business-related exchange, such as payments to employees, sales to customers, or payments to suppliers. Thus, processing business transactions was the first application of computers for most organizations. A **transaction processing system (TPS)** is an organized collection of people, procedures, software, databases, and devices used to record

transaction
Any business-related exchange such as payments to employees, sales to customers, and payments to suppliers.

transaction processing system (TPS)

An organized collection of people, procedures, software, databases, and devices used to record completed business transactions.

completed business transactions. To understand a transaction processing system is to understand basic business operations and functions.

One of the first business systems to be computerized was the payroll system (see Figure 1.7). The primary inputs for a payroll TPS are the numbers of employee hours worked during the week and pay rate. The primary output consists of paychecks. Early payroll systems were able to produce employee paychecks, along with important employee-related reports required by state and federal agencies, such as the Internal Revenue Service. Other routine applications include sales ordering, customer billing and customer relationship management, inventory control, and many other applications. Some automobile companies, for example, use their TPS to buy billions of dollars of needed parts each year through Internet sites. Because these systems handle and process daily business exchanges, or transactions, they are all classified as TPSs.

Figure 1.7

A Payroll Transaction Processing System

The inputs (numbers of employee hours worked and pay rates) go through a transformation process to produce outputs (paychecks).

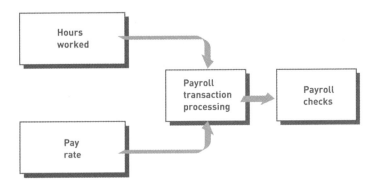

enterprise resource planning (ERP) system

A set of integrated programs capable of managing a company's vital business operations for an entire multisite, global organization.

Enterprise Resource Planning

An **enterprise resource planning (ERP) system** is a set of integrated programs that is capable of managing a company's vital business operations for an entire multisite, global organization. ERP systems can replace many applications with one unified set of programs. Sutter Health, a large network of 33 hospitals with more than 4 million patients in northern California, uses an ERP system to process medical transactions and to exchange information between hospitals, physicians, and employees.[19] Although the scope of an ERP system can vary from company to company, most ERP systems provide integrated software to support the manufacturing and finance business functions of an organization. In such an environment, a forecast is prepared that estimates customer demand for several weeks. The ERP system checks what is already available in finished product inventory to meet the projected demand. Manufacturing must then produce inventory to eliminate any shortcomings. In developing the production schedule, the ERP system checks inventory levels and determines what needs to be ordered to meet the schedule. Most ERP systems also have a purchasing subsystem that orders the needed items. In addition to these core business processes, some ERP systems can support additional business functions, such as human resources, sales, and distribution. Customer relationship management (CRM) features, for example, help organizations manage all aspects of customer interactions, including inquiries, sales, delivery of products and services, and support after the sale.

Information and Decision Support Systems

Although early accounting and financial TPSs were already valuable, companies soon realized that the data stored in these systems could be used to help managers make better decisions in their respective business areas, whether human resource management, marketing, or administration. Satisfying the needs of managers and decision makers continues to be a major factor in developing information systems, discussed later.

Management Information Systems

A **management information system (MIS)** is an organized collection of people, procedures, software, databases, and devices used to provide routine information to managers and decision makers. The focus of an MIS is primarily on operational efficiency. Marketing, production, finance, and other functional areas are supported by MISs and linked through a common database. Management information systems typically provide standard reports generated with data and information from the TPS (see Figure 1.8).

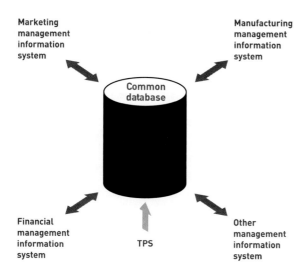

management information system (MIS)
An organized collection of people, procedures, software, databases, and devices used to provide routine information to managers and decision makers.

Figure 1.8

Functional management information systems draw data from the organization's transaction processing system.

Decision Support Systems

By the 1980s, dramatic improvements in technology resulted in information systems that were less expensive but more powerful than earlier systems. People at all levels of organizations began using personal computers to do a variety of tasks; they were no longer solely dependent on the IS department for all their information needs. So, people quickly recognized that computer systems could support additional decision-making activities. A **decision support system (DSS)** is an organized collection of people, procedures, software, databases, and devices used to support problem-specific decision making. The focus of a DSS is on decision-making effectiveness. Whereas an MIS helps an organization "do things right," a DSS helps a manager "do the right thing." Oxford Bookstore, located in Calcutta, uses a DSS and the Internet to allow book lovers in India to purchase their favorite books at Oxford's traditional retail stores or through its Internet site. The Internet site provides a wealth of information to help people make better book-purchasing decisions.[20] Blue Cross of Pennsylvania uses a DSS from InterQual to help it support level-of-care decisions.[21]

decision support system (DSS)
An organized collection of people, procedures, software, databases, and devices used to support problem-specific decision making.

The essential elements of a DSS include a collection of models used to support a decision maker or user (model base), a collection of facts and information to assist in decision making (database), and systems and procedures (dialogue manager) that help decision makers and other users interact with the DSS (see Figure 1.9). Software is often used to manage the database (the database management system, DBMS) and the model base (the model management system, MMS).

In addition to DSSs that support individual decision making, group decision support systems and executive support systems use the same overall approach of a DSS. A group support system, also called a *group decision support system*, includes the DSS elements just described and software, called *groupware*, to help groups make effective decisions. An executive support system, also called an *executive information system*, helps top-level managers, including a firm's president, vice presidents, and members of the board of directors, make better decisions. An executive support system can be used to assist with strategic planning, top-level organizing and staffing, strategic control, and crisis management.

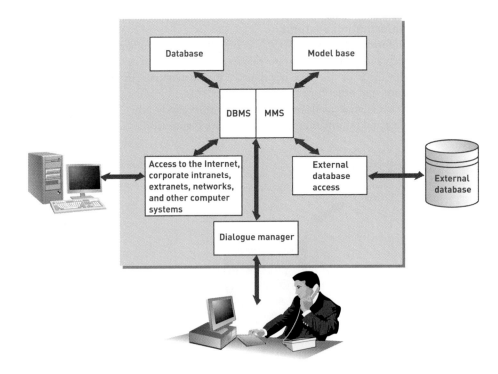

Specialized Business Information Systems: Artificial Intelligence, Expert Systems, and Virtual Reality

artificial intelligence (AI)
A field in which the computer system takes on the characteristics of human intelligence.

In addition to TPSs, MISs, and DSSs, organizations often use specialized systems. One of these systems is based on the notion of **artificial intelligence** (**AI**), in which the computer system takes on the characteristics of human intelligence. The field of artificial intelligence includes several subfields (see Figure 1.10).

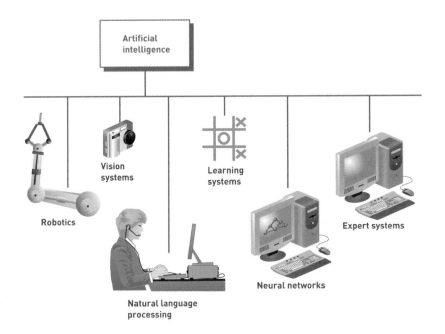

Artificial Intelligence

Robotics is an area of artificial intelligence in which machines take over complex, dangerous, routine, or boring tasks, such as welding car frames or assembling computer systems and components. Vision systems allow robots and other devices to have "sight" and to store and process visual images. Natural language processing involves the ability of computers to understand and act on verbal or written commands in English, Spanish, or other human languages. Learning systems give computers the ability to learn from past mistakes or experiences, such as playing games or making business decisions, and neural networks is a branch of artificial intelligence that allows computers to recognize and act on patterns or trends. Some successful stock, options, and futures traders use neural networks to spot trends and make them more profitable with their investments.

Expert Systems

Expert systems give the computer the ability to make suggestions and act like an expert in a particular field. The unique value of expert systems is that they allow organizations to capture and use the wisdom of experts and specialists. Therefore, years of experience and specific skills are not completely lost when a human expert dies, retires, or leaves for another job. Expert systems can be applied to almost any field or discipline. Expert systems have been used to monitor complex systems such as nuclear reactors, perform medical diagnoses, locate possible repair problems, design and configure IS components, perform credit evaluations, and develop marketing plans for a new product or new investment strategies. The collection of data, rules, procedures, and relationships that must be followed to achieve value or the proper outcome is contained in the expert system's **knowledge base**.

Virtual Reality

Virtual reality is the simulation of a real or imagined environment that can be experienced visually in three dimensions. Virtual worlds can be animated, interactive, and shared.

A variety of input devices such as head-mounted displays (see Figure 1.11), data gloves (see Figure 1.12), joysticks, and handheld wands allow the user to navigate through a virtual environment and to interact with virtual objects. Directional sound, tactile and force feedback devices, voice recognition, and other technologies are used to enrich the immersive experience. Several people can share and interact in the same environment. Because of this ability, virtual reality can be a powerful medium for communication, entertainment, and learning.

expert system
A system that gives a computer the ability to make suggestions and act like an expert in a particular field.

knowledge base
The collection of data, rules, procedures, and relationships that must be followed to achieve value or the proper outcome.

virtual reality
The simulation of a real or imagined environment that can be experienced visually in three dimensions.

Figure 1.11

A Head-Mounted Display

The head-mounted display (HMD) was the first device of its kind providing the wearer with an immersive experience. A typical HMD houses two miniature display screens and an optical system that channels the images from the screens to the eyes, thereby presenting a stereo view of a virtual world. A motion tracker continuously measures the position and orientation of the user's head and allows the image-generating computer to adjust the scene representation to the current view. As a result, the viewer can look around and walk through the surrounding virtual environment.

(Source: Image courtesy of 5DT, Inc., *www.5DT.com*.)

It is difficult to predict where information systems and technology will be in 10 to 20 years. It seems, however, that we are just beginning to discover the full range of their usefulness. Technology has been improving and expanding at an increasing rate; dramatic growth and change are expected for years to come. Without question, a knowledge of the effective use of information systems will be critical for managers both now and in the long term. But how are these information systems created?

Figure 1.12

A Data Glove

Realistic interactions with virtual objects via such devices as a data glove that senses hand position allow for manipulation, operation, and control of virtual worlds. (Source: Image courtesy of 5DT, Inc., *www.5DT.com*.)

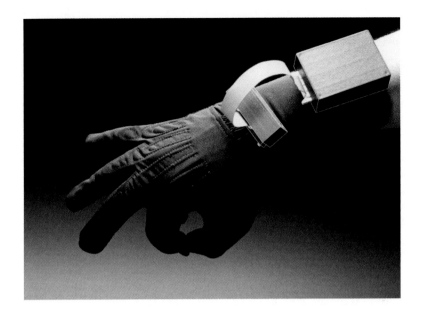

SYSTEMS DEVELOPMENT

systems development
The activity of creating or modifying existing business systems.

Systems development is the activity of creating or modifying existing business systems. People inside a company can develop systems, or companies can use *outsourcing*, hiring an outside company to perform some or all of a systems development project. Outsourcing allows a company to focus on what it does best and delegate other functions to companies with expertise in systems development. Cox Insurance Holdings, for example, outsourced its commercial underwriting operations to another company.[22] Outsourcing enabled Cox Insurance to streamline its operations and reduce costs. Outsourcing, however, is not the best alternative for all companies. Toyota recently stopped outsourcing its financial services and started to perform the financial services function internally.[23] According to the director of Toyota Financial Services, "You depend on that service provider. You worry about whether or not it will be in business next year, and whether or not it will be able to service you consistently throughout the terms of the agreement and beyond." Other companies have used outsourcing for software development, database development, and other aspects of systems development.

The first two steps of systems development are systems investigation and analysis. The goal of the *systems investigation* is to gain a clear understanding of the problem to be solved or opportunity to be addressed. A cruise line company, for example, might launch a systems investigation to determine whether a development project is feasible to automate purchasing at ports around the world. After an organization understands the problem, the next question to be answered is "Is the problem worth solving?" Given that organizations have limited resources—people and money—this question deserves careful consideration. If the decision is to continue with the solution, the next step, *systems analysis*, defines the problems and opportunities of the existing system.

Systems design determines how the new system will work to meet the business needs defined during systems analysis. *Systems implementation* involves creating or acquiring the various system components (hardware, software, databases, etc.) defined in the design step, assembling them, and putting the new system into operation. The purpose of *systems maintenance and review* is to check and modify the system so that it continues to meet changing needs of today's organizations.

ORGANIZATIONS AND INFORMATION SYSTEMS

An **organization** is a formal collection of people and other resources established to accomplish a set of goals. The primary goal of a for-profit organization is to maximize shareholder value, often measured by the price of the company stock. Nonprofit organizations include social groups, religious groups, universities, and other organizations that do not have profit as the primary goal.

An organization is a system. Money, people, materials, machines and equipment, data, information, and decisions are constantly in use in any organization. As shown in Figure 1.13, resources such as materials, people, and money are input to the organizational system from the environment, go through a transformation mechanism, and are output to the environment. The outputs from the transformation mechanism are usually goods or services. The goods or services produced by the organization are of higher relative value than the inputs alone. Through adding value or worth, organizations attempt to achieve their goals.

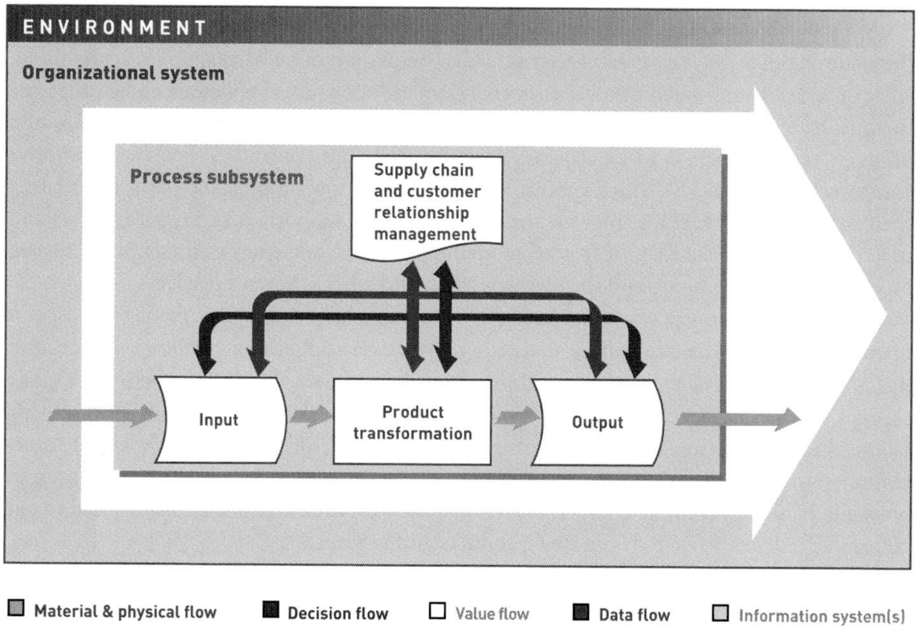

organization
A formal collection of people and other resources established to accomplish a set of goals.

| Material & physical flow | Decision flow | Value flow | Data flow | Information system(s) |

Figure 1.13

A General Model of an Organization

Information systems support and work within all parts of an organizational process. Although not shown in this simple model, input to the process subsystem can come from internal and external sources. Just prior to entering the subsystem, data is external. After it enters the subsystem, it becomes internal. Likewise, goods and services can be output to either internal or external systems.

All business organizations contain a number of processes. Providing value to a stakeholder—customer, supplier, manager, or employee—is the primary goal of any organization. The value chain, first described by Michael Porter, a prominent management theorist, in a 1985 *Harvard Business Review* article, is a concept that reveals how organizations can add value to their products and services. The **value chain** is a series (chain) of activities that includes inbound logistics, warehouse and storage, production, finished product storage, outbound logistics, marketing and sales, and customer service (see Figure 1.14). Each of these activities is investigated to determine what can be done to increase the value perceived by a customer. Depending on the customer, value might mean lower price, better service, higher quality, or uniqueness of product. The value comes from the skill, knowledge, time, and energy invested by the company. By adding a significant amount of value to their products and services, companies will ensure further organizational success. Cessna Aircraft, for example, has used supply chain management to improve the quality of supplies by about 86 percent and increase material availability by 28 percent.[24]

value chain
A series (chain) of activities that includes inbound logistics, warehouse and storage, production, finished product storage, outbound logistics, marketing and sales, and customer service.

The Value Chain of a Manufacturing Company

The management of raw materials, inbound logistics, and warehouse and storage facilities is called *upstream management*, and the management of finished product storage, outbound logistics, marketing and sales, and customer service is called *downstream management*.

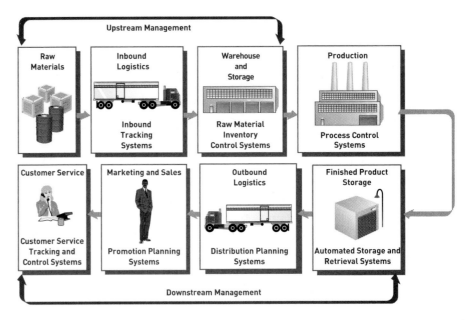

Supply chain and customer relationship management are two key aspects of managing the value chain. *Supply chain management (SCM)* helps determine what supplies are required, what quantities are needed to meet customer demand, how the supplies are to be processed (manufactured) into finished goods and services, and how the shipment of supplies and products to customers is to be scheduled, monitored, and controlled. For an automotive company, for example, SCM is responsible for identifying key supplies and parts, negotiating with supply and parts companies for the best prices and support, ensuring all supplies and parts are available when they are needed to manufacture cars and trucks, and sending finished products to dealerships around the country when and where they are needed.

Customer relationship management (CRM) programs help a company manage all aspects of customer encounters, including marketing and advertising, sales, customer service after the sale, and programs to help keep and retain loyal customers.[25] CRM can help a company collect customer data, contact customers, educate customers on new products, and actively sell products to existing and new customers. Often, CRM software uses a variety of information sources, including sales from different retail stores, surveys, e-mails, and Internet browsing habits, to compile comprehensive customer profiles. CRM can also be used to get customer feedback to help design new products and services.

Organizational Culture and Change

culture
A set of major understandings and assumptions shared by a group.

organizational culture
The major understandings and assumptions for a business, a corporation, or an organization.

organizational change
The responses that are necessary for for-profit and nonprofit organizations to plan for, implement, and handle change.

The internal collective beliefs of and degree of change within organizations greatly affect their employees, processes, and overall effectiveness. **Culture** is a set of major understandings and assumptions shared by a group, for example, within an ethnic group or a country. **Organizational culture** consists of the major understandings and assumptions for a business, a corporation, or an organization. The understandings, which can include common beliefs, values, and approaches to decision making, are often not stated or documented as goals or formal policies. Employees, for example, might be expected to be clean-cut, wear conservative outfits, and be courteous in dealing with all customers. Sometimes, organizational culture is formed over years. In other cases, it can be formed rapidly by top-level managers—for example, implementation of a "casual Friday" dress policy. **Organizational change** deals with how for-profit and nonprofit organizations plan for, implement, and handle change. Change can be caused by internal or external factors. Internal factors include activities initiated by employees at all levels. External factors include activities wrought by competitors, stockholders, federal and state laws, community regulations, natural occurrences (such as hurricanes), and general economic conditions. Many European countries, for example, adopted the euro, a single European currency, which changed how financial companies do business and how they use their information systems.

Change can be sustaining or disruptive.[26] *Sustaining change* can help an organization improve the raw materials supply, the production process, and the products and services offered by the organization. New manufacturing equipment to make disk drives is an example of a sustaining change. The new equipment might reduce the costs of producing the disk drives and improve overall performance. *Disruptive change*, on the other hand, often harms an organization's performance or even puts it out of business. The 3.5-inch hard disk drive was a disruptive technology for companies that produced the 5.25-inch hard disk drive. When it was first introduced, the 3.5-inch drive was slower, had lower capacity, and lower demand than the existing 5.25-inch disk drives. Over time, however, the 3.5-inch drive improved and replaced the 5.25-inch drive in performance and demand. Some companies that produced the 5.25-inch drives that didn't change didn't survive. Today, many of the 5.25-inch drive companies are out of business. In general, disruptive technologies might not originally have good performance, low cost, or even strong demand. Over time, however, they often replace existing technologies. They can cause good, stable companies to fail when they don't change or adopt the new technology.

Technology Diffusion, Infusion, and Acceptance

The use of technology also affects organizational processes. Even if a company buys or develops new computerized systems, managers and employees might never use them. Or, new systems might not be used to their potential. Millions of dollars can be wasted as a result. The extent to which new computerized systems are used throughout an organization can be measured by the amount of technology diffusion, infusion, and acceptance. **Technology diffusion** is a measure of how widely technology is spread throughout an organization. An organization in which computers and information systems are located in most departments and areas has a high level of technology diffusion.[27] Some online merchants, such as Amazon.com, have a high level of diffusion and use computer systems to perform most of their business functions, including marketing, purchasing, and billing. **Technology infusion**, on the other hand, is the extent to which technology permeates an area or department. In other words, it is a measure of how deeply embedded technology is in an area of the organization. Some architectural firms, for example, use computers in all aspects of designing a building or structure. This design area, thus, has a high level of infusion. Of course, it is possible for a firm to have a high level of infusion in one aspect of its operations and a low level of diffusion overall. The architectural firm might use computers in all aspects of design (high infusion in the design area) but might not use computers to perform other business functions, including billing, purchasing, and marketing (low diffusion).

Although an organization might have a high level of diffusion and infusion, with computers throughout the organization, it does not necessarily mean that information systems are being used to their full potential. In fact, the assimilation and use of expensive computer technology throughout organizations varies greatly.[28] One reason is a low degree of acceptance and use of the technology among some managers and employees. Research has attempted to explain the important factors that enhance or hinder the acceptance and use of information systems.[29] A number of possible explanations of technology acceptance and usage have been studied. The **technology acceptance model (TAM)** specifies the factors that can lead to higher acceptance and usage of technology in an organization, including the perceived usefulness of the technology, the ease of its use, the quality of the information system, and the degree to which the organization supports the use of the information system.[30] Companies hope that a high level of diffusion, infusion, and acceptance will lead to greater performance and profitability.[31]

Organizations in a Global Society

Organizations do not operate in a vacuum. The society or societies in which they operate contribute to or detract from their operations and overall success. Increasingly, organizations operate in a global society. An American company, for example, can get inputs for products and services from Europe, assemble them in Asia, and ship them to customers in

Cingular's planned acquisition of AT&T Wireless will combine the strengths of the two companies and is expected to create customer benefits and growth prospects neither company could have achieved on its own. Together, the companies can provide better coverage, improve reliability, enhance call quality, and offer a wide array of new and innovative services for consumers.

(Source: AP/Wide World Photos.)

technology diffusion
A measure of how widely technology is spread throughout the organization.

technology infusion
The extent to which technology is deeply integrated into an area or department.

technology acceptance model (TAM)
A model that describes the factors that can lead to higher acceptance and usage of technology.

Australia and New Zealand. After-the-sale support can be given by a call center in India. The Internet and telecommunications make this trend possible.

There are, however, many challenges to operating in a global society. Some countries have seen high-paying jobs transferred to other countries, where labor and production costs are lower. In addition, every country has a set of customs, cultures, standards, politics, and laws that can make it difficult for businesses operating around the world. Language can also be a potential problem. Some companies that outsourced their call centers to foreign countries are now moving them back because of customer complaints. It can also be more difficult to manage and control operations in different countries. Many of today's organizations operate globally to give them a competitive advantage, discussed next.

COMPETITIVE ADVANTAGE

competitive advantage
A significant and (ideally) long-term benefit to a company over its competition.

A **competitive advantage** is a significant and (ideally) long-term benefit to a company over its competition. Establishing and maintaining a competitive advantage is complex, but a company's survival and prosperity depend on its success in doing so. In his book *Good To Great*, Jim Collins outlined how technology can be used to accelerate companies from good to great.[32] Table 1.4 shows how a few companies accomplished this move. Ultimately, it is not how much a company spends on information systems but how investments in technology are made and managed. Companies can spend less and get more value.

Table 1.4

How Some Companies Used Technology to Move from Good to Great

(Source: Data from Jim Collins, *Good to Great*, Harper Collins Books, 2001, p. 300.)

Company	Business	Competitive Use of Information Systems
Circuit City	Consumer electronics	Developed sophisticated sales and inventory-control systems to deliver a consistent experience to customers
Gillette	Shaving products	Developed advanced computerized manufacturing systems to produce high-quality products at low cost
Walgreens	Drug and convenience stores	Developed satellite communications systems to link local stores to centralized computer systems
Wells Fargo	Financial services	Developed 24-hour banking, ATMs, investments, and increased customer service using information systems

Factors That Lead Firms to Seek Competitive Advantage

A number of factors can lead to the attainment of competitive advantage. Michael Porter, in his research on competitive forces, suggested a now widely accepted model, sometimes called the **five-forces model**. The five forces include (1) the rivalry among existing competitors, (2) the threat of new entrants, (3) the threat of substitute products and services, (4) the bargaining power of buyers, and (5) the bargaining power of suppliers. The more these forces combine in any instance, the more likely firms will seek competitive advantage and the more dramatic the results of such an advantage will be.

five-forces model
A widely accepted model that identifies five key factors that can lead to attainment of competitive advantage, including (1) the rivalry among existing competitors, (2) the threat of new entrants, (3) the threat of substitute products and services, (4) the bargaining power of buyers, and (5) the bargaining power of suppliers.

Rivalry Among Existing Competitors

The rivalry among existing competitors is an important factor leading firms to seek competitive advantage. Typically, highly competitive industries are characterized by high fixed costs of entering or leaving the industry, low degrees of product differentiation, and many competitors. Although all firms are rivals with their competitors, industries with stronger rivalries tend to have more firms seeking competitive advantage. To compete with existing competitors, companies are constantly analyzing how their resources and assets are used. The *resource-based view* is an approach to acquiring and controlling assets or resources that can help the company achieve a competitive advantage.[33] Using the resource-based view, for example, a transportation company might decide to invest in radio-frequency technology to tag and trace products as they move from one location to another.

Threat of New Entrants

The threat of new entrants is another important force leading an organization to seek competitive advantage. A threat exists when entry and exit costs to the industry are low and the technology needed to start and maintain the business is commonly available. For example, consider a small restaurant. The owner does not require millions of dollars to start the business, food costs do not go down substantially for large volumes, and food processing and preparation equipment is commonly available. When the threat of new market entrants is high, the desire to seek and maintain competitive advantage to dissuade new market entrants is usually high.

Threat of Substitute Products and Services

The more consumers are able to obtain similar products and services that satisfy their needs, the more likely firms are to try to establish competitive advantage. For example, consider the photographic industry. When digital cameras started to become more popular, traditional film companies had to respond to stay competitive and profitable. Traditional film companies, such as Kodak and others, started to offer additional products and enhanced services, including digital cameras, the ability to produce digital images from traditional film cameras, and Web sites that could be used to store and view pictures.

Bargaining Power of Customers and Suppliers

Large buyers tend to exert significant influence on a firm. This influence can be diminished if the buyers are unable to use the threat of going elsewhere. Suppliers can help an organization obtain a competitive advantage. In some cases, suppliers have entered into strategic alliances with firms. When they do so, suppliers act as if they were part of the company. Suppliers and companies can use telecommunications to link their computers and personnel to obtain fast reaction times and the ability to get the parts or supplies when they are needed to satisfy customers. Government agencies are also using strategic alliances.[34] The investigative units of the U.S. Customs and Immigration and Naturalization Service entered into a strategic alliance to streamline operations and to place all investigative operations into a single department.

In the restaurant industry, competition is fierce because entry costs are low. So, a small restaurant that enters the market can be a threat to existing restaurants.

(Source: © Owen Franken/CORBIS.)

Strategic Planning for Competitive Advantage

To be competitive, a company must be fast, nimble, flexible, innovative, productive, economical, and customer oriented.[35] It must also align its IS strategy with general business strategies and objectives.[36] Given the five market forces just mentioned, Porter proposed three general strategies to attain competitive advantage: altering the industry structure, creating new products and services, and improving existing product lines and services. Subsequent research into the use of information systems to help an organization achieve a competitive advantage has confirmed and extended Porter's original work to include additional strategies—such as forming alliances with other companies, developing a niche market, maintaining competitive cost, and creating product differentiation.[37]

Altering the Industry Structure

Altering the industry structure is the process of changing the industry to become more favorable to the company or organization. The introduction of low-fare airline carriers, such as Southwest Airlines, has forever changed the airline industry, making it difficult for traditional airline companies to make high profit margins. To fight back, airline companies such as Delta are launching their own low-fare flights.[38] Delta claims that its Song airline will be one of the first "all digital" airlines. The approach will include a host of services on the flights, including entertainment, satellite TV, and many similar services.

A company can also attempt to create barriers to new companies entering the industry. An established organization that acquires expensive new technology to provide better

strategic alliance (strategic partnership)
An agreement between two or more companies that involves the joint production and distribution of goods and services.

products and services can discourage new companies from getting into the marketplace. Creating strategic alliances can also have this effect. A **strategic alliance**, also called a **strategic partnership**, is an agreement between two or more companies that involves the joint production and distribution of goods and services. Samsung Electronics and Echelon Corporation, for example, signed a strategic alliance agreement to develop and market electronic devices that can be connected to each other and the Internet.[39] The alliance is oriented to the home networking market.

Creating New Products and Services

Creating new products and services is always an approach that can help a firm gain a competitive advantage, and it is especially true of the computer industry and other high-tech businesses. If an organization does not introduce new products and services every few months, the company can quickly stagnate, lose market share, and decline. Companies that stay on top are constantly developing new products and services. A large U.S. credit-reporting agency, for example, can use its information system to help it explore new products and services in different markets. Delta Airlines created a new service by installing hundreds of self-service kiosks to reduce customer check-in times. The new kiosks allow Delta customers to check in, get boarding passes, change seats, and sign up for standby flights or upgrades. On average, the kiosks save flyers from 5 to 15 minutes for each check-in at an airport terminal.[40]

Improving Existing Product Lines and Services

Improving existing product lines and services is another approach to staying competitive. The improvements can be either real or perceived. Manufacturers of household products are always advertising new and improved products. In some cases, the improvements are more perceived than real refinements; usually, only minor changes are made to the existing product. Many food and beverage companies are introducing "Healthy" and "Low-Carb" product lines. Some companies are now starting to put radio-frequency ID (RFID) tags on their products to identify them and track their location as they move from one location to another.[41] Customers and managers can instantly locate products as they are shipped from suppliers, to the company, to warehouses, and, finally, to customers. In another case, Metro, the third largest retail store in Europe, used portable computers to show shoppers where products are located in the store and to display discounted prices and any specials.[42]

Using Information Systems for Strategic Purposes

In simplest terms, competitive advantage is usually embodied in either a product or service that has the most added value to consumers and that is unavailable from the competition or in an internal system that delivers benefits to a firm not enjoyed by its competition. Although it can be difficult to develop information systems to provide a competitive advantage, some organizations have done so with success. A classic example is SABRE, a sophisticated computerized reservation system installed by American Airlines and one of the first information systems recognized for providing competitive advantage. Travel agents used this system for rapid access to flight information, offering travelers reservations, seat assignments, and ticketing. The travel agents also achieved an efficiency benefit from the SABRE system. Because SABRE displayed American Airline flights whenever possible, it also gave the airline a long-term, significant competitive advantage. Today, SABRE is aggressively seeking a competitive advantage by investing heavily in e-commerce technology and developing Internet travel sites. It invested more than $200 million in technology recently. Much of the investment was in the company's Travelocity.com site (the second largest online travel agency) and GetThere.com.[43] Increasingly, companies are using e-commerce as part of a strategy to achieve a competitive advantage. Table 1.5 lists several examples of how companies have attempted to gain a competitive advantage.

Table 1.5

Competitive Advantage Factors and Strategies

Factors That Lead to Attainment of a Competitive Advantage	Alter Industry Structure	Create New Products and Services	Improve Existing Product Lines and Services
Rivalry among existing competitors	Netflix changes the industry structure with its use of online ordering for DVDs.	Apple, Dell, and other PC makers develop computers that excel at downloading Internet music and playing the music on high-quality speakers.	Food and beverage companies offer "healthy" and "light" product lines.
Threat of new entrants	HP and Compaq merge to form a large Internet and media company.	Apple Computer introduces an easy-to-use iMac computer that can be used to create and edit home movies.	Starbucks offers new coffee flavors at premium prices.
Threat of substitute products and services	Ameritrade and other discount stockbrokers offer low fees and research on the Internet.	Wal-Mart uses technology to monitor inventory and product sales to determine the best mix of products and services to offer at various stores.	Cosmetic companies add sunscreen to their product lines.
Bargaining power of buyers	Ford, GM, and others require that suppliers locate near their manufacturing facilities.	Investors and traders of the Chicago Board of Trade (CBOT) put pressure on the institution to implement electronic trading.	Retail clothing stores require manufacturing companies to reduce order lead times and improve materials used in the clothing.
Bargaining power of suppliers	American Airlines develops SABRE, a comprehensive travel program used to book airline, car rental, and other reservations.	Broadcom develops a chip for wireless computing used in notebook PCs from Apple, Dell, Hewlett-Packard, and Gateway.	Hayworth, a supplier of office furniture, has a computerized-design tool that helps it design new office systems and products.

PERFORMANCE-BASED INFORMATION SYSTEMS

At least three major stages have occurred in the business use of information systems. The first stage started in the 1960s and was oriented toward cost reduction and productivity. This stage generally ignored the revenue side, not looking for opportunities to increase sales via the use of information systems. The second stage started in the 1980s and was defined by Porter and others. It was oriented toward gaining a competitive advantage. In many cases, companies spent large amounts on information systems and ignored the costs. Today, we are seeing a shift from strategic management to performance-based management in many IS organizations.[44] This third stage carefully considers both strategic advantage and costs. This stage uses productivity, return on investment (ROI), net present value, and other measures of performance. Figure 1.15 illustrates these stages. This balanced approach attempts to reduce costs and increase revenues. Aviall, an aviation parts company, for example, invested over $3 million in a new Web site—Aviall.com—that slashed inventory ordering costs from $9 to $.39 per order.[45]

Productivity

Developing information systems that measure and control productivity is a key element for most organizations. **Productivity** is a measure of the output achieved divided by the input required. A higher level of output for a given level of input means greater productivity; a lower level of output for a given level of input means lower productivity. Consider a tax

productivity
A measure of the output achieved divided by the input required.

Figure 1.15

Three Stages in the Business Use of Information Systems

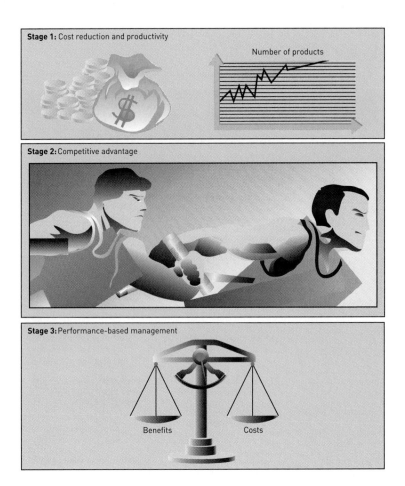

preparation firm, where productivity can be measured by the hours spent on preparing tax returns divided by the total hours the employee worked. For example, in a 40-hour week, an employee may have spent 30 hours preparing tax returns. The productivity is thus equal to 30/40, or 75 percent. With administrative and other duties, a productivity level of 75 might be excellent. The numbers assigned to productivity levels are not always based on labor hours—productivity can be based on factors such as the amount of raw materials used, resulting quality, or time to produce the goods or service. In any case, what is important is not the value of the productivity number but how it compares with other time periods, settings, and organizations.

Productivity = (Output/Input) × 100%

After a basic level of productivity is measured, an information system can monitor and compare it over time to see whether productivity is increasing. Then, corrective action can be taken if productivity drops below certain levels. In addition to measuring productivity, an information system can also be used within a process to significantly increase productivity. Thus, improved productivity can result in faster customer response, lower costs, and increased customer satisfaction. See the "Information Systems @ Work" feature for an example of how one company used information systems to streamline its business processes.

OneBeacon Focuses on Information Systems

"It's not uncommon for insurers to have multiple policy systems that are 15 or 20 years old and don't talk to one another," says Larry Goldberg, senior vice president of Sapiens Americas, a leading insurance IS provider. OneBeacon Insurance Group was in such a predicament. Until recently, the process for policy writing took OneBeacon agents a long, arduous week. Field agents filled out application forms and faxed them to the main office. Data-entry clerks would then input the information into a system. The system would provide a quotation, which was then relayed to the agent via telephone. If the customer decided to accept the quote, she would have to wait one week before the coverage would be issued.

This wasteful system was not only time-consuming but also expensive. OneBeacon was losing $50 million a month due to inefficiencies. The expense of the inefficiencies was passed on to customers in the form of higher premiums. The cost of creating new policies was so high that OneBeacon all but gave up its efforts to insure small businesses; the cost of the quotes didn't justify the return.

Under the management of a new chief information officer (CIO), Mike Natan, OneBeacon restructured its business. The change was focused around a new $15 million Web-based policy administration system purchased from Sapiens International. The system resides on servers at the insurance company's home office and is accessed from Web-based applications downloaded to an agent's desktop or notebook computers. Agents use standard Web forms to send customer information to company headquarters, and the system scores the request for risk and approves or disapproves the policy. Any questionable applications are flagged and forwarded for human inspection. Quotes are provided in a matter of seconds, and a policy is issued within 15 minutes. What used to take a week can now be accomplished in the time it takes to sip a cup of coffee. What's more, the cost of providing a quote was more than halved—from $15 per policy to $7.

The new system has allowed OneBeacon once again to offer insurance coverage to small commercial businesses. Between the money saved by the new system and the additional revenue coming in from small businesses, OneBeacon expects to realize a return on its investment in less than a year.

The change in information systems has led to changes in organizational structure. The company has reduced the number of data-entry clerks—those who collected faxed quotation requests and entered them into the old system. But computers aren't replacing human employees at OneBeacon. The company has been hiring hundreds of agents to keep up with the increased demand for its insurance policies. The end result is that OneBeacon has moved its workforce from its back office to the field, where they can generate more revenue for the company rather than burn it up through inefficiencies.

OneBeacon's project follows an industry trend of automating the underwriting process, says analyst Janie Bisker. "You're going to see more of these automated systems," he says. "The general idea is to reduce cost but also increase accuracy and the convenience for the broker and consumer." Where once brokers waited uncomfortably for quotations while customers lost patience, requests are now filled almost instantaneously. The uninterrupted flow of the sale allows brokers to hold a customer's attention and build momentum for increased sales.

Discussion Questions

1. How does OneBeacon's new system improve the productivity of its field agents?
2. How does the new system improve OneBeacon's standing in terms of return on investment, earnings growth, market share, and customer satisfaction?

Critical Thinking Questions

3. If you were a field agent for OneBeacon, what type of information would you need delivered to your portable computer in order to write a policy?
4. It took a considerable amount of courage for OneBeacon to agree to invest $15 million in a new system and reengineer its business. If you were CIO Mike Natan, how would you have sold the company on your idea for the new system in light of what you have learned about performance-based information systems?

SOURCES: Lucas Mearian, "Sticking to Policy at OneBeacon," *Computerworld*, July 7, 2003, *www.computerworld.com*; Matt Glynn, "OneBeacon Adds 130 Jobs, Doubles Space in Amherst," *The Buffalo News*, August 5, 2003, Business Section, p. B6; OneBeacon Web site, *www.onebeacon.com*, accessed January 23, 2004.

In the late 1980s and early 1990s, overall productivity did not seem to increase with increases in investments in information systems. Often called the *productivity paradox*, this situation troubled many economists who were expecting to see dramatic productivity gains.[46] In the early 2000s, however, productivity again seemed on the rise. According to IDC, a marketing research company, investments in information systems will contribute an estimated 80 percent of the productivity gains from 2002 through 2010.[47]

Return on Investment and the Value of Information Systems

return on investment (ROI)
One measure of IS value that investigates the additional profits or benefits that are generated as a percentage of the investment in information systems technology.

One measure of IS value is **return on investment (ROI)**. This measure investigates the additional profits or benefits that are generated as a percentage of the investment in IS technology. A small business that generates an additional profit of $20,000 for the year as a result of an investment of $100,000 for additional computer equipment and software would have a return on investment of 20 percent ($20,000/$100,000). One study investigated the ROI for computer-related training and certification.[48] According to the study, the Microsoft Certified Solution Developer for Microsoft.NET and the Microsoft Certified Database Administrator received ROI values of 170 percent and 122 percent for large organizations. For smaller organizations, the Check Point Certified Security Administrator received an ROI value of 98 percent.

Because of the importance of ROI, many computer companies provide ROI calculators to potential customers. ROI calculators are typically found on a vendor's Web site and can be used to estimate returns.

Ace Hardware's new Web-based inventory management system benefits customers by speeding checkout time. The new system enables real-time viewing and tracking of sales data, replacing a week-long lag time under the old system.

(Source: AP/Wide World Photos.)

Earnings Growth

Another measure of IS value is the increase in profit, or earnings growth, it brings. For instance, suppose a mail-order company, after installing an order-processing system, had a total earnings growth of 15 percent compared with the previous year. Sales growth before the new ordering system was only about 8 percent annually. Assuming that nothing else affected sales, the earnings growth brought by the system, then, was 7 percent. Aviall, an aviation parts company, invested over $30 million in a new computer system to improve inventory control and earnings growth. According to the vice president of information services, "Our competitors thought we were insane. Some investors asked for my resignation." The investment in the new information system was very successful, however. Net earnings rose almost 75 percent.[49]

Market Share

Market share is the percentage of sales that one company's products or services have in relation to the total market. If installing a new online Internet catalog increases sales, it might help a company increase its market share by 20 percent.

Customer Awareness and Satisfaction

Although customer satisfaction can be difficult to quantify, about half of today's best global companies measure the performance of their information systems based on feedback from internal and external users. Some companies use surveys and questionnaires to determine whether the IS investment has increased customer awareness and satisfaction.

Total Cost of Ownership

total cost of ownership (TCO)
The measurement of the total cost of owning computer equipment, including desktop computers, networks, and large computers.

In addition to such measures as return on investment, earnings growth, market share, and customer satisfaction, some companies are also tracking total costs. One measure, developed by the Gartner Group is the **total cost of ownership (TCO)**. This approach breaks total costs into such areas as the cost to acquire the technology, technical support, administrative costs, and end-user operations. Other costs in TCO include retooling and training costs. TCO can be used to get a more accurate estimate of the total costs for systems that range from small PCs to large mainframe systems. Market research groups often use TCO to compare different products and services. For example, a survey of large global enterprises ranked messaging and collaboration software products using the TCO model.[50] The survey analyzed acquisition, maintenance, administration, upgrading, downtime, and training costs. In this survey, Oracle had the lowest TCO of about $65 per user for messaging and collaboration.

CAREERS IN INFORMATION SYSTEMS

Realizing the benefits of any information system requires competent and motivated information systems personnel, and many companies offer excellent job opportunities. Numerous schools have degree programs with such titles as information systems, computer information systems, and management information systems. These programs are typically in business schools and within computer science departments. Degrees in information systems have provided high starting salaries for many students after graduation from college. In addition, students are increasingly looking at business degrees with a global or international orientation.[51]

Many companies, such as FedEx, are joining with colleges and universities to help prepare students for careers.[52] FedEx has opened its $23 million four-story FedEx Technology Institute in Memphis, Tennessee, part of the University of Memphis. Programs include Managing Emerging Technology, Supply Chain Management, Multimedia Arts, Digital Economic and Regional Development, Spatial Analysis, Artificial Intelligence, Life Sciences, and Advanced Learning.

Today, companies are rebounding and looking for IS talent. Online job listings for IS positions, for example, increased in 2003.[53] Demand for IS professionals has grown also in nonprofit organizations and in government. In addition to salary, IS workers seek paid vacation, health insurance, stock options, and flexible hours as important job factors. A study done by *Computerworld* listed the top places to work in information systems (see Table 1.6).

Company Name	Business
Hershey Foods Corp.	Maker of candies
Harley-Davidson, Inc.	Motorcycle manufacturer
University of Miami	Florida university
Network Appliances, Inc.	Hardware storage company
Vision Service Plan	Eye-care provider
Harrah's Entertainment	Casino operator
Saint Luke's Health System	Nonprofit health organization, including eight hospitals
Rich Products, Inc.	Family-owned food company
Discover Financial Services	Credit card company
Software Performance Systems, Inc.	Provider of financial management software

Table 1.6

The Top Places to Work in Information Systems

(Source: Data from Steve Ulfelder, "100 Best Places to Work in IT," *Computerworld*, June 9, 2003, p. 23.)

On the job, computer systems are also making IS professionals' work easier. Called *autonomics* by some, the use of advanced computer systems can help IS professionals spend less time maintaining existing systems and more time solving problems or looking for new opportunities. Colgate-Palmolive Co., for example, uses autonomics to keep its computer systems running well in more than 50 countries.

Opportunities in information systems are not confined to single countries. Some companies seek skilled IS employees from foreign countries, including Russia and India. The U.S. H-1B and L-1 visa programs seek to allow skilled employees from foreign lands into the United States. But not everyone is happy with the H-1B program. Some companies may be firing U.S. workers and hiring less-expensive workers under the H-1B program. Because of a recent difficult economy, some companies may be abusing the H-1B visa program.[54] The L-1 visa program is often used for intracompany transfers for multinational companies. Some people fear, however, that the L-1 visa program could also be used to bring cheap IS personnel into the United States to replace more expensive American workers.[55] The Internet also makes it easier to export IS jobs to other countries.[56] Procter & Gamble estimates that it has reduced costs by about $1 billion by exporting IS jobs to Costa Rica, the Philippines, and Great Britain.

Roles, Functions, and Careers in the IS Department

Information systems personnel typically work in an IS department that employs Web developers, computer programmers, systems analysts, computer operators, and a number of other personnel. They may also work in other functional departments or areas in a support capacity. In addition to technical skills, IS personnel also need skills in written and verbal communication, an understanding of organizations and the way they operate, and the ability to work with people. According to George Voutes, enterprise technology programs manager for Deutsche Asset Management Technology, "We have to get away from strict programming and systems development. Those are skills to get into the field, but we have to train our technology people more like business people and arm them with strong communications skills."[57] IS personnel also need skills to work in groups.[58] Today, many good business and computer science schools require business and communications skills of their graduates. In general, IS personnel are charged with maintaining the broadest perspective on organizational goals. For most medium- to large-sized organizations, information resources are typically managed through an IS department. In smaller businesses, one or more people might manage information resources, with support from outside services—outsourcing. Outsourcing is also popular with larger organizations. According to a study by Gartner, Inc., a technology consulting company, "By 2004, 80 percent of U.S. executive boardrooms will have discussed offshore outsourcing, and more than 40 percent will have completed some type of pilot."[59] As shown in Figure 1.16, the IS organization has three primary responsibilities: operations, systems development, and support.

Operations

The operations component of a typical IS department focuses on the use of information systems in corporate or business unit computer facilities. It tends to focus more on the *efficiency* of information system functions rather than their effectiveness.

The primary function of a system operator is to run and maintain IS equipment. System operators are responsible for starting, stopping, and correctly operating mainframe systems, networks, tape drives, disk devices, printers, and so on. System operators are typically trained at technical schools or through on-the-job experience. Other operations include scheduling, hardware maintenance, and preparation of input and output. Data-entry operators convert data into a form the computer system can use. They can use terminals or other devices to enter business transactions, such as sales orders and payroll data. Increasingly, data entry is being automated—captured at the source of the transaction rather than being entered later. In addition, companies might have local area network and Web or Internet operators who are responsible for running the local network and any Internet sites the company might have.

Systems Development

The systems development component of a typical IS department focuses on specific development projects and ongoing maintenance and review. Systems analysts and programmers, for example, focus on these concerns. The role of a systems analyst is multifaceted. Systems analysts help users determine what outputs they need from the system and construct the plans needed to develop the necessary programs that produce these outputs. Systems analysts then work with one or more programmers to ensure that the appropriate programs are purchased, modified from existing programs, or developed. The major responsibility of a computer programmer is to use the plans developed by

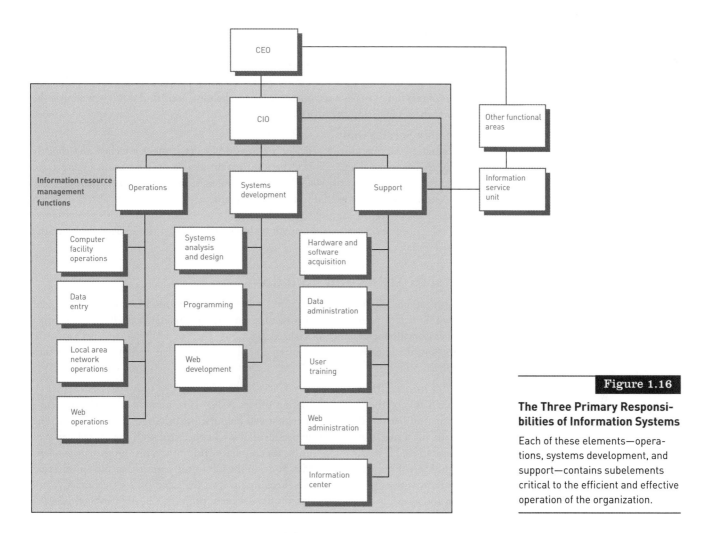

Figure 1.16

The Three Primary Responsibilities of Information Systems

Each of these elements—operations, systems development, and support—contains subelements critical to the efficient and effective operation of the organization.

the systems analyst to develop or adapt one or more computer programs that produce the desired outputs. The main focus of systems analysts and programmers is to achieve and maintain IS effectiveness. To help companies select the best analysts and programmers, companies such as TopCoder offer tests to evaluate the proficiency and competence of existing IS employees or job candidates. Some companies, however, are skeptical of the usefulness of these types of tests.[60]

Support

The support component of a typical IS department focuses on providing user assistance in the areas of hardware and software acquisition and use, data administration, user training and assistance, and Web administration. In many cases, the support function is delivered through an information center.

A database administrator focuses on planning, policies, and procedures regarding the use of corporate data and information. For example, database administrators develop and disseminate information about the corporate databases for developers of IS applications. In addition, the database administrator is charged with monitoring and controlling database use.

User training is a key to get the most from any information system. The support area ensures that appropriate training is available to users. Training can be provided by internal staff or from external sources.

Web administration is another key area of the support function. With the increased use of the Internet and corporate Web sites, Web administrators are sometimes asked to regulate and monitor Internet use by employees and managers to ensure that it is authorized and appropriate. Web administrators also are responsible for maintaining the corporate Web site. Keeping corporate Web sites accurate and current can require substantial resources.

information center
A support function that provides users with assistance, training, application development, documentation, equipment selection and setup, standards, technical assistance, and troubleshooting.

information service unit
A miniature IS department.

The support component typically operates the information center. An **information center** provides users with assistance, training, application development, documentation, equipment selection and setup, standards, technical assistance, and troubleshooting. Although many firms have attempted to phase out information centers, others have changed the focus of this function from technical training to helping users find ways to maximize the benefits of the information resource.

Information Service Units

An **information service unit** is basically a miniature IS department attached and directly reporting to a functional area. Notice the information service unit shown in Figure 1.16. Even though this unit is usually staffed by IS professionals, the project assignments and the resources necessary to accomplish these projects are provided by the functional area to which it reports. Depending on the policies of the organization, the salaries of IS professionals staffing the information service unit might be budgeted to either the IS department or the functional area.

Typical IS Titles and Functions

The organizational chart shown in Figure 1.16 is a simplified model of an IS department in a typical medium-sized or large organization. Many organizations have even larger departments, with increasingly specialized positions, such as librarian, quality assurance manager, and the like. Smaller firms often combine the roles depicted in Figure 1.16 into fewer formal positions.

The Chief Information Officer

The overall role of the chief information officer (CIO) is to employ an IS department's equipment and personnel in a manner that will help the organization attain its goals. The CIO is usually a manager at the vice-presidential level concerned with the overall needs of the organization. He or she is responsible for corporate-wide policy, planning, management, and acquisition of information systems. Some of the CIO's top concerns include integrating IS operations with corporate strategies, keeping up with the rapid pace of technology, and defining and assessing the value of systems development projects. The high level of the CIO position is consistent with the idea that information is one of the organization's most important resources. This individual works with other high-level officers of an organization, including the chief financial officer (CFO) and the chief executive officer (CEO), in managing and controlling total corporate resources. CIOs must work closely with advisory committees, stressing effectiveness and teamwork and viewing information systems as an integral part of the organization's business processes—not an adjunct to the organization. Thus, CIOs need both technical and business skills. For federal agencies, the Clinger-Cohen Act of 1996 required the establishment of a CIO to coordinate the purchase and management of information systems.[61]

Depending on the size of the IS department, there might be several people at senior IS managerial levels. Some of the job titles associated with IS management are the CIO, vice president of information systems, manager of information systems, and chief technology officer (CTO). A central role of all these individuals is to communicate with other areas of the organization to determine changing needs. Often, these individuals are part of an advisory or steering committee that helps the CIO and other IS managers with their decisions about the use of information systems. Together they can best decide what information systems will support corporate goals. The CTO, for example, typically works under a CIO and specializes in hardware and related equipment and technology.[62] According to John Voeller, a CTO for the Black & Veatch engineering and construction company, "I don't just look at the technology of my enterprise. I look far beyond information technology at nanotechnologies, biotech, and other domains." The CTO position also exists in federal agencies. According to Debra Stouffer, CTO of the Environmental Protection Agency (EPA), "I can carve out what the CTO position will be. I was very attracted to the mission—protecting human health and safeguarding the environment."[63]

LAN Administrators

Local area network (LAN) administrators set up and manage the network hardware, software, and security processes. They manage the addition of new users, software, and devices to the network. They isolate and fix operations problems. LAN administrators are in high demand and often solve both technical and nontechnical problems.

Internet Careers

The recent bankruptcy of some Internet start-up companies, called the *dot-gone era* by some, has resulted in layoffs for some firms. Some executives of these bankrupt start-up Internet companies lost hundreds of millions of dollars in a few months. Yet, the growth in the use of the Internet to conduct business continues and has caused a steady need for skilled personnel to develop and coordinate Internet usage. As seen in Figure 1.16, these careers are in the areas of Web operations, Web development, and Web administration. As with other areas in information systems, there are a number of top-level administrative jobs related to the Internet. These career opportunities are with traditional companies and companies that specialize in the Internet.

Internet jobs within a traditional company include Internet strategists and administrators, Internet systems developers, Internet programmers, and Internet or Web site operators. The Internet has become so important to some companies that some have suggested a new position, chief Internet officer, with responsibilities and salary similar to the CIO's.

In addition to traditional companies, many exciting career opportunities exist in companies that offer products and services over the Internet. These companies include Amazon.com, Yahoo!, eBay, and many others. Systest, for example, specializes in finding and eliminating digital bugs that could halt the operation of a computer system.

A number of Internet sites, such as Monster.com, post job opportunities for Internet careers and more traditional careers. Most large companies have job opportunities listed on their Internet sites. These sites allow prospective job hunters to browse job opportunities, job locations, salaries, benefits, and other factors. In addition, some of these sites allow job hunters to post their résumé.

Internet job sites such as Monster.com allow job hunters to browse job opportunities and post their résumés.

certification
A process for testing skills and knowledge that results in a statement by the certifying authority that says an individual is capable of performing a particular kind of job.

Often, the people filling IS roles have completed some form of certification. **Certification** is a process for testing skills and knowledge resulting in an endorsement by the certifying authority that an individual is capable of performing a particular job. Certification frequently involves specific, vendor-provided or vendor-endorsed coursework. A number of popular certification programs are available, including Novell Certified Network Engineer, Microsoft Certified Professional Systems Engineer, Certified Project Manager, and others. Microsoft, for example, offers Certified Systems Engineer: Security on Microsoft Windows that requires passing six exams.[64] The Certified Information Systems Security Professional (CISSP) is also becoming increasingly important to companies. The federal government is helping military personnel get IS certification. Some GI bill beneficiaries, for example, can now be reimbursed for technology certification through the Computing Technology Industry Association.

Other IS Careers

Many other exciting IS careers are also available. With the increase in computer attacks, there are new and exciting careers in security and fraud detection and prevention. Insurance fraud and vehicle theft are no longer perpetrated primarily by small-time crooks but by organized crime rings using computers to falsify claim receipts, ship stolen vehicles throughout the world, and commit identity theft and fraud. The National Insurance Crime Bureau, a nonprofit organization supported by roughly 1,000 property and casualty insurance companies, uses computers to join forces with special investigation units and law enforcement agencies, as well as to conduct online fraud-fighting training to investigate and prevent these types of crimes.

In addition to working for an IS department in an organization, IS personnel can work for one of the large consulting firms, such as Accenture, IBM, EDS, and others. These jobs often entail a large amount of travel, because consultants are assigned to work on various projects wherever the client is. Such roles require excellent people and project management skills in addition to IS technical skills.

Other IS career opportunities also exist, including being employed by a hardware or software vendor developing or selling products. Such a role enables an individual to work on the cutting edge of technology, which can be extremely challenging and exciting! As some computer companies cut their services to customers, new companies are being formed to fill the need. With names such as Speak With a Geek and the Geek Squad, these companies are helping people and organizations with their computer-related problems that computer vendors are no longer solving.

SUMMARY

Principle

The value of information is directly linked to how it helps decision makers achieve the organization's goals.

Data consists of raw facts; information is data transformed into a meaningful form. The process of defining relationships between data requires knowledge. Knowledge is an awareness and understanding of a set of information and the ways that information can be made useful to support a specific task. To be valuable, information must have several characteristics: It should be accurate, complete, economical to produce, flexible, reliable, relevant, simple to understand, timely, verifiable, accessible, and secure. The value of information is directly linked to how it helps people achieve their organization's goals.

Information systems are sets of interrelated elements that collect (input), manipulate (process) and store, and disseminate (output) data and information. Input is the activity of capturing and gathering new data; processing involves converting or transforming data into useful outputs; and output involves producing useful information. Feedback is the output that is used to make adjustments or changes to input or processing activities. The components of a computer-based information system include hardware, software, databases, telecommunications and networks, people, and procedures.

Principle

Knowing the potential impact of information systems and having the ability to put this knowledge to work can result in a successful personal career, organizations that reach their goals, and a society with a higher quality of life.

Information systems play an important role in today's businesses and society. The key to understanding the existing variety of systems begins with learning their fundamentals. The types of systems used within organizations can be classified into four basic groups: (1) e-commerce and m-commerce, (2) TPS and ERP, (3) MIS and DSS, and (4) specialized business information systems.

E-commerce involves any business transaction executed electronically between parties such as companies (business-to-business, B2B), consumers and other consumers (consumer-to-consumer, C2C), companies and consumers (business-to-consumer, B2C), business and the public sector, and consumers and the public sector. The major volume of e-commerce and its fastest-growing segment is business-to-business transactions. Mobile commerce (m-commerce) is transactions conducted anywhere, anytime using handheld computers, portable phones, laptop computers connected to a network, and other mobile devices.

The most fundamental system is the transaction processing system (TPS). A transaction is any business-related exchange. The TPS handles the large volume of business transactions that occur daily within an organization. TPSs include order processing, purchasing, accounting, and related systems. An enterprise resource planning (ERP) system is a set of integrated programs that is capable of managing a company's vital business operations for an entire multisite, global organization.

A management information system (MIS) uses the information from a TPS to generate information useful for management decision making. The focus of an MIS is primarily on operational efficiency. A decision support system (DSS) is an organized collection of people, procedures, databases, and devices used to support problem-specific decision making. The DSS differs from an MIS in the support given to users, the decision emphasis, the development and approach, and the system components, speed, and output. The specialized business information systems include artificial intelligence systems, expert systems, and virtual reality systems.

Principle

System users, business managers, and information systems professionals must work together to build a successful information system.

Systems development is the activity of creating or modifying existing business systems. The goal of the systems investigation is to gain a clear understanding of the problem to be solved or opportunity to be addressed. If the decision is to continue with the solution, the next step, systems analysis, defines the problems and opportunities of the existing system. Systems design determines how the new system will work to meet the business needs defined during systems analysis. Systems implementation involves creating or acquiring the various system components (hardware, software, databases, etc.) defined in the design step, assembling them, and putting the new system into operation. The purpose of systems maintenance and review is to check and modify the system so that it continues to meet changing business needs.

Principle

The use of information systems to add value to the organization can also give an organization a competitive advantage.

An organization is a formal collection of people and various other resources established to accomplish a set of goals. The primary goal of a for-profit organization is to maximize shareholder value. Nonprofit organizations include social groups, religious groups, universities, and other organizations that do not have profit as the primary goal. Organizations are systems with inputs, transformation mechanisms, and outputs.

Value-added processes increase the relative worth of the combined inputs on their way to becoming final outputs of the organization. The value chain is a series (chain) of activities that includes (1) inbound logistics, (2) warehouse and storage, (3) production, (4) finished product storage, (5) outbound logistics, (6) marketing and sales, and (7) customer service.

Supply chain management (SCM) helps determine what supplies are required, what quantities are needed to meet customer demand, how the supplies are to be processed (manufactured) into finished goods and services, and how the shipment of supplies and products to customers is to be scheduled, monitored, and controlled. Customer relationship management (CRM) programs help a company manage all aspects of customer encounters, including marketing and advertising, sales, customer service after the sale, and programs to help keep and retain loyal customers.

Organizations use information systems to support organizational goals. Because information systems typically are designed to improve productivity, methods for measuring the system's impact on productivity should be devised.

Organizational culture and change are important internal issues that affect most organizations. Organizational culture consists of the major understandings and assumptions for a business, a corporation, or an organization. Organizational change deals with how for-profit and nonprofit organizations plan for, implement, and handle change.

The extent to which technology is used throughout an organization is a function of technology diffusion, infusion, and acceptance. Technology diffusion is a measure of how widely technology is in place throughout an organization. Technology infusion is the extent to which technology permeates an area or department. The technology acceptance model (TAM) investigates factors, such as the perceived usefulness of the technology, the ease of use of the technology, the quality of the information system, and the degree to which the organization supports the use of the information system, to predict IS usage and performance.

Competitive advantage is usually embodied in either a product or service that has the most added value to consumers and that is unavailable from the competition or in an internal system that delivers benefits to a firm not enjoyed by its competition. The five-forces model covers factors that lead firms to seek competitive advantage: the rivalry among existing competitors, the threat of new market entrants, the threat of substitute products and services, the bargaining power of buyers, and the bargaining power of suppliers. Three strategies to address these factors and to attain competitive advantage include altering the industry structure, creating new products and services, and improving existing product lines and services.

Developing information systems that measure and control productivity is a key element for most organizations. A useful measure of the value of an IS project is return on investment (ROI). This measure investigates the additional profits or benefits that are generated as a percentage of the investment in IS technology. Total cost of ownership (TCO) can also be a useful measure.

Principle

Information systems personnel are the key to unlocking the potential of any new or modified system.

Information systems personnel typically work in an IS department that employs a chief information officer, systems analysts, computer programmers, computer operators, and a number of other people. The overall role of the chief information officer (CIO) is to employ an IS department's equipment and personnel in a manner that will help the organization attain its goals. Systems analysts help users determine what outputs they need from the system and construct the plans for developing the necessary programs that produce these outputs. Systems analysts then work with one or more programmers to ensure that the appropriate programs are purchased, modified from existing programs, or developed. The major responsibility of a computer programmer is to use the plans developed by the systems analyst to develop or adapt one or more computer programs that produce the desired outputs. Computer operators are responsible for starting, stopping, and correctly operating mainframe systems, networks, tape drives, disk devices, printers, and so on. LAN administrators set up and manage the network hardware, software, and security processes. Trained personnel are also increasingly needed to set up and manage a company's Internet site. Information systems personnel might also work in other functional departments or areas in a support capacity for one of the large consulting firms. Another IS career opportunity is to be employed by a hardware or software vendor developing or selling products.

CHAPTER 1: SELF-ASSESSMENT TEST

The value of information is directly linked to how it helps decision makers achieve the organization's goals.

1. A (An) _____ is a set of interrelated components that collect, manipulate, and disseminate data and information and provide a feedback mechanism to meet an objective.
2. The value of data is measured by the increase in revenues. True or False?

Knowing the potential impact of information systems and having the ability to put this knowledge to work can result in a successful personal career, organizations that reach their goals, and a society with a higher quality of life.

3. A (An) _____ consists of hardware, software, databases, telecommunications, people, and procedures.
4. Computer programs that govern the operation of a computer system are called
   a. feedback
   b. feedforward

   c. software
   d. transaction processing system

5. Payroll and order processing are examples of a computerized management information system. True or False?

System users, business managers, and information systems professionals must work together to build a successful information system.

6. What involves creating or acquiring the various system components (hardware, software, databases, etc.) defined in the design step, assembling them, and putting the new system into operation?

   a. systems implementation
   b. systems review
   c. systems development
   d. systems design

7. _____ involves anytime, anywhere commerce that uses wireless communications.

8. _____ involves contracting with outside professional services to meet specific business needs.

The use of information systems to add value to the organization can also give an organization a competitive advantage.

9. _____ change can help an organization improve the raw materials supply, the production process, and the products and services offered by the organization.

10. Technology diffusion is a measure of how widely technology is spread throughout an organization. True or False?

Information systems personnel are the key to unlocking the potential of any new or modified system.

11. Who is involved in helping users determine what outputs they need and constructing the plans needed to produce these outputs?

a. the CIO
b. the applications programmer
c. the systems programmer
d. the systems analyst

12. The systems development component of a typical IS department focuses on specific development projects and ongoing maintenance and review. True or False?

13. The _____ is typically in charge of the information systems department or area in a company.

CHAPTER 1: SELF-ASSESSMENT TEST ANSWERS

(1) information system (2) False (3) computer-based information system (CBIS) (4) c (5) False (6) a (7) Mobile commerce (m-commerce) (8) Outsourcing (9) Sustaining (10) True (11) d (12) True (13) chief information officer (CIO)

REVIEW QUESTIONS

1. What are the components of any information system?
2. How would you distinguish data and information? Information and knowledge?
3. Identify at least six characteristics of valuable information.
4. What is a computer-based information system? What are its components?
5. What are the most common types of computer-based information systems used in business organizations today? Give an example of each.
6. What is the difference between e-commerce and m-commerce?
7. What are some of the benefits organizations seek to achieve through using information systems?
8. What is a value-added process? Give several examples.

9. What is the technology acceptance model (TAM)?
10. What are some general strategies employed by organizations to achieve competitive advantage?
11. Define the term *productivity*. Why is it difficult to measure the impact that investments in information systems have on productivity?
12. What is the productivity paradox?
13. What is the total cost of ownership?
14. What is the role of the systems analyst? What is the role of the programmer?
15. What is the operations component of a typical IS department?
16. What is the role of the chief information officer?

DISCUSSION QUESTIONS

1. Describe the "ideal" automated auto license plate renewal system for the drivers in your state. Describe the input, processing, output, and feedback associated with this system.
2. You have decided to open an Internet site to buy and sell used music CDs to other students. Describe the value chain for your new business.
3. How is it that useful information can vary widely from the quality attributes of valuable information?
4. What is the difference between an MIS and a DSS?
5. Discuss the potential use of virtual reality to enhance the learning experience for new automobile drivers. How

might such a system operate? What are the benefits and potential drawbacks of such a system?
6. Discuss how information systems are linked to the business objectives of an organization.
7. You have been hired to work in the IS area of a manufacturing company that is starting to use the Internet to order parts from its suppliers and offer sales and support to its customers. What types of Internet positions would you expect to see at the company?
8. You have been asked to participate in the preparation of your company's strategic plan. Specifically, your task is to analyze the competitive marketplace using Porter's five-forces model. Prepare your analysis, using your knowledge

of a business you have worked for or have an interest in working for.

9. Based on the analysis you performed in the preceding discussion question, what possible strategies could your organization adopt to address these challenges? What role could information systems play in these strategies? Use Porter's strategies as a guide.

10. You have been hired as a sales representative for a sporting goods store. You would like the IS department to develop new software to give you reports on which customers are spending the most at your store. Describe your role in getting the new software developed. Describe the roles of the systems analysts and the computer programmers.

11. Imagine that you are the CIO for a large, multinational company. Outline a few of your key responsibilities.

12. What sort of IS position would be most appealing to you—working as a member of an IS organization, being a consultant, or working for an IS hardware or software vendor? Why?

13. What are your career goals and how can a computer-based information system be used to achieve them?

PROBLEM-SOLVING EXERCISES

1. Prepare a data disk and a backup disk for the problem-solving exercises and other computer-based assignments you will complete in this class. Create one directory or folder for each chapter in the textbook (you should have nine directories or folders). As you work through the problem-solving exercises and complete other work using the computer, save your assignments for each chapter in the appropriate directory or folder. On the label of each disk, be certain to include your name, course, and section. On one disk, write "Working Copy"; on the other, write "Backup."

2. Do some research to obtain estimates of the rate of growth of e-commerce and m-commerce. Use the plotting capabilities of your spreadsheet or graphics software to produce a bar chart of that growth over a number of years. Share your findings with the class.

 3. For an industry of your choice, find the number of employees, total sales, total profits, and earnings growth rate of 15 firms. Using a database program, enter this information for the last year. Use the database to generate a report of the three companies with the highest earnings growth rate. Use your word processor to create a document that describes the 15 firms. What other measures would you use to determine which is the best company in terms of future profit potential?

TEAM ACTIVITIES

 1. Before you can do a team activity, you need a team! The class members may self-select their teams, or the instructor may assign members to groups. After your group has been formed, meet and introduce yourselves to each other. You will need to find out the first name, hometown, major, and e-mail address and phone number of each member. Find out one interesting fact about each member of your team, as well. Come up with a name for your team. Put the information about each team member into a database and print enough copies for each team member and your instructor.

2. Have your team interview a company that recently introduced new technology. Write a brief report that describes the extent of technology infusion and diffusion.

3. Have your team research a firm that has achieved a competitive advantage. Write a brief report that describes how the company was able to achieve its competitive advantage.

WEB EXERCISES

1. Throughout this book, you will see how the Internet provides a vast amount of information to individuals and organizations. We will stress the World Wide Web, or simply the Web, which is an important part of the Internet. Most large universities and organizations have an address on the Internet, called a Web site or home page. The

address of the Web site for the publisher of this text is *www.course.com*. You can gain access to the Internet through a browser, such as Internet Explorer or Netscape. Using an Internet browser, go to the Web site for this publisher. What did you find? Try to obtain information on this book. You may be asked to develop a report or send an e-mail message to your instructor about what you found.

2. Go to an Internet search engine, such as *www.yahoo.com*, and search for information about a company, including its Web site. Write a report that summarizes the size of the company, the number of employees, its products, the location of its headquarters, and its profits (or losses) for last year. Would you want to work for this company?

3. Use the Internet to research a career in an IS area, such as programmer, systems analyst, Web developer, or CIO. Write a brief report describing the career area, including salaries and job opportunities. You may be asked to send an e-mail message to your instructor about what you found.

CAREER EXERCISES

1. In the Career Exercises found at the end of every chapter, you will explore how material in the chapter can help you excel in your college major or chosen career. Write a brief report on the career that appeals to you the most. Do the same for two other careers that interest you.

2. Pick the five best companies for your career. Describe how each company uses information systems to help achieve a competitive advantage.

VIDEO QUESTIONS

Watch the video clip **Go Inside Krispy Kreme** and answer these questions:

1. Provide a description of how Krispy Kreme is using each of the elements of an information system: hardware, software, databases, telecommunications, people, and procedures to provide services for its employees.

2. How have information systems assisted the many Krispy Kreme franchises in providing consistent products and services for their customers?

CASE STUDIES

Case One

The Queen Mary 2 and Partner

The Queen Mary 2 (QM2) is the largest and most expensive cruise ship ever built. It includes five swimming pools, a planetarium, a two-story theater that seats 1,000, a casino, a gym, luxurious kennels, a nursery staffed with British nannies, and the largest ballroom, library, and wine collection at sea. Of all its amenities, the one considered most valuable to the crew and management and key to the functioning of the vessel is the integrated network and information system accessible in every cabin.

The $800 million QM2, constructed in the shipyard at Chantiers de l'Atlantique, France, and owned by Miami-based

Cunard Line Ltd., made her maiden voyage in early 2004. Passengers in each of her 1,310 cabins had access to digital entertainment such as on-demand movies and interactive television. Each cabin is also wired with Internet access and network services. For example, passengers use the network to make shore excursion reservations and dinner plans.

Upon checking in, passengers are presented with a plastic bar-coded card. The card is used while on board to make purchases, which are then billed to the customer's account. It is also swiped as guests leave and return to the ship to track passenger location. The ship's massive data network brings order where there once was chaos. Ship managers can run reports showing which passengers are on board, how many will be attending the morning exercise class, and which entrée was most popular at last night's dinner. The network and database are backed up by redundant systems that automatically take over if the primary system fails.

The information system, called The Ship Partner, is used to track security, billing, telephone service, onboard television, and other operations. It was designed by Discovery Travel Systems LP (DTS). John Broughan, president of DTS, says that the IT needs of cruise ship operators differ from those of typical hotel property management companies, so specialized systems had to be created to better serve cruise companies.

The Queen Mary 2 provides yet another example of how information systems assist with management functions, providing valuable information and offering services to customers.

Discussion Questions

1. What conveniences does The Ship Partner information system provide to passengers of the Queen Mary 2? What entertainment services could be made available to passengers through this digital network?

2. How does The Ship Partner information system assist ship managers with their duties and responsibilities?

Critical Thinking Questions

3. How does The Ship Partner information system assist Cunard in competing in the travel industry? What other travel and leisure industries would benefit from a system like The Ship Partner?

4. Why is it important for The Ship Partner to have a backup system? How would a systemwide failure affect the functioning of the ship?

SOURCES: Todd R. Weiss, "New Queen Mary 2 Offers High Tech on the High Seas," *Computerworld*, January 12, 2004, *www.computerworld.com*; Eric Thomas, "Queen Mary 2, World's Biggest Liner, Awaits Its Champagne Moment," *Agence France Presse*, January 8, 2004; The Chantiers de l'Atlantique Web site, *www.chantiersatlantique.com/UK/index_UK.htm*, accessed January 18, 2004.

Case Two
MyFamily Comforts Its Members

MyFamily.com, Inc., is a leading online subscription business for researching family history, and the site also allows families to set up their own Web site and share photos with other family members.

MyFamily.com was one of the rare companies that survived the hardships of the dot-com bust. Through smart business management and providing a highly valued service to its customers, MyFamily.com actually grew its business when many others lost theirs. From 1999 through 2003, MyFamily.com doubled its subscribers each year, finishing 2003 with 1.6 million customers.

The rapid growth of the company presented MyFamily.com with challenges in customer relationship management (CRM). The company was hiring many customer service representatives just to respond to customer e-mail. Much of the e-mail involved simple questions that employees answered repeatedly day in and day out. What MyFamily.com needed was a system to help organize its customer support function and allow the company to make better use of its employees. The solution lay in a self-service CRM application from RightNow Technologies called eService Center.

The eService Center provides Web-based customer support for routine customer inquiries, freeing up customer service representatives to handle more difficult problems. It uses artificial intelligence contained in a single self-learning knowledge base that can be accessed from the Web, e-mail, chat room, or telephone. The system makes it easy for customers to find answers to questions by presenting the most successful solutions first and refining the solution based on customer responses. The eService Center also includes ana-

lytics and the ability to measure customer satisfaction through surveys.

Within 30 days after MyFamily.com implemented RightNow's system, the number of e-mails that employees had to answer fell 30 percent, according to Mary Kay Evans, spokeswoman for MyFamily.com. Calculating the savings in employee time, MyFamily.com has received over two and a half times as much as it invested in the system over nine months—a 260 percent return on investment (ROI). The new system earned MyFamily.com two awards in 2003: SearchCRM honored MyFamily.com with the Customer Touch award, and *CRM Magazine* awarded MyFamily its 2003 CRM Elite award.

Discussion Questions

1. What type of information system is RightNow's eService Center, a TPS, MIS, DSS, or some other specialized system? Present the rationale for your answer.

2. Besides cost savings, what other benefits does the eService Center provide for the upper-level managers of MyFamily.com?

Critical Thinking Questions

3. Have you had any experience with automated customer service systems? Do you think that these services benefit the company or the customer more? Why?

4. The types of questions that this automated system assists customers with are described as typical customer inquiries. Do you think handling frequently asked questions (FAQs) is a job better suited for man or machine? Why?

SOURCES: Linda Rosencrance, "CRM with a Family Touch," *Computerworld*, April 7, 2003, *www.computerworld.com*; "RightNow Customer MyFamily.com Wins SearchCRM.com's Customer Touch Award for Effective Service & Sup- port," *PR Newswire*, September 11, 2003; MyFamily.com Web site, *www.myfamily.com*, accessed January 17, 2004.

NOTES

Sources for the opening vignette: Marc Songini, "Case Study: Boehringer Cures Slow Reporting," *Computerworld*, July 21, 2003, *www.computer-world.com*; "Boehringer Ingelheim Deploys BackWeb's Offline Solution for the Plumtree Corporate Portal," *PR Newswire*, December 15, 2003; the Boehringer Ingelheim Web site, *www.boehringer-ingelheim.com/corporate/home/home.asp*, accessed January 22, 2004.

1. Booth-Thomas, Cathy, "The See-It-All Chip," *Time* magazine special technology section, October 2003, p. A12.
2. Nussbaum, Bruce, "Technology: Just Make It Simpler," *Business Week,* September 8, 2003, p. 38.
3. Wildstrom, Stehen, "Tablet PCs," *Business Week,* August 4, 2003, p. 22.
4. Heun, Christopher, "Marine Mouse Takes IT to New Depths," *InformationWeek,* November 5, 2002, p. 20.
5. Clark, Don, "A 64-bit Bet On Its Future," *The Wall Street Journal,* April 21, 2003, p. B1.
6. Goldsmigh, Charles, "German Visual Image Firm Is Honored for Film Graphics," *The Wall Street Journal,* February 26, 2003, p. B1.
7. Wildstrom, Stephen, "A Dana for Every Schoolkid," *Business Week,* April 21, 2003, p. 26.
8. Brandel, S. "35 Years of Leadership," *Computerworld,* September 30, 2003, p. 55
9. Beauprez, Jennifer, "State Urged to Think Small," *The Denver Post,* July 13, 2003, p. K1.
10. Cowley, Stacy, "Software Market Hit By Purchasing Delays," *Computerworld,* July 14, 2003, p. 12.
11. Brandel, S. "35 Years of Leadership," *Computerworld,* September 30, 2003, p. 55.
12. King, Julia, "Open for Inspection," *Computerworld,* July 21, 2003, p. 39.
13. Hamblen, Matt, "Compression Relieves Congestion," *Computerworld,* March 10, 2003, p. 30.
14. Hamblen, Matt, "Hotel Goes Wireless," *Computerworld,* July 14, 2003, p. 16.
15. Sitch, Stephane, "Invasion of the Drones," *Forbes,* March 17, 2003, p. 52.
16. Kelly, Lisa, "Virgin Sets Up Global Intranet," *Computing,* November 6, 2003, p. 15.
17. Anthes, Gary, "Corporate Express Goes Direct," *Computerworld,* September 1, 2003, p. 17.
18. Chabrow, Eric, "Online Ad Sales Rebounding," *Information Week,* January 6, 2003, p.16.
19. Cuneo, Wileen Colkin, "Uptick in Care," *InformationWeek,* November 3, 2003, p. H18.
20. Vijayan, Jaikumar, "Bookseller Expands Its Reach with Integrated Internet Platform," *Computerworld,* June 2, 2003, p. 29.
21. Staff, "IBC Supports Decision Support," *Health Management Technology,* September, 2003, p.10.
22. Staff, "The Rise of Outsourcing," *Insurance Day,* November 5, 2003.
23. Nash, Emma, "Toyota Puts Brakes on Outsourcing," *Computing,* November 6, 2003, p. 4.
24. Avery, Susan, "Cessna Soars," *Purchasing,* September 4, 2003, p. 25.
25. Gentle, Michael, "CRM: Ready or Not," *Computerworld,* August 18, 2003, p. 40.

26. Christensen, Clayton, "The Innovator's Dilemma," *Harvard Business School Press,* 1997, p. 225 and "The Inventor's Solution," *Harvard Business School Press,* 2003.
27. Loch, Christoph, and Huberman, Berndao, "A Punctuated-Equilibrium Model of Technology Diffusion," *Management Science,* February, 1999, p. 160.
28. Armstrong, Curtis, and Sambamurthy, V. "Information Technology Assimilation in Firms," *Information Systems Research,* December, 1999, p. 304.
29. Agarwal, Ritu, and Prasad, Jayesh, "Are Individual Differences Germane to the Acceptance of New Information Technology?" *Decision Sciences,* Spring 1999, p. 361.
30. Kwon et al., "A Test of the Technology Acceptance Model," *Proceedings of the Hawaii International Conference on System Sciences,* January 4-7, 2000.
31. Watts, Stephanie, et al., "Informational Influence in Organizations," *Information Systems Research,* March 2003, p. 47.
32. Collins, Jim *Good to Great,* Harper Collins Books, 2001, p. 300.
33. Slotegraaf, Rebecca, et al., "The Role of Firm Resources," *Journal of Marketing Research,* August 2003, p. 295.
34. Saccomano, Ann, "Bureau of Immigration and Customs Reorganizes," *The Journal of Commerce Online Edition,* May 21, 2003, p. 1.
35. Kurtz, et al., "The New Dynamics of Strategy," *IBM Systems Journal,* Vol. 42, No. 3, 2003, p. 462.
36. Kearns, et al., "A Resource-Based View of Strategic IT Alignment," *Decision Sciences,* Winter 2003, p. 1.
37. M. Porter and V. Millar, "How Information Systems Give You Competitive Advantage," *Journal of Business Strategy*, Winter 1985. See also M. Porter, *Competitive Advantage* (New York: Free Press, 1985).
38. Melymuka, Kathleen, "Delta's New All-Digital Song," *Computerworld,* August 8, 2003, p. 37.
39. Staff, "Echelon and Samsung Sign Strategic Alliance," *Business Wire,* June 23, 2003.
40. Rosencrane, Linda, "Best In Class: Self-Service Check-In Kiosks Give Travelers More Control," *Computerworld,* February 24, 2003, p. 48.
41. Khermouch, Gerry, "Bar Codes Better Watch Their Backs," *Business Week,* July 14, 2003, p. 42.
42. Moes, Annick, "Germans Put Stock in Store of the Future," *The Wall Street Journal,* June 19, 2003, p. A10.
43. Trottman, Melanie, "Sabre Looks to Web Ventures," *The Wall Street Journal,* June 24, 2003, p. B6.
44. Kohli, Rajiv, et al., "Measuring Information Technology Performance," *Information Systems Research,* June 2003, p. 127.
45. Alexander, Steve, "Best In Class: Web Site Adds Inventory Control and Forecasting," *Computerworld,* February 24, 2003, p. 45.
46. Staff, "Paradox Lost," *The Economist,* September 13, 2003, www.economist.com.
47. Park, Andrew, "Computers Get Their Groove Back," *Business Week,* January 12, 2004, p. 96.
48. Staff, "Certification—Return On Investment," *Computer Reseller News,* August 25, 2003, p. 54.
49. Alexander, Steve, "Best In Class: WebSite Adds Inventory Control and Forecasting," *Computerworld,* February 24, 2003, p.45.

50. Staff, "Detailed Comparison of a Range of Factors Affecting Total Cost of Ownership (TCO) for Messaging and Collaboration," *M2 Presswire,* September 5, 2003.

51. Staff, "An International Internet Research Asignment," *Journal of Education for Business,* January 2003, p. 158.

52. Brandel, Mary, "Home-Schooling—IT Talent," *Computerworld,* January 27, 2003, p. 36.

53. Hoffman, Thomas, "Online Tech Job Positions Increase, But IT Execs Don't Expect a Jump in Hiring," *Computerworld,* September 8, 2003, p. 11.

54. Grow, Brian, "Skilled Workers or Indentured Servants," *Business Week,* June 16, 2003, p. 54.

55. Grow, Brian, "A Loophole as Big as a Mainframe," *Business Week,* March 10, 2003, p. 82.

56. Kirkpatrick, David, "The Net Makes It All Easier," *Fortune,* May 26, 2003, p. 146.

57. Hoffman, Thomas, "Preparing Generation Z," *Computerworld,* August 25, 2003, p. 41.

58. Tesch, et al., "The Impact of Information System Personnel Skill Discrepancies on Stakeholder Satisfaction," *Decision Science,* Winter 2003, p. 107.

59. Gongloff, Mark, "U.S. Jobs Jumping Ship," *CNN Money Online,* May 2, 2003.

60. Hoffman, Thomas, "Programmer Testing Services Get Wary Reception from IT," *Computerworld,* April 14, 2003, p. 6.

61. Staff, "The Clinger-Cohen Act of 1996," *The Governments Accounts Journal,* Winter 1997, p. 8.

62. Melymuks, K. "So You Want to Be a CTO," *Computerworld,* March 24, 2003, p. 42.

63. Prencipe, Loretta, " Chief Technology Officers," *InfoWorld,* January 6, 2003, p. 44.

64. Sliwa, Carol, "Microsoft to Introduce Security Certifications," *Computerworld,* June 2, 2003, p. 6.

PART
· 2 ·

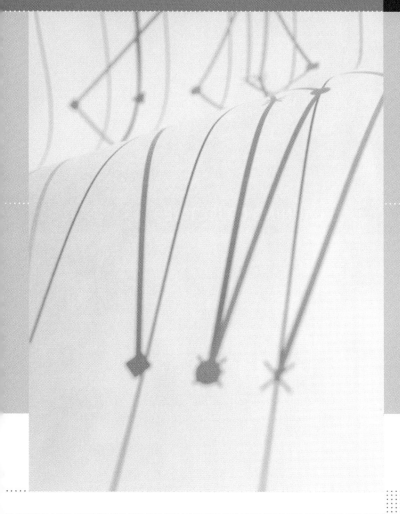

Technology

Chapter 2 Hardware and Software

Chapter 3 Organizing Data and Information

Chapter 4 Telecommunications, the Internet, Intranets, and Extranets

CHAPTER
• 2 •

Hardware and Software

PRINCIPLES	LEARNING OBJECTIVES
▪ Users must work closely with IS professionals to define business needs, evaluate options, and select the hardware and software that provide a cost-effective solution to those needs.	▪ Identify and discuss the role of the essential hardware components of a computer system. ▪ List and describe popular classes of computer systems and discuss the role of each.
▪ When selecting an operating system, you must consider the current and future requirements for application software to meet the needs of the organization. In addition, your choice of a particular operating system must be consistent with your choice of hardware.	▪ Identify and briefly describe the functions of the two basic kinds of software. ▪ Outline the role of the operating system and identify the features of several popular operating systems.
▪ Do not develop proprietary application software unless doing so will meet a compelling business need that can provide a competitive advantage.	▪ Discuss how application software can support personal, workgroup, and enterprise business objectives. ▪ Identify three basic approaches to developing application software and discuss the pros and cons of each.
▪ Choose a programming language whose functional characteristics are appropriate for the task at hand, taking into consideration the skills and experience of the programming staff.	▪ Outline the overall evolution of programming languages and clearly differentiate among the five generations of programming languages.
▪ The software industry continues to undergo constant change; users need to be aware of recent trends and issues to be effective in their business and personal life.	▪ Identify several key issues and trends that have an impact on organizations and individuals.

INFORMATION SYSTEMS IN THE GLOBAL ECONOMY
PORSCHE AG, GERMANY

Auto Manufacturer Upgrades Its Computer Hardware and Software

Porsche is known for manufacturing world-class sports cars, among them its famous 911 and Boxster automobiles. The company built on its reputation when it recently introduced the four-wheel-drive Cayenne, the world's fastest SUV, and the Carerra GT, a super sports car with a powerful V10 engine. Porsche's headquarters are in Stuttgart, Germany; it maintains production facilities in Stuttgart and Leipzig, and it has dealerships and sales offices worldwide. Annual sales are at record levels—$7 billion—and its employees number just under 10,000.

To maintain its market leadership, Porsche must employ information systems as cutting edge as its cars. The company was one of the first auto makers to introduce enterprise resource planning (ERP) software to support its accounting, finance, purchasing, and material-management business processes. It implemented its ERP system in the early 1990s and chose then state-of-the-art Hewlett-Packard (HP) V-class servers as the underlying computer hardware to handle some 1.5 million daily transactions quickly and reliably. Over the years, however, the volume of transactions its system must handle more than doubled. This increase comes from four factors: (1) Porsche plans to use new ERP software modules to support additional business processes, (2) the company introduced new automobile models, (3) it built a new production plant for the Cayenne and Carerra GT, and (4) the number of users of the ERP system exceeded 4,000.

Porsche Information Kommunikation Services (PIKS) GmbH is a wholly owned subsidiary of Porsche AG, with headquarters in Stuttgart. Its 86 employees are responsible for planning and operating the IS infrastructure for the entire Porsche group, including software, networks, servers, security systems, and storage systems. For more than a year, PIKS evaluated several different computer manufacturers and hardware options to meet the new processing requirements. It was critical that the new hardware not increase the company's hardware budget. Obviously, the new hardware must work well with existing components of the infrastructure (software, network, and other computer hardware). More important, the new hardware must be extremely reliable and available.

To meet the company's computing needs, PIKS decided to replace its HP V-class servers with two HP Superdome servers, each with 24 processors and 28 GB of RAM. This hardware upgrade also provides for increased processing power to meet future needs; the PA-8700 processors in the current version of the Superdome server can be upgraded to processors from the Intel Itanium processor family to double the processing capability.

As you read this chapter, consider the following:

- How are companies using computer hardware and software to compete and meet their business objectives?
- How do organizations go about selecting computer hardware and software and what must you know to assist in this process?

Why Learn About Hardware and Software?

Organizations invest in computer hardware and software to improve worker productivity, increase revenue, reduce costs, and provide better customer service. Those that don't may be stuck with outdated hardware and software that often fail and cannot take advantage of the latest advances. As a result, obsolete hardware and software can place an organization at a competitive disadvantage. Managers, no matter what their career field and educational background, are expected to know enough about hardware and software to ask tough questions to invest wisely for their area of the business. Managers in marketing, sales, and human resources often help IS specialists assess opportunities to apply hardware and software and evaluate the various options and features. Managers in finance and accounting especially must also keep an eye on the bottom line, guarding against overspending, yet be willing to invest in computer hardware and software when and where business conditions warrant it.

Porsche needed cost-effective, reliable, powerful computers on which to run its ERP software.

(Source: Getty Images.)

In building a car, manufacturers try to match the intended use of the vehicle to its components. Racing cars require special types of engines, transmissions, and tires. The selection of a transmission for a racing car, then, requires not only consideration of how much of the engine's power can be delivered to the wheels (efficiency and effectiveness) but also how expensive the transmission is (cost), how reliable it is (control), and how many gears it has (complexity). Similarly, organizations assemble the hardware components of a computer system so that they are effective, efficient, and well suited to the tasks that need to be performed. Users and IS professionals often need to make these decisions together, combining their knowledge of systems and business functions as well as forecasting their future needs.

Because the business needs and their importance vary at different companies, the IS solutions they choose can be quite different.

- ARZ Allgemeines Rechenzentrum GmbH (ARZ) is one of Austria's leading providers of information services, processing more than 8 million transactions per day for financial and medical institutions. ARZ is keenly interested in reducing processing costs, maintaining sufficient capacity to handle an increasing workload, and providing highly reliable processing. Peter Gschirr, information technology director at ARZ, selected two large, extremely powerful IBM zSeries mainframe computers to handle the workload.[1]
- Specialized Bicycles is a pioneer in designing and manufacturing high-performance bicycles, helmets, and other cycling accessories. When Specialized Bicycles wanted to offer its products over the Web, it selected computer hardware that could rapidly increase computing capacity, provide high reliability so that the Web site was always available, and easily be integrated with the rest of the organization's hardware. To meet these requirements, Ron Pollard, chief information officer, went with Sun Microsystems Enterprise 450 Server.[2]
- Sullivan Street Bakery, started in 1994 by an art student and anthropologist, is today one of the most popular Italian bakeries in New York. The owners chose Apple's Mac computers because they are efficient and easy to use. The Macs make managing production almost effortless, and they support billing and inventory control. Workers don't have to worry about how to use the computers and can concentrate instead on maintaining the quality of their hand-crafted traditional Italian-style breads.[3]

As each of these examples demonstrates, assembling the right computer hardware requires an understanding of its relationship to the information system and the needs of the organization. Remember that the computer hardware objectives are subordinate to, but supportive of, the information system and the needs of the organization.

HARDWARE COMPONENTS

Computer system hardware components include devices that perform the functions of input, processing, data storage, and output (see Figure 2.1).

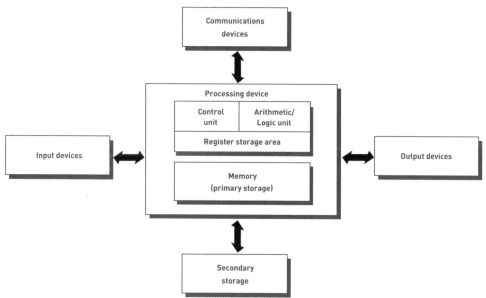

Figure 2.1

Computer System Components

These components include input devices, output devices, communications devices, primary and secondary storage devices, and the central processing unit (CPU). The control unit, the arithmetic/logic unit (ALU), and the register storage area constitute the CPU.

central processing unit (CPU)
The part of the computer that consists of two associated elements: the arithmetic/logic unit and the control unit.

arithmetic/logic unit (ALU)
The portion of the CPU that performs mathematical calculations and makes logical comparisons.

control unit
The part of the CPU that sequentially accesses program instructions, decodes them, and coordinates the flow of data in and out of the ALU, primary storage, and even secondary storage and various output devices.

The ability to process (organize and manipulate) data is a critical aspect of a computer system, in which processing is accomplished by an interplay between one or more of the central processing units and primary storage. Each **central processing unit (CPU)** consists of two primary elements: the arithmetic/logic unit and the control unit. The **arithmetic/logic unit (ALU)** performs mathematical calculations and makes logical comparisons. The **control unit** sequentially accesses program instructions, decodes them, and coordinates the flow of data in and out of the ALU, primary storage, and even secondary storage and various output devices. Primary memory, which holds program instructions and data, is closely associated with the CPU.

Now that you have learned about the basic hardware components and the way they function, we turn to an examination of processing power, speed, and capacity. These three attributes determine the capabilities of a hardware device.

PROCESSING AND MEMORY DEVICES: POWER, SPEED, AND CAPACITY

The components responsible for processing—the CPU and memory—are housed together in the same box or cabinet, called the *system unit*. All other computer system devices, such as the monitor and keyboard, are linked either directly or indirectly into the system unit housing. As discussed previously, achieving IS objectives and organizational goals should be the primary consideration in selecting processing and memory devices. In this section, we investigate the characteristics of these important devices.

Processing Characteristics and Functions

Because efficient processing and timely output are important, organizations use a variety of measures to gauge processing speed. These measures include the time it takes to complete a machine cycle, clock speed, and others.

Machine Cycle Time

The time it takes to execute the instruction phase and the execution phase is the machine cycle time. Machine cycle time is one measure of processing speed.

Clock Speed

clock speed
A series of electronic pulses produced at a predetermined rate that affect machine cycle time.

Each CPU produces a series of electronic pulses at a predetermined rate, called the **clock speed**, which affects machine cycle time. The control unit executes an instruction in accordance with the electronic cycle, or pulses of the CPU "clock." Each instruction takes at least the same amount of time as the interval between pulses. The shorter the interval between pulses, the faster each instruction can be executed. Clock speed is often measured in megahertz (MHz), or millions of cycles per second. The clock speed for personal computers is in the multiple gigahertz (GHz), or billions of cycles per second, range.

Wordlength

wordlength
The number of bits the CPU can process at any one time.

Data is moved within a computer system in units called *bits*. A bit is a binary digit—0 or 1. Another factor affecting overall system performance is the number of bits the CPU can process at one time, or the **wordlength** of the CPU. Early computers were built with CPUs that had a wordlength of 4 bits, meaning that the CPU was capable of processing 4 bits at one time. The 4 bits could be used to represent actual data, an instruction to be processed, or the address of data to be accessed. The 4-bit limitation was quite confining and greatly constrained the power of the computer. Over time, CPUs have evolved to 8-, 16-, 32-, and 64-bit machines with dramatic increases in power and capability. Computers with larger wordlengths can transfer more data between devices in the same machine cycle. They can also use the larger number of bits to address more memory locations and, hence, are a requirement for systems with certain large memory requirements. The 64-bit chip allows the CPU to directly address 16 quintillion (billion billion) unique address locations compared with 4.3 billion for a 32-bit processor. The ability to directly access a larger address space is critical for multimedia, imaging, and database applications.

Physical Characteristics of the CPU

CPU speed is also limited by physical constraints. Most CPUs are collections of digital circuits imprinted on silicon wafers, or chips, each no bigger than the tip of a pencil eraser. To turn a digital circuit within the CPU on or off, electrical current must flow through a medium (usually silicon) from point A to point B. The speed at which it travels between points can be increased by either reducing the distance between the points or reducing the resistance of the medium to the electrical current.

Memory Characteristics and Functions

Located physically close to the CPU (to decrease access time), memory provides the CPU with a working storage area for program instructions and data. The chief feature of memory is that it rapidly provides the data and instructions to the CPU.

Storage Capacity

byte (B)
Eight bits that together represent a single character of data.

Like the CPU, memory devices contain thousands of circuits imprinted on a silicon chip. Each circuit is either conducting electrical current (on) or not (off). By representing data as a combination of on or off circuit states, the data is stored in memory. Usually eight bits are used to represent a character, such as the letter *A*. Eight bits together form a **byte**. Table 2.1 summarizes commonly used measurements. Storage capacity is measured in bytes, abbreviated with the letter *B*, with one byte usually equal to one character. The contents of the Library of Congress, with more than 126 million items and 530 miles of bookshelves, would require about 20 petabytes of digital storage.

Name	Abbreviation	Exact Number of Bytes	Approximate Number of Bytes
Byte	B	1	1
Kilobyte	KB	1,024 Bytes	1 thousand
Megabyte	MB	1,024 Kilobytes	1 million
Gigabyte	GB	1,024 Megabytes	1 billion
Terabyte	TB	1,024 Gigabytes	1 trillion
Petabyte	PB	1,024 Terabytes	1 quadrillion

Types of Memory

Several forms of memory are available, Instructions or data can be temporarily stored in **random access memory (RAM)**. RAM is temporary and volatile—RAM chips lose their contents if the current is turned off or disrupted (as in a power surge, brownout, or electrical noise generated by lightning or nearby machines). RAM chips are mounted directly on the computer's main circuit board or in chips mounted on peripheral cards that plug into the computer's main circuit board. These RAM chips consist of millions of switches that are sensitive to changes in electric current.

Another type of memory, **ROM**, an acronym for **read-only memory**, is usually non-volatile. In ROM, the combination of circuit states is fixed, and, therefore, its contents are not lost if the power is removed. ROM provides permanent storage for data and instructions that do not change, such as programs and data from the computer manufacturer.

Multiprocessing

There are a number of forms of **multiprocessing**, which involves the simultaneous execution of two or more instructions. One form of multiprocessing involves **coprocessors**. A coprocessor speeds processing by executing specific types of instructions while the CPU works on another processing activity. Coprocessors can be internal or external to the CPU and can have different clock speeds than the CPU. Each type of coprocessor best performs a specific function. For example, a math coprocessor chip can be used to speed mathematical calculations, and a graphics coprocessor chip decreases the time it takes to manipulate graphics.

Parallel Processing

Another form of multiprocessing, called **parallel processing**, speeds processing by linking several processors to operate at the same time, or in parallel. The most frequent business uses for parallel processing are modeling, simulation, and analysis of large amounts of data. In today's marketplace, consumers demand quick response and customized service, so companies are gathering and reporting more information about their customers. Collecting and organizing the enormous amount of customer data is no easy task, but parallel processing can help companies organize data on existing consumer buying patterns and process them more quickly to build an effective marketing program. As a result, a company can gain a competitive advantage. Ford Motor Company uses crash dummies to determine whether a certain type of crash is survivable. It is exploring the use of parallel processing to predict driver and passenger injuries in accident scenarios to actually predict the damage to various body parts and organs. Such occupant injury analysis takes much more computing power than is currently available.[4]

Grid Computing

Grid computing is the use of a collection of computers, often owned by multiple individuals or organizations, to work in a coordinated manner to solve a common problem. Grid computing is one low-cost approach to parallel processing. The grid can include dozens, hundreds, or even thousands of computers that run collectively to solve extremely large parallel processing problems. Key to the success of grid computing is a central server that

Table 2.1

Number of Bytes

random access memory (RAM)
A form of memory in which instructions or data can be temporarily stored.

read-only memory (ROM)
A nonvolatile form of memory.

multiprocessing
The simultaneous execution of two or more instructions at the same time.

coprocessor
The part of the computer that speeds processing by executing specific types of instructions while the CPU works on another processing activity.

parallel processing
A form of multiprocessing that speeds processing by linking several processors to operate at the same time, or in parallel.

grid computing
The use of a collection of computers, often owned by multiple individuals or organizations, to work in a coordinated manner to solve a common problem.

acts as the grid leader and traffic monitor. This controlling server divides the computing task into subtasks and assigns the work to computers on the grid that have (at least temporarily) surplus processing power. The central server also monitors the processing, and if a member of the grid fails to complete a subtask, it restarts or reassigns the task. When all the subtasks are completed, the controlling server combines the results and advances to the next task until the whole job is completed.

By installing the SETI@home screen saver program on their personal computers, millions of people worldwide contribute their idle CPU time to analyzing radio data from space for signs of intelligent life. With SETI, a central server doles out the work by giving each computer on the grid radio data from a tiny slice of the sky to analyze.

(Source: Copyright ©2003 SETI@home.)

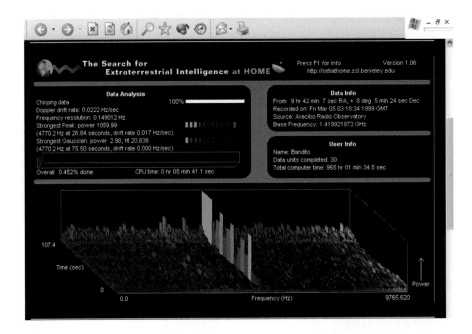

SECONDARY STORAGE AND INPUT AND OUTPUT DEVICES

As we have seen, memory is an important factor in determining overall computer system power. However, memory provides only a small amount of storage area for the data and instructions the CPU requires for processing. Computer systems also need to store larger amounts of data, instructions, and information more permanently than main memory allows. **Secondary storage**, also called *permanent storage*, serves this purpose.

Compared with memory, secondary storage offers the advantages of nonvolatility, greater capacity, and greater economy. Most forms of secondary storage are considerably less expensive than memory (see Table 2.2). Because of the electromechanical processes involved in using secondary storage, however, it is considerably slower than memory. The selection of secondary storage media and devices requires an understanding of their primary characteristics—access method, capacity, and portability.

Secondary Storage Access Methods

secondary storage (permanent storage)
The devices that store larger amounts of data, instructions, and information more permanently than allowed with main memory.

Data and information access can be either sequential or direct. **Sequential access** means that data must be accessed in the order in which it is stored. For example, inventory data stored sequentially might be stored by part number, such as 100, 101, 102, and so on. If you want to retrieve information on part number 125, you need to read and discard all the data relating to parts 001 through 124.

sequential access
The retrieval method in which data must be accessed in the order in which it is stored.

direct access
The retrieval method in which data can be retrieved without the need to read and discard other data.

Direct access means that data can be retrieved directly, without having to pass by other data in sequence. With direct access, it is possible to go directly to and access the needed data—for example, part number 125—without reading through parts 001 through 124. For this reason, direct access is usually faster than sequential access. The devices used to

Data Storage Media	Capacity	Cost/GB
Data tape cartridge	500 GB	$.16
DVD-RW	4.7 GB	$.29
CD-ROM	740 MB	$.34
Floppy diskette	1.44 MB	$347
Compact flash memory	128 MB	$367
SDRAM	128 MB	$874

Table 2.2

Cost Comparison for Various Forms of Data Storage

All forms of secondary storage cost considerably less per megabyte of capacity than SDRAM, although they have slower access times. A data tape cartridge costs about $.16 per gigabyte, whereas SDRAM can cost around $900 per gigabyte.

sequentially access secondary storage data are simply called **sequential access storage devices (SASDs)**; those used for direct access are called **direct access storage devices (DASDs)**.

Secondary Storage Devices

The most common forms of secondary storage include magnetic tapes, magnetic disks, and optical discs. Some of these media (magnetic tape) allow only sequential access, whereas others (magnetic and optical discs) provide direct and sequential access. Figure 2.2 shows some different secondary storage media.

sequential access storage device (SASD)
The device used to sequentially access secondary storage data.

direct access storage device (DASD)
The device used for direct access of secondary storage data.

Figure 2.2

Types of Secondary Storage

Secondary storage devices such as magnetic tapes and disks, optical discs, CD-ROMs, and DVDs are used to store data for easy retrieval at a later date.

(Source: Courtesy of Imation Corp.)

Magnetic Tapes

One common secondary storage medium is **magnetic tape**. Similar to the kind of tape found in audio- and videocassettes, magnetic tape is a Mylar film coated with iron oxide. Portions of the tape are magnetized to represent bits. Magnetic tape is a sequential access storage medium. Although access is slower, magnetic tape is usually less expensive than disk storage. In addition, magnetic tape is often used to back up disk drives and to store data off-site for recovery in case of disaster. Technology is improving to provide tape storage devices with greater capacities and faster transfer speeds. In addition, the large, bulky tape drives used to read and write on large diameter reels of tapes in the early days of computing have been replaced with much smaller tape cartridge devices measuring a few millimeters in diameter that take up much less floor space and allow hundreds of tape cartridges to be stored in a small area.

magnetic tape
A common secondary storage medium; Mylar film coated with iron oxide with portions of the tape magnetized to represent bits.

Magnetic Disks

Magnetic disks are also coated with iron oxide; they can be thin steel platters (see Figure 2.3) or Mylar film (disks). As with magnetic tape, magnetic disks represent bits by small magnetized areas. When reading from or writing onto a disk, the disk's read/write head can go directly to the desired piece of data. Thus, the disk is a direct access storage medium and allows for fast data retrieval. For example, if a manager needs information on the credit history of a customer, the information can be obtained in a matter of seconds if the data is stored on a direct access storage device. Magnetic disk storage varies widely in capacity and portability.

magnetic disk
A common secondary storage medium, with bits represented by magnetized areas.

RAID

Companies' data storage needs are expanding rapidly. Today's storage configurations routinely entail many hundreds of gigabytes. However, putting the company's data online involves a serious business risk—the loss of critical business data can put a corporation out of operation. The concern is that the most critical mechanical components inside a disk storage device—the disk drives, the fans, and other input/output devices—can break.

Organizations now require their data storage devices to be fault tolerant—the ability to continue with little or no loss of performance in the event of a failure of one or more key components. **Redundant array of independent/inexpensive disks (RAID)** is a method of storing data so that if a hard drive fails, the lost data on that drive can be rebuilt. With this approach, data is stored redundantly on different physical disk drives using a technique called *stripping* to evenly distribute the data. Quicken Loans, Inc., an online mortgage-lending company, moved to a centralized RAID storage system with automated management software from EMC Corp. and reduced the number of employees devoted to managing storage from 15 to 3. The easier access to information also helps cut the time to create detailed financial reports, from one to three days to two to eight hours.[5]

SAN

A **storage area network (SAN)** uses computer servers, distributed storage devices, and networks to tie everything together, as shown in Figure 2.4. To increase the speed of storing and retrieving data, fiber-optic channels are often used. Although SAN technology is relatively new, a number of companies are using SANs to successfully and efficiently store critical data. Wildman, Harrold, Allen & Dixon LLP employs 550 people, and the amount of litigation data it must store has been growing rapidly—from 90 GB to 600 GB in just one year. As a result, the firm decided to change its storage infrastructure, moving from a group of high-capacity disk drives to an automated storage area network. The firm spent less than $80,000 on implementing the SAN.[6]

redundant array of independent/inexpensive disks (RAID)
A method of storing data that generates extra bits of data from existing data, allowing the system to create a "reconstruction map" so that if a hard drive fails, the system can rebuild lost data.

storage area network (SAN)
The technology that provides high-speed connections between data-storage devices and computers.

Figure 2.4

Storage Area Network

A SAN provides high-speed connections between data-storage devices and computers over a network.

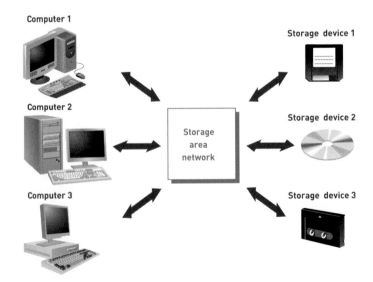

A common form of optical disc is called **compact disc read-only memory (CD-ROM)**. After data has been recorded on a CD-ROM, it cannot be modified—the disc is "read only." CD-recordable (CD-R) discs allow data to be written once to a CD disc. CD-rewritable (CD-RW) technology allows personal computer users to replace their 3.5-inch disks with high-capacity CDs that can be written on and edited. The CD-RW disc can hold roughly 500 times the capacity of a 1.4-MB 3.5-inch disk. A popular use of recordable and rewritable CD technology is to enable users to burn a CD of their favorite music for their later listening pleasure.

Digital Versatile Disc

A **digital versatile disc (DVD)** is a five-inch diameter CD-ROM look-alike with the ability to store about 135 minutes of digital video or several gigabytes of data (see Figure 2.5). Software programs, video games, and movies are common uses for this storage medium. At a data transfer rate of 1.25 MB/second, the access speed of a DVD drive is faster than that of the typical CD-ROM drive.

DVDs are replacing recordable and rewritable CD discs (CD-R and CD-RW) as the preferred format for sharing movies and photos. Whereas a CD can hold about 740 MB of data, a single-sided DVD can hold 4.7 GB, with double-sided DVDs having a capacity of 9.4 GB. Unfortunately, DVD manufacturers haven't agreed on a standard, so there are several types of recorders and discs. Recordings can be made on record-once discs (DVD-R and DVD+R) or on rewritable discs (DVD-RW, DVD+RW, and DVD-RAM). Rewritable discs are less widely compatible than others. Dell and Hewlett-Packard use DVD+RW; Apple, Gateway, and IBM offer DVD-RW.

Memory Cards

A group of computer manufacturers formed the Personal Computer Memory Card International Association (PCMCIA) to create standards for a peripheral device known as a *PC memory card*. These PC memory cards are credit card-sized devices that can be installed in an adapter or slot in many personal computers. To the rest of the system, the PC memory card functions as though it were a fixed hard disk drive. Although the cost per megabyte of storage is greater than for traditional hard disk storage, these cards are less prone to fail than hard disks, are portable, and are relatively easy to use. Software manufacturers often store the instructions for their program on a memory card for use with laptop computers.

Expandable Storage

Expandable storage devices use removable disk cartridges (see Figure 2.6). Expandable storage devices can be internal or external, and a few personal computers now include internal expandable storage devices as standard equipment. CD-RW drives by Hewlett-Packard, Iomega, and others can also be used for expandable storage. These expandable storage devices are ideal for backups of the critical data on your hard drive. The storage capacity can range from less than 100 MB to several gigabytes per cartridge. The access speed of some expandable devices is about as fast as an internal disk drive.

The overall trend in secondary storage is toward more direct-access methods, higher capacity, and increased portability. The business needs and needs of individual users should be considered when selecting a specific type of storage. In general, the ability to store large amounts of data and information and access it quickly can increase organizational effectiveness and efficiency. Table 2.3 lists the most common secondary storage devices and their capacities for easy reference.

---

compact disc read-only memory (CD-ROM)
A common form of optical disc on which data, after it has been recorded, cannot be modified.

digital versatile disc (DVD)
A storage medium used to store digital video or computer data over a network.

Figure 2.5

Digital Versatile Disc Player

DVD discs look like CDs but have a much greater storage capacity and can transfer data at a much faster rate.

(Source: Courtesy of Sony Electronics.)

---

expandable storage devices
Expandable storage devices use removable disk cartridges to provide additional storage capacity.

Figure 2.6

Expandable Storage

Expandable storage drives allow users to add storage capacity by simply inserting a removable disk or cartridge. The disks can be used to back up hard disk data or to transfer large files to colleagues.

(Source: Courtesy of Iomega.)

Table 2.3

Comparison of Secondary Storage Devices

Storage Device	Year First Introduced	Maximum Capacity
3.5-inch diskette	1987	1.44 MB
CD-ROM	1990	650 MB
Zip	1995	100–250 MB
DVD	1996	17 GB

Input Devices

A user's first experience with computers is usually through input and output devices. Through these devices—the gateways to the computer system—people provide data and instructions to the computer and receive results from it. Input and output devices are part of the overall user interface, which includes other hardware devices and software that allow humans to interact with a computer system.

As with other computer system components, the selection of input and output devices depends on the needs of the users and business objectives. For example, many restaurant chains use handheld input devices or computerized terminals that let waiters enter orders quickly and accurately. These systems also cut costs by making inventory tracking more efficient and marketing to customers more effective.

Literally hundreds of devices can be used for data input, ranging from special-purpose devices used to capture specific types of data to more general-purpose input devices. The following sections discuss several data input devices.

Personal Computer Input Devices

A keyboard and a computer mouse are the most common devices used for entry and input of data, such as characters, text, and basic commands. Some companies are developing newer keyboards that are more comfortable, adjustable, and faster to use. These keyboards, such as the split keyboard by Microsoft and others, are designed to avoid wrist and hand injuries caused by hours of keyboarding. Using the same keyboard, you can enter sketches on the touchpad and text using the keys.

A keyboard and mouse are two of the most common devices for computer input. Wireless mouses and keyboards are now readily available.

(Source: Courtesy of Gateway, Inc.)

A computer mouse is used to point to and click on symbols, icons, menus, and commands on the screen. The computer responds by taking a number of actions, such as copying data into the computer system or opening files.

Voice-Recognition Devices

Another type of input device can recognize human speech. Called **voice-recognition devices**, these tools use microphones and special software to record and convert the sound of the human voice into digital signals. Speech recognition can be used on a factory floor to allow equipment operators to give basic commands to machines while their hands perform other operations. Voice recognition can also be used by security systems to allow only authorized personnel into restricted areas. Voice-recognition systems now available on many makes of automobiles and trucks allow a driver to activate radio programs and CDs. It can even tell you the time. Asking "What time is it?" will get a response such as, "Eleven thirty-four."

Burlington Northern and Santa Fe Railway Co. (BNSF) recently launched a project to use its voice radios as the interface to an interactive voice response (IVR) system to provide input directly to the company's databases. The new system automatically integrates radio calls with its back-end systems, providing it with current and accurate information on its trains and individual cars. BNSF is now able to provide customers with more frequent information on car movement and better estimates of expected time of arrival.[7]

Terminals

Inexpensive and easy to use, terminals are input devices that perform data input. A terminal is connected to a complete computer system, including a processor, memory, and secondary storage. General commands, text, and other data are entered via a keyboard or mouse, converted into machine-readable form, and transferred to the processing portion of the computer system. Terminals are normally connected directly to the computer system by telephone lines or cables and can be placed in offices, in warehouses, and on factory floors.

Touch-Sensitive Screens

Advances in screen technology allow display screens to function as input as well as output devices. By touching certain parts of a sensitive screen, you can execute a program or cause the computer to take an action. Touch-sensitive screens are frequently used at gas stations for customers to select grades of gas and request a receipt, at fast-food restaurants for order clerks to enter customer choices, at information centers in hotels to allow guests to request facts about local eating and drinking establishments, and at amusement parks to provide directions to patrons. They also are used in kiosks at airports and department stores.

Bar-Code Scanners

A bar-code scanner employs a laser scanner to read a bar-coded label. This form of input is used widely in grocery store checkouts and in warehouse inventory control. Often, bar-code technology is combined with other forms of technology to create innovative ways for capturing data.

Optical Data Readers

A special scanning device called an *optical data reader* can also be used to scan documents. The two categories of optical data readers are for optical mark recognition (OMR) and optical character recognition (OCR). People completing OMR forms use pencils to fill in boxes on OMR paper, which is also called a "mark sense form." OMR is used in standardized tests (including SAT and GMAT tests), surveys, and cenus taking. In comparison, most OCR readers use reflected light to recognize various characters. With special software, OCR readers can convert handwritten or typed documents into digital data.

Point-of-Sale (POS) Devices

Point-of-sale (POS) devices are terminals used in retail operations to enter sales information into the computer system. The POS device then computes the total charges, including tax. Many POS devices also use other types of input and output devices, such as keyboards, bar-code readers, printers, and screens. A large portion of the money that businesses spend on computer technology involves POS devices.

voice-recognition device
An input device that recognizes human speech.

point-of-sale (POS) device
A terminal used in retail operations to enter sales information into the computer system.

Automatic Teller Machine (ATM) Devices

Another type of special-purpose input/output device, the automatic teller machine (ATM), is a terminal most bank customers use to perform withdrawals and other transactions for their bank accounts. The ATM, however, is no longer used only for cash and bank receipts. Companies use various ATM devices to support their business processes. Some ATMs dispense tickets for airlines, concerts, and soccer games. Some colleges use them to output transcripts. For this reason, the input and output capabilities of ATMs are quite varied. Like POS devices, ATMs may combine other types of input and output devices. Unisys, for example, has developed an ATM kiosk that allows bank customers to make cash withdrawals and pay bills, and also receive advice on investments and retirement planning.[8]

Magnetic Ink Character Recognition (MICR) Devices

In the 1950s, the banking industry became swamped with paper checks, loan applications, bank statements, and so on. To remedy this overload and process documents more quickly, the industry developed *magnetic ink character recognition (MICR)*, a system for reading this data quickly. With MICR, data is placed on the bottom of a check or other form using a special magnetic ink. Data printed with this ink using a special character set can be read by both people and computers. Read the "Information Systems @ Work" special-interest feature to learn more about how banks are using IS technology to speed up the check-clearing process.

Radio-Frequency Identification

radio-frequency identification (RFID)
A technology that employs a microchip with an antenna that broadcasts its unique identifier and location to receivers.

Radio-frequency identification (RFID) technology employs a microchip, called a *smart tag*, with an antenna that broadcasts its unique 96-bit identifier and location to corresponding receivers. The receiver relays the data to a computer, which decodes the information and processes it. One application of RFID is to place a microchip on retail items and install in-store readers that constantly count the inventory on the shelves.

Smart tags can be also embedded in items that are difficult to bar code, such as a bunch of grapes. They can even be embedded in raw materials, extending their benefits to manufacturers and suppliers. The difference between bar-code scanners and RFID scanners is that the latter records data from each item at a distance of 4 to 5 feet. It does not require a person to manually pass the item over a scanner, so transactions can be completed faster. As a result, RFID can produce more accurate inventory counts than bar codes.

Output Devices

Computer systems provide output to decision makers at all levels of an organization to solve a business problem or capitalize on a competitive opportunity. In addition, output from one computer system can be used as input into another computer system within the same information system. The desired form of this output might be visual, audio, and even digital. Whatever the output's content or form, output devices function to provide the right information to the right person in the right format at the right time.

Display Monitors

The display monitor is a TV-screen-like device on which output from the computer is displayed. Because traditional monitors use a cathode ray tube to display images, they are sometimes called *CRTs*. The monitor works in much the same way as a traditional TV screen—one or more electron beams are generated from cathode ray tubes. As the beams strike a phosphorescent compound (phosphor) coated on the inside of the screen, a dot on the screen called a *pixel* lights up. The electron beam sweeps back and forth across the screen so that as the phosphor starts to fade, it is struck and lights up again.

A monitor's ability to display color is a function of the quality of the monitor, the amount of RAM in the computer system, and the monitor's graphics adapter card. The color graphics adapter (CGA) was one of the first technologies to display color images on the screen. Today, super video graphics array (SVGA) displays are standard, providing vivid colors and superior resolution.

Banks Weigh Move to Improved Check-Clearing Process

The effective use of information systems is critical to the success of the banking industry. Basic transaction processing, debit card processing, automated teller machines, online bill payment, automated bill payment, and check imaging and processing all rely on information systems for efficiency and high reliability.

Until recently, the check-clearing and settlement process had changed little since the 1950s, when banks introduced magnetic-ink encoding. The clearance and settlement of checks took from one to four days and required paper checks to be shipped by plane and truck to the banks that issued them. This process also involved error-prone manual data entry.

Recent technological advances, such as image-capture hardware and high-speed networks, created an opportunity to store and exchange check images in lieu of actual checks. However, banks could not be forced to accept electronic check images. So, until recently, banks that didn't invest in imaging systems could force other banks to continue the costly practice of hiring air couriers to transport bundles of checks across the country.

The Check Clearing for the 21st Century Act, or Check 21, advocates the use of a device called an *image replacement document (IRD)*, a paper facsimile that is the legal equivalent of an original check. Here's how it works. Suppose that Bank of America wants to transmit a check image to a bank in North Dakota, but the North Dakota bank isn't equipped to receive digital images. Bank of America would transmit the image to a processing site near the bank in North Dakota, which would print out the image as an IRD and ship it to the North Dakota bank. From then on, the process would proceed exactly as in the past. This process enables the Bank of America to reduce its check transportation costs, but the North Dakota bank is not forced to invest in imaging technology.

Use of this improved process speeds the collection of checks, improves the availability of funds, reduces fraud losses, and lowers collection costs. According to Small Value Payment Company, a bank-owned provider of electronic-payment services, banks can achieve a net savings of $2.1 billion a year through the use of substitute checks. In addition, customers receive the convenience of viewing cleared checks via a PC connected to the Internet or from a link on an institution's Web page. The use of IRDs allows banks to use more efficient delivery channels for statements via CDs, DVDs, and e-mail.

Most large banks already have much of the basic computer hardware required for check imaging and processing. Many smaller banks, however, instead depend on their larger counterparts to perform the check-image processing. Another approach that some might take is to outsource check clearing and settlement, relying on a third-party provider to perform these functions.

Imagine that you are the manager in charge of the check-clearing and settlement operation in a midsized bank that is considering investing in information systems and hardware to take advantage of Check 21. Senior bank managers have asked for your recommendation on whether the bank should make the necessary investments to convert to this new process.

Discussion Questions

1. What are the potential benefits for your bank to convert to the new process?
2. What are some of the issues and factors that may complicate this move?

Critical Thinking Questions

3. How would you decide whether your bank should convert to the new Check 21 process?
4. What factors should you consider in deciding whether you should outsource the check-clearing and settlement functions rather than continue to perform them in-house?

SOURCES: Steven Marlin, "Check Clearing to Get Electronic Overhaul," *InformationWeek*, June 10, 2003, *www.informationweek.com*; Stessa B. Cohen, "New Sterling Spinoff Focuses on Check 21 Support," Gartner Group, September 17, 2003, at *www4.gartner.com*; Lucas Mearian, "Check 21 Becomes Law, Allows Speedier Electronic Settlements," *Computerworld*, November 23, 2003, *www.computerworld.com*; Carey Richardson, "Check 21: Check Clearing for the 21st Century," de novo banks.com, January 22, 2004, *www.denovobanks.com*.

Liquid Crystal Displays (LCDs)

A different technology, flat-panel display, is used for portable personal computers and laptops. One common technology used for flat-screen displays is the same liquid crystal display technology used for pocket calculators and digital watches. LCD monitors are flat displays that use liquid crystals—organic, oil-like material placed between two polarizers—to form characters and graphic images on a backlit screen.

CRT monitors are large and bulky in comparison with LCD monitors (flat displays).

(Source: Courtesy of ViewSonic Corporation.)

LCD technology is also being used to create thin and extremely high-resolution monitors for desktop computers. Although the screen might measure just 13 inches from corner to corner, the display's extremely high resolution—1,280×1,280 pixels—lets it show as much information as a conventional 20-inch monitor. In addition, although cramming more into a smaller area causes text and images to shrink, you can comfortably sit much closer to an LCD screen than the conventional CRT monitor.

Organic Light-Emitting Diodes

Organic light-emitting diode (OLED) technology is based on research done by Eastman Kodak Co. and is just reaching the market in small electronic devices. OLEDs use the same base technology as LCDs, with one key difference: Whereas LCD screens contain a fluorescent backlight and the LCD acts as a shutter to selectively block that light, OLEDs directly emit light. OLEDs can provide sharper and brighter colors than LCDs and CRTs, and because they don't require a backlight, the displays can be half the thickness of OCDs and can be used in flexible displays. Another big advantage is that OLEDs don't break when dropped. OLEDs are currently limited to use in cell phones, car radios, and digital cameras but might be used in computer displays—if the average display lifetime can be extended beyond the current 8,000 hours.[9]

Printers and Plotters

Printers with different speeds, features, and capabilities are available. Some can be set up to accommodate different paper forms, such as blank check forms, invoice forms, and so forth. Newer printers allow businesses to create customized printed output for each customer from standard paper and data input using full color.

The speed of the printer is typically measured by the number of pages printed per minute (ppm). Like a display screen, the quality, or resolution, of a printer's output depends on the number of dots printed per inch. A 600-dpi (dots-per-inch) printer prints more clearly than a 300-dpi printer. A recurring cost of using a printer is the ink-jet or laser cartridge that must be replaced every few thousand pages of output. Figure 2.7 shows a laser printer.

Plotters are a type of hard-copy output device used for general design work. Businesses typically use these devices to generate paper or acetate blueprints, schematics, and drawings of buildings or new products onto paper or transparencies. Standard plot widths are 24 inches and 36 inches, and the length can be whatever meets the need—from a few inches to several feet.

Figure 2.7

Laser Printer

Laser printers, available in a wide variety of speeds and price ranges, have many features, including color capabilities. They are the most common solution for outputting hard copies of information.

(Source: Courtesy of Lexmark International.)

COMPUTER SYSTEM TYPES

Computer systems can range from desktop (or smaller) portable computers to massive supercomputers that require housing in large rooms. Let's examine the types of computer systems in more detail. Table 2.4 shows general ranges of capabilities for various types of computer systems.

Table 2.4

Types of Computer Systems

Factor	Single-User Systems					Multiuser Systems (Servers)		
	Handheld	Portable	Thin Client	Desktop	Workstation	Server	Mainframe	Supercomputer
Cost Range	$200 to $1,500	$1,000 to $3,500	$250 to $1,000	$600 to $3,500	$4,000 to $40,000	$500 to $50,000	> $100,000	> $250,000
Weight	< 24 oz.	< 7 lbs.	< 15 lbs.	< 25 lbs.	< 25 lbs.	> 25 lbs.	> 200 lbs.	> 200 lbs.
Typical Size	Palm size	Size of a three-ring notebook	Fits on desktop	Fits on desktop	Fits on desktop	Three-drawer filing cabinet	Refrigerator	Refrigerator and larger
CPU Speed	> 200 MHz	> 2 GHz	> 200 MHz	> 3 GHz	> 3 GHz	> 2 GHz	> 300 MIPS	> 2 teraflops
Typical Use	Personal organizer	Improvement of worker productivity	Data entry and Internet access	Improvement of worker productivity	Engineering, CAD, software development	Support for network and Internet applications	Computing for large organization; provides massive data storage	Scientific applications; intensive number crunching
Example	Handspring Treo 600 smart phone	Motion Computing M1300 Mainstream Tablet PC	Max-speed Max-term 8400	iMac Power PC G4	Sun Microsystems Sun Blade 2500 Workstation	Hewlett-Packard HP ProLiant BL	Unisys ES5000	IBMs RS/6000 SP

Handheld Computers

Handheld computers are single-user computers that provide ease of portability because of their small size—some are as small as a credit card. These systems often include a wide variety of software and communications capabilities. Most are compatible with and able to communicate with desktop computers over wireless networks. Some even add a built-in global positioning system receiver with software that can integrate the location data into the

handheld computer
A single-user computer that provides ease of portability because of its small size.

application. For example, if you click on an entry in the address book, the device displays a map and directions from your current location. Such a computer can also be mounted in your car and serve as a navigation system. One of the shortcomings of handheld computers is that they require lots of power relative to their size.

Smart phones combine the functions of a telephone, PDA, game console, and a wireless-data device—and sometimes digital cameras. They enable a user to browse the Internet while on the move, store and play music, jot down brief messages, and place and receive phone calls. With their greater functionality, smart phones are expected to outsell PDAs very soon.[10]

Portable Computers

<div class="glossary">

portable computer
A computer small enough to be carried easily.

</div>

A variety of **portable computers**, those that can be carried easily, are now available—from laptops, to notebooks, to subnotebooks, to tablet computers. A *laptop computer* is a small, lightweight PC about the size of a three-ring notebook. The even smaller and lighter *notebook* and *subnotebook* computers offer similar computing power. Some notebook and subnotebook computers fit into docking stations of desktop computers to provide additional storage and processing capabilities. *Tablet PCs* are portable, lightweight computers that allow users to roam the office, home, or factory floor carrying the device like a clipboard. Such devices are quite popular in the health services field. HealthSouth, a company that provides outpatient surgery and other healthcare services, ordered 5,000 tablet PCs equipped with wireless LAN connections from Motion Computing, Inc. The tablet devices are used by physical therapists at the company's 1,400 rehabilitation centers to provide access to patient records and enable them to document clinical progress.[11]

The ViewSonic V1250 has a 12.1-inch screen and comes with a small navigation pad that protrudes from the right of the screen, with buttons for tasks such as scrolling, toggling between applications, and launching Internet Explorer. It also comes with a docking station that doubles as a battery charger.

(Source: Courtesy of ViewSonic Corporation.)

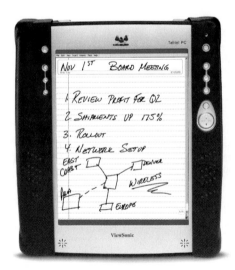

Thin Client

<div class="glossary">

thin client
A low-cost, centrally managed computer with essential but limited capabilities that is devoid of a DVD player, internal disk drive, and expansion slots.

</div>

A **thin client** is a low-cost, centrally managed computer that is devoid of a DVD player, internal disk drive, and expansion slots. These computers have limited capabilities and perform only essential applications, so they remain "thin" in terms of the client applications they include. These stripped-down versions of desktop computers do not have the storage capacity or computing power of typical desktop computers, nor do they need it for the role they play. With no hard disk, they never pick up viruses or experience a hard disk crash. Unlike personal computers, thin clients download software from a network when needed, making support, distribution, and updating of software applications much easier and less expensive. Their primary market is small businesses and educational institutions.

Desktop Computers

Desktop computers are relatively small, inexpensive single-user computer systems that are highly versatile. Named for their size—they are small enough to fit on an office desk—desktop computers can provide sufficient memory and storage for most business computing tasks. Desktop computers have become standard business tools; more than 30 million are in use in large corporations.

In addition to traditional PCs that use Intel processors and Microsoft software, other options are available. One of the most popular is the iMac by Apple Computer.

Workstations

Workstations are more powerful than personal computers but still small enough to fit on a desktop. They are used to support engineering and technical users who perform heavy mathematical computing, computer-aided design (CAD), and other applications requiring a high-end processor. Such users need very powerful CPUs, large amounts of main memory, and extremely high-resolution graphic displays to meet their needs.

Servers

A computer **server** is a computer used by many users to perform a specific task, such as running network or Internet applications. Servers typically have large memory and storage capacities, along with fast and efficient communications abilities. A Web server is used to handle Internet traffic and communications. An enterprise server stores and provides access to programs that meet the needs of an entire organization. A file server stores and coordinates program and data files. Server systems consist of multiuser computers, including supercomputers, mainframes, and servers.

Servers offer great **scalability**, the ability to increase the processing capability of a computer system so that it can handle more users, more data, or more transactions in a given period. Scalability is increased by adding more, or more powerful, processors. *Scaling up* adds more powerful processors, and *scaling out* adds many more equal (or even less powerful) processors to increase the total data-processing capacity.

Mainframe Computers

A **mainframe computer** is a large, powerful computer shared by dozens or even hundreds of concurrent users connected to the machine over a network. The mainframe computer must reside in a data center with special heating, venting, and air-conditioning (HVAC) equipment to control the environment around the computer. In addition, most mainframes are kept in a secure data center with limited access. The construction and maintenance of such a controlled-access room with HVAC can add hundreds of thousands of dollars to the cost of owning and operating a mainframe computer. Mainframe computers also require specially trained individuals (called *system engineers* and *system programmers*) to care for them.

The role of the mainframe is undergoing some remarkable changes as lower-cost, single-user computers become increasingly powerful. Many computer jobs that used to run on mainframe computers have migrated onto these smaller, less expensive computers. This information processing migration is called *computer downsizing*. The new role of the mainframe is as a large information processing and data-storage utility for a corporation—running jobs too large for other computers, storing files and databases too large to be stored elsewhere, and storing backups of files and databases created elsewhere. The mainframe is capable of handling the millions of daily transactions associated with airline, automobile, and hotel/motel reservation systems. It can process the tens of thousands of daily queries necessary to provide data to decision support systems. Its massive storage and input/output capabilities enable it to play the role of a video computer, providing full-motion video to multiple, concurrent users.

CartaSi S.p.A is one of Europe's largest credit card companies, with 7.5 million credit cards outstanding. Its 350 call-center service agents must process some 50,000 customer

---

desktop computer
A relatively small, inexpensive single-user computer that is highly versatile.

workstation
A more powerful personal computer that is used for technical computing, such as engineering, but still fits on a desktop.

server
A computer designed for a specific task, such as network or Internet applications.

scalability
The ability to increase the capability of a computer system to process more transactions in a given period by adding more, or more powerful, processors.

mainframe computer
A large, powerful computer often shared by hundreds of concurrent users connected to the machine via terminals.

Mainframe computers have been the workhorses of corporate computing for more than 50 years. They can support hundreds of users simultaneously and handle all of the core functions of a corporation.

(Source: Courtesy of IBM Corporation.)

calls per day for 16 banks in Italy. The agents retrieve customer data located in databases in the company's mainframe computers to answer questions related to a customer's account.[12] The processing speed and large data-storage capacity of these machines make this an efficient and cost-effective solution.

Supercomputers

supercomputers
The most powerful computer systems, with the fastest processing speeds.

Supercomputers are the most powerful computer systems, with the fastest processing speeds. They are designed for applications that require extensive and rapid computational capabilities. With an entry-level cost of $250,000, not all organizations can afford such a computer.

Originally, supercomputers were used primarily by government agencies to perform the high-speed number crunching needed in weather forecasting and military applications. With recent improvements in the cost and performance (lower cost and faster speeds) of these machines, they are being used more broadly for commercial purposes today. For example, France's Compagnie Générale de Géophysique SA is a global oil-services company that has linked more than 3,000 Dell PowerEdge servers to act as a supercomputing cluster to analyze seismic data to identify new oil and gas reservoirs, as well as to model existing reservoirs to optimize production.[13]

This 512-processor supercomputer at NASA's Ames Research Center is used for performing scalability studies, benchmarking, and solving large-scale problems for the Information Power Grid. The system has 192 GB of main memory, contains 2 TB of disk storage, and can reach a peak processing speed of 307 GFLOPS.

(Source: Tom Trower, NASA Ames Research Center.)

We now turn to the other critical component of effective computer systems—software. Like hardware, software has made technological leaps in a relatively short time span.

OVERVIEW OF SOFTWARE

In the 1950s, when computer hardware was relatively rare and expensive, software costs were a comparatively small percentage of total IS costs. Today, that situation has dramatically changed. Software can represent 75 percent or more of the total cost of an information system for three major reasons: Advances in hardware technology have dramatically reduced hardware costs, increasingly complex software requires more time to develop and so is more costly, and salaries for software developers have increased because the demand for these workers far exceeds the supply. In the future, software is expected to make up an even greater portion of the cost of the overall information system. The critical functions that software serves, however, make it a worthwhile investment.

One of software's most important functions is to direct the workings of the computer hardware. **Computer programs** are sequences of instructions for the computer. **Documentation** describes the program functions and helps the user operate the computer system. A program displays some of its documentation on screen, and other forms appear in external resources, such as printed manuals. Two basic types of software are available: systems software and application software. **Systems software** is the set of programs designed to coordinate the activities and functions of the hardware and various programs throughout the computer system. A particular systems software package is designed for a specific CPU design and class of hardware. The combination of a particular hardware configuration and systems software package is known as a **computer system platform**. **Application software** consists of programs that help users solve particular computing problems.

computer programs
The sequences of instructions for the computer.

documentation
The text that describes the program functions to help the user operate the computer system.

systems software
The set of programs designed to coordinate the activities and functions of the hardware and various programs throughout the computer system.

computer system platform
The combination of a particular hardware configuration and systems software package.

application software
The programs that help users solve particular computing problems.

sphere of influence
The scope of problems and opportunities addressed by a particular organization.

Supporting Individual, Group, and Organizational Goals

Every organization relies on the contributions of individuals, groups, and the entire enterprise to achieve business objectives. To help them achieve these objectives, the organization provides them with specific application software and information systems. One useful way of classifying the many potential uses of information systems is to identify the scope of the problems and opportunities addressed by a particular information system, called the **sphere of influence**. These spheres of influence are personal, workgroup, and enterprise, as shown in Table 2.5.

Software	Personal	Workgroup	Enterprise
Systems software	Personal computer and workstation operating systems	Network operating systems	Midrange computer and mainframe operating systems
Application software	Word processing, spreadsheet, database, graphics	Electronic mail, group scheduling, shared work	General ledger, order entry, payroll, human resources

Information systems that operate within the *personal sphere of influence* serve the needs of an individual user. These information systems enable users to improve their personal effectiveness, increasing the amount of work that can be done and its quality. Such software is often referred to as **personal productivity software**. Many examples of such

Table 2.5

Classifying Software by Type and Sphere of Influence

personal productivity software
The software that enables users to improve their personal effectiveness, increasing the amount of work they can do and its quality.

applications operating within the personal sphere of influence exist—a word processing application to enter, check spelling of, edit, copy, print, distribute, and file text material; a spreadsheet application to manipulate numeric data in rows and columns for analysis and decision making; a graphics application to perform data analysis; and a database application to organize data for personal use.

A *workgroup* is two or more people who work together to achieve a common goal. A workgroup might be a large, formal, permanent organizational entity such as a section or department or a temporary group formed to complete a specific project. The human resource department of a large firm is an example of a formal workgroup. It consists of several people, is a formal and permanent organizational entity, and appears on a firm's organization chart. An information system that operates in the *workgroup sphere of influence* supports a workgroup in the attainment of a common goal. Users of such applications are operating in an environment in which communication, interaction, and collaboration are critical to the success of the group. Applications include systems that support information sharing, group scheduling, group decision making, and conferencing. These applications enable members of the group to communicate, interact, and collaborate.

Information systems that operate within the *enterprise sphere of influence* support the firm in its interaction with its environment. The surrounding environment includes customers, suppliers, shareholders, competitors, special-interest groups, the financial community, and government agencies. Every enterprise has many applications that operate within the enterprise sphere of influence. The input to these systems is data about or generated by basic business transactions with someone outside the business enterprise. These transactions include customer orders, inventory receipts and withdrawals, purchase orders, freight bills, invoices, and checks. One of the results of processing transaction data is that the records of the company are updated. The order entry, finished product inventory, and billing information systems are examples of applications that operate in the enterprise sphere of influence. These applications support interactions with customers and suppliers.

SYSTEMS SOFTWARE

Controlling the operations of computer hardware is one of the most critical functions of systems software. Systems software also supports the application programs' problem-solving capabilities. Different types of systems software include operating systems and utility programs.

Operating Systems

operating system (OS)
A set of computer programs that controls the computer hardware and acts as an interface with application programs.

An **operating system (OS)** is a set of computer programs that control the computer hardware and act as an interface with application programs (see Figure 2.8). Operating systems can control one computer or multiple computers, or they can allow multiple users to interact with one computer. The various combinations of OSs, computers, and users include the following:

- *A Single Computer with a Single User.* This system is commonly used in a personal computer or a handheld computer that allows one user at a time.
- *A Single Computer with Multiple Users.* This system is typical of larger, mainframe computers that can accommodate hundreds or thousands of people, all using the computer at the same time.
- *Multiple Computers.* This system is typical of a network of computers, such as a home network that has several computers attached or a large computer network with hundreds of computers attached around the world.
- *Special-Purpose Computers.* This system is typical of a number of special-purpose systems that control sophisticated military aircraft, the space shuttle, some home appliances, and a variety of other special-purpose computers.

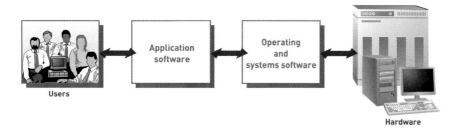

Figure 2.8

The role of the operating system and other systems software is as an interface or buffer between application software and hardware.

The OS, which plays a central role in the functioning of the complete computer system, is usually stored on disk. After a computer system is started, or "booted up," portions of the OS are transferred to memory as they are needed. The group of programs, collectively called the *operating system*, executes a variety of activities, including the following:

- Performing common computer hardware functions
- Providing a user interface
- Providing a degree of hardware independence
- Managing system memory
- Managing processing tasks
- Providing networking capability
- Controlling access to system resources
- Managing files

Common Hardware Functions

All application programs must perform certain tasks—for example, getting input from the keyboard or some other input device, retrieving data from disks, storing data on disks, and displaying information on a monitor or printer. Each of these basic functions requires a more detailed set of instructions. The OS converts a simple, basic instruction into the set of detailed instructions the hardware requires. In effect, the OS acts as an intermediary between the application program and the hardware. The typical OS performs hundreds of such functions, each of which is translated into one or more instructions for the hardware. The OS notifies the user if input/output devices need attention, if an error has occurred, or if anything abnormal occurs in the system.

User Interface

One of the most important functions of any OS is providing a **user interface**. A user interface allows individuals to access and command the computer system. The first user interfaces for mainframe and personal computer systems were command based.

A **command-based user interface** requires text commands to be given to the computer to perform basic activities. For example, the command ERASE 00TAXRTN would cause the computer to erase or delete a file called 00TAXRTN. RENAME and COPY are other examples of commands used to rename files and copy files from one location to another.

A **graphical user interface (GUI)** uses pictures called *icons* and menus displayed on screen to send commands to the computer system. Many people find that GUIs are easier to use because users intuitively grasp the functions. Today, the most widely used graphical user interface is Windows by Microsoft. As the name suggests, Windows is based on the use of a window, or a portion of the display screen dedicated to a specific application. The screen can display several windows at once. The use of GUIs has contributed greatly to the increased use of computers because users no longer need to know command-line syntax to accomplish a task.

Hardware Independence

The applications use the OS by making requests for services through a defined **application program interface (API)**, as shown in Figure 2.9. Programmers can use APIs to create application software without understanding the inner workings of the operating system.

Memory Management

The memory management feature of OSs converts a user's request for data or instructions (called a *logical view* of the data) to the physical location where the data or instructions are stored. A computer understands only the *physical view* of data—that is, the specific

user interface
The element of the operating system that allows individuals to access and command the computer system.

command-based user interface
A user interface that requires that text commands be given to the computer to perform basic activities.

graphical user interface (GUI)
An interface that uses icons and menus displayed on screen to send commands to the computer system.

application program interface (API)
The interface that allows applications to make use of the operating system.

Application Program Interface Links Application Software to the Operating System

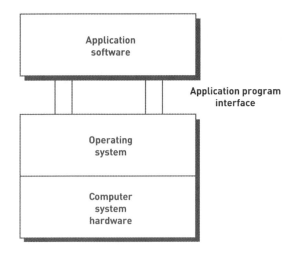

location of the data in storage or memory and the techniques needed to access it. This concept is described as logical versus physical access. For example, the current price of an item, for instance, a Texas Instruments BA-35 calculator with an item code of TIBA35, might always be found in the logical location "TIBA35$." If the CPU needed to fetch the price of TIBA35 as part of a program instruction, the memory management feature of the OS would translate the logical location "TIBA35$" into an actual physical location in memory or secondary storage (see Figure 2.10).

An Example of the Operating System Controlling Physical Access to Data

The user prompts the application software for specific data. The OS translates this prompt into instructions for the hardware, which finds the data the user requested. Having successfully completed this task, the OS then relays the data back to the user via the application software.

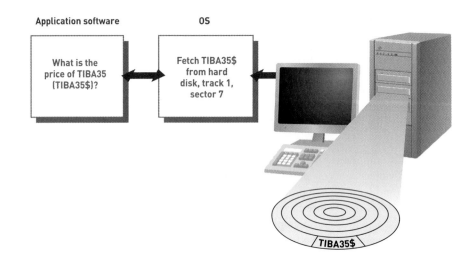

Processing Tasks

Task management features of today's OSs manage all processing activities. Task management allocates computer resources to make the best use of each system's assets. Task management software can permit one user to run several programs or tasks at the same time (multitasking) and allow several users to use the same computer at the same time (time-sharing). With **multitasking**, a user can run more than one application at the same time. Without having to exit a program, you can work in one application, easily pop into another, and then jump back to the first program, picking up where you left off. Better still, while you're working in the *foreground* in one program, one or more other applications can be churning away, unseen, in the *background*—sorting a database, printing a document, or performing other lengthy operations that otherwise would monopolize your computer and leave you staring at the screen unable to get other work done. Multitasking can save users a considerable amount of time and effort.

multitasking
The capability that allows a user to run more than one application at the same time.

Time-sharing allows more than one person to use a computer system at the same time. For example, 15 customer service representatives might be entering sales data into a computer system for a mail-order company simultaneously. In another case, thousands of people might be simultaneously using an online computer service to get stock quotes and valuable business news. Time-sharing works by dividing time into small CPU processing time slices, which can be a few milliseconds or less in duration. During a time slice, some tasks are done for the first user. The computer then goes from that user to the next and completes some tasks for that user during that time slice. This process continues for each user and cycles back to the first user. Because the CPU processing time slices are small, it appears that tasks for all users are being completed at the same time. In reality, each user is sharing the time of the computer with other users.

The ability of the computer to handle an increasing number of concurrent users smoothly is called *scalability*. Scalability is a critical feature for systems that must handle a large number of users, such as a mainframe computer or a Web server. Because personal computer OSs usually are oriented toward single users, management of multiple-user tasks often is not needed.

Networking Capability

The OS can provide features and capabilities that aid users in connecting to a computer network. For example, Apple computer users have built-in network access through the AppleTalk feature, and the Microsoft Windows OSs come with the capability to link users to the Internet.

Access to System Resources

Computers often handle sensitive data that can be accessed over networks, so the OS needs to provide a high level of security against unauthorized access to users' data and programs. Typically, the OS establishes a log-on procedure that requires users to enter an identification code and a matching password. If the identification code is invalid or if the password does not match the identification code, the user cannot gain access to the computer. The OS also requires that user passwords be changed frequently—for example, every 30 days. If the user is successful in logging on to the system, the OS records who is using the system and for how long. In some organizations, such records are also used to bill users for time spent using the system. The OS also reports any attempted breaches of security.

File Management

The OS also performs a file management function to ensure that files in secondary storage are available when needed and that they are protected from unauthorized access. Many computers support multiple users who store files on centrally located disks or tape drives. The OS keeps track of where each file is stored and who may access it. The OS must also be able to determine what to do if more than one user requests access to the same file at the same time.

Current Operating Systems

Early OSs were very basic. Today, however, more advanced OSs have been developed, incorporating some features previously available only with mainframe OSs. Table 2.6 classifies a number of current OSs by sphere of influence.

Microsoft PC Operating Systems

Ever since a then-small company called Microsoft developed PC-DOS and MS-DOS to support the IBM personal computer introduced in the 1970s, there has been a continuous and steady evolution of personal computer OSs. *PC-DOS* and *MS-DOS* had command-driven interfaces that were difficult to learn and use. Each new version of OS has improved the ease of use, processing capability, reliability, and ability to support new computer hardware devices.

Windows XP (XP reportedly stands for the wonderful *ex*perience that you will have with your personal computer) was released in fall of 2001. With XP, Microsoft hopes to bring reliability to the consumer. Its redesigned icons, task bar, and window borders make for more pleasant viewing. The Start menu is two columns wide with recently used programs in the left column and everything else (e.g., My Documents, My Computer, and Control

time-sharing
The capability that allows more than one person to use a computer system at the same time.

Table 2.6

Popular Operating Systems Across All Three Spheres of Influence

Personal	Workgroup	Enterprise
Windows XP, Windows Mobile, and Windows Embedded	Windows NT Server	Windows NT Server
Mac OS	Windows 2003 Server	Windows 2003 Server
Mac OS X	Mac OS Server	Windows Advanced Server, Limited Edition
UNIX	UNIX	UNIX
Solaris	Solaris	Solaris
Linux	Linux	Linux
Red Hat Linux	Red Hat Linux	Red Hat Linux
Palm OS	NetWare	
	IBM OS/390	IBM OS/390
	IBM z/OS	IBM z/OS
	HP MPE/iX	HP MPE/iX

Panel) in the right column. It comes with Internet Explorer 6 browser software, which boasts improved security and reliability features, including a one-way firewall that blocks hacker invasions coming in from the Internet. Radio Shack is using Windows XP in more than 5,000 stores to help run point-of-sale terminals and other devices.[14] Today, Microsoft has about 93 percent of the PC OS market.[15] Apple has 3 percent of the market, Linux has 3 percent, and other companies account for about 1 percent of the PC OS market.

Apple Computer Operating Systems

Although IBM system platforms traditionally use one of the Windows OSs and Intel microprocessors (often called *WINTEL* for this reason), Apple computers typically use non-Intel microprocessors designed by Apple, IBM, and Motorola and a proprietary Apple OS—the Mac OS. Although IBM and IBM-compatible computers hold the largest share of the business PC market, Apple computers are also quite popular, especially in the fields of publishing, education, graphic arts, music, movies, and media.

The Apple OSs have also evolved over a number of years and often provide features not available from Microsoft. The Mac OS 9 had a Multiple Users feature, which allowed you to safely share your Macintosh computer with other people. The Mac OS X includes an entirely new user interface, which provides a new visual appearance for users—including luminous and semitransparent elements, such as buttons, scroll bars, windows, and fluid animation to enhance the user's experience. OS X has been upgraded with additional releases, nicknamed Jaguar (OS X.2) and Panther (OS X.3). Cline, Davis & Mann, Inc., a New York advertising agency, specializes in healthcare industry campaigns with major pharmaceutical clients. About 90 percent of the computers for the ad agency's 360 employees are Macs. The firm is moving to Mac OS X to take advantage of the improved reliability, 3-D effects, and better management of windows on a display screen.[16] Fewer virus attacks, compared with Microsoft's Windows OS, are another advantage of using Mac OS X.[17]

Linux

Linux is an OS developed by Linus Torvalds in 1991 as a student in Finland. The OS is under the GNU General Public License, and its source code is freely available to everyone. This doesn't mean, however, that Linux and its assorted distributions are free—companies and developers can charge money for it as long as the source code remains available. Linux is actually only the *kernel* of an OS, the part that controls hardware, manages files, separates processes, and so forth. Several combinations of Linux are available, with various sets of capabilities and applications to form a complete OS. Each of these combinations is called a *distribution* of Linux. Read the "Ethical and Societal Issues" special-interest feature to learn more about the pros and cons of open source.

Linux is available over the Internet and from other sources, including Red Hat Linux and SCO OpenLinux. Many individuals and organizations are starting to use Linux. Galileo, the large travel and airline ticketing company uses Linux to run its Internet site.[18] Panasonic is using Linux to run one of its high-speed television tuners.[19] A survey revealed that many CIOs are considering switching to Linux and open-source software because of security concerns with Microsoft software.[20]

Workgroup Operating Systems

To keep pace with today's high-tech society, the technology of the future must support a world in which network usage, data-storage requirements, and data-processing speeds increase at a dramatic rate. Powerful and sophisticated OSs are needed to run the servers that meet these business needs for workgroups. Small businesses, for example, often use workgroup OSs to run networks and perform critical business tasks.[21]

Windows Server

Microsoft designed *Windows Server 2003* to do a host of new tasks that are vital for Web sites and corporate Web applications on the Internet.[22] For example, Microsoft Windows Server 2003 can be used to coordinate large data centers. The OS also works with other Microsoft products. It can be used to prevent unauthorized disclosure of information by blocking text and e-mails from being copied, printed, or forwarded to other people.[23] Besides being more reliable than Windows NT, this OS is capable of handling extremely demanding computer tasks, such as order processing. It can be tuned to run on machines with up to 32 microprocessors—satisfying the needs of all but the most demanding of Web operators. Four machines can be clustered to prevent service interruptions, which are disastrous for Web sites.

Microsoft *Windows Advanced Server, Limited Edition*, was the first 64-bit version of the Windows Server family. Introduced in August 2001, it was designed to run on the 64-bit Itanium processor from Intel (also known as the IA64). This OS enables Microsoft to begin competing with rival Linux vendors (Red Hat, SCO, SuSE, and TurboLinux), which already have 64-bit Itanium versions of their Linux distributions. In addition, Sun Microsystems and IBM have had 64-bit UNIX OSs for years.

UNIX

UNIX is a powerful OS originally developed by AT&T for minicomputers. UNIX can be used on many computer system types and platforms, from personal computers to mainframe systems. UNIX also makes it much easier to move programs and data among computers or to connect mainframes and personal computers to share resources. Many variants of UNIX are available—including HP/UX from Hewlett-Packard, AIX from IBM, UNIX SystemV from SCO, Solaris from Sun Microsystems, and SCO from Santa Cruz Operations.

NetWare

NetWare is a network OS sold by Novell that can support end users on Windows, Macintosh, and UNIX platforms. NetWare provides directory software to track computers, programs, and people on a network, making it easier for large companies to manage complex networks. NetWare users can log on from any computer on the network and still get their own familiar desktop with all their applications, data, and preferences.

Red Hat Linux

Red Hat Software offers a Linux network OS that taps into the talents of tens of thousands of volunteer programmers who generate a steady stream of improvements for the Linux OS. The *Red Hat Linux* network OS is very efficient at serving up Web pages and can manage a cluster of up to eight servers. The film *Lord of the Rings* used Linux and hundreds of servers to deliver many of the special effects seen in the finished film.[24] Linux environments typically have fewer virus and security problems than other OSs. Red Hat Linux has proven to be a very stable and efficient OS.

Weighing the Benefits of Open Source

Recently, the city of Munich decided to switch its 14,000 computers used by local government employees from Microsoft Windows to Linux, a free open-source OS. Open-source software is distributed along with its source code so that users can edit and change it as desired. When Microsoft's CEO, Steve Ballmer, heard about Munich's plans, he flew there to lobby the mayor. In the negotiations, Ballmer even matched Linux's price—essentially giving the city Microsoft Windows for free—and still the mayor turned him down. The rationale? The municipality wanted to control its technological destiny and avoid an OS from a commercial vendor that is accountable to shareholders rather than to citizens.

This sentiment is growing among governments around the world. Brazil also recommended that all its government agencies and state enterprises buy open-source software. China is working to adopt Linux to become self-sufficient and secure. India, Japan, and South Korea are also aggressively pursuing open-source alternatives to Microsoft software. In the United States, state, county, and city governments are equal in their interest in open-source software. In Florida, Miami-Dade County is considering swapping out its 15,000 copies of Microsoft Windows for Linux.

Some in private industry are following government's lead. Such big players in the technology industry as IBM, Novell, and Sun Microsystems are switching or have switched their internal operations to Linux. Many nontech companies are also converting. Ford Motor Company is switching to Linux for much of its server computing.

Why all this fuss over software? As we saw in Munich, it isn't just a matter of price. The reason governments and businesses like open-source software is its accessibility. It can be tweaked, edited, and dissected and has no secrets. Traditional, proprietary software is just the opposite—it is one big secret. Using software packages such as Microsoft Windows is similar to owning a car with no hood to access the engine. You are provided with a slick user interface, but if you are curious about how it works or need to repair it, you are out of luck.

Supporters of open-source software development have a simple philosophy: When programmers can read, redistribute, and modify the source code, the software evolves for everyone's benefit. People improve it, people adapt it, and people fix bugs. Development can happen at a speed that seems astonishing.

Opponents of the open-source movement, such as Microsoft, have sought to discredit the software. They have suggested that the very openness of the software makes it vulnerable to hackers and terrorists. Microsoft has also funded studies that have found that Windows has a lower cost of ownership than Linux.

Still, Microsoft has opened up some of its code under an initiative called "shared source," which allows certain approved governments and large corporate clients to access most of the Windows source code. This limited access helps Microsoft assure purchasers that Windows doesn't contain secret security backdoors. Microsoft has also made available portions of the source code for Windows CE, the OS for handheld PCs and mobile phones, so others can more easily develop applications for it.

The one major drawback of open-source software—and the major challenge for the Linux OS—is the lack of application programs for personal computing. Because Microsoft holds a monopoly in the desktop PC OS market, the vast majority of software is written for Microsoft Windows. The software developed for Linux is much smaller in quantity, variety, and some may argue in quality. Although inexpensive alternatives are available, most users are more comfortable with the familiar Microsoft software. Because a solid business model for open-source development is still in the blue-print stages, developers have little incentive to create software that will be given away for free. Companies such as IBM, however, are finding success in distributing Linux to their customers and profiting from the design, training, and support aspects of Linux-based information systems.

Critical Thinking Questions

1. Why do you think Ford is using Linux on its back-end servers but not on its employee's desktop PCs?
2. How can software companies like Microsoft compete against the trend toward open-source software?

What Would You Do?

As head of the technology group at Seabreeze Security Systems Corporation, Tom Gaskins had felt a lot of pressure to cut costs over the past two years. His CIO just asked Tom to reduce the technology budget by one-third over the next year. But this was the year the organization was scheduled to upgrade to the latest version of Microsoft Windows and Office. That investment would put him 20 percent over last year's budget. Tom could put off the upgrade for another, but it didn't look as though the budget would be any better then. How could he reduce costs? Should he lay off staff? Should he investigate alternatives to Microsoft? Changing platforms would mean quite a bit of work and significant staff training.

3. If you were Tom, how would you calculate the cost of changing to a platform such as Linux against the cost of the typical Windows/Office upgrade?
4. How would the costs of changing platforms differ over the next five years from multiple upgrades of Windows and Office?

SOURCES: "Microsoft at the Power Point," *Economist.com*, September 11, 2003, *www.economist.com*, John Lettice, "Motor Giant Ford to Move to Linux," *The Register*, September 9, 2003, *www.theregister.com*, Linux Online, *www.linux.org/*, accessed February 8, 2004; Open Source Initiative Web site, *www.opensource.org/*, accessed February 8, 2004; Sun Microsystems Star Office Web site, *wwws.sun.com/*, accessed February 8, 2004.

Mac OS X Server

The *Mac OS X Server* is the first modern server OS from Apple Computer. It provides UNIX-style process management. Protected memory puts each service in its own well-guarded chunk of dynamically allocated memory, preventing a single process from going awry and bringing down the system or other services. Under preemptive multitasking, a computer OS uses some criteria to decide how long to allocate to any one task before giving another task a turn to use the OS. Preempting is the act of taking control of the OS from one task and giving it to another. A common criterion for preempting is simply elapsed time. In more sophisticated OSs, certain applications can be given higher priority than other applications, giving the higher-priority programs longer processing times. Preemptive multitasking ensures that each process gets the right amount of CPU time and the system resources it needs for optimal efficiency and responsiveness.

Enterprise Operating Systems

The new generation of mainframe computers provides the computing and storage capacity to meet massive data-processing requirements and provide a large number of users with high performance and excellent system availability, strong security, and scalability. In addition, a wide range of application software has been developed to run in the mainframe environment, making it possible to purchase software to address almost any business problem. As a result, mainframe computers remain the computing platform of choice for mission-critical business applications for many companies. *z/OS* from IBM, *MPE/iX* from Hewlett-Packard, and *Red Hat Linux* are examples of mainframe operating systems.

z/OS

The *z/OS* operating system supports IBM's z900 and z800 lines of mainframes that can come with up to sixteen 64-bit processors. (The z stands for zero downtime.) It provides several new capabilities to make it easier and less expensive for users to run large mainframe computers. The OS has improved workload management and advanced e-commerce security. The IBM zSeries mainframe, like previous generations of IBM mainframes, lets users subdivide a single computer into multiple smaller servers, each of which is capable of running a different application. In recognition of the widespread popularity of a competing OS, z/OS allows partitions to run a version of the Linux OS.

MPE/iX and HP-UX

Multiprogramming Executive with integrated POSIX (MPE/iX) is the Internet-enabled OS for the Hewlett-Packard e3000 family of computers. MPE/iX is a robust OS designed to handle a variety of business tasks, including online transaction processing and Web applications. It runs on a broad range of HP e3000 servers—from entry-level to workgroup and enterprise servers within the data centers of large organizations. *HP-UX* is a mainframe OS from Hewlett-Packard. The OS is designed to support Internet, database, and a variety of business applications. It can work with Java programs and Linux applications.

Linux

Red Hat Software announced the availability of *Red Hat Linux for IBM* mainframe computers in December 2001. This version of Red Hat Linux means that the company now has Linux versions for everything from handheld devices to the largest enterprise mainframes.

Operating Systems for Small Computers and Special-Purpose Devices

New OSs and other software are changing the way users interact with personal digital assistants (PDAs), cell phones, digital cameras, TVs, and other appliances. These OSs are also called *embedded operating systems* because they are typically embedded in a computer chip. Embedded software is a $21 billion industry.[25] Some of these OSs allow handheld devices to be synchronized with PCs using cradles, cables, and wireless connections. Cell phones also use embedded OSs (see Figure 2.11). In addition, there are OSs for special-purpose devices, such as TV set-top boxes, computers on the space shuttle, computers in military weapons, and computers in some home appliances. Here are some of the more popular OSs for such devices.

Figure 2.11

Many cell phones such as this one from Nokia also have an integrated imaging device. Point, use the color display as a viewfinder, and snap a picture. Images can be stored on the device and sent to a friend.

(Source: Courtesy of Nokia.)

Palm OS

PalmSource makes the Palm operating system that is used on more than 30 million handheld computers and smart phones by PalmOne and other companies. The company also develops and supports applications, including business, multimedia, games, productivity, reference and education, hobbies and entertainment, travel, sports, utilities, and a variety of wireless applications.

Windows Embedded

Windows Embedded is a family of Microsoft OSs included with or embedded into small computer devices. Windows Embedded includes Windows CE.Net and Windows XP Embedded. Windows Embedded OSs can be used in mobile devices, such as smart phones and PDAs, and also in a variety of other devices, such as digital cameras, thin clients, TV set-top boxes, and automotive computers.

Windows Mobile

Windows Mobile is a family of Microsoft OSs for mobile or portable devices. Windows Mobile includes Pocket PC, Pocket PC Phone Edition, and SmartPhone.[26] These OSs have many features, including handwriting recognition, the ability to beam information to devices running either Pocket PC or PalmSource's competing OS, Microsoft's instant messaging technology, and support for more secure Internet connections. Motorola, Samsung, Dell, Hewlett-Packard, and Toshiba have products that run Windows Mobile. Wireless services from Verizon, Cingular, and others are often available with devices that use Windows Mobile.

APPLICATION SOFTWARE

The primary function of application software is to apply the power of a computer to give individuals, workgroups, and the entire enterprise the ability to solve problems and perform specific tasks. Application programs perform those specific computer tasks by interacting with systems software to direct the computer hardware. Programs that complete sales orders, control inventory, pay bills, write paychecks to employees, and provide financial and marketing information to managers and executives are examples of application software. Most of the computerized business jobs and activities discussed in this book involve the use of application software.

Types and Functions of Application Software

The key to unlocking the potential of any computer system is application software. A company can either develop a one-of-a-kind program for a specific application (called **proprietary software**) or purchase and use an existing software program (sometimes called **off-the-shelf software**). It is also possible to modify some off-the-shelf programs, giving a blend of off-the-shelf and customized approaches. These different sources of software are shown in Figure 2.12. The relative advantages and disadvantages of proprietary software and off-the-shelf software are summarized in Table 2.7.

proprietary software
A one-of-a-kind program for a specific application, usually developed and owned by a single company.

off-the-shelf software
An existing software program that is purchased.

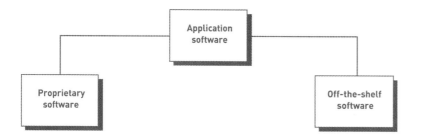

Figure 2.12

Types of Application Software

Some off-the-shelf software can be modified to allow some customization.

Proprietary Application Software

Software to solve a unique or specific problem is called *proprietary application software*. This type of software is usually built, but it can also be purchased from an outside company. If an organization has the time and IS talent, it might opt for *in-house development* for all aspects of the application programs. Alternatively, an organization might obtain customized software from external vendors. For example, a third-party software firm, often called a *value-added software vendor*, might develop or modify a software program to meet the needs of a particular industry or company. A specific software program developed for a particular company is called **contract software**.

contract software
The software developed for a particular company.

Off-the-Shelf Application Software

Software can also be purchased, leased, or rented from a software company that develops programs and sells them to many computer users and organizations. Software programs developed for a general market are called off-the-shelf software packages because they can literally be purchased "off the shelf" in a store. Many companies use off-the-shelf software to support business processes.

Customized Package

In some cases, companies use a blend of external and internal software development. That is, off-the-shelf software packages are modified or customized by in-house or external personnel. For example, a software developer might write a collection of programs to be used in an auto body shop that includes features to generate estimates, order parts, and process

Proprietary Software		Off-the-Shelf Software	
Advantages	**Disadvantages**	**Advantages**	**Disadvantages**
You can get exactly what you need in terms of features, reports, and so on.	It can take a long time and significant resources to develop required features.	The initial cost is lower since the software firm is able to spread the development costs over a large number of customers.	An organization might have to pay for features that are not required and never used.
Being involved in the development offers a further level of control over the results.	In-house system development staff may become hard pressed to provide the required level of ongoing support and maintenance because of pressure to get on to other new projects.	There is a lower risk that the software will fail to meet the basic business needs—you can analyze existing features and the performance of the package.	The software may lack important features, thus requiring future modification or customization. This can be very expensive because users must adopt future releases of the software as well.
There is more flexibility in making modifications that may be required to counteract a new initiative by one of your competitors or to meet new supplier and/or customer requirements. A merger with another firm or an acquisition also will necessitate software changes to meet new business needs.	There is more risk concerning the features and performance of the software that has yet to be developed.	Package is likely to be of high quality because many customer firms have tested the software and helped identify many of its bugs.	Software may not match current work processes and data standards.

Table 2.7

A Comparison of Proprietary and Off-the-Shelf Software

application service provider (ASP)
A company that provides software, end-user support, and the computer hardware on which to run the software from the user's facilities.

insurance. Designed properly—and allowing for minor tailoring for each body shop—the same software package can be sold to many businesses. However, because each body shop has slightly different requirements, software vendors would probably provide a wide range of services, including installation of their standard software, modifications for unique customer needs, training of end users, and other consulting services.

Another approach to obtaining a customized software package is to use an **application service provider (ASP)**—a company that provides software, end user support, and the computers on which to run the software from the user's facilities. They can also simplify a complex corporate software package for users so that it is easier to set up and manage. ASPs also provide contract customization of off-the-shelf software, assist in speeding deployment of new applications, and help IS managers avoid implementation headaches, reducing the need for skilled IS staff members and reducing project start-up expenses. Perhaps the biggest advantage of employing an ASP is that it frees in-house corporate resources from staffing and managing complex computing projects so that they can focus on more important things.

Using an ASP is not without risks—sensitive information could be compromised in a number of ways, including unauthorized access by employees or computer hackers, the ASP being unable to keep its computers and network up and running as consistently as is needed, or a disaster disabling the ASP's data center, temporarily putting an organization out of business. These are legitimate concerns that the ASP must address.

Personal Application Software

Literally hundreds of computer applications can help individuals at school, home, and work. The features of personal application software are summarized in Table 2.8. In addition to these general-purpose programs, there are literally thousands of other personal computer applications to perform specialized tasks: to help you do your taxes, get in shape, lose weight, get medical advice, write wills and other legal documents, make repairs to your computer, fix your car, write music, and edit your pictures and videos (see Figures 2.13 and 2.14). This type of software, often called *user software* or *personal productivity software*, includes the general-purpose tools and programs that support individual needs.

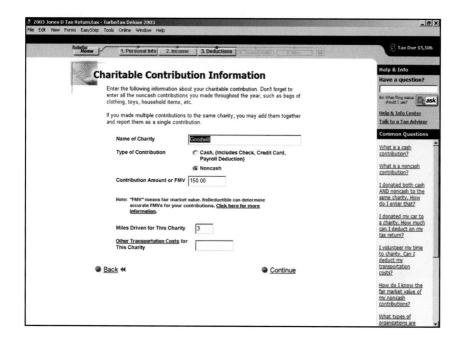

TurboTax

Tax-preparation programs can save hours of work and are typically more accurate than doing a tax return by hand. Programs can check for potential problems and give you help and advice about what you may have forgotten to deduct.

(Source: Turbo Tax Deluxe 2003 screenshot courtesy of Intuit.)

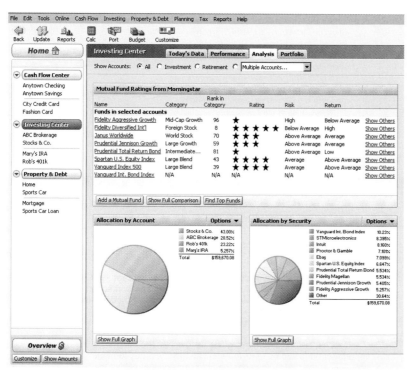

Figure 2.14

Quicken

Off-the-shelf financial-management programs are useful for paying bills and tracking expenses.

(Source: Courtesy of Intuit.)

Word Processing

If you write reports, letters, or term papers, word processing applications can be indispensable. The majority of personal computers in use today have word processing applications installed. Such applications can be used to create, edit, and print documents. Most come with a vast array of features, including those for checking spelling, creating tables, inserting formulas, creating graphics, and much more (see Figure 2.15). This book (and most like it) was entered into a word processing application using a personal computer.

**Examples of Personal
Productivity Software**

Type of Software	Explanation	Example	Vendor
Word processing	Create, edit, and print text documents	Word WordPerfect	Microsoft Corel
Spreadsheet	Provide a wide range of built-in functions for statistical, financial, logical, database, graphics, and date and time calculations	Excel Lotus 1-2-3	Microsoft Lotus/IBM
Database	Store, manipulate, and retrieve data	Access Approach dBASE	Microsoft Lotus/IBM Borland
Online information services	Obtain a broad range of information from commercial services	America Online CompuServe MSN	America Online CompuServe Microsoft
Graphics	Develop graphs, illustrations, and drawings	Illustrator FreeHand	Adobe Macromedia
Project management	Plan, schedule, allocate, and control people and resources (money, time, and technology) needed to complete a project according to schedule	Project for Windows On Target Project Schedule Time Line	Microsoft Symantec Scitor Symantec
Financial management	Provide income and expense tracking and reporting to monitor and plan budgets (some programs have investment portfolio management features)	Managing Your Money Quicken	Meca Software Intuit
Desktop publishing (DTP)	Work with personal computers and high-resolution printers to create high-quality printed output, including text and graphics; various styles of pages can be laid out; art and text files from other programs can also be integrated into "published" pages	QuarkXPress Publisher PageMaker Ventura Publisher	Quark Microsoft Adobe Corel
Creativity	Help generate innovative and creative ideas and problem solutions. The software does not propose solutions, but provides a framework conducive to creative thought. The software takes users through a routine, first naming a problem, then organizing ideas and "wishes," and offering new information to suggest different ideas or solutions	Organizer Notes	Macromedia Lotus

Word Processing Program

Word processing applications can be used to write letters, holiday greeting cards, work reports, and term papers.

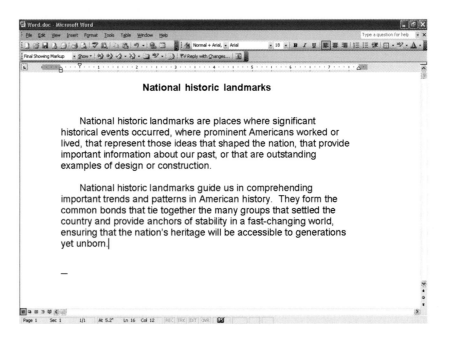

Word processing programs can be used with a team or group of people collaborating on a project. The authors and editors who developed this book, for example, used the "track changes" and "reviewing" features of Microsoft Word to track and make changes to chapter files. You can insert comments in or make revisions to a document that a coworker can review and either accept or reject.

Spreadsheet Analysis

People use spreadsheets to prepare budgets, forecast profits, analyze insurance programs, summarize income tax data, and analyze investments. Whenever numbers and calculations are involved, spreadsheets should be considered. Woodward Aircraft Engine Systems, for example, uses spreadsheets to compute the inventory levels it needs to manufacture engine parts.[27] The calculations made in the spreadsheet have helped the company reduce inventory levels to one-third of their original values for some parts and components, reducing inventory costs. Features of spreadsheets include graphics, limited database capabilities, statistical analysis, built-in business functions, and much more (see Figure 2.16). The business functions include calculation of depreciation, present value, internal rate of return, and the monthly payment on a loan, to name a few. Optimization is another powerful feature of many spreadsheet programs. *Optimization* allows the spreadsheet to maximize or minimize a quantity subject to certain constraints.

	Catalog No.	Description	Price	Inventory	Value	Reorder
2	0-59088-473-X	A Guide to National Parks	$29.95	4320	$129,384.00	4000
3	1-55858-024-7	Tracing Your Genealogy	$99.95	184	$18,390.80	600
4	1-55858-092-1	History of Our Flag	$17.95	996	$17,878.20	200
5	1-55858-265-3	Women in History	$59.95	213	$12,769.35	100
6	0-61767-628-2	Computer Security	$27.95	427	$11,934.65	300
7	1-87263-823-3	Modern Décor	$22.95	543	$12,461.85	400
8	0-93747-288-1	Nutrition	$43.95	1497	$65,705.25	900
9	1-33778-945-7	Engineering Marvels	$62.95	344	$21,654.80	300
10	1-33778-653-8	Managing Your Finances	$21.95	567	$12,445.65	200
11	0-59088-274-3	Music of the 21st Century	$21.95	960	$21,072.00	800
12	0-14183-903-2	Creating a Winning Webpage	$19.95	745	$14,862.75	300
13	0-61767-274-9	Time Management Skills	$28.95	367	$10,624.65	600
14	1-87263-782-2	Networking Know-How	$32.95	211	$6,952.45	1000
15	0-61767-743-7	Natural Fractals	$39.95	976	$38,991.20	400
16	1-55858-229-0	El Ninol	$29.95	321	$946.95	500
17	0-31984-794-X	The History of Dance	$17.95	402	$7,215.90	700
18	1-93372-256-1	Gothic Architecture	$65.95	566	$37,327.70	500
19	0-59088-833-0	Touring Finland	$17.95	560	$10,052.00	300

Figure 2.16

Spreadsheet Program

Spreadsheet programs should be considered when calculations are required.

Database Applications

Database applications are ideal for storing, manipulating, and retrieving data. These applications are particularly useful when you need to manipulate a large amount of data and produce reports and documents. Database manipulations include merging, editing, and sorting data. The uses of a database application are varied. You can keep track of a CD collection, items in your apartment, tax records, and expenses. A student club can use a database to store names, addresses, phone numbers, and dues paid. In business, a database application can help process sales orders, control inventory, order new supplies, send letters to customers, and pay employees. A database can also be a front end to another application. For example, a database application can be used to enter and store income tax information; the stored results can then be exported to other applications, such as a spreadsheet or tax-preparation application (see Figure 2.17).

Figure 2.17

Database Program

After being entered into a database application, information can be manipulated and used to produce reports and documents.

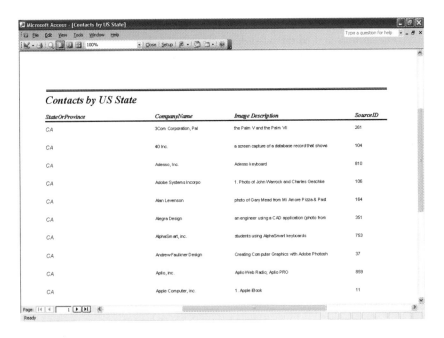

Graphics Programs

With today's graphics programs, it is easy to develop attractive graphs, illustrations, and drawings. Graphics programs can be used to develop advertising brochures, announcements, and full-color presentations. If you are asked to make a presentation at school or work, you can use a graphics program to develop and display slides while you are making your speech. A graphics program can be used to help you make a presentation, a drawing, or an illustration (see Figure 2.18). Most presentation graphics programs come with many pieces of *clip art*, such as drawings and photos of people meeting, medical equipment, telecommunications equipment, entertainment, and much more.

Figure 2.18

Graphics Program

Graphics programs can help you make a presentation at school or work. They can also be used to develop attractive brochures, illustrations, drawings, and maps.

(Source: Courtesy of Adobe Systems Incorporated.)

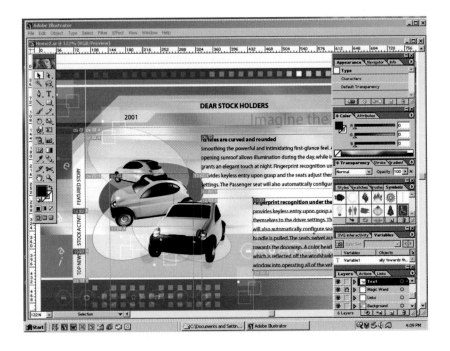

Personal Information Managers

Personal information managers (PIMs) help individuals, groups, and organizations store useful information, such as a list of tasks to complete or a list of names and addresses. They usually provide an appointment calendar and a place to take notes. In addition, information in a PIM can be linked. For example, you can link an appointment with a sales manager that appears in the calendar with information on the sales manager in the address book. When you click the appointment in the calendar, information on the sales manager from the address book is automatically opened and displayed on the computer screen. Microsoft Outlook is an example of a PIM software package.[28]

Software Suites and Integrated Software Packages

A **software suite** is a collection of single application programs packaged in a bundle.[29] Software suites can include word processors, spreadsheets, database management systems, graphics programs, communications tools, organizers, and more. Some suites support the development of Web pages, note taking, and speech recognition, in which applications in the suite can accept voice commands and record dictation.[30] There are a number of advantages to using a software suite. The software programs have been designed to work similarly, so after you learn the basics of one application, the other applications are easy to learn and use. Buying software in a bundled suite is cost-effective; the programs usually sell for a fraction of what they would cost individually.

Microsoft Office, Corel's WordPerfect Office, Lotus SmartSuite, and Sun Microsystems's StarOffice are examples of popular general-purpose software suites for personal computer users (see Figure 2.19).[31] The Free Software Foundation offers software similar to Sun Microsystems's StarOffice that includes word processing, spreadsheet, database, presentation graphics, and e-mail applications for the Linux OS.[32] OpenOffice is another Office suite for Linux.[33] Each of these software suites includes a spreadsheet program, word processor, database program, and graphics package with the ability to move documents, data, and diagrams among them (see Table 2.9). Thus, a user can create a spreadsheet and then cut and paste that spreadsheet into a document created using the word processing application.

software suite
A collection of single application programs packaged in a bundle.

Table 2.9

Major Components of Leading Software Suites

Personal Productivity Function	Microsoft Office	Lotus SmartSuite Millennium Edition	Corel WordPerfect Office	Sun Microsystems
Word Processing	Word	WordPro	WordPerfect	Writer
Spreadsheet	Excel	Lotus 1-2-3	Quattro Pro	Calc
Presentation Graphics	PowerPoint	Freelance Graphics	Presentations	Impress
Database	Access	Lotus Approach	Paradox	

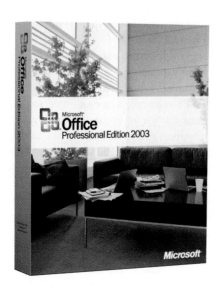

Figure 2.19

Software Suite

A software suite, such as Microsoft Office 2003, offers a collection of powerful programs, including word processing, spreadsheet, database, graphics, and other programs. The programs in a software suite are designed to be used together. In addition, the commands, icons, and procedures are the same for all programs in the suite.

(Source: Courtesy of Microsoft Corporation.)

Workgroup Application Software

Workgroup application software is designed to support teamwork, whether people are in the same location or dispersed around the world. This support can be accomplished with software known as *groupware* that helps groups of people work together more efficiently and effectively. Microsoft Exchange Server 2003, for example, has groupware and e-mail features.[34] Monsanto, a chemical manufacturing company, used groupware to let members of its agricultural chemical group work together to develop new products.[35] According to one company official, "We determined that this application has the potential to significantly increase the productivity of our engineering group by making programs more accessible to our young engineers." Also called *collaborative software*, the approach allows a team of managers to work on the same production problem, letting them share their ideas and work via connected computer systems. The "Three Cs" rule for successful implementation of groupware is summarized in Table 2.10.

Table 2.10

Ernst & Young's "Three Cs" Rule for Groupware

Convenient	If it's too hard to use, it doesn't get used; it should be as easy to use as the telephone.
Content	It must provide a constant stream of rich, relevant, and personalized content.
Coverage	If it isn't close to everything you need, it may never get used.

Workgroup application software is found in most industries. IBM, for example, is developing software to help Sesame Street Productions convert thousands of episodes of the popular *Sesame Street* TV show from the old analog film format to a digital format.[36] The project "will save both time and money while it opens up new avenues for improving our abilities to create and generate revenues," says Sherra Pierre of Sesame Workshop.

Enterprise Application Software

Software that benefits an entire organization can also be developed or purchased. A fast-food chain, for example, might develop a materials ordering and distribution program to ensure that each fast-food franchise gets the necessary raw materials and supplies during the week. This materials ordering and distribution program can be developed internally using staff and resources in the IS department or purchased from an external software company. Table 2.11 lists a number of applications that can be addressed with enterprise software. Many organizations are moving to integrated enterprise software that supports supply chain management (movement of raw materials from suppliers through shipment of finished goods to customers), as shown in Figure 2.20.

Organizations can no longer respond to market changes using nonintegrated information systems based on overnight processing of yesterday's business transactions, conflicting data models, and obsolete technology. As a result, many corporations are turning to **enterprise resource planning (ERP)** software, a set of integrated programs that manage a company's vital business operations for an entire multisite, global organization. Thus, an ERP system must be able to support multiple legal entities, multiple languages, and multiple currencies. Although the scope of an ERP system might vary from vendor to vendor, most ERP systems provide integrated software to support manufacturing and finance. The primary benefits of implementing ERP include eliminating inefficient systems, easing adoption of improved work processes, improving access to data for operational decision making, standardizing technology vendors and equipment, and enabling the implementation of supply chain management.

Table 2.11

Examples of Enterprise Application Software

Accounts receivable	Sales ordering
Accounts payable	Order entry
Airline industry operations	Payroll
Automatic teller systems	Human resource management
Cash-flow analysis	Check processing
Credit and charge card administration	Tax planning and preparation
Manufacturing control	Receiving
Distribution control	Restaurant management
General ledger	Retail operations
Stock and bond management	Invoicing
Savings and time deposits	Shipping
Inventory control	Fixed asset accounting

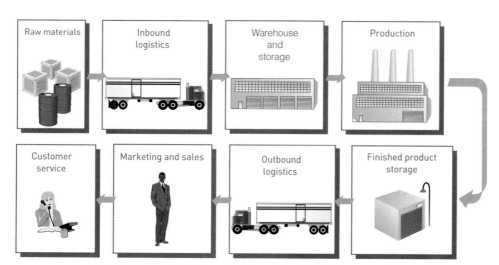

Integrated Enterprise Software to Support Supply Chain Management

Figure 2.20

Use of Integrated Supply Chain Management Software

Application Software for Information, Decision Support, and Specialized Purposes

Specialized application software for information, decision support, and other purposes is available in every industry. Sophisticated decision support software is now being used to increase the cure rate for cancer by analyzing about 100 different scans of a cancer tumor to create a 3-D view. Software can then consider thousands of angles and doses of radiation to determine the best radiation program. The software analysis takes only minutes, but the results can save years or decades of life for the patient. As you will see in future chapters, information, decision support, and specialized systems are used in businesses of all sizes and types to increase profits or reduce costs. But how are all these systems actually developed or built? The answer is through the use of programming languages, discussed next.

PROGRAMMING LANGUAGES

programming languages
The sets of keywords, symbols, and a system of rules for constructing statements by which humans can communicate instructions to be executed by a computer.

syntax
A set of rules associated with a programming language.

Table 2.12

Summary of Programming Languages

Both systems and application software are written in coding schemes called **programming languages**. The primary function of a programming language is to provide instructions to the computer system so that it can perform a processing activity. IS professionals work with programming languages, which are sets of symbols and rules used to write program code. Programming involves translating what a user wants to accomplish into instructions that the computer can understand and execute. Like writing a report or a paper in English, writing a computer program in a programming language requires that the programmer follow a set of rules. Each programming language uses a set of symbols that have special meaning. Each language also has its own set of rules, called the **syntax** of the language. The language syntax dictates how the symbols should be combined into statements capable of conveying meaningful instructions to the CPU. The desire to use the power of information processing efficiently in problem solving has pushed the development of literally thousands of programming languages, but only a few dozen are commonly used today. A brief summary of the various programming language generations is provided in Table 2.12.

Generation	Key Characteristics	Examples
First	Computer instructions written in machine code (0s and 1s) directly executable by the computer. Very tedious and time-consuming to create.	Machine language
Second	Abbreviations used for common operations (such as ADD, MOV, JMP). Each operation was followed by the address of one or more parameters. Recognizable names replaced numerical addresses. Assemblers were required to convert code into the machine language of a specific manufacturer's computer.	Assembly language
Third	Computer instructions became even more English-like and easier to read. Software programs called compilers or interpreters were required to convert instructions into machine language.	FORTRAN COBOL
Fourth	Language was easy enough for nonprogrammers to use. Computer instructions became very powerful, telling the computer what was to be done, without requiring step-by-step instructions on how to do it.	Focus Powerhouse SAS Essbase
Visual programming languages	Enabled developers to create applications using point-and-click and drag-and-drop techniques instead of text-oriented methods.	Visual Basic Visual C++
Object-oriented languages	Objects contain the data, instructions, and procedures. The same objects can be easily reused in different programs for a variety of applications, greatly increasing program quality and programmer productivity.	Eiffel C++ C# Java SmallTalk

The various languages have characteristics that make them appropriate for particular types of problems or applications. Among the third-generation languages, COBOL has excellent file-handling and database-handling capabilities for manipulating large volumes of business data, whereas FORTRAN is better suited for scientific applications. Java is an obvious choice for Web developers. End users will chooses one of the fourth- or fifth-generation languages to develop programs. Although many programming languages are used to write new business applications, more lines of code are written in COBOL in existing business applications than any other programming language.

SOFTWARE ISSUES AND TRENDS

Because software is such an important part of today's computer systems, issues such as software bugs, licensing, and global software support have received increased attention.

Software Bugs

A **software bug** is a defect in a computer program that keeps it from performing in the manner intended. Some software bugs are obvious and cause the program to terminate unexpectedly. Other bugs are subtler and allow errors to creep into your work. Computer and software vendors tell us that because humans design and program hardware and software, bugs are inevitable. In fact, according to the Pentagon and the Software Engineering Institute at Carnegie Mellon University, there are typically 5 to 15 bugs in every 1,000 lines of code—the software instructions that make sense only to computers and programmers.

Most software bugs arise because manufacturers release new software as early as possible instead of waiting until all bugs are identified and removed. They are under intense pressure from customers to deliver the software they have announced and from shareholders to begin selling the new product to increase sales. Meanwhile, the software manufacturer's quality-assurance people fight a losing battle for more testing time to identify and remove bugs. Although the decision of when to release new software is based on a fine line, the industry clearly favors releasing software early and with defects. After all, software companies make money on upgrades, so there is little incentive to achieve a perfect first release.

software bug
A defect in a computer program that keeps it from performing in the manner intended.

Copyrights and Licenses

Most software products are protected by law using copyright or licensing provisions. Those provisions can vary, however. In some cases, you are given unlimited use of software on one or two computers. This is typical with many applications developed for personal computers. In other cases, you pay for your usage—if you use the software more, you pay more. This approach is becoming popular with software placed on networks or larger computers. Most of these protections prevent you from copying software and giving it to others without restrictions. Some software now requires that you *register* or *activate* it before it can be fully used. Registration and activation sometimes put software on your hard disk that monitors activities and changes to your computer system.

Global Software Support

Large, global companies have little trouble persuading vendors to sell them software licenses for even the most far-flung outposts of their company. But can those same vendors provide adequate support for their software customers in all locations? Supporting local operations is one of the biggest challenges IS teams face when putting together standardized, company-wide systems. In slower technology growth markets, such as Eastern Europe and Latin America, there may be no official vendor presence at all. Instead, large vendors such as Sybase, IBM, and Hewlett-Packard typically contract out support for their software to local providers.

One approach that has been gaining acceptance in North America is to outsource global support to one or more third-party distributors. The software-user company might still negotiate its license with the software vendor directly, but it then hands over the global support contract to a third-party supplier. The supplier acts as a middleman between software vendor and user, often providing distribution, support, and invoicing. American Home Products Corporation handles global support for both Novell NetWare and Microsoft Office applications this way—throughout the 145 countries in which it operates. American Home Products negotiated the agreements directly with the vendors for both purchasing and maintenance, but fulfillment of the agreement is handled exclusively by Philadelphia-based Softsmart, an international supplier of software and services.

In today's computer systems, software is an increasingly critical component. Whatever approach individuals and organizations take to acquire software, it is important for everyone to be aware of the current trends in the industry. Informed users are wiser consumers, and they can make better decisions.

SUMMARY

Principle

Users must work closely with IS professionals to define business needs, evaluate options, and select the hardware and software that provide a cost-effective solution to those needs.

Hardware devices work together to perform input, processing, data storage, and output. Processing is performed by an interplay between the central processing unit (CPU) and memory. Primary storage, or memory, provides working storage for program instructions and data to be processed and provides them to the CPU.

Processing that uses several processing units is called *multiprocessing*. One form of multiprocessing uses coprocessors; coprocessors execute one type of instruction while the CPU works on others. Parallel processing involves linking several processors to work together to solve complex problems. Grid computing is the use of a collection of computers, often owned by multiple individuals or organizations, to work in a coordinated manner to solve a common problem.

Computer systems can store large amounts of data and instructions in secondary storage, which is less volatile and has greater capacity than memory. Storage media can be either sequential access or direct access. Common forms of secondary storage include magnetic tape, magnetic disk, optical disc storage, and PC memory cards. Redundant array of independent/inexpensive disks (RAID) is a method of storing data that allows the system to more easily recover data in the event of a hardware failure. A storage area network (SAN) uses computer servers, distributed storage devices, and networks to provide fast and efficient storage.

Input and output devices allow users to provide data and instructions to the computer for processing and allow subsequent storage and output. These devices are part of a user interface through which humans interact with computer systems.

Point-of-sale (POS) devices are terminals with scanners that read and enter codes into computer systems. Automatic teller machines (ATMs) are terminals with keyboards used for transactions. Radio-frequency identification (RFID) technology employs a microchip, called a *smart tag*, with an antenna that broadcasts its unique 96-bit identifier and location to corresponding receivers.

Output devices provide information in different forms, from hard copy to sound to digital format. Display monitors are standard output devices; monitor quality is determined by size, number of colors that can be displayed, and resolution. Other output devices include printers and plotters.

The main computer system types are handheld computers, portable computers, thin clients, desktop computers, workstations, servers, mainframe computers, and supercomputers. Personal computers (PCs) are small, inexpensive computer systems.

Principle

When selecting an operating system, you must consider the current and future requirements for application software to meet the needs of the organization. In addition, your choice of a particular operating system must be consistent with your choice of hardware.

Software consists of programs that control the workings of the computer hardware. Two main categories of software are available: systems software and application software. Systems software is a collection of programs that interacts between hardware and application software. Application software enables people to solve problems and perform specific tasks. Application software can be proprietary or off the shelf.

An operating system (OS) is a set of computer programs that controls the computer hardware to support users' computing needs. OSs function by converting an instruction from an application into a set of instructions needed by the hardware. An OS also provides a user interface, which allows users to access and command the computer. The OS also serves as an intermediary between application programs and hardware, allowing hardware independence. Memory management involves controlling storage access and use by converting logical requests into physical locations and by placing data in the best storage space. Task management allocates computer resources through multitasking and time-sharing.

The ability of a computer to handle an increasing number of concurrent users smoothly is called *scalability*. Other functions of the OS are to provide networking capabilities, control access to system resources, and file management.

Over the years, several popular OSs have been developed. These include several proprietary OSs used primarily on mainframes. MS-DOS is an early OS for IBM-compatibles. Windows XP is the current Windows OS. Apple computers use proprietary OSs like the Mac OS and Mac OS X. UNIX is a powerful OS that can be used on many computer system types and platforms, from personal computers to mainframe systems. Linux is the kernel of an OS whose source code is freely available to everyone. Several variations of Linux are available, with sets of capabilities and applications to form a complete OS. A number of OSs have been developed to support consumer appliances such as Palm OS, Windows XP Embedded, Pocket PC, and variations of Linux.

Principle

Do not develop proprietary application software unless doing so will meet a compelling business need that can provide a competitive advantage.

Application software applies the power of the computer to solve problems and perform specific tasks. One useful way of classifying the many potential uses of information systems is to identify the scope of problems and opportunities addressed by a particular organization or its sphere of influence. For most companies, the spheres of influence are personal, workgroup, and enterprise.

User software, or personal productivity software, includes general-purpose programs that enable users to improve their personal effectiveness. Software that helps groups work together is often referred to as workgroup application software. Enterprise software that benefits the entire organization can also be developed or purchased. Many organizations are turning to enterprise resource planning software, a set of integrated programs that manage a company's vital business operations for an entire multisite, global organization.

Three approaches to developing application software are as follows: Build proprietary application software, buy existing programs off the shelf, or use a combination of customized and off-the-shelf application software.

An application service provider (ASP) is a company that can provide software, end-user support, and the computer hardware on which to run the software from the user's facilities.

Although there are literally hundreds of computer applications that can help individuals at school, home, and work, the primary applications are word processing, spread sheet analysis, database, graphics, and personal information managers. A software suite, such as SmartSuite, WordPerfect, StarOffice, or Office, offers a collection of powerful programs.

Principle

Choose a programming language whose functional characteristics are appropriate for the task at hand, taking into consideration the skills and experience of the programming staff.

All software programs are written in coding schemes called *programming languages*, which provide instructions to a computer to perform some processing activity. Several classes of programming languages are available, including machine, assembly, high-level, query and database, object-oriented, and visual programming languages.

Programming languages have gone through changes since their initial development in the early 1950s. In the first generation, computers were programmed in machine language, or binary code, a series of statements written in 0s and 1s. The second generation of languages was termed *assembly languages*; these languages support the use of symbols and words rather than 0s and 1s. The third generation consists of many high-level programming languages that use English-like statements and commands. These languages include BASIC, COBOL, FORTRAN, and others. A fourth-generation language is less procedural and more English-like than third-generation languages. The fourth-generation languages include database and query languages such as SQL.

Visual and object-oriented programming languages—such as Smalltalk, C++, and Java—use groups of related data, instructions, and procedures called *objects*, which serve as reusable modules in various programs. These languages can reduce program development and testing time. Java can be used to develop applications on the Internet.

Principle

The software industry continues to undergo constant change; users need to be aware of recent trends and issues to be effective in their business and personal life.

Software bugs, software licensing and copyrighting, and global software support are all important software issues and trends.

A software bug is a defect in a computer program that keeps it from performing in the manner intended. Software bugs are common, even in key pieces of business software.

Software manufacturers are developing new approaches to licensing their software to lock in a steady, predictable stream of revenue from their customers.

Global software support is an important consideration for large, global companies putting together standardized, company-wide systems. A common solution is outsourcing global support to one or more third-party software distributors.

CHAPTER 2: SELF-ASSESSMENT TEST

Users must work closely with IS professionals to define business needs, evaluate options, and select the hardware and software that provide a cost-effective solution to those needs.

1. Non-IS managers have little need to understand computer hardware. True or False?
2. _____ is any machinery (most of which uses digital circuits) that assists in the input, processing, storage, and output activities of an information system.
3. Which of the following performs mathematical calculations and makes logical comparisons?

    a. control unit
    b. register
    c. ALU
    d. main memory

4. A form of main memory that does not lose its contents if power is lost is called _____.

    a. ROM
    b. RAM
    c. CD-ROM
    d. DVD

5. The relative clock speed of two CPUs from different manufacturers is a good indicator of their relative processing speed. True or False?

When selecting an operating system, you must consider the current and future requirements for application software to meet the needs of the organization. In addition, your choice of a particular operating system must be consistent with your choice of hardware.

6. Name an operating system whose source code is freely available to everyone.

    a. Windows XP
    b. UNIX
    c. Linux
    d. Mac OS X

7. A command-based user interface requires that text commands be given to the computer to perform basic activities. True or False?
8. The file manager component of the operating system controls how memory is accessed and maximizes available memory and storage. True or False?

Do not develop proprietary application software unless doing so will meet a compelling business need that can provide a competitive advantage.

9. The primary function of application software is to apply the power of the computer to give individuals, workgroups, and the entire enterprise the ability to solve problems and perform specific tasks. True or False?
10. Software that enables users to improve their personal effectiveness, increasing the amount of work they can do and its quality is called _____.

    a. personal productivity software
    b. operating system software
    c. utility software
    d. graphics software

11. Software used to solve a unique or specific problem that is usually built in-house, but can also be purchased from an outside company is called _____.
12. What type of software has the greatest potential to affect the processes that add value to a business because it is designed for specific organizational activities and functions?

    a. personal productivity software
    b. operating system software
    c. utility software
    d. application software

13. Off-the-shelf software is never customized. True or False?
14. A class of applications software that helps groups work together and collaborate is called _____.

Choose a programming language whose functional characteristics are appropriate to the task at hand, taking into consideration the skills and experience of the programming staff.

15. Each programming language has its own set of rules, called the _____ of the language.

The software industry continues to undergo constant change; users need to be aware of recent trends and issues to be effective in their business and personal life.

16. A defect in a computer program that keeps it from performing as intended is called a _____.

CHAPTER 2: SELF-ASSESSMENT TEST ANSWERS

(1) False (2) Hardware (3) c (4) a (5) False (6) c (7) True (8) False (9) True (10) a (11) proprietary software (12) d (13) False (14) workgroup application software (15) syntax (16) software bug

REVIEW QUESTIONS

1. What role does the mainframe computer play in today's large organization?
2. How would you distinguish between a desktop computer and a workstation?
3. What is RFID technology? Identify three practical uses for this technology.
4. Identify the three components of the CPU and explain the role of each.
5. What is the difference between secondary storage and main memory?
6. What is the overall trend in secondary storage devices?
7. What is the difference between systems and application software? Give four examples of personal productivity software.
8. What are the two basic types of software? Briefly describe the role of each.
9. What is multitasking?
10. Identify the two primary sources for acquiring application software.
11. What is an application service provider? What issues arise in considering the use of one?
12. What is open-source software? What is the biggest stumbling block with the use of open-source software?
13. What does the acronym API stand for? What is the role of an API?
14. Describe the term *enterprise resource planning* (*ERP*) system. What functions does such a system perform?

DISCUSSION QUESTIONS

1. Imagine that you are the business manager for your university. What type of computer would you recommend for broad deployment in the university's computer labs—a standard desktop personal computer or a thin client? Why?
2. Which would you rather have—a PDA or smart phone? Why?
3. If cost were not an issue, describe the characteristics of your ideal computer. What would you use it for? Would you choose a handheld, portable, desktop, or workstation computer? Why?
4. Identify the three spheres of influence and briefly discuss the software needs of each.
5. Identify the three fundamental types of application software. Discuss the advantages and disadvantages of each type.
6. Describe how the operating system can manage the computer's memory.
7. Contrast and compare two popular operating systems for personal computers.
8. If you were the IS manager for a large manufacturing company, what issues might you have with the use of open-source software? What advantages might there be for use of such software?

PROBLEM-SOLVING EXERCISES

1. Some believe that the information technology industry has driven the economy and in large measure determines stock market prices—not just technology stocks but other stocks as well. Do some research to find an index that measures the stock performance of the largest technology companies. Plot that index versus the index for the S&P 500 for the past five years. What is your conclusion?

2. Use a spreadsheet package to prepare a simple monthly budget and forecast your cash flow—both income and expenses for the next six months (make up numbers rather than use actual). Now use a graphics package to plot the total monthly income and monthly expenses for six months. Cut and paste both the spreadsheet and the graph into a word processing document that summarizes your financial condition.

TEAM ACTIVITIES

1. Form a group of three or four classmates. Identify and contact an individual with a local firm. Interview the individual and describe the application software the company uses and the importance of the software to the organization. Write a brief report summarizing your findings.

2. With one or two of your classmates, visit several local car dealers to identify what makes and models of automobiles come equipped with on-board computer-based navigation systems. Ask to see a demo of the capabilities of these systems. Write a brief report summarizing your findings, including costs and features.

WEB EXERCISES

1. Do research on the Web to identify the current status of the use of radio-frequency ID chips in the consumer goods industry. Write a brief report summarizing your findings.

2. Do research on the Web to find the current status of implementing the new 13-digit bar code. Are companies struggling to do this or has it already been successfully accomplished? Write a brief report summarizing your findings.

3. Microsoft, IBM/Lotus, and Corel are the important providers of personal productivity software suites. Do research to assess the relative success of these three products in terms of sales of their software suites. Do you think it is possible that Microsoft will become the only provider of such software? Would this be good or bad? Why? Write a brief report summarizing your findings and conclusions.

CAREER EXERCISES

1. Think of your ideal job. Imagine that you are going to buy some sort of handheld computer device to support you in this job. What tasks could it help you perform? What sort of features would you look for in this device? Visit a computer store or a consumer electronics store and see whether you can purchase such a device for under $400.

2. Think of your ideal job. Describe five software packages that could help you advance in your career. If the software package doesn't exist, describe the kinds of software packages that could help you in your career.

VIDEO QUESTIONS

Watch the video clip **Getting Started** and answer these questions:

1. What reason is given for buying a good high-quality keyboard and monitor?

2. After seeing what it takes to build your own computer, do you think that it is more economical to build, or buy a pre-built computer? What do you think are the advantages and disadvantages of both?

CASE STUDIES

Case One

PACS at North Bronx Healthcare Network

"I have never done an implementation of any system that has so dramatically impacted the way we do business," says Dan Morreale, CIO at North Bronx Healthcare Network, one of six regional networks established by New York City Health and Hospitals Corp. Morreale is referring to the North Bronx's new $6 million picture archiving and communications software (PACS). The software captures, stores, and displays patient x-rays and other images in digital form.

You might wonder how a medical facility justifies a $6 million technology investment. Morreale could clearly show how the system would pay for itself in the matter of a few years. In his calculations, he showed that the new software would save North Bronx Healthcare Network almost $2 million per year by eliminating costs associated with film- and paper-based reports, such as:

- $1 million from reduced film-processing costs
- $400,000 from eliminating labor costs for film processing and storage
- $130,000 from lower real estate costs, because the organization no longer needs 5,000 square feet of floor space to store film
- $400,000 from eliminating manually produced reports

Not only could the software pay for itself in three years, but North Bronx would enjoy additional rewards that are more difficult to calculate.

The PACS implementation was the final step in digitizing all of North Bronx Healthcare Network's record keeping. Being able to advertise itself as fully digital helps a hospital to attract the best medical personnel. This advantage is particularly important in tight labor markets experiencing a shortage of good radiologists. The best professionals want to work with PACS because it allows them to be more effective in their work.

Digitizing records and images has many other advantages. Digital images can be sent to multiple physicians simultaneously. Traditional x-rays travel by courier from doctor to doctor, often keeping doctors and patients waiting for days or weeks for results. Physicians can manipulate and zoom in on the images to focus on medical problems. Digital x-rays are easy to copy and less likely to be lost. Also, through computer networks, doctors can share x-rays with experts around the world and benefit from consultations.

Although PACS have been used since the mid-1990s, about two-thirds of U.S. hospitals haven't purchased the system yet, estimates Jocelyn Young, a healthcare industry analyst at market research firm IDC. But the advantages of such software for cost savings, improved workflow, better patient diagnoses, and a competitive advantage over other hospitals are sure to make these high priced, advanced software systems increasingly popular in years to come.

Discussion Questions

1. How can medical facilities save money by moving to digital x-rays?
2. How does switching from a traditional x-ray system to PACS improve the effectiveness of a medical facility?

Critical Thinking Questions

3. What other photography and imaging industries could benefit from the digitization of traditional paper-based processing?
4. How can PACS affect the medical services patients receive from multiple hospitals, doctors, and specialists?

SOURCES: John Webster, "Hospital Imaging Systems: A Tough Sell." *Computerworld*, February 2, 2004, *www.computerworld.com*; "North Bronx Healthcare Network to Reduce Digital Medical Image Management Costs with EMC Centera," techinfocenter.com, June 9, 2003, *www.techinfocenter.com*; AGFA Radiology Solutions Web site, *www.afga.com*, accessed February 8, 2004.

Case Two

UPS Provides Shipping Management Software to Customers

Because of the popularity of the Web, and thanks to recent innovations in software development tools, many of today's software programs are written to run on Web servers using Web browsers as the user interface. Businesses are particularly fond of Web-delivered services because they provide universal accessibility and are easy to deploy and maintain. UPS provides many Web-based software applications for its customers, the most recent of which is called Quantum View Manage.

Quantum View Manage is a component of the larger Quantum View system, which includes three Web-based applications that help businesses control their supply chains. Quantum View Manage is an application that allows UPS small package shippers to track package movements within their own supply chains. It is designed for customers who ship, on average, 30 or more packages per week and want to track packages coming in and going out.

Quantum View Manage offers more convenience than traditional tracking software by putting shipping status information at users' fingertips. UPS spokeswoman Laurie Mallis says that one of the primary benefits of the software is that customers don't have to dig to find or enter tracking information. The software maintains and displays all shipping information for an organization in a number of useful report formats. Customers can configure the software to view shipment information for multiple accounts, so they can see when a package is processed, where it is in transit, when it arrives, or whether and why it is delayed, according to Mallis.

The software provides many features for ultimate customer flexibility. It shows both an inbound view of packages being shipped by your vendors and suppliers to designated locations and an outbound view about packages billed to your UPS account. "Customers can customize the information that they're receiving, because different departments within different companies have different needs," Mallis explained. For example, a customer service department might want to know whether a package was shipped or, if there's a delay, what the cause is. Shippers can be notified via e-mail, so they can see problems before their customers do, enabling better service. In addition, a shipper's finance department could use the confirmed delivery notice to automatically trigger an invoice and then view all outstanding C.O.D. orders. Some customers use Quantum View Manage to help manage inventory.

UPS customers seem to love this new free service. All 622 customers who tested the product have adopted it. In addition, another 300 customers are asking for it.

Quantum View Manage is an excellent example of how a company can leverage software to win a competitive advantage. UPS has invested heavily in software over the years and maintains its own international computer network. UPS employs more than 4,450 technology employees and was voted as one of the best places to work in IT by *Computerworld* magazine.

Using automated and handheld bar-code scanners, UPS is able to track the 13 million-plus packages it handles each day. With this sophisticated tracking system and network in place, it was a simple matter for UPS to organize and deliver pertinent information to customers and assist them in managing the flow of goods through their supply chains.

Discussion Questions

1. What advantage does Web-based software provide to businesses such as UPS?
2. What risks are involved for companies that rely on Web-based software such as Quantum View Manage in managing their own businesses?

Critical Thinking Questions

3. UPS advertises its overall goal as "enabling commerce around the globe." How does Quantum View Manage assist UPS in achieving that goal?
4. Consider the components of a supply chain for a typical manufactured product. How could Quantum View Manage improve the management of shipping between supply chain units?

SOURCES: Linda Rosencrance, "UPS Launches Quantum View Manage," *Computerworld*, February 4, 2004, *www.computerworld.com*; "UPS Opens Supply Chain 'Window' with Quantum View Manage." *Business Wire*, February 4, 2004, www.lexis-nexis.com.

NOTES

Sources for the opening vignette: "Porsche's CEO Talks Shop, *BusinessWeek Online*, December 22, 2003, accessed at *www.businessweekonline.com*; Daren Fonda, "A New Porsche for Purists," *Time Magazine*, July 28, 2003, accessed at *www.time.com*; "About Porsche," "Finances," "Corporate Divisions," and "Cayenne" accessed at the Porsche Web site at *www2.USporsche*, January 19, 2004; "Porsche AG, SAP Customer Success Story," accessed at *www.sap.com*, January 19, 2004; "Porsche, HP Success Story," March 10, 2003, accessed at *www.hp.com*, January 19, 2004.

1. "ARZ Allgemeines Reschenzentrum Enjoys Reduced Costs through Workload License Charges for IBM eServer," the IBM Web site—Success Stories, *www-306.ibm.com/software/success/cssdb.nsf/cs/DNSD-5UZKY7?OpenDocument&Site=eserverzseries*, accessed January 19, 2004.

2. "Specialized Bicycles Manufacturing," the Sun Web site, *www.sun.com/desktop/success/specialized.html*, accessed January 19, 2004.

3. Levy, David, "A Bakery on the Rise," the Apple Web site, *www.apple.com/business/profiles/sullivanstreet/index2.html,* accessed January 19, 2004.

4. Anthes, Gary H., "U.S. Losing Lead in Supercomputing, User Says," *Computerworld*, September 29, 2003, accessed at *www.computerworld.com*.

5. Dunn, Darrell, "Automating the IT Factory," *InformationWeek*, July 12, 2004, accessed at *www.informationweek.com*.

6. Garvey, Martin J., "Legal Storage Upgrade," *InformationWeek*, June 28, 2004, accessed at *www.informationweek.com*.

7. Brewin, Bob, "A Railroad Finds Its Voice," *Computerworld*, January 26, 2004, accessed at *www.computerworld.com*.

8. Unsys Web site, *www.unisys.com*, accessed January 16, 2002.

9. Robb, Drew, "Displays Go for Sharper Image," *Computerworld*, January 12, 2004, accessed at *www.computerworld.com*.

10. Kessler, Michelle, "Sales of Smart Phones Leave PDAs in the Dust," *USA Today*, January 29, 2004, p. B1.

11. Krazit, Tom, "Motion Computing Adds Celeron to Low-Cost Tablet PC," *Computerworld*, December 4, 2003, accessed at *www.computerworld.com*.

12. Babcock, Charles, "Italian Company Finds Web Service," *InformationWeek,* January 7, 2004, accessed at *www.informationweek.com*.

13. Greenemeier, Larry, "Dell Adds French Accent to Supercomputing Clusters," *InformationWeek*, September 10, 2003, accessed at *www.informationweek.com*.

14. Bacheldor, Beth, "Retail Innovation Starts at Store," *InformationWeek*, January 19, 2004, p. 45.

15. Spanbauer, Scott, "After Antitrust," *PC World*, May 2004, p. 30.

16. Wildstrom, Stephen, "Apple Gets the Little Things Right," *BusinessWeek*, November 17, 2003, p. 30.

17. Mossberg, Walter, "If You're Getting Tired of Fighting Viruses, Consider a New Mac," *The Wall Street Journal*, October 23, 2003, p. B1.

18. Staff, "Travel Web Services Take an Open Source Route," *Computer Weekly*, April 8, 2003, p. 4.

19. Staff, "Panasonic to Supply Linux Development Platform and Operating System for Its Latest Broadband TV Tuner," *Electronic News*, April 21, 2003.

20. Staff, "CIOs Eyeing Open-Source Software," *Information Security*, January 2004, p. 22.

21. Janowske, Davis, et al., "Taking Care of Small Business," *PC Magazine*, February 3, 2004, p. 12.

22. Dragan, Richard, "Windows Server 2003 Delivers Improvements All Around," *PC Magazine*, May 27, 2003, p. 28.

23. Clark, Don, "Microsoft Offers New Lock for Files," *The Wall Street Journal*, February 24, 2003.

24. Grimes, Brad, "Linux Goes to the Movies," *PC Magazine*, May 27, 2003, p. 70.

25. Krishnadas, K. C. "India Pursues Global Role in Embedded Software," *Electronic Engineering Times*, April 14, 2003, p. 23.

26. Brown, Bruce and Magre, "Redone Pocket PC OS Gets a New Name," *PC Magazine*, August 5, 2003, p. 32.

27. Srinivasan, M. et al., "Woodward Aircraft Engine Systems Sets Work-In-Process Levels," *Interfaces*, July–August, 2003, p. 61.

28. Dragan, Richard, "Microsoft Outlook 2003," *PC Magazine*, February 17, 2004, p. 64.

29. Mendelson, Edward, "Microsoft Office 2003," *PC Magazine*, April 22, 2003, p. 34.

30. Sliwa, Carol, "Microsoft Expands Its Office Family by Two," *Computerworld*, March 10, 2003, p. 6.

31. O'Reilly, "The Other Office Suite," *PC World*, July 2003, p. 60.

32. Mendelson, Edward, "StarOffice 7 Makes A Run At Office," *PC Magazine*, February 3, 2004, p. 40.

33. Albro, Edward, "The Linux Experiment," *PC World*, February 2004, p. 105.

34. Lipschutz, Robert, "Exchange 2003: More Approachable, More Affordable," *PC Magazine*, August 19, 2003, p. 37.

35. Staff, "Monsanto Puts a Friendly Face on Hard Software," *KMWorld*, February 2003, p. 28.

36. Crupi, Anthony, "Turning Muppets into Digital Assets," *Cable World*, April 14, 2003, p. 15.

CHAPTER
• 3 •

Organizing Data and Information

PRINCIPLES

- The database approach to data management provides significant advantages over the traditional file-based approach.

- A well-designed and well-managed database is an extremely valuable tool in supporting decision making.

- The number and types of database applications will continue to evolve and yield real business benefits.

LEARNING OBJECTIVES

- Define general data management concepts and terms, highlighting the advantages of the database approach to data management.

- Describe the relational database model and outline its basic features.

- Identify the common functions performed by all database management systems and identify popular end-user database management systems.

- Identify and briefly discuss current database applications.

INFORMATION SYSTEMS IN THE GLOBAL ECONOMY
DDB WORLDWIDE, UNITED STATES

Using Databases to Understand Gut Feelings

DDB is a global advertising agency based in New York, with 206 offices in 96 countries. As an advertising agency, DDB's primary objective is to generate ideas to sell its clients' products. To be successful, an advertiser must thoroughly understand prospective customers—what interests and motivates buyers. One of the founders of DDB, Bill Burnboch (the B in DDB), put it this way: "You can say the right thing about a product and nobody will listen. You've got to say it in such a way that people will feel it in their gut. Because if they don't feel it, nothing will happen."

To understand what customers feel "in their gut," every two years DDB conducts marketing research on 21,000 consumers in 23 countries. The participants respond to 600 questions relating to brand interests and purchases, lifestyle attitudes, interests, and opinions—even seemingly irrelevant information such as how well they think they'd do in a fist fight.

"Traditionally, a lot of advertising strategy was prepared based on things that were easy to manage, such as man-on-the-street interviews and focus groups," explains Janice Riggs, a project director at DDB. "While those are interesting and useful, they're not necessarily statistically valid." Riggs is in charge of DDB's Brand Capital project. Brand Capital is DDB's worldwide brand perception and consumer lifestyle project designed to help clients position and market their brands globally. Brand Capital generates mountains of data, which can only be analyzed and managed with a powerful database management system. DDB uses a database management system from leading database company SAS to "mine" the 12.5 million consumer responses for nuggets of valuable information.

By using sophisticated data-mining tools, DDB is able to turn terabytes of consumer data into useful business intelligence. An example of the types of valuable insights produced by DDB's Brand Capital is the discovery of a correlation between sports and greeting cards. DDB discovered that people who attend at least a dozen professional sporting events a year are also very likely to send greeting cards—useful information for a card company. DDB used this information as a springboard to design different greeting card advertising strategies to build on customers' loyalty with team sporting events, the team's city, and even local historical events to generate excitement about a product—something that's going to be talked about tomorrow.

The advertising industry has historically shied away from such intense and thorough marketing research because agencies don't have the time, talent, or money to turn raw data into intelligence and to present results in an easy-to-read report. Systems such as those provided by SAS are making reliable and thorough research convenient and affordable. The SAS system makes the centralized database available to DDB's 300 strategists, analysts, and client-services directors worldwide through an easy-to-use Web interface that requires no knowledge of code or complicated queries. "We've taken out the pain-in-the-neck part of it, and we free people up to apply creativity and imagination to the information they're given," Riggs says.

Because all DDB employees are working from the same information, they speak to clients with a unified voice. The easily accessible huge data store has significantly altered the way DDB employees work. When they sign a new client, rather than rushing off to create consumer focus groups to collect their opinions, DDB employees sit down at their computers and run queries on the database containing the responses that consumers have already provided in the target market and location.

Before implementing the SAS database system, a single market study request could take two days and cost a few hundred thousand dollars. "But now we turn those requests around instantaneously and at no extra cost," says Denny Merritt, the DDB software developer who helps develop software tools for interacting with brand and lifestyle data.

Being able to collect, store, and evaluate large quantities of consumer opinions has given DDB rare insight into human behavior and an edge over competing agencies, which are simply unable to provide the level of intelligence that DDB offers. "There is never a time when we do not find cool stuff in all of this," Riggs says. "People really are interesting."

As you read this chapter, consider the following:

- How do databases and database management systems allow businesses to do things that they couldn't previously do?
- What considerations and precautions are necessary when developing databases to ensure that they fully support an organization's requirements and lead, rather than mislead, the decision-making process?

Why Learn About Database Systems?

A huge amount of data is entered into computer systems every day. Where does all this data go and how is it used? How can it help you on the job? In this chapter, you learn about database management systems and how they can help you. If you become a marketing manager, you can have access to a vast store of data on existing and potential customers from surveys, their Web habits, and their past purchases from different stores. This information can help you sell products and services. If you become a corporate lawyer, you will have access to past cases and legal opinions from sophisticated legal databases. This information can help you win cases and protect your organization legally. If you become a human resource manager, you will be able to use databases to analyze the impact of raises, employee insurance benefits, and retirement contributions on long-term costs to your company. Regardless of your major in school, you likely will find that using database management systems will be a critical part of your job. You discover in this chapter how you can use data mining to get valuable information to help you succeed. This chapter begins with an introduction to basic concepts of database management systems.

Like other components of a computer-based information system, the overall objective of a database is to help an organization achieve its goals. A database can contribute to organizational success in a number of ways, including the ability to provide managers and decision makers with timely, accurate, and relevant information based on data. Databases also help companies generate information to reduce costs, increase profits, track past business activities, and open new market opportunities. For example, to achieve better customer satisfaction, Hilton Hotels is using its vast database system to customize service.[1] The new database system provides detailed information about Hilton customers. A receptionist at a Hilton Hotel in New York, for example, might apologize to a customer for not having her room cleaned as she wanted during a recent stay at a Hilton in Orlando. The new system, called OnQ, allows Hilton to store and retrieve a tremendous amount of detailed customer satisfaction information. OnQ will also help Hilton make better decisions to meet customer needs. As a result of the data contained in OnQ, the receptionist at the Hilton in New York could decide to offer a customer a special rate or provide additional service based on information from the customer's stay in Orlando. Hilton hopes that this information support will improve customers' stays and increase profits in the long run.

A **database management system (DBMS)** consists of a group of programs that manipulate the database and provide an interface between the database and its users and other application programs. A DBMS is normally purchased from a database company. A DBMS provides a single point of management and control over data resources, which can be critical to maintain the integrity and security of the data. A database, a DBMS, and the application programs that utilize the data in the database make up a database environment. A **database administrator (DBA)** is a skilled and trained IS professional who directs all activities related to an organization's database, including providing security from intruders.[2]

database management system (DBMS)
A group of programs that manipulate the database and provide an interface between the database and its users and other application programs.

database administrator (DBA)
A skilled IS professional who directs all activities related to an organization's database.

DATA MANAGEMENT

Without data and the ability to process it, an organization would not be able to successfully complete most business activities. It would not be able to pay employees, send out bills, order new inventory, or produce information to assist managers in decision making. As you recall, data consists of raw facts, such as employee numbers and sales figures. For data to be transformed into useful information, it must first be organized in a meaningful way.

The Hierarchy of Data

Data is generally organized in a hierarchy that begins with the smallest piece of data used by computers (a bit) and progresses through the hierarchy to a database. As discussed in Chapter 2, a bit (a binary digit) represents a circuit that is either on or off. Bits can be organized into units called *bytes*. A byte is typically eight bits. Each byte represents a **character**, which is the basic building block of information. A character can consist of uppercase letters (A, B, C, . . . , Z), lowercase letters (a, b, c, . . . , z), numeric digits (0, 1, 2, . . . , 9), or special symbols (.![+][-]/ . . .).

Characters are put together to form a field. A **field** is typically a name, number, or combination of characters that describes an aspect of a business object (e.g., an employee, a location, a truck) or activity (e.g., a sale). In addition to being entered into a database, fields can be computed from other fields. *Computed fields* include the total, average, maximum, and minimum values. A collection of related data fields is a **record**. By combining descriptions of various aspects of an object or activity, a more complete description of the object or activity is obtained. For instance, an employee record is a collection of fields about one employee. One field would be the employee's name, another her address, and still others her phone number, pay rate, earnings made to date, and so forth. A collection of related records is a **file**—for example, an employee file is a collection of all company employee records. Likewise, an inventory file is a collection of all inventory records for a particular company or organization. Some database software refers to files as tables.

At the highest level of this hierarchy is a *database*, a collection of integrated and related files. Together, bits, characters, fields, records, files, and databases form the **hierarchy of data** (see Figure 3.1). Characters are combined to make a field, fields are combined to make a record, records are combined to make a file, and files are combined to make a database. A database stores not only all these levels of data, but also the relationships among them.

character
A basic building block of information, consisting of uppercase letters, lowercase letters, numeric digits, or special symbols.

field
Typically a name, number, or combination of characters that describes an aspect of a business object or activity.

record
A collection of related data fields.

file
A collection of related records.

hierarchy of data
Bits, characters, fields, records, files, and databases.

Data Entities, Attributes, and Keys

Entities, attributes, and keys are important database concepts. An **entity** is a generalized class of people, places, or things (objects) for which data is collected, stored, and maintained. Examples of entities include employees, inventory, and customers. Most organizations organize and store data as entities.

entity
A generalized class of people, places, or things for which data is collected, stored, and maintained.

Figure 3.1

The Hierarchy of Data

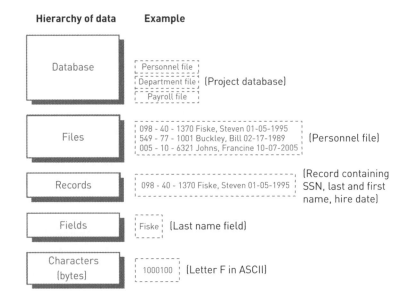

attribute
A characteristic of an entity.

data item
The specific value of an attribute.

An **attribute** is a characteristic of an entity. For example, employee number, last name, first name, hire date, and department number are attributes for an employee (see Figure 3.2). Inventory number, description, number of units on hand, and the location of the inventory item in the warehouse are examples of attributes for items in inventory. Customer number, name, address, phone number, credit rating, and contact person are examples of attributes for customers. Attributes are usually selected to capture the relevant characteristics of entities such as employees or customers. The specific value of an attribute, called a **data item**, can be found in the fields of the record describing an entity.

Figure 3.2

Keys and Attributes

The key field is the employee number. The attributes include last name, first name, hire date, and department number.

Employee #	Last name	First name	Hire date	Dept. number
005-10-6321	Johns	Francine	10-07-2005	257
549-77-1001	Buckley	Bill	02-17-1989	632
098-40-1370	Fiske	Steven	01-05-1995	598

Entities (records)

Key field

Attributes (fields)

key
A field or set of fields in a record that is used to identify the record.

primary key
A field or set of fields that uniquely identifies the record.

As discussed, a collection of fields about a specific object is a record. A **key** is a field or set of fields in a record that is used to identify the record. A **primary key** is a field or set of fields that uniquely identifies the record. No other record can have the same primary key. The primary key is used to distinguish records so that they can be accessed, organized, and manipulated. For an employee record such as the one shown in Figure 3.2, the employee number is an example of a primary key.

Locating a particular record that meets a specific set of criteria might require the use of a combination of secondary keys. For example, a customer might call a mail-order company to place an order for clothes. If the customer does not know his primary key (such as a customer number), a secondary key (such as last name) can be used. In this case, the order clerk enters the last name, such as Adams. If there are several customers with a last name of Adams, the clerk can check other fields, such as address, first name, and so on, to find the correct customer record. After the correct customer record is obtained, the order can be completed, and the clothing items can be shipped to the customer.

The Traditional Approach Versus the Database Approach

Since the first use of computers to perform routine business functions in the 1950s, companies have used the traditional approach to process their transactions. This approach uses separate files for each application, such as payroll and billing. Although some companies still use the traditional approach, most organizations today use the database approach, which utilizes a unified and integrated database for most or all of a company's transactions. This section explores both the traditional and database approach.

The Traditional Approach

One of the most basic ways to manage data is via files. Because a file is a collection of related records, all records associated with a particular application (and, therefore, related by the application) can be collected and managed together in an application-specific file. At one time, most organizations had numerous application-specific data files; for example, customer records often were maintained in separate files, with each file relating to a specific process completed by the company, such as shipping or billing. This approach to data management, whereby separate data files are created and stored for each application program, is called the **traditional approach to data management**. For each particular application, one or more data files is created (see Figure 3.3).

traditional approach to data management
An approach whereby separate data files are created and stored for each application program.

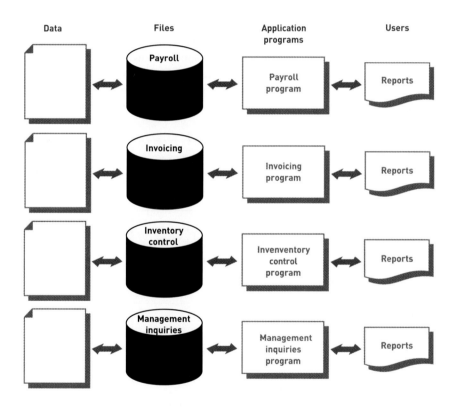

Figure 3.3

The Traditional Approach to Data Management

With the traditional approach, one or more data files is created and used for every application. For example, the inventory control program would have one or more files containing inventory data, such as the inventory item, number on hand, and item description. Likewise, the invoicing program can have files on customers, inventory items being shipped, and so on. With the traditional approach to data management, it is possible to have the same data, such as inventory items, in several different files used by different applications.

One of the flaws in this traditional file-oriented approach to data management is that much of the data—for example, customer name and address—is duplicated in two or more files. This duplication of data in separate files is known as **data redundancy**. The problem with data redundancy is that changes to the data (e.g., a new customer address) might be made in one file and not in another. The order-processing department might have updated its file to the new address, but the billing department is still sending bills to the old address.

data redundancy
A duplication of data in separate files.

data integrity
The degree to which the data
in any one file is accurate.

Data redundancy, therefore, conflicts with **data integrity**—the degree to which the data in any one file is accurate. Data integrity follows from the control or elimination of data redundancy. Keeping a customer's address in only one file decreases the possibility that the customer will have two different addresses stored in different locations. The efficient operation of a business requires a high degree of data integrity, which is one of the advantages of the database approach.

The Database Approach

Because of the problems associated with the traditional approach to data management, many managers wanted a more efficient and effective means of organizing data. The result was the **database approach to data management**. In a database approach, a pool of related data is shared by multiple application programs. Rather than having separate data files, each application uses a collection of data that is either joined or related in the database.

database approach to data management
An approach whereby a pool of related data is shared by multiple application programs.

The database approach offers significant advantages over the traditional file-based approach. For one, by controlling data redundancy, the database approach can use storage space more efficiently and increase data integrity. The database approach can also provide an organization with increased flexibility in the use of data. Because data once kept in two files is now located in the same database, it is easier to locate and request data for many types of processing. A database also offers the ability to share data and information resources. This can be a critical factor in coordinating organization-wide responses across diverse functional areas of a corporation.

To use the database approach to data management, additional software—a database management system (DBMS)—is required. As previously discussed, a DBMS consists of a group of programs that can be used as an interface between a database and the user or the database and the application programs. Typically, this software acts as a buffer between the application programs and the database itself. Figure 3.4 illustrates the database approach.

Figure 3.4

The Database Approach to Data Management

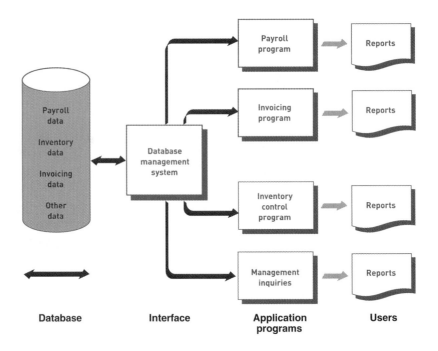

Table 3.1 lists some of the primary advantages of the database approach, whereas Table 3.2 lists the primary disadvantages of the database approach.

Table 3.1

Advantages of the Database Approach

Advantages	Explanation
Improved strategic use of corporate data	Accurate, complete, up-to-date data can be made available to decision makers where, when, and in the form they need it.
Reduced data redundancy	The database approach can reduce or eliminate data redundancy. Data is organized by the DBMS and stored in only one location. This results in more efficient utilization of system storage space.
Improved data integrity	With the traditional approach, some changes to data were not reflected in all copies of the data kept in separate files. This is prevented with the database approach because there are no separate files that contain copies of the same piece of data.
Easier modification and updating	With the database approach, the DBMS coordinates updates and data modifications. Programmers and users do not have to know where the data is physically stored. Data is stored and modified once. Modification and updating are also easier because the data is stored at only one location in most cases.
Data and program independence	The DBMS organizes the data independently of the application program. With the database approach, the application program is not affected by the location or type of data. Introduction of new data types not relevant to a particular application does not require the rewriting of that application to maintain compatibility with the data file.
Better access to data and information	Most DBMSs have software that makes it easy to access and retrieve data from a database. In most cases, simple commands can be given to get important information. Relationships between records can be more easily investigated and exploited, and applications can be more easily combined.
Standardization of data access	A primary feature of the database approach is a standardized, uniform approach to database access. This means that the same overall procedures are used by all application programs to retrieve data and information.
A framework for program development	Standardized database access procedures can mean more standardization of program development. Because programs go through the DBMS to gain access to data in the database, standardized database access can provide a consistent framework for program development. In addition, each application program need address only the DBMS, not the actual data files, reducing application development time.
Better overall protection of the data	The use of and access to centrally located data is easier to monitor and control. Security codes and passwords can ensure that only authorized people have access to particular data and information in the database, thus ensuring privacy.
Shared data and information resources	The cost of hardware, software, and personnel can be spread over a large number of applications and users. This is a primary feature of a DBMS.

Table 3.2

Disadvantages of the Database Approach

Disadvantages	Explanation
More complexity	Database management systems can be difficult to set up and operate. Many decisions must be made correctly for the database management system to work effectively. In addition, users have to learn new procedures to take full advantage of a database management system.
More difficult to recover from a failure	With the traditional approach to file management, a failure of a file only affects a single program. With a database management system, a failure can shut down the entire database.
More expensive	Database management systems can be more expensive to purchase and operate. The expense includes the cost of the database and specialized personnel, such as a database administrator, who is needed to design and operate the database.

DATA MODELING AND THE RELATIONAL DATABASE MODEL

▼

Because there are so many elements in today's businesses, it is critical to keep data organized so that it can be used effectively. A database should be designed to store all data relevant to the business and provide quick access and easy modification. Moreover, it must reflect the business processes of the organization. When building a database, an organization must carefully consider these questions:

- *Content.* What data should be collected and at what cost?
- *Access.* What data should be provided to which users and when?
- *Logical structure.* How should data be arranged so that it makes sense to a given user?
- *Physical organization.* Where should data be physically located?

Data Modeling

Key considerations in organizing data in a database include determining what data is to be collected in the database, who will have access to it, and how they might want to use the data. Based on these determinations, a database can then be created. Building a database requires two different types of designs: a logical design and a physical design. The *logical design* of a database shows an abstract model of how the data should be structured and arranged to meet an organization's information needs. The logical design of a database involves identifying relationships among the different data items and grouping them in an orderly fashion. Because databases provide both input and output for information systems throughout a business, users from all functional areas should assist in creating the logical design to ensure that their needs are identified and addressed. *Physical design* starts from the logical database design and fine-tunes it for performance and cost considerations (e.g., improved response time, reduced storage space, lower operating cost). The person who fine-tunes the physical design must have an in-depth knowledge of the DBMS to implement the database. For example, the logical database design might need to be altered so that certain data entities are combined, summary totals are carried in the data records rather than calculated from elemental data, and some data attributes are repeated in more than one data entity. These are examples of **planned data redundancy**, which is done to improve the system performance so that user reports or queries can be created more quickly.

One of the tools database designers use to show the logical relationships among data is a data model. A **data model** is a diagram of entities and their relationships. Data modeling usually involves understanding a specific business problem and analyzing the data and information needed to deliver a solution. When done at the level of the entire organization, this is called *enterprise data modeling*. **Enterprise data modeling** is an approach that starts by investigating the general data and information needs of the organization at the strategic level and then examining more specific data and information needs for the various functional areas and departments within the organization. Various models have been developed to help managers and database designers analyze data and information needs. An entity-relationship diagram is an example of such a data model.

Entity-relationship (ER) diagrams use basic graphical symbols to show the organization of and relationships between data. In most cases, boxes are used in ER diagrams to indicate data items or entities contained in data tables, and diamonds show relationships between data items and entities. In other words, ER diagrams are used to show data items in tables (entities) and the ways they are related.

ER diagrams help ensure that the relationships among the data entities in a database are correctly structured so that any application programs developed are consistent with business operations and user needs. In addition, ER diagrams can serve as reference documents after a database is in use. If changes are made to the database, ER diagrams help design them.

planned data redundancy
A way of organizing data in which the logical database design is altered so that certain data entities are combined, summary totals are carried in the data records rather than calculated from elemental data, and some data attributes are repeated in more than one data entity to improve database performance.

data model
A diagram of data entities and their relationships.

enterprise data modeling
The data modeling done at the level of the entire enterprise.

entity-relationship (ER) diagram
The data models that use basic graphical symbols to show the organization of and relationships between data.

Figure 3.5 shows an ER diagram for an order database. In this database design, one salesperson serves many customers. This is an example of a one-to-many relationship, as shown by the one-to-many symbol ("crow's-foot") shown in Figure 3.5. The ER diagram also shows that each customer can place one-to-many orders, each order includes one-to-many line items, and many line items can specify the same product (a many-to-one relationship). There can also be one-to-one relationships. For example, one order generates one invoice.

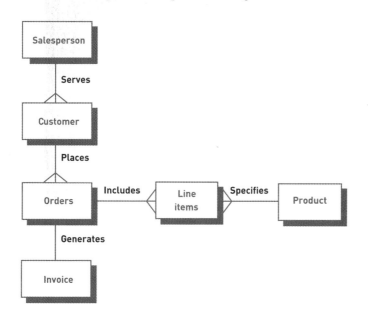

Figure 3.5

An Entity-Relationship (ER) Diagram for a Customer Order Database

Development of ER diagrams helps ensure that the logical structure of application programs are consistent with the data relationships in the database.

The Relational Database Model

Although there are a number of different database models, including flat files, hierarchical, and network models, the **relational model** has become the most popular database model, and use of this model will increase in the future. The relational model describes data using a standard tabular format. In a database structured according to the relational model, all data elements are placed in two-dimensional tables, called *relations*, that are the logical equivalent of files. The tables in relational databases organize data in rows and columns, simplifying data access and manipulation. It is normally easier for managers to understand the relational model (see Figure 3.6) than other database models. Databases based on the relational model include DB2 from IBM, Oracle, Sybase, SQL Server and Access from Microsoft, and MySQL.

relational model
A database model that describes data in which all data elements are placed in two-dimensional tables, called *relations*, that are the logical equivalent of files.

In the relational database model, each row of a table represents a data entity, with the columns of the table representing attributes. Each attribute can take on only certain values. The allowable values for these attributes are called the **domain**. The domain for a particular attribute indicates what values can be placed in each of the columns of the relational table. For instance, the domain for an attribute such as gender would be limited to male or female. A domain for pay rate would not include negative numbers. Defining a domain can increase data accuracy. For example, a pay rate of −$5.00 could not be entered into the database because it is a negative number and not in the domain for pay rate.

domain
The allowable values for data attributes.

Manipulating Data

After data has been placed into a relational database, users can make inquiries and analyze data. Basic data manipulations include selecting, projecting, and joining. **Selecting** involves eliminating rows according to certain criteria. Suppose a project table contains the project number, description, and department number for all projects being performed by a company. The president of the company might want to find the department number for Project 226, a sales manual project. Using selection, the president can eliminate all rows but the one for Project 226 and see that the department number for the department completing the sales manual project is 598.

selecting
The data manipulation that eliminates rows according to certain criteria.

Figure 3.6

A Relational Database Model

In the relational model, all data elements are placed in two-dimensional tables, or relations. As long as they share at least one common element, these relations can be linked to output useful information.

Data table 1: Project table

Project number	Description	Dept. number
155	Payroll	257
498	Widgets	632
226	Sales manual	598

Data table 2: Department table

Dept.	Dept. name	Manager SSN
257	Accounting	005-10-6321
632	Manufacturing	549-77-1001
598	Marketing	098-40-1370

Data table 3: Manager table

SSN	Last name	First name	Hire date	Dept. number
005-10-6321	Johns	Francine	10-07-2005	257
549-77-1001	Buckley	Bill	02-17-1989	632
098-40-1370	Fiske	Steven	01-05-1995	598

projecting
The data manipulation that eliminates columns in a table.

joining
The data manipulation that combines two or more tables.

linking
The data manipulation that relates or links two or more tables using common data attributes.

Projecting involves eliminating columns in a table. For example, we might have a department table that contains the department number, department name, and Social Security number (SSN) of the manager in charge of the project. The sales manager might want to create a new table with only the department number and the Social Security number of the manager in charge of the sales manual project. Projection can be used to eliminate the department name column and create a new table containing only department number and SSN.

Joining involves combining two or more tables. For example, we can combine the project table and the department table to get a new table with the project number, project description, department number, department name, and Social Security number for the manager in charge of the project.

As long as the tables share at least one common data attribute, the tables in a relational database can be **linked** to provide useful information and reports. Being able to link tables to each other through common data attributes is one of the keys to the flexibility and power of relational databases. Suppose the president of a company wants to find out the name of the manager of the sales manual project and the length of time the manager has been with the company. Assume that the company has the manager, department, and project tables shown in Figure 3.6. A simplified ER diagram showing the relationship between these tables is shown in Figure 3.7. Note the crow's-foot by the project table. This indicates that a department can have many projects. The president would make the inquiry to the database, perhaps via a personal computer. The DBMS would start with the project description and search the project table to find out the project's department number. It would then use the department number to search the department table for the manager's Social Security number. The department number is also in the department table and is the common element that allows the project table and the department table to be linked. The DBMS would then use the manager's Social Security number to search the manager table for the manager's hire date.

The manager's Social Security number is the common element between the department table and the manager table. The final result would be: The manager's name and hire date are presented to the president as a response to the inquiry (see Figure 3.8).

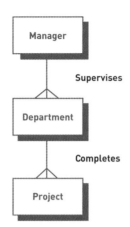

Figure 3.7

A Simplified ER Diagram Showing the Relationship Between the Manager, Department, and Project Tables

One of the primary advantages of a relational database is that it allows tables to be linked, as shown in Figure 3.8. This linkage is especially useful when information is needed from multiple tables, as in our example. The manager's Social Security number, for example, is maintained in the manager table. If the Social Security number is needed, it can be obtained by linking to the manager table.

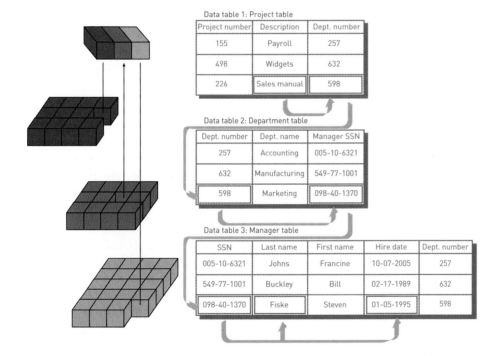

 Figure 3.8

Linking Data Tables to Answer an Inquiry

In finding the name and hire date of the manager working on the sales manual project, the president needs three tables: project, department, and manager. The project description (Sales Manual) leads to the department number (598) in the project table, which leads to the manager's SSN (098-40-1370) in the department table, which leads to the manager's name (Fiske) and hire date (01-05-1985) in the manager table.

The relational database model is by far the most widely used. It is easier to control, more flexible, and more intuitive than other approaches because it organizes data in tables. As seen in Figure 3.9, a relational database management system, such as Microsoft Access, provides a number of tips and tools for building and using database tables. This figure shows the database displaying information about data types and indicating that additional help is available. The ability to link relational tables also allows users to relate data in new ways without having to redefine complex relationships. Because of the advantages of the relational model, many companies use it for large corporate databases, such as in marketing and accounting.

The relational model can be used with personal computers and mainframe systems. A travel reservation company, for example, can develop a fare-pricing system by using relational database technology that can handle millions of daily queries from online travel companies, such as Expedia, Travelocity, and Orbitz.

Building and Modifying a Relational Database

Relational databases provide many tools, tips, and tricks to simplify the process of creating and modifying a database.

DATABASE MANAGEMENT SYSTEMS (DBMS)

Creating and implementing the right database system ensures that the database will support both business activities and goals. But how do we actually create, implement, use, and update a database? The answer is found in the database management system. As discussed earlier, a database management system (DBMS) is a group of programs used as an interface between a database and the application programs or a database and the user. The capabilities and types of database systems, however, vary considerably.

Overview of Database Types

Database management systems can range from small, inexpensive software packages to sophisticated systems costing hundreds of thousands of dollars. The following sections detail a few popular alternatives.

Flat File

A flat file is a simple database program that has no relationship between its records and is often used to store and manipulate a single table or file. Flat files do not use any of the data-base models discussed previously, such as the relational model. Many spreadsheet and word processing programs have flat file capabilities. These software packages can sort tables and make simple calculations and comparisons. OneNote, developed by Microsoft in 2003, was designed to let people put ideas, thoughts, and notes into a computer file.[3] Each can be placed anywhere on a page in a OneNote file. Ideas, thoughts, and notes can also be placed in a box on a page, called a *container*. Pages are organized into sections and subsections that appear as colored tabs. After being entered, the ideas, thoughts, and items in a OneNote file can be retrieved, copied, and pasted into other applications, such as word processing and spreadsheet programs.

Single User

Databases for personal computers are most often meant for a single user. Only one person can use the database at a time. Access and Quicken are examples of popular single-user DBMSs, through which users store and manipulate financial data. Microsoft's InfoPath is another example of a single-user database.[4] The database is part of the Microsoft Office suite, and it helps people collect and organize information from a variety of sources. InfoPath has built-in forms that can be used to enter expense information, time-sheet data, and a variety of other information.

Multiple Users

Large mainframe computer systems need multiuser DBMSs. These more powerful, expensive systems allow dozens or hundreds of people to access the same database system at the same time. Popular vendors for multiuser database systems include Oracle, Sybase, and IBM.

All DBMSs share some common functions, such as providing a user view, physically storing and retrieving data in a database, allowing for database modification, manipulating data, and generating reports. These DBMSs are capable of handling the most complex of data-processing tasks. A medical clinic, for example, can use a database like IBM's DB2 to develop an all-inclusive database containing patient records, physician notes, demographic data, genetic data, and proteomic (protein-related) data for millions of patients.

Providing a User View

Because the DBMS is responsible for access to a database, one of the first steps in installing and using a database involves telling the DBMS the logical and physical structure of the data and relationships among the data in the database. This description is called a **schema** (as in schematic diagram). A schema can be part of the database or a separate schema file. The DBMS can refer to a schema to find where to access the requested data in relation to another piece of data.

A DBMS also acts as a user interface by providing a view of the database. A user view is the portion of the database a user can access. To create different user views, subschemas are developed. A **subschema** is a file that contains a description of a subset of the database and identifies which users can view and modify the data items in that subset. Whereas a schema is a description of the entire database, a subschema shows only some of the records and their relationships in the database. Normally, programmers and managers need to view or access only a subset of the database. For example, a sales representative might need only data describing customers in his region, not the sales data for the entire nation. A subschema could be used to limit his view to data from his region. With subschemas, the underlying structure of the database can change, but the view the user sees might not change. For example, even if all the data on the southern region changed, the northeastern region's sales representative's view would not change if he accessed data on his region.

A number of subschemas can be developed for different users and the various application programs. Typically, the database user or application will access the subschema, which then accesses the schema (see Figure 3.10). Subschemas can also provide additional security because programmers, managers, and other users are typically allowed to view only certain parts of the database.

Creating and Modifying the Database

Schemas and subschemas are entered into the DBMS (usually by database personnel) via a data definition language. A **data definition language** (DDL) is a collection of instructions and commands used to define and describe data and data relationships in a specific database. A DDL allows the database's creator to describe the data and the data relationships that are to be contained in the schema and the many subschemas. In general, a DDL describes logical access paths (LAPs) and logical records in the database. Figure 3.11 shows a simplified example of a DDL used to develop a general schema. The *Xs* in Figure 3.11 reveal where specific information concerning the database is to be entered. File description, area description, record description, and set description are terms the DDL defines and uses in this example. Other terms and commands can be used, depending on the particular DBMS employed.

---

schema
A description of the entire database.

subschema
A file that contains a description of a subset of the database and identifies which users can view and modify the data items in the subset.

data definition language (DDL)
A collection of instructions and commands used to define and describe data and data relationships in a specific database.

Figure 3.10

**The Use of Schemas
and Subschemas**

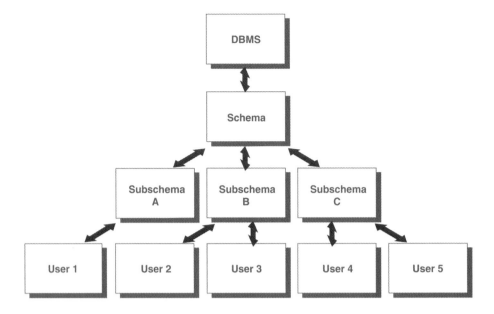

Figure 3.11

**Using a Data Definition
Language to Define a Schema**

data dictionary
A detailed description of all the data
used in the database.

Another important step in creating a database is to establish a **data dictionary**, a detailed description of all data used in the database. The data dictionary contains the name of the data item, the aliases or other names that can be used to describe the item, the range of values that can be used, the type of data (such as alphanumeric or numeric), the amount of storage needed for the item, a notation of the person responsible for updating it and the various users who can access it, and a list of reports that use the data item. A data dictionary can also include a description of data flows, the way records are organized, and the data-processing requirements. Figure 3.12 shows a typical data dictionary entry.

For example, the information in a data dictionary for the part number of an inventory item can include the name of the person who made the data dictionary entry (D. Bordwell), the date the entry was made (August 4, 2005), the name of the person who approved the

Figure 3.12

A Typical Data Dictionary Entry

NORTHWESTERN MANUFACTURING

PREPARED BY: D. BORDWELL
DATE: 04 AUGUST 2005
APPROVED BY: J. EDWARDS
DATE: 13 OCTOBER 2005
VERSION: 3.1
PAGE: 1 OF 1

DATA ELEMENT NAME: PARTNO
DESCRIPTION: INVENTORY PART NUMBER
OTHER NAMES: PTNO
VALUE RANGE: 100 TO 5000
DATA TYPE: NUMERIC
POSITIONS: 4 POSITIONS OR COLUMNS

entry (J. Edwards), the approval date (October 13, 2005), the version number (3.1), the number of pages used for the entry (1), the part name (PARTNO), the other part names that can be used (PTNO), the range of values (part numbers can range from 100 to 5,000), the type of data (numeric), and the storage required (four positions are required for the part number).

Storing and Retrieving Data

As just described, one function of a DBMS is to be an interface between an application program and the database. When an application program needs data, it requests that data through the DBMS. Suppose that to calculate the total price of a new car, an auto dealer pricing program needs price data on the engine option—six cylinders instead of the standard four cylinders. The application program thus requests this data from the DBMS. In doing so, the application program follows a logical access path. Next, the DBMS, working in conjunction with various system software programs, accesses a storage device, such as disk or tape, where the data is stored. When the DBMS goes to this storage device to retrieve the data, it follows a path to the physical location (physical access path or PAP) where the price of this option is stored. In the pricing example, the DBMS might go to a disk drive to retrieve the price data for six-cylinder engines. This relationship is shown in Figure 3.13.

Figure 3.13

Logical and Physical Access Paths

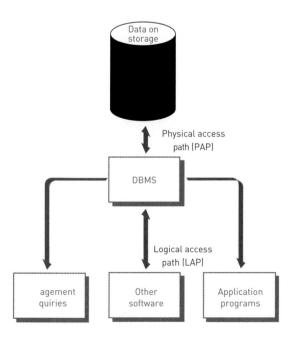

When two or more people or programs attempt to access the same record in the same database at the same time, a problem can occur. For example, an inventory control program might attempt to reduce the inventory level for a product by ten units because ten units were just shipped to a customer. At the same time, a purchasing program might attempt to increase the inventory level for the same product by 200 units because more inventory was just received. Without proper database control, one of the inventory updates might not be correctly made, resulting in an inaccurate inventory level for the product. **Concurrency control** can be used to avoid this potential problem. One approach is to lock out all other application programs from access to a record if the record is being updated or used by another program.

Manipulating Data and Generating Reports

After a DBMS has been installed, employees and managers can use it to generate reports and obtain important information. Some databases use *Query-By-Example (QBE)*, which is a visual approach to developing database queries or requests. Like Windows and other GUI operating systems, you can perform queries and other database tasks by opening windows and clicking on the data or features you want (see Figure 3.14).

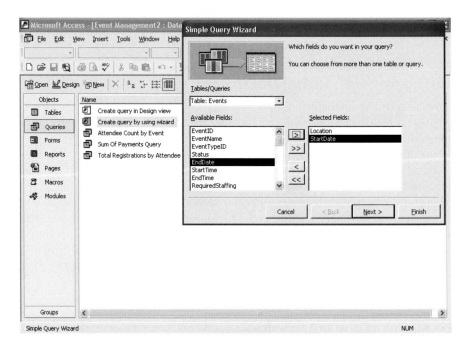

In other cases, database commands can be used in a programming language. For example, COBOL commands can be used in simple programs that will access or manipulate certain pieces of data in the database. Here's another example of a DBMS query: SELECT * FROM EMPLOYEE WHERE JOB_CLASSIFICATION = "C2". The * tells the program to include all columns from the EMPLOYEE table. In general, the commands that are used to manipulate the database are part of the **data manipulation language (DML)**. This specific language, provided with the DBMS, allows managers and other database users to access, modify, and make queries about data contained in the database to generate reports. Again, the application programs go through the subschemas, schemas, and DBMS before actually getting to the physically stored data on a device such as a disk.

In the 1970s, D. D. Chamberlain and others at the IBM Research Laboratory in San Jose, California, developed a standardized data manipulation language called *Structured Query Language (SQL)*, pronounced like the word *sequel* or simply spelled out as *SQL*. The EMPLOYEE query shown earlier is written in SQL. In 1986, the American National Standards Institute (ANSI) adopted SQL as the standard query language for relational databases. Since ANSI's acceptance of SQL, interest in making SQL an integral part of relational databases on both mainframe and personal computers has increased. SQL has many built-in functions, such as the average (AVG), the largest value (MAX), the smallest value (MIN), and others.

SQL lets programmers learn one powerful query language and use it on systems ranging from PCs to the largest mainframe computers (see Figure 3.15). Programmers and database users also find SQL valuable because SQL statements can be embedded into many programming languages, such as the widely used C++ and COBOL languages. Because SQL uses standardized and simplified procedures for retrieving, storing, and manipulating data in a database system, the popular database query language can be easy to understand and use.

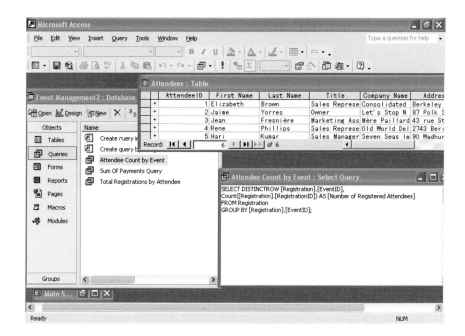

Figure 3.15

Structured Query Language

SQL has become an integral part of most relational database packages, as shown by this screen from Microsoft Access.

After a database has been set up and loaded with data, it can produce desired reports, documents, and other outputs (see Figure 3.16). These outputs usually appear in screen displays or hard-copy printouts. The output-control features of a database program allow you to select the records and fields to appear in reports. You can also make calculations specifically for the report by manipulating database fields. Formatting controls and organization options (such as report headings) help you to customize reports and create flexible, convenient, and powerful information-handling tools.

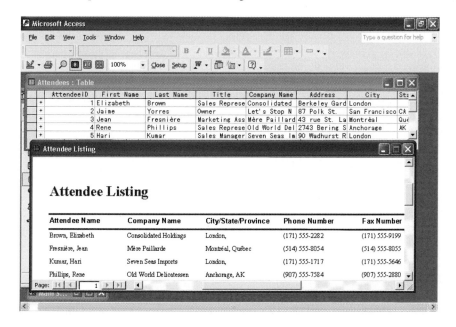

Figure 3.16

Database Output

A database application offers sophisticated formatting and organization options to produce the right information in the right format.

Reports generated from the database that highlight unusual, problematic, or significant events might require urgent management attention. Database programs can produce literally hundreds of documents and reports. A few examples include the following:

- Form letters with address labels
- Payroll checks and reports
- Invoices
- Orders for materials and supplies
- A variety of financial performance reports

Database Administration

As mentioned earlier, a database administrator (DBA) is a highly skilled systems professional who directs or performs all activities to maintain a successful database environment. The DBA's responsibilities include designing, implementing, and maintaining the database system and the DBMS; establishing policies and procedures pertaining to the management, security, maintenance, and use of the database management system; and training employees in database management and use.

A DBA is expected to have a clear understanding of the fundamental business of the organization, be proficient in the use of selected database management systems, and stay abreast of emerging technologies and new design approaches. Typically, a DBA has either a degree in computer science or management information systems and some on-the-job training with a particular database product or more extensive experience with a range of database products.

The DBA works with users to decide the content of the database—to determine exactly what entities are of interest and what attributes are to be recorded about those entities. Thus, it is important for non-IS personnel to have some idea of what the DBA does and why this function is important. The DBA can play a crucial role in the development of effective information systems to benefit the organization, its employees, and the managers.

The DBA also works with programmers as they build applications to ensure that their programs comply with database management system standards and conventions. After the database is built and operating, the DBA monitors operations logs for security violations. Database performance is also monitored to ensure that the system's response time meets users' needs and that it operates efficiently. If there is a problem, the DBA attempts to correct it before it becomes serious.

Popular Database Management Systems

A key to making DBMSs more usable for some databases is the incorporation of "wizards" that walk you through how to build customized databases, modify ready-to-run applications, use existing record templates, and quickly locate the data you want. These applications also include powerful new features, such as help systems and Web-publishing capabilities. For example, users can create a complete inventory system and then instantly post it to the Web, where it does double duty as an electronic catalog. Some of the more popular DBMSs for end users include Access and Corel's Paradox. The complete database management software market encompasses software used by professional programmers and that runs on midrange, mainframe, and supercomputers. The entire market generates $10 billion per year in revenue, including IBM, Oracle, and Microsoft. Although Microsoft rules the desktop PC software market, its share of database software on bigger computers is small.

Like other software products, a number of open-source database systems are available, including PostgreSQL and MySQL.[5] MySQL is the most popular open-source database management system used by travel agencies, manufacturing companies, and other companies.[6] The use of MySQL has increased by more than 30 percent in some years, compared with a 6 percent growth for other popular database packages.[7] According to Chares Gary of the MetaGroup, "I get calls now from customers, not [asking,] 'Should I use an open-source database?' but instead it's, 'Should I use PostgreSQL or MySQL?'"[8] In addition, many traditional database programs are now available on open-source operating systems. The popular DB2 relational database from IBM, for example, is available on the Linux operating system.[9] The Sybase IQ database is also available on the Linux operating system.[10] S&H Co., which offers S&H Greenpoints to reward loyal retail customers, is considering the use of Sybase IQ running on the Linux operating system.[11] "We find Linux to be very cost effective and powerful for what we want to do. It's the best use of financial resources to go with an open-source platform," says Frank Lundy, database director for S&H Co.

Special-Purpose Database Systems

In addition to the popular database management systems just discussed, a number of specialized database packages are used for specific purposes or in specific industries. Summation and Concordance, for example, is a special-purpose database system used in law firms to organize legal documents.[12] CaseMap organizes information about a case, and LiveNote is used to display and analyze transcripts. These databases help law firms develop and execute good litigation strategies. The Scottish Intelligence Database (SID) is used by the Scottish Drug Enforcement Agency to share crime reports and up-to-date information about crime and criminals.[13] According to Detective Superintendent Ian McCandish, "SID will allow officers access to data from the Shetlands to Gretna Green. Criminals may travel the length and breadth of Scotland, but if we hold the intelligence on them it will be available to front-line officers irrespective of where they are." GlobalSpec is a specialized database for engineers and product designers that contains more than 45 million parts from about 9,500 parts catalogues.[14] Architects Wimberly, Allison, Gong & Goo use a special-purpose database to store and manipulate three-dimensional drawings.[15] J.P. Morgan Chase & Co., uses an "in-memory" database to speed trade orders from pension funds, hedge funds, and various institutional investors.[16] In-memory databases use a computer's memory instead of a hard disk to store and manipulate important data. The database can process more than 100,000 queries per second.

Selecting a Database Management System

The database administrator often selects the best database management system for an organization. The process begins by analyzing database needs and characteristics. The information needs of the organization affect the type of data that is collected and the type of database management system that is used. Important characteristics of databases include the size of the database, the number of concurrent users, the performance, the ability of the DBMS to integrate with other systems, the features of the DBMS, the vendor considerations, and the cost of the system.

Database Size

Database size depends on the number of records or files in the database. The size determines the overall storage requirement for the database. Most database management systems can handle relatively small databases of less than 100 million bytes; fewer can manage terabyte-size databases.

Today, companies are trimming the size of their databases to maintain good performance and reduce costs.[17] According to a project manager at Kennametal, "Our overweight database was months away from crashing due to exceeding our production diskspace capacity." The company was able to save about $700,000 in additional hardware and storage costs by trimming its database.

Number of Concurrent Users

The number of simultaneous users who can access the contents of the database is also an important factor. Clearly, a database that is used by a large workgroup must be able to support a number of concurrent users; if it cannot, the efficiency of the members is lowered. The term *scalability* is sometimes used to describe how well a database performs as the size of the database or the number of concurrent users is increased. A highly scalable database management system is desirable to provide flexibility. Unfortunately, many companies make a poor DBMS choice in this regard and then later are forced to convert to a new DBMS when the original does not meet expectations.

Performance

How fast the database is able to update records can be the most important performance criterion for some organizations. Credit card and airline companies, for example, must have database systems that can update customer records and check credit or make a plane reservation in seconds, not minutes. Other applications, such as payroll, can be done once a week or less frequently and do not require immediate processing. If an application demands immediacy, it also demands rapid recovery facilities in the event the computer system shuts down temporarily. Other performance considerations include the number of concurrent users who can be supported and the amount of memory that is required to execute the database management program. One database was able to process 1,184,893 transactions per minute on average, setting a new world record for speed.[18] Organizations often undergo *performance tuning* to increase database speed and storage efficiency. Performance tuning involves making adjustments to the database to enhance overall performance.

A database management system used by online stores must be able to support a large number of concurrent users by quickly checking a customer's credit card and processing his or her order for merchandise.

Integration

A key aspect of any database management system is its ability to integrate with other applications and databases. A key determinant here is what operating systems it can run under—such as Linux, UNIX, or Windows. Some companies use several databases for different applications at different locations. A manufacturing company with four plants in three different states might have a separate database at each location. The ability of a database program to import data from and export data to other databases and applications can be a critical consideration.

Features

The features of the database management system can also make a big difference. Most database programs come with security procedures, privacy protection, and a variety of tools. Other features can include the ease of use of the database package and the availability of manuals and documentation to help the organization get the most from the database package. Additional features such as wizards and ready-to-use templates help improve the ease of use. Because of the pressure to reduce IS budgets and the scarcity of experienced database administrators, organizations are demanding database software that comes with features that simplify database management tasks.

The Vendor

The size, reputation, and financial stability of the vendor should also be considered in making any database purchase. Some vendors are well respected in the IS industry and have a large support staff to give assistance, if necessary. A well-established and financially secure database company is more likely to remain in business than others.

Cost

Database packages for personal computers can cost a few hundred dollars, but large database systems for mainframe computers can cost hundreds of thousands of dollars. In addition to the initial cost of the database package, monthly operating costs should be considered. Some database companies rent or lease their database software. Monthly rental or lease costs, maintenance costs, additional hardware and software costs, and personnel costs can be substantial.

Using Databases with Other Software

Database management systems are often used in conjunction with other software packages or the Internet. A database management system can act as a front-end application or a back-end application. A *front-end application* is one that directly interacts with people or users. Marketing researchers often use a database as a front-end to a statistical analysis program. The researchers enter the results of market questionnaires or surveys into a database. The data is then transferred to a statistical analysis program to determine the potential for a new product or the effectiveness of an advertising campaign. A *back-end application* interacts with other programs or applications; it only indirectly interacts with people or users. When people request information from a Web site on the Internet, the Web site can interact with a database (the back end) that supplies the desired information. For example, you can connect to a university Web site to find out whether the university's library has a book you want to read. The Web site then interacts with a database that contains a catalog of library books and articles to determine whether the book you want is available.

DATABASE APPLICATIONS

Today, there is a shift from simply accessing the data contained in a database to managing and manipulating the content of a database to produce useful information. Common manipulations are searching, filtering, synthesizing, and assimilating the data contained in a database using a number of database applications. These applications allow users to link the company databases to the Internet, set up data warehouses and marts, use databases for strategic business intelligence, place data at different locations, use online processing and open connectivity standards for increased productivity, develop databases with the object-oriented approach, and search for and use unstructured data, such as graphics, audio, and video.

Linking the Company Database to the Internet

Customers, suppliers, and company employees must be able to access corporate databases through the Internet, intranets, and extranets to meet various business needs. For example, Internet customers need to access the corporate product database to obtain product information, including size, color, type, and price details. Suppliers use the Internet and corporate extranets to view inventory databases to check the levels of raw materials and the current production schedule to determine when and how much of their products must be delivered to support just-in-time inventory management. Company employees need to be able to access corporate databases to support decision making even when they are located remotely. In such cases, they might use laptop computers and access the data via the Internet and company intranet. Read the "Information Systems @ Work" feature to see how one cruise line has helped its employees access and use its data.

Web-Based DBMS Empowers Cruise Line Personnel

In these days of value-driven, performance-based information systems, many businesses are turning to new database technologies to streamline operations and save money. Recently, Holland America Cruise Lines traded its old, complex mainframe database system—one that only a computer programmer could understand—for a new system with which ordinary employees could interact directly.

The goal of the upgrade was to increase revenues by $1 million annually by speeding information to employees for more efficient sales, marketing, and revenue management. Prior to the upgrade, database programmers and tech staff would prepare and distribute weekly reports to address other employees' ad hoc inquiries and would generate scheduled reports to assess the company's revenue and inventory, according to Paul Grigsby, senior revenue manager at Holland America. The new reporting system, called WebFocus, connects to the same mainframe database as the old DBMS but provides more powerful analysis, querying, and reporting tools that are accessed through an intuitive Web-based user interface.

Fulfilling employees' report requests previously took up to two days, but with the new system, the IS staff can fine-tune requests and give reports to end users almost immediately. With a bit of training, some end users are interacting with the system directly to create their own reports. Those who work with the data most, such as revenue management personnel, began using the system first. The new system had its skeptics at the beginning, says Grigsby. "We are yield managers, not computer programmers, and I was frankly suspicious about putting the reporting function in our hands," he explains. "However, after training and a goodly amount of trial and error, we began to see the rewards of empowering the end users." After running a query and producing a report, employees can alter the view of the data by re-sorting it or introducing new fields into the inquiry—without formally requesting a new report from the IS department. Grigsby says, "It makes both my time and the information systems department's time more efficient."

Teresa Tennant is the manager of online communications for the shore excursion department at Holland America. She plans to become less dependent on the IS staff with training on the new system. With the new system, she can log on to a Web site to access reports in whatever format she wants—an Excel spreadsheet, Word document, or portable document file (PDF). "It's very handy. I can drill down into complete details of the booking by travel agency and individual," she says.

The company is limited, however, because of the size of the existing mainframe DBMS. It plans to load the information into a streamlined data warehouse, where it can be accessed more quickly. Speed and accessibility of information access, after all, are the attributes that will increase productivity and revenues. "The real value of the new system at Holland America," Bill Hostmann, an analyst at Gartner Inc., says, "is how it makes it easier to develop and distribute business management information to more users in a timely fashion around the world than ever before and thereby more fully leverage existing IT investments."

Discussion Questions

1. How does the new reporting system at Holland America Cruise Lines empower revenue management personnel?
2. How does the new system allow the IS staff to work more efficiently?

Critical Thinking Questions

3. The change in information access at Holland America Cruise Lines is indicative of a general trend in many industries: Non-IS employees are assuming traditional IS staff responsibilities. Do you think that this trend evolved purely out of efforts to save money by reducing IS staff, or are there substantial benefits to bringing IS power to the people? What might those benefits be?
4. Is it realistic to expect nontechnical employees to acquire higher-level technical skills? Will the nontechnical staff be willing and able to assume the task?

SOURCES: Mark L Sangini, "Cruise Line Changes BI Tack," *Computerworld*, October 6, 2003, *www.computerworld.com*; Information Builder's Web site, *www.informationbuilders.com*, accessed March 5, 2004; Holland America Line Web site, *www.hollandamerica.com*, accessed March 5, 2004.

Developing a seamless integration of traditional databases with the Internet is often called a *semantic Web*.[19] According to Tim Berners-Lee, creator of the World Wide Web (WWW), "The Semantic Web is about taking the relational database and webbing it." A semantic Web allows people to access and manipulate a number of traditional databases at the same time through the Internet. Many software vendors—including IBM, Oracle, Microsoft, Macromedia, Inline Internet Systems, and Netscape Communications—are incorporating the capability of the Internet into their products. Such databases allow companies to create an Internet-accessible catalog, which is nothing more than a database of items, descriptions, and prices. Simplest-Shop, for example, uses the Web to sell compact discs on the Internet.[20] The company offers a wealth of information on each compact disc, including product reviews. The Web site makes it appear as though the company has a large number of employees, but the Internet has allowed one person, Calin Uioreanu from Romania, to set up the site as a side business. He is a full-time software engineer. AutoTradeCenter, Inc., has linked its Web site to an Oracle database.[21] The database supports about 24,000 franchise dealerships and 80,000 independent dealerships. The database helps AutoTradeCenter customers, including Honda, DaimlerChrysler, Volkswagen, Subaru, and others.

In addition to the Internet, organizations are gaining access to databases through networks to get good prices and reliable service. Catholic Health, for example, has developed a network to allow physicians to access patient information from remote locations.[22] Perry Manufacturing in North Carolina uses a network to allow its managers to have access to e-mail and other information stored on its database. Connecting databases to corporate Web sites and networks can lead to potential problems, however. One database expert believes that up to 40 percent of Web sites that connect to corporate databases are susceptible to hackers taking complete control of the database.[23] By typing certain characters in a form on some Web sites, a hacker is able to give SQL commands to control the corporate database.

Data Warehouses, Data Marts, and Data Mining

The raw data necessary to make sound business decisions is stored in a variety of locations and formats. This data is initially captured, stored, and managed by transaction processing systems that are designed to support the day-to-day operations of the organization. For decades, organizations have collected operational, sales, and financial data with their online transaction processing (OLTP) systems. The data can be used to support decision making using data warehouses, data marts, and data mining.

Data Warehouses

A **data warehouse** is a database that holds business information from many sources in the enterprise, covering all aspects of the company's processes, products, and customers. The data warehouse provides business users with a multidimensional view of the data they need to analyze business conditions. Data warehouses allow managers to *drill down* to get more detail or *roll up* to take detailed data and generate aggregate or summary reports. A data warehouse is designed specifically to support management decision making, not to meet the needs of transaction processing systems. Ace Hardware Corporation, for example, uses a data warehouse to analyze pricing trends.[24] According to the company's data warehouse designer, "We had one store that only sold one wheelbarrow a year, but when he lowered the price, he sold four in one month." The price reduction was suggested by the data warehouse. Office Depot developed a powerful data warehouse to give employees better information on employee and store performance. According to a company spokesperson, "Sometimes data warehousing can be ignored by the stores unless it provides direct metrics on how the store is performing."[25] A data warehouse stores historical data that has been extracted from operational systems and external data sources (see Figure 3.17). This operational and external data is "cleaned up" to remove inconsistencies and integrated to create a new information database that is more suitable for business analysis.

Data warehouses typically start out as very large databases, containing millions and even hundreds of millions of data records. As this data is collected from the various production systems, a historical database is built that business analysts can use. To remain fresh and accurate, the data warehouse receives regular updates. Old data that is no longer needed is

data warehouse
A database that collects business information from many sources in the enterprise, covering all aspects of the company's processes, products, and customers.

Figure 3.17

Elements of a Data Warehouse

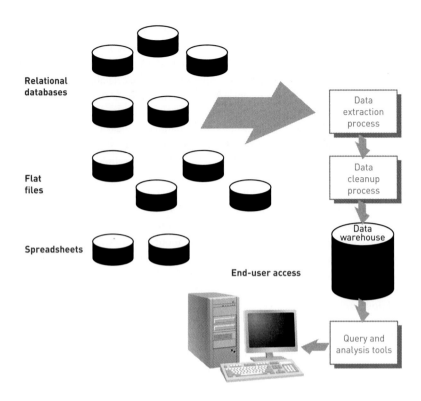

purged from the data warehouse. Updating the data warehouse must be fast, efficient, and automated, or the ultimate value of the data warehouse is sacrificed. It is common for a data warehouse to contain from 3 to 10 years of current and historical data. Data-cleaning tools can merge data from many sources into one database, automate data collection and verification, delete unwanted data, and maintain data in a database management system. Data warehouses can also get data from unique sources. Oracle's Warehouse Management software, for example, can accept information from radio-frequency identification (RFID) technology, which is being used to tag products as they are shipped or moved from one location to another.[26]

The primary advantage of data warehousing is the ability to relate data in new, innovative ways. However, a data warehouse can be extremely difficult to establish, with the typical cost exceeding $2 million.

Data Marts

data mart
A subset of a data warehouse.

A **data mart** is a subset of a data warehouse. Data marts bring the data warehouse concept—online analysis of sales, inventory, and other vital business data that has been gathered from transaction processing systems—to small and medium-sized businesses and to departments within larger companies. Rather than store all enterprise data in one monolithic database, data marts contain a subset of the data for a single aspect of a company's business—for example, finance, inventory, or personnel. In fact, a specific area in the data mart might contain more detailed data than the data warehouse would provide.

Data marts are most useful for smaller groups who want to access detailed data. A warehouse is used for summary data that can be used by an entire company. Because data marts typically contain tens of gigabytes of data, as opposed to the hundreds of gigabytes in data warehouses, they can be deployed on less-powerful hardware with smaller secondary storage devices, delivering significant savings to an organization. Although any database software can be used to set up a data mart, some vendors deliver specialized software designed and priced specifically for data marts. Already, companies such as Sybase, Software AG, Microsoft, and others have announced products and services that make it easier and cheaper to deploy these scaled-down data warehouses. The selling point: Data marts put targeted business information into the hands of more decision makers.

Data Mining

Data mining is an information-analysis tool that involves the automated discovery of patterns and relationships in a data warehouse or a data mart. The FBI, for example, is using the ClearForest database package to support data mining of the Terrorism Intelligence Database.[27] According to FBI director Robert Mueller, "We are now focused on implementing a data warehousing capability that can bring together our information into databases that can be accessed by agents throughout the world as well as our analysts as soon as a piece of information is developed." Data mining has also been used in the airline-passenger profiling system used to block suspected terrorists from flying.[28] Organizations are also investing in systems for data mining to meet new government regulations, as the "Ethical and Societal Issues" feature discusses.

Data mining's objective is to extract patterns, trends, and rules from data warehouses to evaluate (i.e., predict or score) proposed business strategies, which, in turn, will improve competitiveness, improve profits, and transform business processes. It is used extensively in marketing to improve customer retention; cross-selling opportunities; campaign management; market, channel, and pricing analysis; and customer segmentation analysis (especially one-to-one marketing). In short, data-mining tools help end users find answers to questions they never even thought to ask.

Predictive analysis is a form of data mining that combines historical data with assumptions about future conditions to predict outcomes of events such as future product sales or the probability that a customer will default on a loan. Retailers use predictive analysis to upgrade occasional customers into frequent purchasers by predicting what products they will buy if offered an appropriate incentive. Genalytics, Magnify, NCR Teradata, SAS Institute, Sightward, SPSS, and Quadstone have developed predictive analysis tools. Predictive analysis software can be used to analyze a company's customer list and a year's worth of sales data to find new market segments that could be profitable.

data mining
An information-analysis tool that involves the automated discovery of patterns and relationships in a data warehouse or a data mart.

predictive analysis
A form of data mining that combines historical data with assumptions about future conditions to predict outcomes of events such as future product sales or the probability that a customer will default on a loan.

Halfords, a U.K. retailer of car parts, bikes, and accessories, used SPSS's Clementine, a predictive analysis data-mining tool, to identify the most profitable sites to build new stores as part of its expansion program.

Traditional DBMS vendors are well aware of the great potential of data mining. Thus, companies such as Oracle, Sybase, Tandem, and Red Brick Systems are all incorporating data-mining functionality into their products. Table 3.3 summarizes a few of the most frequent applications for data mining.

The Growing Cost of Data-Related Regulations

Increasing government regulations are forcing many businesses to scrutinize their data, database systems, and storage technologies. Government agencies such as the Internal Revenue Service have been concerned about the accuracy and security of records storage since the invention of the computer. Recently, the mounting concerns of varying government agencies have been transformed into laws—laws that are costing businesses big bucks to implement.

In the wake of corporate scandals and in an effort to make investment brokers and dealers accountable for their practices, the Securities and Exchange Commission (SEC) supported passage of the Sarbanes-Oxley Act of 2002. Among the act's many provisions is a requirement for brokers and dealers to log and record all electronic communications, such as e-mail, using "write-once, read many," or WORM, technology. WORM technology ensures that stored data cannot be tampered with later.

The SEC regulatory requirements compelled Jay Cohen, corporate compliance officer at The Mony Group, to install an EMC Centera storage server system and database application called AXS-One Email and Instant Messaging Management Solution software suite to track e-mails. Cohen explains. "All the external e-mail from the sales force, either incoming or outgoing, goes into the AXS-One Email archival system." The system permanently stores each communication as a record with a unique identifier in a database. Administrators can then run queries on the e-mail data for content surveillance, security, and audit trails.

While financial industries buckle down, medical industries have an equal share of new government regulations to meet. In 1996, the U.S. Department of Health and Human Services established national standards to protect the privacy of personal health information with HIPAA—the Health Insurance Portability and Accountability Act. HIPAA caused health and medical organizations to scramble to secure medical records in private database systems.

More recently, the Food and Drug Administration issued a ruling that requires pharmaceutical companies to apply bar codes to thousands of prescription and over-the-counter drugs dispensed in hospitals. The FDA believes the move will save lives by reducing medical errors but estimated that it would hit the nation's 6,000-plus hospitals with a $7 billion technology bill for bar-code readers, databases, and management tools.

Information system companies are designing systems to help businesses and organizations comply with the many new laws and regulations. IBM recently unveiled an integrated system to help users preserve electronic documents for regulations such as the Sarbanes-Oxley Act and HIPAA. The system, called IBM TotalStorage Data Retention 450, combines server, storage, and software components in a secure cabinet. At a cost starting at $141,600 for a 3.5-TB configuration, systems such as this are costly but necessary to comply with government regulations.

With budgets a constant worry, financial and medical organizations are complaining that regulation should be backed with financial assistance to help them comply with the regulations. In the meantime, the costs for e-mail archiving, prescription bar-code readers, and complex and expensive systems that support secure and private recordkeeping will most likely be passed along to customers in the form of higher prices.

Critical Thinking Questions

1. Why has the government stepped up efforts to regulate recordkeeping in financial and medical industries?
2. With many medical organizations already strapped for cash, especially hospitals, is it fair for the government to force them to comply with expensive new regulations? Who should bear the financial burden of secure and private recordkeeping?

What Would You Do?

As CIO of the Prime Market Finance Corp., you are responsible for complying with new laws that require you to permanently store all electronic communications. Prime Market employees currently send and receive 200,000 e-mail messages per day. One-third of the e-mail received is junk mail—spam. One-third of the e-mail sent is not related to the business. You are considering the IBM TotalStorage Data Retention system, but first you need to determine how much data storage will be required. Estimating that 100 e-mail messages equal 1 MB of storage space, you calculate that 200,000 e-mail messages per day amount to 2,000 MB, or 2 GB. That amounts to 730 GB per year, and more than 7 TB over 10 years. It appears that IBM's low-end system, which supports 3.5 TB, will not meet your storage needs over the long haul.

3. What policies and technologies might you implement to reduce the overall amount of e-mail sent and received?
4. Who should bear the financial burden of archiving data that might or might not be needed as evidence in court?

SOURCES: Lucas Mearian, "SEC Holds Fast on WORM Standard for Securities Firms," *Computerworld*, March 9, 2003, *www.computerworld.com*; Mitch Betts, "Editor's Note: The New Rules of Storage," *Computerworld*, November 17, 2003, *www.computerworld.com*; Lucas Mearian, "Sidebar: Regulations, Volume and Capacity Add Archiving Pressure," *Computerworld*, February 16, 2004, *www.computerworld.com*; Lucas Mearian, "Compliance Laws Vex IT: The USA Patriot Act Is Keeping Financial Firms Busy," *Computerworld*, September 8, 2003, *www.computerworld.com*; Thomas Hoffman, "IBM Tailors Bundle for Preserving Corporate Data: Integrated System to Aid in Regulatory Compliance Efforts," *Computerworld*, February 23, 2004, *www.computerworld.com*; Bob Brewin, "FDA Mandates Bar Codes on Drugs Used in Hospitals," *Computerworld*, February 26, 2004, *www. computerworld.com*.

Application	Description
Branding and positioning of products and services	Enable the strategist to visualize the different positions of competitors in a given market using performance (or importance) data on dozens of key features of the product in question and then to condense all that data into a perceptual map of just two or three dimensions
Customer churn	Predict current customers who are likely to go to a competitor
Direct marketing	Identify prospects most likely to respond to a direct marketing campaign (such as a direct mailing)
Fraud detection	Highlight transactions most likely to be deceptive or illegal
Market basket analysis	Identify products and services that are most commonly purchased at the same time (e.g., nail polish and lipstick)
Market segmentation	Group customers based on who they are or on what they prefer
Trend analysis	Analyze how key variables (e.g., sales, spending, promotions) vary over time

Business Intelligence

Closely linked to the concept of data mining is use of databases for business intelligence purposes. **Business intelligence (BI)** is the process of gathering enough of the right information in a timely manner and usable form and analyzing it so that it can have a positive impact on business strategy, tactics, or operations. Business intelligence turns data into useful information that is then distributed throughout an enterprise. Companies use this information to make improved strategic decisions about which markets to enter, how to select and manage key customer relationships, and how to select and effectively promote products to increase profitability and market share.

Today, a number of companies use the business intelligence approach. BankFinancial Corporation of Chicago uses it to help target promotions to bank customers.[29] Owens & Minor, a large medical supplies company, uses business intelligence software from Business Objects to analyze sales data.[30] According to Scott Wiener, chief technology officer at Certive Corporation, "Today, the most innovative business-intelligence technology is able to recommend the optimal course of action based on business rules, representing the first step in automated decision making."[31] Wiener predicts that business-intelligence software may eliminate the need for a large number of midlevel managers. Companies like Ben and Jerry's ice cream need to store and process huge amounts of data.[32] The company collects data on all 190,000 pints it produces in its factories each day, with all the data being shipped to the company's headquarters in Burlington, Vermont, which is a few miles from its first store that was opened more than 25 years ago. In the marketing department, the massive amount of data is analyzed. Using business-intelligence software, the company is able to cut costs and improve customer satisfaction. The software allows Ben and Jerry's to match the over 200 calls and e-mails received each week with ice cream products and supplies. Today, the company can quickly determine whether a bad batch of milk or eggs was used in production or whether sales of its Chocolate Chip Cookie Dough is gaining on the No. 1 selling Cherry Garcia. An insurance and financial services company can use business intelligence to gain an in-depth understanding of its customers—from the profits they generate, their rate of retention, and the opportunities they offer to cross-sell the company's products. Employees can obtain the data needed to zero in on problem areas, obtain a detailed picture of the profitability of any customer, and see which products are selling and which are not. If sales decline, they can track those declines to specific offices or even to individual sales reps to pinpoint problems and take immediate steps to remedy them.

Table 3.3

Common Data-Mining Applications

business intelligence (BI)
The process of gathering enough of the right information in a timely manner and usable form and analyzing it to have a positive impact on business strategy, tactics, or operations.

competitive intelligence
A continuous process involving the legal and ethical collection of information about competitors, its analysis, and controlled dissemination of information to decision makers.

Competitive intelligence is one aspect of business intelligence and is limited to information about competitors and the ways that knowledge affects strategy, tactics, and operations. Competitive intelligence is a critical part of a company's ability to see and respond quickly and appropriately to the changing marketplace. Competitive intelligence is not espionage—the use of illegal means to gather information. In fact, almost all the information a competitive intelligence professional needs can be collected by examining published information sources, conducting interviews, and using other legal, ethical methods. Using a variety of analytical tools, a skilled competitive intelligence professional can by deduction fill the gaps in information already gathered.

counterintelligence
The steps an organization takes to protect information sought by "hostile" intelligence gatherers.

The term **counterintelligence** describes the steps an organization takes to protect information sought by "hostile" intelligence gatherers. One of the most effective counterintelligence measures is to define "trade secret" information relevant to the company and control its dissemination.

knowledge management
The process of capturing a company's collective expertise wherever it resides—in computers, on paper, in people's heads—and distributing it wherever it can help produce the biggest payoff.

Knowledge management is the process of capturing a company's collective expertise wherever it resides—in computers, on paper, or in people's heads—and distributing it wherever it can help produce the biggest payoff. The goal of knowledge management is to get people to record knowledge (as opposed to data) and then share it. Although a variety of technologies can support it, knowledge management is really about changing people's behavior to make their experience and expertise available to others. Knowledge management had its start in large consulting firms and has expanded to nearly every industry. Pharmaceuticals, for example, must have access to various databases from different biotechnology companies to ensure they make informed decisions.

Distributed Databases

distributed database
A database in which the data can be spread across several smaller databases connected via telecommunications devices.

Distributed processing involves placing processing units at different locations and linking them via telecommunications equipment. A **distributed database**—a database in which the data can be spread across several smaller databases connected via telecommunications devices—works on much the same principle. A user in the Milwaukee branch of a clothing manufacturer, for example, might make a request for data that is physically located at corporate headquarters in Milan, Italy. The user does not have to know where the data is physically stored. The user makes a request for data, and the DBMS determines where the data is physically located and retrieves it (see Figure 3.18).

Distributed databases give corporations more flexibility in how databases are organized and used. Local offices can create, manage, and use their own databases, and people at other offices can access and share the data in the local databases. Giving local sites more direct access to frequently used data can improve organizational effectiveness and efficiency significantly.

Despite its advantages, distributed processing creates additional challenges in maintaining data security, accuracy, timeliness, and conformance to standards. Distributed databases allow more users direct access at different sites; thus, controlling who accesses and changes data is sometimes difficult. Also, because distributed databases rely on telecommunications lines to transport data, access to data can be slower.

replicated database
A database that holds a duplicate set of frequently used data.

To reduce telecommunications costs, some organizations build a replicated database. A **replicated database** holds a duplicate set of frequently used data. At the beginning of the day, the company sends a copy of important data to each distributed processing location. At the end of the day, the different sites send the changed data back to update the main database. This process, often called *data synchronization*, is used to ensure that replicated databases are accurate, up-to-date, and consistent with each other. A railroad, for example, can use a replicated database to increase punctuality, safety, and reliability. The primary database can hold data on fares, routings, and other essential information. The data can be continually replicated and downloaded on a read-only basis from the master database to hundreds of remote servers across the country. The remote locations can send back the latest figures on ticket sales and reservations to the main database.

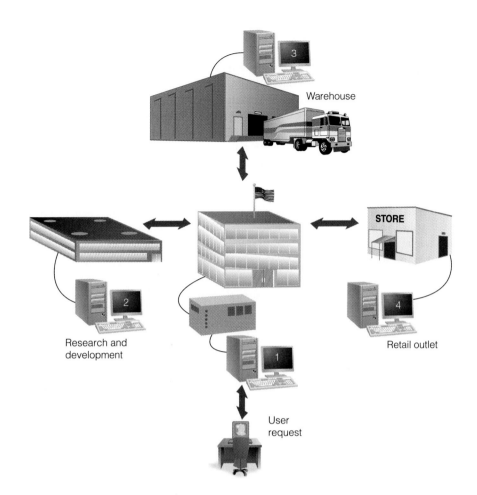

Figure 3.18

The Use of a Distributed Database

For a clothing manufacturer, computers might be located at corporate headquarters, in the research and development center, in the warehouse, and in a company-owned retail store. Telecommunications systems link the computers so that users at all locations can access the same distributed database no matter where the data is actually stored.

Online Analytical Processing (OLAP)

For nearly two decades, multidimensional databases and their analytical information display systems have provided flashy sales presentations and trade show demonstrations. All you have to do is ask where a certain product is selling well, for example, and a colorful table showing sales performance by region, product type, and time frame automatically pops up on the screen. Called **online analytical processing (OLAP)**, these programs are now being used to store and deliver data warehouse information efficiently.[33] OLAP allows users to explore corporate data from a number of different perspectives.

OLAP servers and desktop tools support high-speed analysis of data involving complex relationships, such as combinations of a company's products, regions, channels of distribution, reporting units, and time periods. Speed is essential as businesses grow and accumulate more and more data in their operational systems and data warehouses. Long popular with financial planners, OLAP is now being put in the hands of other professionals. The leading OLAP software vendors include Cognos, Comshare, Hyperion Solutions, Oracle, MineShare, WhiteLight, and Microsoft.

Consumer goods companies use OLAP to analyze the millions of consumer purchase records captured by scanners at the checkout stand. This data is used to spot trends in purchases and to relate sales volume to promotions and store conditions, such as displays and even the weather. OLAP tools let managers analyze business data using multiple dimensions, such as product, geography, time, and salesperson. The data in these dimensions, called *measures*, is generally aggregated—for example, total or average sales in dollars or units, or budget dollars or sales forecast numbers. Rarely is the data studied in its raw, unaggregated form. Each dimension also can contain some hierarchy. For example, in the time dimension, users can examine data by year, by quarter, by month, by week, and even by day. A geographic dimension might compile data from city, state, region, country, and even hemisphere.

online analytical processing (OLAP)
The software that allows users to explore data from a number of different perspectives.

The value of data ultimately lies in the decisions it enables. Powerful information-analysis tools in areas such as OLAP and data mining, when incorporated into a data warehousing architecture, bring market conditions into sharper focus and help organizations deliver greater competitive value. OLAP provides top-down, query-driven data analysis; data mining provides bottom-up, discovery-driven analysis. OLAP requires repetitive testing of user-originated theories; data mining requires no assumptions and instead identifies facts and conclusions based on patterns discovered. OLAP, or multidimensional analysis, requires a great deal of human ingenuity and interaction with the database to find information in the database. A user of a data-mining tool does not need to figure out what questions to ask; instead, the approach is, "Here's the data, tell me what interesting patterns emerge." For example, a data-mining tool in a credit card company's customer database can construct a profile of fraudulent activity from historical information. Then, this profile can be applied to all incoming transaction data to identify and stop fraudulent behavior, which might otherwise go undetected. Table 3.4 compares the OLAP and data-mining approaches to data analysis.

Table 3.4

Comparison of OLAP and Data Mining

Characteristic	OLAP	Data Mining
Purpose	Supports data analysis and decision making	Supports data analysis and decision making
Type of analysis supported	Top-down, query-driven data analysis	Bottom-up, discovery-driven data analysis
Skills required of user	Must be very knowledgeable of the data and its business context	Must trust in data-mining tools to uncover valid and worthwhile hypothesis

Object-Oriented and Object-Relational Database Management Syems

object-oriented database
The database that stores both data and its processing instructions.

An **object-oriented database** uses the same overall approach of object-oriented programming, first discussed in Chapter 2. With this approach, both the data and the processing instructions are stored in the database. For example, an object-oriented database could store both monthly expenses and the instructions needed to compute a monthly budget from those expenses. A traditional DBMS might only store the monthly expenses. In an object-oriented database, a *method* is a procedure or action. A sales tax method, for example, could be the procedure to compute the appropriate sales tax for an order or sale—for example, multiplying the total amount of an order by 5 percent, if that is the local sales tax. A *message* is a request to execute or run a method. For example, a salesclerk could issue a message to the object-oriented database to compute sales tax for a new order. Many object-oriented databases have their own query language, called *object query language (OQL)*, which is similar to SQL, discussed previously.

object-oriented database management system (OODBMS)
A group of programs that manipulate an object-oriented database and provide a user interface and connections to other application programs.

An object-oriented database uses an **object-oriented database management system (OODBMS)** to provide a user interface and connections to other programs. A number of computer vendors sell or lease OODBMSs, including eXcelon, Versant, Poet, and Objectivity. Object-oriented databases are used by a number of organizations. Versant's OODBMS, for example, is being used by companies in the telecommunications, financial services, transportation, and defense industries.[34] J.D. Edwards is using an object-oriented database to help its customers make fast and efficient forecasts of future sales and to determine whether they have enough materials and supplies to meet future demand for products and services.[35] With an object-oriented database, customers can quickly get a variety of reports on inventory and supplies. The *Object Data Standard* is a design standard by the *Object Database Management Group* (*www.odmg.org*) for developing object-oriented database systems.

An **object-relational database management system (ORDBMS)** provides a complete set of relational database capabilities plus the ability for third parties to add new data types and operations to the database. These new data types can be audio, images, unstructured text, spatial, or time series data that require new indexing, optimization, and retrieval features. Each of the vendors offering ORDBMS facilities provides a set of application programming interfaces to allow users to attach external data definitions and methods associated with those definitions into the database system. They are essentially offering a standard socket into which users can plug special instructions. DataBlades, Cartridges, and Extenders are the names applied by Oracle and IBM to describe the plug-ins to their respective products. Other plug-ins serve as interfaces to Web servers.

> **object-relational database management system (ORDBMS)**
> A DBMS capable of manipulating audio, video, and graphical data.

Visual, Audio, and Other Database Systems

In addition to raw data, organizations are increasingly finding a need to store large amounts of visual and audio signals in an organized fashion. Credit card companies, for example, input pictures of charge slips into an image database using a scanner. The images can be stored in the database and later sorted by customer, printed, and sent to customers along with their monthly statements. Image databases are also used by physicians to store x-rays and transmit them to clinics away from the main hospital. Financial services, insurance companies, and government branches are also using image databases to store vital records and replace paper documents. Music companies need the ability to store and manipulate sound from recording studios.

Visual databases can be stored in some object-relational databases or special-purpose database systems. Purdue University has developed an audio database and processing software to give singers a voice makeover.[36] The database software can correct pitch errors and modify voice patterns to introduce vibrato and other voice characteristics. According to the project director, "We look at the results from good singers and those of bad singers, and try to understand those differences." Drug companies often need the ability to analyze a large number of visual images from laboratories.[37] The PetroView database and analysis tool allows petroleum engineers to analyze geographic information to help them determine where to drill for oil and gas.[38] A visual-fingerprint database was used to solve a 40-year-old murder case in California.[39] The fingerprint database was a $640 million project started in 1995.

Combining and analyzing data from separate and totally different databases is an increasingly important database challenge. Global businesses, for example, sometimes need to analyze sales and accounting data stored around the world in different database systems. Companies such as IBM are developing *virtual database systems* to allow different databases to work together as a unified database system.[40] Using an IBM virtual database system, a Canadian bioresearch firm was able to integrate data from different databases that used different file formats and types. DiscoveryLink, one of IBM's projects, is allowing biomedical data from different sources to be integrated. According to Raimond Winslow, professor of biomedical engineering and computer science and director of the Center for Cardiovascular Bioinformatics and Modeling at Johns Hopkins University, "Information tools [such as DiscoveryLink] provide ways in which you can mine these huge data sets." The Centers for Disease Control (CDC) also has the problem of integrating different databases.[41] The CDC has more than 100 databases on various diseases. Searching these databases for data and information on diseases like SARS (severe acute respiratory syndrome) can be difficult.

In addition to visual, audio, and virtual databases, a number of other special-purpose database systems also exist.[42] *Spatial data technology* involves the use of a database to store and access data according to the locations it describes and to permit spatial queries and analysis. MapExtreme is spatial technology software from MapInfo that extends a user's database so it can store, manage, and manipulate location-based data. Police departments, for example, can use this type of software to bring together crime data and map the data visually so that patterns are easier to analyze. Police officers can select and work with spatial data at a specified location, within a rectangle, a given radius, or a polygon such as a precinct. For example, a police officer can request a list of all liquor stores within a two-mile radius of the precinct. Builders and insurance companies use spatial data to make decisions related to natural hazards. Spatial data can even be used to improve financial risk management with information stored by investment type, currency type, interest rates, and time.

Spatial data technology is used by NASA to store data from satellites and Earth stations. Location-specific information can be accessed and compared.

(Source: Courtesy of NASA.)

SUMMARY

Principle

The database approach to data management provides significant advantages over the traditional file-based approach.

Data is one of the most valuable resources a firm possesses. It is organized into a hierarchy that builds from the smallest element to the largest. The smallest element is the bit, a binary digit. A byte (a character such as a letter or numeric digit) is made up of eight bits. A group of characters, such as a name or number, is called a *field* (an object). A collection of related fields is a *record*; a collection of related records is called a *file*. The database, at the top of the hierarchy, is an integrated collection of records and files.

The traditional approach to data management has been from a file perspective. Separate files are created for each application, which can create problems over time: As more files are created for new applications, data that is common to the individual files becomes redundant. Also, if data is changed in one file, those changes might not be made to other files, reducing data integrity.

Traditional file-oriented applications are often characterized by program-data dependence, meaning that their data cannot be read by other programs. In contrast, benefits of the database approach include reduced data redundancy, improved data consistency and integrity, easier modification and updating, data and program independence, standardization of data access, and more efficient program development.

One of the tools database designers use to show the relationships among data is a data model. A data model is a map or diagram of entities and their relationships. Enterprise data modeling involves analyzing the data and information needs of an entire organization. Entity-relationship (ER) diagrams can be employed to show the relationships between entities in the organization.

The newest, most flexible structure is the relational model, in which data is set up in two-dimensional tables. Tables can be linked by common data elements, which are used to access data when the database is queried. Each row represents a record. Columns of the tables are called attributes, and allowable values for these attributes are called the *domain*. Basic data manipulations include selecting, projecting, and joining. The relational model is easier to control, more flexible, and more intuitive than the other models because it organizes data in tables.

Principle

A well-designed and well-managed database is an extremely valuable tool in supporting decision making.

A DBMS is a group of programs used as an interface between a database and its users and other application programs. When an application program requests data from the database, it follows a logical access path. The actual retrieval of the data follows a physical access path. Records can be considered in the same way: A logical record is what the record contains; a physical record is where the record is stored on storage devices. Schemas are used to describe the entire database, its record types, and their relationships to the DBMS.

A database management system provides four basic functions: providing user views, creating and modifying the database, storing and retrieving data, and manipulating data and generating reports. Schemas and subschemas are entered into the computer via a data definition language, which describes the data and relationships in a specific database. Subschemas are used to define a user view, the portion of the database a user can access and manipulate. Another tool used in database management is the data dictionary, which contains detailed descriptions of all data in the database.

After a DBMS has been installed, the database can be accessed, modified, and queried via a data manipulation language. A more specialized data manipulation language is the query language, the most common being Structured Query Language (SQL). SQL is used in several popular database packages today and can be installed on PCs and mainframes.

Selecting a database management system begins by analyzing the information needs of the organization. Important characteristics of databases include the size of the database, the number of concurrent users, the performance, the ability of the DBMS to integrate with other systems, the features of the DBMS, the vendor considerations, and the cost of the database management system.

Principle

The number and types of database applications will continue to evolve and yield real business benefits.

Traditional online transaction processing (OLTP) systems put data into databases very quickly, reliably, and efficiently, but they do not support the types of data analysis needed today. So, organizations are building data warehouses, which are relational database management systems specifically designed to support management decision making. Data marts are subdivisions of data warehouses that are commonly devoted to specific purposes or functional business areas.

Data mining, which is the automated discovery of patterns and relationships in a data warehouse, is emerging as a practical approach to predict future behavior. Predictive analysis is a form of data mining that combines historical data with assumptions about future conditions to forecast outcomes of events such as future product sales or the probability that a customer will default on a loan.

Business intelligence is the process of getting enough of the right information in a timely manner and usable form and analyzing it so that it can help form business strategy, create tactics, or modify operations. Competitive intelligence is one aspect of business intelligence limited to information about competitors and the ways that information affects strategy, tactics, and operations. Competitive intelligence is not espionage—illegally gathering information. Counterintelligence describes the steps an organization takes to protect information sought by "hostile" intelligence gatherers. Knowledge management is the process of capturing a company's collective expertise wherever it resides—in computers, on paper, or in people's heads—and distributing it wherever it can help produce the biggest payoff. The goal of knowledge management is to get people to record knowledge (as opposed to data) and then share it.

Distributed databases, which allow multiple users and different sites access to data that might be stored in different locations, are gaining in popularity. To reduce telecommunications costs, some organizations build replicated databases, which hold a duplicate set of frequently used data.

Multidimensional databases and online analytical processing (OLAP) programs are being used to store data and allow users to explore the data from a number of different perspectives.

An object-oriented database uses the same overall approach of object-oriented programming, first discussed in Chapter 2. With this approach, both the data and the processing instructions are stored in the database. An object-relational database management system (ORDBMS) provides a complete set of relational database capabilities, plus the ability for third parties to add new data types and operations to the database. These new data types can be audio, video, and graphical data that require new indexing, optimization, and retrieval features.

In addition to raw data, organizations are increasingly finding a need to store large amounts of visual and audio signals in an organized fashion. In addition, a number of special-purpose database systems are also available. Spatial data technology involves the use of an object-relational database to store and access data according to the locations it describes and to permit spatial queries and analysis.

CHAPTER 3: SELF-ASSESSMENT TEST

The database approach to data management provides significant advantages over the traditional file-based approach.

1. A group of programs that manipulate the database and provide an interface between the database and the user of the database and other application programs is called a(an) _____.

   a. GUI
   b. operating system

c. DBMS

d. productivity software

2. A(An) _____ has no relationship between its records and is often used to store and manipulate a single table or file.

3. A primary key is a field or set of fields that uniquely identifies the record. True or False?

4. The duplication of data in separate files is known as _____.

    a. data redundancy

    b. data integrity

    c. data relationships

    d. data entities

5. _____ is a data-modeling approach that starts by investigating the general data and information needs of the organization at the strategic level and then examining more specific data and information needs for the various functional areas and departments within the organization.

6. The most popular database model is _____.

    a. relational

    b. network

    c. normalized

    d. hierarchical

A well-designed and well-managed database is an extremely valuable tool in supporting decision making.

7. A(An)_____ is a highly skilled and trained systems professional who directs or performs all activities related to maintaining a successful database environment.

8. After data has been placed into a relational database, users can make inquiries and analyze data. Basic data manipulations include selecting, projecting, and optimization. True or False?

9. Because the DBMS is responsible for access to a database, one of the first steps in installing and using a database involves telling the DBMS the logical and physical structure of the data and relationships among the data in the database. This description is called a(an) _____.

10. The commands that are used to access and report information from the database are part of the _____.

    a. data definition language

    b. data manipulation language

    c. data normalization language

    d. subschema

11. Access is a popular DBMS for _____.

    a. personal computers

    b. graphics workstations

    c. mainframe computers

    d. supercomputers

12. The ability of a vendor to provide global support for large, multinational companies or companies outside the United States is becoming increasingly important. True or False?

The number and types of database applications will continue to evolve and yield real business benefits.

13. A(An) _____ holds business information from many sources in the enterprise, covering all aspects of the company's processes, products, and customers.

14. An information-analysis tool that involves the automated discovery of patterns and relationships in a data warehouse is called _____.

    a. a data mart

    b. data mining

    c. predictive analysis

    d. business intelligence

15. _____ is a continuous process involving the legal and ethical collection of information, analysis that doesn't avoid unwelcome conclusions, and controlled dissemination of that information to decision makers.

CHAPTER 3: SELF-ASSESSMENT TEST ANSWERS

(1) c (2) flat file (3) True (4) a (5) Enterprise data modeling (6) a (7) database administrator (8) False (9) schema (10) b (11) a (12) True (13) data warehouse (14) b (15) Competitive intelligence

REVIEW QUESTIONS

1. What is an attribute? How is it related to an entity?

2. Define the term *database*. How is it different from a database management system?

3. What is a flat file?

4. How would you describe the traditional approach to data management? How does it differ from the database approach?

5. What is data modeling? What is its purpose? Briefly describe three commonly used data models.

6. What is a database schema and what is its purpose?

7. Identify important characteristics in selecting a database management system.

8. What is the difference between a data definition language (DDL) and a data manipulation language (DML)?

9. What is a distributed database system?
10. What is a data warehouse, and how is it different from a traditional database used to support OLTP?
11. What is data mining? What is OLAP? How are they different?

12. What is an ORDBMS? What kind of data can it handle?
13. What is business intelligence? How is it used?
14. Give an example of a visual database.

DISCUSSION QUESTIONS

1. You have been selected to represent the student body on a project to develop a new student database for your school. What actions might you take to fulfill this responsibility to ensure that the project meets the needs of students and is successful?

2. Your company wants to increase revenues from its existing customers. How can data mining be used to accomplish this objective?

3. You are going to design a database for your cooking club to track its recipes. Identify the database characteristics most important to you in choosing a DBMS. Which of the database management systems described in this chapter would you choose? Why? Is it important for you to know what sort of computer the database will run on? Why or why not?

4. Make a list of the databases in which data about you exists. How is the data in each database captured? Who updates each database and how often? Is it possible for you to request a printout of the contents of your data record from each database? What data privacy concerns do you have?

5. You are the vice president of information technology for a large, multinational, consumer packaged goods company (e.g., Procter & Gamble, Unilever, or Gillette). You must make a presentation to persuade the board of directors to invest $5 million to establish a competitive intelligence organization—including people, data-gathering services, and software tools. What key points do you need to make in favor of this investment? What arguments can you anticipate others might make?

6. Briefly describe how visual and audio databases can be used by companies today.

7. Briefly discuss what impact data privacy legislation might have on the building and use of customer and employee data warehouses.

PROBLEM-SOLVING EXERCISES

1. Develop a simple data model for the members of a student club, where each row is a student. For each row, what attributes should you capture? What will be the unique key for the records in your database? Describe how you might use the database.

2. A video movie rental store is using a relational database to store information on movie rentals to answer customer questions. Each entry in the database contains the following items: Movie ID No. (primary key), Movie Title, Year Made, Movie Type, MPAA Rating, Number of Copies on Hand, and Quantity Owned. Movie types are comedy, family, drama, horror, science fiction, and western. MPAA ratings are G, PG, PG-13, R, X, and NR (not rated). Use an end-user database management system to build a data-entry screen to enter this data. Build a small database with at least ten entries.

3. To improve service to their customers, the salespeople at the video rental store have proposed a list of changes being considered for the database in the previous exercise. From this list, choose two database modifications and modify the data-entry screen to capture and store this new information.

Proposed changes:
   a. Add the date that the movie was first available to help locate the newest releases.
   b. Add the director's name.
   c. Add the names of three primary actors in the movie.
   d. Add a rating of one, two, three, or four stars.
   e. Add the number of Academy Award nominations.

TEAM ACTIVITIES

1. In a group of three or four classmates, interview a database administrator (DBA) for a company in your area. Describe this person's duties and responsibilities. What are the career opportunities of a DBA?

2. As a team of three or four classmates, interview business managers from three different businesses that use databases to help them in their work. What data entities and data attributes are contained in each database? How do they access the database to perform analysis? Have they received training in any query or reporting tools? What do they like about their database and what could be improved? Do any of them use data mining or OLAP techniques? Weighing the information obtained, select one of these databases as being most strategic for the firm and briefly present your selection and the rationale for the selection to the class.

3. Imagine that you and your classmates are a research team developing an improved process for evaluating auto loan applicants. The goal of the research is to predict which applicants will become delinquent or forfeit their loan. Those who score well on the application will be accepted;

those who score exceptionally well will be considered for lower-rate loans. Prepare a brief report for your instructor addressing these questions:

   a. What data do you need for each loan applicant?
   b. What data might you need that is not typically requested on a loan application form?
   c. Where might you get this data?
   d. Take a first cut at designing a database for this application. Using the chapter material on designing a database, show the logical structure of the relational tables for this proposed database. In your design, include the data attributes you believe are necessary for this database and show the primary keys in your tables. Keep the size of the fields and tables as small as possible to minimize required disk drive storage space. Fill in the database tables with the sample data for demonstration purposes (ten records). After your design is complete, implement it using a relational DBMS.

WEB EXERCISES

1. Use a Web search engine to find information on one of the following topics: business intelligence, knowledge management, predictive analysis. Find a definition of the term, an example of a company using the technology, and three companies that provide such software. Cut graphics and text material from the Web pages and paste them into a word processing document to create a two-page report on your selected topic. At the home page of each software

company, request further information from the company about its products.

2. Use a Web search engine to find three companies in an industry that is of interest to you that use a database management system. Describe how databases are used in each company. Could the companies survive without the use of a database management system? Why?

CAREER EXERCISE

1. For a career area of interest to you, describe three databases that could help you on the job.

2. How could you use data mining to help you make better decisions at work? Give five specific examples.

VIDEO QUESTIONS

Watch the video clip **Predicting Huge Surf** and answer these questions:

1. The Maverick's Surf Contest has been hailed by Sports Illustrated as "the Super Bowl of big wave

surfing." What role do databases play in Jeff Clark's ability to predict the best day and time for the contest, when the surf will be at its peak?

2. The reporter claims that the surf contest is sometimes called with only hours of notice. How, would you guess, does Jeff get the word out?

CASE STUDIES

Case One

Brazilian Grocer Gets Personal with Customers

Grupo Pão de Açúcar is Brazil's largest retail company, and its brand Pão de Açúcar (Portuguese for "Sugarloaf," Rio de Janeiro's famous mountain) is Brazil's most popular supermarket. Pão de Açúcar attributes its popularity to the personalized service it offers its millions of customers from its 433 stores. You may wonder how such a large company can possibly offer personalized service to so many customers. Pão de Açúcar found the solution in databases, data mining, and database management systems.

Through a study on customer loyalty, Pão de Açúcar found that customers today are going to fewer grocery stores than they used to and becoming more loyal to a specific store. Based on this information, Miriam Salomão, market-intelligence manager for Pão de Açúcar, set out to analyze customer purchasing data so that the chain could deliver more personalized services. However, the customer data being captured and stored in the company database was not customer specific. "Although we had a lot of data from our checkouts—detailed information by product, day, and store—this data didn't bring us a true understanding of our customers," says Salomão. "We needed to create a view of the data that consolidated each one of the tickets [customer receipts] and associated those tickets with people."

The store launched a customer relationship program that centered around a "loyalty card." Customers interested in receiving personalized discounts fill out a short registration form to receive a loyalty card, which is swiped at the checkout to apply additional discounts. Within no time, Pão de Açúcar had 1.73 million households signed up for the program. The loyalty card provided customer identification, a database key to group purchases by customer, and a foundation on which to build powerful information systems and services.

Pão de Açúcar called in database specialists to assist it with the project. The specialists implemented a data-mining solution to analyze customer activity data. They began by developing a data mart that organized the data so thatindividual transactions of each customer could be viewed.The data mart is used to create statistical models related to customer segmentation, consumption profiles, buying propensity, and other meaningful reports. This information provides valuable customer information and insight into their interests and needs.

The new program, formally named Mais, Portuguese for "More," allowed Pão de Açúcar to relate with customers individually. Customers receive a personalized mailing indicating sales on items that they have previously purchased or offers for new products related to those the customer uses. Mais members also receive special treatment and perks while shopping. A survey indicated that customers appreciate the personal recognition even more than the financial benefits. Miriam Salomão believes that getting to know customers as individuals has provided the store with a competitive advantage over its competition. "We know our customers, and we know they prefer a more personal, customized relationship. I'm sure they recognize us as a different store. This perception, although not easily measurable, is very important—maybe the most important perception of all."

Discussion Questions

1. What types of information can retailers such as Pão de Açúcar gather from checkout data that is not associated with a customer?
2. What additional information can be acquired when customer identity is associated with checkout data?

Critical Thinking Questions

3. What privacy issues come into play when customer identity is associated with checkout data?
4. What additional benefits and issues arise when customer data is combined from several different types of retail stores, such as the information associated with a typical credit card?

SOURCES: "Pão de Açúcar Making Millions of Customers Feel Like Family with SAS," *www.sas.com/success/paodeacucar.html*, accessed March 6, 2004; Grupo Pão de Açúcar Web site, *www.cbd-ri.com.br/eng/home/index.asp*, accessed March 6, 2004; SAS Data and Text Mining, *www.sas.com/technologies/analytics/datamining*, accessed March 6, 2004.

Case Two

DBMS Upgrade Faces Employee Opposition

Upgrading a database management system (DBMS) often leads to changes in business procedures, and change is sometimes met with resistance by the workforce. This was the result when Huntington Bancshares Inc., a $28 billion regional bank holding company based in Columbus, Ohio, moved to a new DBMS.

Prior to updating its system, Huntington Bancshares delivered a paper report, called the balance sheet income report, to its hundreds of offices every month. The report included 200,000 pages—the equivalent of 40 cartons of paper—of detailed financial information. The company decided it was time to move to a paperless, Web-based system to save on costs related to processing, printing, and delivering 2.4 million pages per year. Employees should celebrate such a decision, shouldn't they? Unfortunately, Huntington Bancshares employees had a different reaction. Here is a sample of their comments:

- "My manager says I have to have these [paper] reports for my file every month."
- "I'm not going to be able to do my job anymore."
- "I can't possibly ask my people to learn this. I'll have to do it for them every month."
- "You may have saved paper, but you have just doubled my workload."
- "Who made this decision?"

In response to the employee feedback, Raymond Heizer, IS project leader for corporate profitability systems at Huntington, says, "As far as our users were concerned, the sun came up every morning, and they got their balance sheet and income statement delivered to their desk every month end."

Despite employee opposition, Huntington Bancshares went ahead with the upgrade. It implemented an Oracle Corp. database running on a UNIX server and purchased a reporting system from Crystal Decisions Inc. The rollout took four months and cost a little more than $1 million. The bank is now saving $30,000 per year in paper costs alone, says Al Werner, vice president of corporate systems. But the return on investment has been realized in areas other than paper savings.

Cost center managers can view up-to-date balance sheet income reports immediately online, rather than waiting for the first of the month. They can also more easily see and resolve exceptions—items in an account balance that don't match the credits. "We've always had a lot of data in the bank, but it was always seen in rows and columns. Now we can have bar charts of mismatches," says Al Werner, vice president of corporate systems. Huntington now has a standard set of metrics on a single report that "each branch can use to determine the 'health' of his/her branch."

Huntington Bancshares' success in reducing its piles of paper has not been a universal experience, though. With the invention of the computer, many dreamed of a paperless society. In many cases, the opposite has been true. Computers generate amazing amounts of useful information, and printed reports are the norm in most corporations. Paper consumption by U.S. companies is growing 6 to 8 percent annually, according to document technology user group Xplor International in Torrance, California.

Still, new government programs and regulations are encouraging banks to go paperless. In June 2003, Congress passed the Check Clearing for the 21st Century Act, also known as Check 21, which allows banks to voluntarily exchange electronic images over networks instead of using paper checks. Huntington is riding the crest of a wave of banks trying to go paperless with online banking statements, to save money, and to comply with new regulations, says Avivah Litan, an analyst at Gartner Inc. "The trend started two years ago, but in the last nine months, it's really been moving ahead," Litan says.

As for the disgruntled Huntington Bancshares employees, they have stopped mourning the loss of their beloved paper reports now that they see the benefits of the database and reporting tools, Werner says. Now they're saying, "Wow, these are pretty nice features."

Discussion Questions

1. What are the benefits of viewing database reports on paper?
2. What are the benefits of being able to interact with the data online?

Critical Thinking Questions

3. Why do you think Huntington employees reacted so strongly to news of the new paperless system?
4. What will it take to turn businesses away from their dependency on paper?

SOURCES: Lucas Mearian, "Bank Tries to Break the Paper Habit," *Computerworld*, August 4, 2003, *www.computerworld.com*; Huntington Bancshares Inc. Web site, *www.huntngton.com*, accessed March 6, 2004; Lucas Mearian, "Huntington Bancshares Moves to AIX for Scalability," *Computerworld*, July 7, 2003, *www.computerworld.com*.

NOTES

Sources for the opening vignette: "The Secret of Their Success: DDB Uses SAS® to Explore Why, How Consumers Respond to Certain Brands," SAS Web site, *www.sas.com/success/ddb.html*, accessed March 5, 2004; DDB press kit, *www.ddb.com/5_media/downloads/presskit.pdf*, accessed

March 5, 2004; "Bill Burnbach Said...," *www.ddb.com/5_media/downloads/bb_quotes.pdf*, accessed March 5, 2004.

1. Binkley, Christina, "Soon, the Desk Clerk Will Know All About You," *The Wall Street Journal*, May 8, 2003, p. D4.
2. Fonseca, Brian, "DBA Boundaries Blurring," *eWeek,* January 26, 2004, p. 995.
3. Mossberg, Walter, "Microsoft's OneNote Turns Scribbled Ideas into Computer Files," *The Wall Street Journal,* October 30, 2003, p. B1.
4. McAmis, David, "Introducing InfoPath," *Intelligent Enterprise,* February 7, 2004, p. 36.
5. Perez, Jeanette, "Open-Source DBs Go Big Time," *Intelligent Enterprise,* February 7, 2004, p. 8.
6. Hall, Mark, "MySQL Breaks Into The Data Center," *Computerworld,* October 13, 2003, p. 32.
7. Whiting, Rick, "Open-Source Database Gaining," *InformationWeek,* January 12, 2004, p. 10.
8. Fonseca, Brian, "Database Opening Up," *eWeek,* January 12, 2004, p. 12.
9. Langley, Nick, "DB2 vies for Top Database Position," *Computer Weekly,* January 27, 2004, p. 40.
10. Whiting, Rick, "Sybase and IBM Ready Databases for Linux," *InformationWeek,* January 26, 2004, p. 49.
11. Whiting, Rick, "Sybase and IBM Ready Databases for Linux," *InformationWeek,* January 26, 2004, p. 49.
12. Voorhees, Mark, "Ready to Rumble," *The American Lawyer,* February 2004.
13. Arnott, Sarah, "Scottish Police Forces to Share Data," *Computing,* January 8, 2004, p. 8.
14. Staff, "Partners in Supply," *Design News,* January 12, 2004, p. 50.
15. Foley, John, "Blue Print for Change," *InformationWeek,* January 26, 2004, p. 22.
16. Whiting, Rick, "Stock Trades Get a Boost," *InformationWeek,* January 12, 2004, p. 47.
17. Robb, Drew, "The Database Diet," *Computerworld,* March 8, 2004, p. 32.
18. Staff, "Oracle and HP Set World Record," *VARBusines,* January 26, 2004, p. 64.
19. Thibodeau, Patrick, "The Web's Nest Leap," *Computerworld,* April 21, 2003, p. 34.
20. Loftus, Peter, "Smooth Talk," *The Wall Street Journal,* March 31, 2003, p. R9.
21. Fonseca, Brian, "ATC Database Upgrade Supports Growth, Improves Reliability," *eWeek,* January 26, 2004, p. 33.
22. Radding, Alan, "SSL Virtual Private Networks are Simpler to Set Up," *Computerworld,* April 28, 2003, p. 28.
23. Saran, Cliff, "Code Issue Affects 40% of Websites," *Computer Weekly,* January 13, 2004, p. 5.
24. Betts, Mitch, "Unexpected Insights," *Computerworld,* April 14, 2003, p. 34.
25. Anthes, Gary, "Best In Class: Data Warehouse Boosts Profits by Empowering Sales Force," *Computerworld,* February 24, 2003, p. 46.
26. Sullivan, Laurie, "Oracle Embraces RFID," *Information Week,* February 2, 2004, p. 8.
27. Verton, Dan, "FBI Begins Knowledge Management Face-Lift," *Computerworld,* April 21, 2003, p. 10.
28. Davis, Ann, "Data Collection Is Up Sharply Following 9/11," *The Wall Street Journal,* May 22, 2003, p. B1.
29. Anthes, Gary, "The Forrest Is Clear," *Computerworld,* April 14, 2003, p. 31.
30. Leon, Mark, "Keys to the Kingdom," *Computerworld,* April 14, 2003, p. 42.
31. Siener, Scott, "Total Automation," *Computerworld,* April 14, 2003, p. 52.
32. Scholsser, Julie, "Looking for Intelligence in Ice Cream," *Fortune,* March 17, 2003, p. 114.
33. Staff, "The State of Business Intelligence," *Computer Weekly,* February 3, 2004, p. 24.
34. Vaas, Lisa, "Tools Give Insights Into Databases," *eWeek,* January 27, 2003, p. 9.
35. Bacheldor, Beth, "Object-Oriented Database Speeds Queries," *InformationWeek,* March 10, 2003, p. 49.
36. Johnson, Colin, "Tech Gives Tone-Deaf a Voice Makeover," *Electronic Engineering Times,* May 5, 2003, p. 51.
37. Derra, Skip, "Image Analysis Software Shows Its Flexibility," *Drug Discovery and Development,* March 1, 2003, p. 61.
38. Staff, "Software System Allows Geographic Display, Analysis of Upstream Activity," *Offshore,* February 2003, p. 58.
39. Worthen, Ben, "Database Cracks Murder Case," *CIO Magazine,* May 1, 2003.
40. Vaas, Lisa, "Virtual Databases Make Sense Out of Our Varied Data," *eWeek,* March 31, 2003, p. 12.
41. Dignan, Larry, "Diagnosis Disconnected," *Baseline,* May 5, 2003.
42. Thomas, Daniel, "Online Age Verification," *Computer Weekly,* May 06, 2003, p. 14.

CHAPTER
· 4 ·

Telecommunications, the Internet, Intranets, and Extranets

PRINCIPLES	LEARNING OBJECTIVES
▪ Effective communications are essential to organizational success.	▪ Define the term *telecommunications* and describe the function of the components of a telecommunications system. ▪ Identify the three types of telecommunications carriers and discuss the services they provide. ▪ Name three distributed processing alternatives and outline their basic features.
▪ The Internet is like many other technologies—it provides a wide range of services, some of which are effective and practical for use today, others are still evolving, and still others will fade away from lack of use.	▪ Briefly describe how the Internet works, including alternatives for connecting to it and the role of Internet service providers.
▪ Originally developed as a document-management system, the World Wide Web is a menu-based system that is easy to use for personal and business applications.	▪ Describe the World Wide Web and the way it works, including the use of Web browsers, search engines, and other Web tools.
▪ Because the Internet and the World Wide Web are becoming more universally used and accepted for business, management, service and speed, privacy and security issues must continually be addressed and resolved.	▪ Identify and briefly describe the applications associated with the Internet and the Web. ▪ Define the terms *intranet* and *extranet* and discuss how organizations are using them. ▪ Identify several issues associated with the use of networks.

INFORMATION SYSTEMS IN THE GLOBAL ECONOMY
THE BARILLA GROUP, ITALY

Pasta Giant Uses Wireless Networking to Liberate Employees

In 1877, Pietro Barilla opened a bread and pasta bakery in Parma, Italy. Today, 128 years later, Pietro's three great-grandsons, Guido, Luca, and Paolo, control what has become Italy's largest food-processing company: Barilla. With 29 production plants distributed around the world, providing 1,265,000 tons of food products to more than 100 countries, Barilla is undisputedly the world's leading name in pasta. Barilla is a great family run, multinational corporation, squarely set on the shoulders of four generations of the Barilla family.

Barilla's business philosophy incorporates the metaphor of an iceberg: The visible part of a product is what is eaten, or directly perceived by the consumer. But like the tip of the iceberg, it is only a part of the entire product. The hidden part is enormous, made up of research, intuition, design, control, and study—the activities that are part of the Barilla Quality System. Barilla considers its human resources its greatest asset—the key to its success—and fosters employee talent and promotes their leadership with a managerial style based on integrity in decision making and conduct.

Recently, the American branch of Barilla constructed a new facility in Bannockburn, Illinois. Barilla designed a state-of-the-art communications and computer network system to improve information flow throughout the large company campus. The new network connects the two sprawling wings of the main facility and also a research kitchen located in a separate building a mile away.

With Barilla's emphasis on people, teamwork is an important component of its organizational culture, and the new facility's design supports teamwork with several "huddle" rooms, as well as a central area where people can hold discussions or planning sessions in a congenial, open atmosphere. The new computer network adds yet another benefit, freeing employees to travel wherever they are needed within the company campus through mobile technology.

Barilla America worked with Cisco Systems to implement a wireless computer network that spanned the entire campus. Devices called "access points," which send and receive communications signals through the air, were distributed about the organization and connected to the company's servers. Special wireless adapter cards were installed in employees' notebook and handheld computers so that wherever they roamed on the corporate campus, they would be connected to the network. Cisco also installed powerful long-range wireless connections to link the headquarters and the test kitchen. Today's technology provides top-notch speed in wireless networks for a fraction of the installation expense of wired networks.

Perhaps the most interesting of the new technologies implemented at Barilla are its Internet phones. They offer standard phone services, as well as additional services that interact with computer systems. Internet phones use the same technology as the regular Internet—the Internet Protocol (IP)—to communicate over the local data network. IP phones enable voice, video, and data transmissions to converge in one device. An additional bonus of IP phones is that they are less expensive than traditional leased lines and provide a company with full control over the telephone network.

The combination of wireless computing and IP phones provides Barilla employees the ability to access any information, including database reports from the corporate database server, e-mail, and the Web, and communicate with colleagues one at a time or in groups through voice or video from any location on the campus. Such flexibility frees Barilla employees to work and interact in any number of environments and settings. "People can sit in a huddle room or anywhere else, enter data, or send and receive e-mail," says Vince Danca, infrastructure manager. "I see people in our corporate 'living room' with laptops checking sales data, discussing trends, doing research. Everybody loves using wireless." Executives use wireless-equipped PDAs to move throughout the office while staying connected to the network. They can use the PDA to check e-mail or conduct calls. "I can take calls anywhere at head-quarters without running to the phone," says Danca. "It makes me and any-body else using the system immediately more responsive to customers and other callers."

Wireless networking has liberated Barilla employees from their desks and offices and allows them to work more naturally together. Productivity and employee morale is at an all-time high at Barilla's U.S. offices. It's easy to understand why so many businesses are moving to untether their employees through wireless networking.

As you read this chapter, consider the following:

- Having instant access to valuable information is the primary goal of an information system. How do computer networks support this goal?
- How are wireless networks affecting the way people do their jobs?

Why Learn About Telecommunications, the Internet, Intranet, and Extranets?

Today's decision makers need the ability to access data wherever it resides. They must be able to establish fast, reliable connections to exchange messages; download data, software, and updates; route transactions to processors; and send output to printers. Regardless of your chosen major or future career field, you will need the communications capabilities provided by telecommunications, the Internet, intranets, and extranets. Among all business functions, perhaps supply chain management emphasizes the use of telecommunications and networks the most because it encompasses inbound logistics, warehouse and storage, production, finished product storage, outbound logistics, marketing and sales, and customer service. All members of the supply chain must work together efficiently and effectively to increase the value perceived by the customer, so they must be able to communicate with key business partners as well. Other employees in managerial, human resource, finance, research and development, marketing, and sales positions must also use communications technology to communicate with people both inside and outside the organization. As a member of any organization, you must be able to take advantage of the capabilities that these technologies offer you to be successful. This chapter begins by discussing the importance of effective communications.

In today's high-speed business world, effective communication is critical to organizational success, as it is with The Barilla Group. Often, what separates good management from poor management is the ability to identify problems and solve them with available resources. Efficient communication is one of the most valuable of these resources because it enables a company to keep in touch with its operating divisions, customers, suppliers, and stockholders. For example, Wells Fargo & Company, the global financial services firm, is investing in a major upgrade of its online application called Commercial Electronic Office (CEO) to make the services easier to use. CEO is accessed over telecommunications networks by some

140,000 end users within Wells Fargo's corporate clients. They use the system to view account information and bank statements and to access money-transfer services and an automated clearinghouse that processes nearly $6 trillion in electronic payments annually. "We are an information-rich company, but there's so much information that it's sometimes hard to find what you need," said Steve Ellis, executive vice president of Wells Fargo's wholesale services group. "So we're grabbing it and pushing it up front to the person who needs it, so they can do their job faster."[1] Forward-thinking companies such as Wells Fargo hope to save billions of dollars, reduce time to market, and enable collaboration with their business partners through the use of telecommunications systems.

AN OVERVIEW OF TELECOMMUNICATIONS AND NETWORKS

Telecommunications refers to the electronic transmission of signals for communications, and it has the potential to create profound changes in business because it lessens the barriers of time and distance. Telecommunications not only is changing the way businesses operate but also is altering the nature of commerce itself. As networks are connected with one another and information is transmitted more freely, a competitive marketplace is making excellent quality and service imperative for success.

Figure 4.1 shows a general model of telecommunications. The model starts with a sending unit (1), such as a person, a computer system, a terminal, or another device, that originates the message. The sending unit transmits a signal (2) to a telecommunications device (3). The telecommunications device performs a number of functions, which can include converting the signal into a different form or from one type to another. A telecommunications device is a hardware component that allows electronic communication to occur or to occur more efficiently. The telecommunications device then sends the signal through a medium (4). A **telecommunications medium** is anything that carries an electronic signal and interfaces between a sending device and a receiving device. The signal is received by another telecommunications device (5) that is connected to the receiving computer (6). This chapter explores the components of the telecommunications model shown in Figure 4.1. An important characteristic of telecommunications is the speed at which information is transmitted, measured in bits per second (bps). Common speeds are in the range of thousands of bits per second (Kbps) to millions of bits per second (Mbps) and even billions of bits per second (Gbps).

telecommunications medium
Anything that carries an electronic signal and interfaces between a sending device and a receiving device.

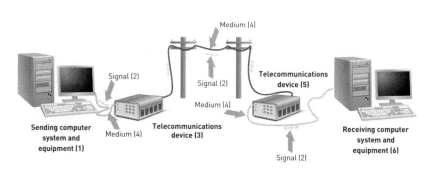

Figure 4.1

Elements of a Telecommunications System

Telecommunications devices relay signals between computer systems and transmission media.

Advances in telecommunications technology allow us to communicate rapidly with clients and coworkers almost anywhere in the world. Telecommunications also reduces the amount of time needed to transmit information that can drive and conclude business actions. A manufacturing sales representative, for example, can use telecommunications technology to get new product prices and availability information from the central sales office while working at a customer's location. This empowers the sales representative and often results in faster, higher-quality customer service. Telecommunications technology also helps businesses coordinate activities and integrate various departments to increase operational efficiency and support effective decision making. The far-reaching developments of telecommunications have a profound effect on business information systems and on society in general.

Telecommunications

The use of telecommunications can help businesses solve problems and maximize opportunities. Using telecommunications effectively requires careful analysis of telecommunications media, devices, and carriers and services.

Telecommunications technology enables businesspeople to communicate with coworkers and clients from remote locations.

(Source: © Corbis.)

Transmission Media

Various types of communications media are available. Each type exhibits its own characteristics, including cost, capacity, and speed. In developing a telecommunications system, the selection of media depends on the purpose of the overall information and organizational systems, the purpose of the telecommunications subsystems, and the characteristics of the media. As with other components, the media should be chosen to support the goals of the information and organizational systems at the least cost and to allow for possible modification of system goals over time. Transmission media can be divided into two broad categories: guided transmission, in which communications signals are guided along a solid medium, and wireless media, in which communications signals are sent over airwaves. Various media types are summarized in Table 4.1, and common types of wiring and cabling are shown in Figures 4.2a through c.

Figure 4.2

Common Wiring and Cable Types

(a) Twisted-pair wire cable (Source: Fred Bodin.) (b) Coaxial cable (Source: Fred Bodin.) (c) Fiber-optic cable (Source: © Greg Pease/Getty Images.)

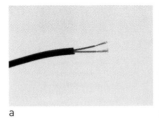

a

b

c

Guided Media Types			
Media Type	**Description**	**Advantages**	**Disadvantages**
Twisted-pair wire cable	Twisted pairs of copper wire, shielded or unshielded	Used for telephone service; widely available	Transmission speed and distance limitations
Coaxial cable	Inner conductor wire surrounded by insulation	Cleaner and faster data transmission than twisted-pair wire	More expensive than twisted-pair wire
Fiber-optic cable	Many extremely thin strands of glass bound together in a sheathing; uses light beams to transmit signals	Diameter of cable much smaller than coaxial; less distortion of signal; capable of high transmission rates	Expensive to purchase and install
Wireless Media Types			
Media Type	**Description**	**Advantages**	**Disadvantages**
Microwave	High-frequency radio signal sent through atmosphere and space (often involves use of communications satellites)	Avoids cost and effort to lay cable or wires; capable of high-speed transmission	Must have unobstructed line of sight between sender and receiver; signal highly susceptible to interception
Cellular	Divides coverage area into cells; each cell has mobile telephone subscriber office	Supports mobile users; costs are dropping	Signal highly susceptible to interception
Infrared	Signals sent through air as light waves	Devices can be moved, removed, and installed without expensive wiring	Must have unobstructed line of sight between sender and receiver; transmission effective only for short distances

Table 4.1

Transmission Media Types

Telecommunications Devices

A telecommunications device is a hardware device that allows electronic communication to occur or to occur more efficiently. Almost every telecommunications system uses one or more of these devices to transmit or receive signals. Table 4.2 summarizes some of the more common telecommunications devices.

Device	Function
Modem	Translates data from a digital form (as it is stored in the computer) into an analog signal that can be transmitted over ordinary telephone lines. This process is called *modulation*. Also performs a demodulation function to convert the analog signal received back into digital form.
Fax modem	Facsimile devices, commonly called *fax devices*, allow businesses to transmit text, graphs, photographs, and other digital files via standard telephone lines. A fax modem is a very popular device that combines a fax with a modem, giving users a powerful communications tool.
Multiplexer	Allows several telecommunications signals to be transmitted over a single communications medium at the same time, thus saving expensive long-distance communications costs.
PBX	A communications system that manages both voice and data transfer within a building and to outside lines. In a PBX system, switching equipment routes phone calls and messages within the building. PBXs can be used to connect hundreds of internal phone lines to a few phone company lines.

Table 4.2

Common Telecommunications Devices

Carriers and Services

Telecommunications carriers organize communications channels, networks, hardware, software, people, and business procedures to provide individuals and businesses with valuable communications services. The types of carriers can be divided into three broad categories: local exchange carriers, competitive local exchange carriers, and long-distance carriers.

local exchange carrier (LEC)
A public telephone company in the United States that provides service to homes and businesses within its defined geographical area called its *local access and transport area* (*LATA*).

Local Exchange Carriers   A **local exchange carrier** (**LEC, also called** *telco*) is a public telephone company in the United States that provides service to homes and businesses within its defined geographical area called its *local access and transport area* (*LATA*). Homes and businesses from within the LEC's LATA are connected to the local exchange via what is called a *local loop*—typically a pair of copper wires called a twisted pair (see Figure 4.3). This local loop is also called the "last mile." After the subscriber reaches the LEC, calls can be routed literally anyplace in the world using the telco's switching equipment.

Figure 4.3

Local Exchange Carriers

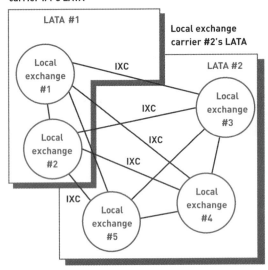

Each LEC has many local exchanges. Each local exchange (outside the United States, the term *public exchange* is used) has equipment that can switch calls locally to subscribers connected to the same local exchange, connect to other local exchanges within the same LATA, or connect to interexchange carriers (IXCs) that carry traffic between LECs such as long-distance carriers AT&T, MCI, and Sprint. The current rules for permitting a company to provide intraLATA or interLATA service (or both) are based on the Telecommunications Act of 1996.

competitive local exchange carrier (CLEC)
A company that is allowed to compete with the LECs, such as a wireless, satellite, or cable service provider.

Competitive Local Exchange Carriers   A **competitive local exchange carrier** (**CLECs**) is a company allowed to compete with the LECs. They include wireless service providers, satellite TV service providers, cable TV companies, and even power companies—the same power lines that bring electricity to our homes and businesses may become the next pathway for high-speed Internet access. This development could turn every power plug into a broadband connection. This technology is not yet commercially feasible. However, utility companies PPL Corporation in Allentown, Pennsylvania, and Ameren Corporation in St. Louis are conducting trial programs with consumers.[2] Cable companies have offered residential phone service since 1998. The number of subscribers is approaching 10 million.[3] The competitive local exchange carriers provide valuable backup capability over the "last mile." In many cases, they also offer faster and cheaper rates than the traditional local exchange carriers.

long-distance carrier
A traditional long-distance phone provider, such as AT&T, Sprint, or MCI.

Long-Distance Carriers   In the past few years, the three established **long-distance carriers**—AT&T, Sprint, and MCI—have lost market share under the provisions of the Telecommunications Act of 1996. The act allowed BellSouth, Qwest Communications, SBC Communications, and Verizon Communications to offer long-distance service within the regions they serve on a state-by-state basis after receiving approval from each state's regulators and the Federal Communications Commission (FCC). These formerly local companies now bundle their long-distance service with traditional local service in a comprehensive package that they advertise aggressively.

Telecommunications networks require state-of-the-art computer software technology to continuously monitor the flow of voice, data, and image transmission over billions of circuit miles worldwide.

(Source: © Roger Tully/Getty Images.)

Telecommunications carriers are providing more and more phone and dialing services to home and business users. Automatic number identification (ANI), or caller ID, equipment can be installed on a phone system to identify and display the number of an incoming call. In a business setting, ANI can be used to identify the caller and link that caller with information stored in a computer. For example, when a customer calls Federal Express, the customer service representative uses ANI to identify the name and address of the customer, thus saving time when handling a request for a pickup. Common carriers offer even more services to extend the capabilities of the typical phone system, such as intelligent dialing (when a busy signal is received, the phone redials the number when your line and the line of the party you are trying to reach are both free) and access codes to screen out junk calls, wrong numbers, and unwanted phone calls.

All the major cellular carriers offer some sort of wireless mobile data services. Cingular employs Global System for Mobile communication (GSM) technology, which is the de facto wireless telephone standard in Europe, with more than 120 million users worldwide in 120 countries. Because many GSM network operators have roaming agreements with foreign operators, users can often continue to use their mobile phones when they travel to other countries. Adoption of cellular data services is still in its early stages; many business users are taking a trial approach. Wireless data communications will be broadly adopted once providers can offer users enough bandwidth and connectivity to enable business users to use cellular as their sole connection. Table 4.3 lists some of the costs, advantages, and disadvantages of different lines and services offered by telecommunications carriers.

Genex Services is one of the nation's largest healthcare management service providers, with hundreds of case-management workers who visit homebound patients across the country to coordinate their care and relay information to employers, doctors, and insurance agencies. Before the caseworkers had an advanced telecommunications system, they would take handwritten notes during a patient visit or conversation with an insurance company or care provider, and later write the information up, make a conventional modem call, and upload their patient information. Genex decided to try wireless technology that employs PC cards that use the same communications protocol as cell phones. Caseworkers can now get online anywhere they can receive a cellular signal—and access patient information literally anywhere and anytime. The result is that data is more accurate and up to date, and the process is streamlined.[4]

Line/Service	Speed	Cost per Month	Advantages	Disadvantages
Standard phone service	56 Kbps	$10–$40	Low cost and broadly available	Too slow for video and downloads of large files
ISDN	64 Kbps–128 Kbps	$50–$150	Fast for video and other applications	Higher costs and not available everywhere
DSL	500 Kbps–1.544 Mbps	$20–$120 in addition to standard phone service	Fast, and the service comes over standard phone lines	Slightly higher costs and not available everywhere
Cable modem	500 Kbps–1.544 Mbps	$20–$120	Fast and uses existing cable that comes into the home	Slightly higher costs and not available everywhere
T1	1.544 Mbps	$600–$1,200	Very fast broadband service, typically used by corporations and universities	Very expensive, high installation fee, and users pay a monthly fee based on distance
Wireless data communications	70 Kbps–2Mbps	$35–$80	Provides network access for mobile worker	Areas of country may not have service yet

Networks

computer network
The communications media, devices, and software needed to connect two or more computer systems or devices.

A **computer network** consists of communications media, devices, and software needed to connect two or more computer systems or devices. The computers and devices on the networks are also called *network nodes*. After connecting, the nodes can share data, information, and processing jobs. More and more businesses are linking computers in networks to streamline work processes and allow employees to collaborate on projects. The effective use of networks can turn a company into an agile, powerful, and creative organization, giving it a long-term competitive advantage. Networks can be used to share hardware, programs, and databases across the organization. They can transmit and receive information to improve organizational effectiveness and efficiency. They enable geographically separated workgroups to share documents and opinions, which fosters teamwork, innovative ideas, and new business strategies. To take full advantage of networks, it is important to understand strategies, network concepts and considerations, network types, and related topics.

Basic Processing Strategies

centralized processing
The processing alternative in which all processing occurs in a single location or facility.

When an organization needs to use two or more computer systems, one of three basic processing strategies can be followed: centralized, decentralized, or distributed. With **centralized processing**, all processing occurs in a single location or facility. This approach offers the highest degree of control because all data processing is done on a single centrally managed computer. 7-Eleven has implemented a centralized processing strategy to manage its 5,800 stores. It uses a proprietary Retail Information System (RIS) that runs on a single mainframe computer and a centralized processing network to enable it to keep all its stores operating efficiently and to share information among all its suppliers. The RIS provides store managers with daily, weekly, and monthly sales tallies, which they use to create their orders. Store managers enter orders into workstations or handheld computers by 10 A.M. daily. By 11 A.M., orders have been transmitted to a central database, consolidated, and dispatched to 7-Eleven's suppliers. The centralized processing network also connects the stores to McLane Company, 7-Eleven's primary wholesale distributor, and to the commissaries and bakeries that provide fresh-food products so that all can view the same sales and shipment information.[5]

decentralized processing
The processing alternative in which processing devices are placed at various remote locations.

With **decentralized processing**, processing devices are placed at various remote locations. The individual computer systems are isolated and do not communicate with each other. Decentralized systems are suitable for companies that have independent operating divisions. Some drugstore chains, for example, operate each location as a completely separate entity; each store has its own computer system that works independently of the computers at other stores.

With **distributed processing**, computers are placed at remote locations but connected to each other via a network. One benefit of distributed processing is that processing activity can be allocated to the location(s) where it can most efficiently occur. Cooper Tire and Rubber Co.'s tire division plans to move more of its manufacturing offshore, with a goal of having its own manufacturing facility just outside Shanghai by 2007. Meanwhile, Cooper's tire division is implementing a two-year business plan that calls for establishing a global distributed processing network to connect the computers and decision makers at all its plants.[6]

The September 11, 2001, terrorist attacks sparked many companies to distribute their workers, operations, and systems much more widely, a reversal of the recent trend toward centralization. The goal is to minimize the consequences of a catastrophic event at one location while ensuring uninterrupted systems availability.

distributed processing
The processing alternative in which computers are placed at remote locations but connected to each other via a network.

Terminal-to-Host, File Server, and Client/Server Systems

If an organization chooses distributed information processing, it can connect computers in several ways, including terminal-to-host, file server, and client/server architectures. With **terminal-to-host** architecture, the application and database reside on one host computer, and the user interacts with the application and data using a "dumb" terminal. (Even if you use a PC to access the application, you run terminal emulation software on the PC to make it act as if it were a dumb terminal with no processing capacity.) Because a dumb terminal has no data-processing capability, all computations, data accessing and formatting, and data display are done by an application that runs on the host computer (see Figure 4.4).

terminal-to-host
An architecture in which the application and database reside on one host computer, and the user interacts with the application and data using a "dumb" terminal.

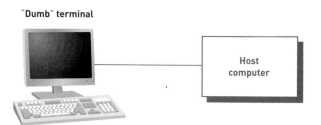

Figure 4.4

Terminal-to-Host Connection

In **file server** architecture, the application and database reside on the one host computer, called the *file server*. The database management system runs on the end user's personal computer or workstation. If the user needs even a small subset of the data that resides on the file server, the file server sends the user the entire file that contains the data requested, including a lot of data the user does not want or need. The downloaded data can then be analyzed, manipulated, formatted, and displayed by a program that runs on the user's personal computer (see Figure 4.5).

file server
An architecture in which the application and database reside on the one host computer, called the *file server*.

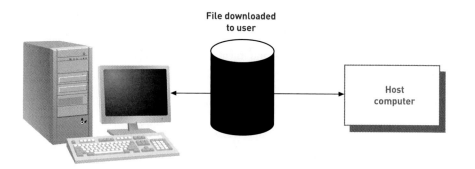

Figure 4.5

File Server Connection

The file server sends the user the entire file that contains the data requested. The downloaded data can then be analyzed, manipulated, formatted, and displayed by a program that runs on the user's personal computer.

client/server

An architecture in which multiple computer platforms are dedicated to special functions, such as database management, printing, communications, and program execution.

In **client/server** architecture, multiple computer platforms are dedicated to special functions, such as database management, printing, communications, and program execution. These platforms are called *servers*. Each server is accessible by all computers on the network. Servers can be computers of all sizes; they store both application programs and data files and are equipped with operating system software to manage the activities of the network. The server distributes programs and data files to the other computers (clients) on the network as they request them. An application server holds the programs and data files for a particular application, such as an inventory database. Processing can be done at the client or server. A client is any computer (often an end user's personal computer) that sends messages requesting services from the servers on the network. A user at a personal computer initiates a request to extract data that resides in a database somewhere on the network. A data request server intercepts the request and determines on which data server the data resides. The server then formats the user's request into a message that the database server will understand. Upon receipt of the message, the database server extracts and formats the requested data and sends the results to the client. Only the data needed to satisfy a specific query is sent—not the entire file (see Figure 4.6). As with the file server approach, after the downloaded data is on the user's machine, it can then be analyzed, manipulated, formatted, and displayed by a program that runs on the user's personal computer.

Figure 4.6

Client/Server Connection

Multiple computer platforms, called *servers*, are dedicated to special functions, such as database management, data storage, printing, communications, network security, and program execution. Each server is accessible by all computers on the network. A server distributes programs and data files to the other computers (clients) on the network as they request them. The client requests services from the servers, provides a user interface, and presents results to the user. After data is moved from a server to the client, the data can be processed on the client.

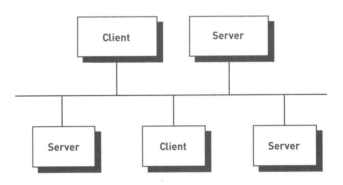

Network Types

Depending on the physical distance between nodes on a network and the communications and services provided by the network, networks can be classified as personal area, local area, metropolitan area, wide area, or international.

Personal Area Networks A **personal area network** (PAN) is a wireless network that supports the interconnection of information technology devices within a range of 33 feet or so. With a PAN, a person with a laptop, digital camera, and portable printer could connect them without having to hardwire anything. Digital image data could be downloaded from the camera to the laptop and then printed on a high-quality printer—all wirelessly.

personal area network (PAN)

A network that supports the interconnection of information technology within a range of 33 feet or so.

Local Area Networks A network that connects computer systems and devices within the same geographic area is a **local area network** (LAN). Typically, local area networks are wired into office buildings and factories (see Figure 4.7). They can be built to connect personal computers, laptop computers, or powerful mainframe computers.

local area network (LAN)

A network that connects computer systems and devices within the same geographic area.

Metropolitan Area Networks A **metropolitan area network** (MAN) is a telecommunications network that connects users and their computers in a geographical area larger than that covered by a LAN but smaller than the area covered by a WAN. Most MANs have a range of roughly 30 miles. An example of a MAN would be redefining the many networks within a city into a single larger network or connecting several LANs into a single campus LAN.

metropolitan area network (MAN)

A telecommunications network that connects users and their computers within a geographical area larger than that covered by a LAN, but smaller than the area covered by a WAN, such as a city or college campus.

Clark County, Nevada (which encompasses Las Vegas), is the nation's sixth-largest school district, with 289 schools. The county spent $15 million to build an IP-based metropolitan area network. An additional $16 million will be spent to outfit the district offices and every classroom with about 27,000 phone sets that can operate in both digital and IP modes. The MAN and the dual-mode phone system are designed to support the school system's explosive growth. The district, which serves 268,000 students and has 30,000 workers, is adding new schools at the rate of one per month.[7]

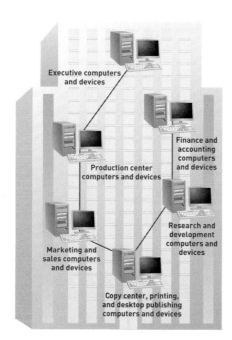

Figure 4.7

A Typical LAN

All network users within an office building can connect to each other's devices for rapid communication. For instance, a user in research and development could send a document from her computer to be printed at a printer located in the desktop publishing center.

Wide Area Networks A **wide area network (WAN)** is a telecommunications network that ties together large geographic regions. A WAN can be privately owned or rented and includes the use of public (shared users) networks. When you make a long-distance phone call or access the Internet, you are using a WAN. WANs usually consist of computer equipment owned by the user, together with data communications equipment and telecommunications links provided by various carriers and service providers (see Figure 4.8).

wide area network (WAN)
A network that ties together large geographic regions.

North America

Figure 4.8

A Wide Area Network

Wide area networks are the basic long-distance networks used by organizations and individuals around the world. The actual connections between sites, or nodes (shown by dashed lines), can be any combination of satellites, microwave, or cabling. When you make a long-distance telephone call or access the Internet, you are using a WAN.

International Networks Networks that link systems between countries are called **international networks.** However, international telecommunications comes with special problems. In addition to requiring sophisticated equipment and software, global area networks must meet specific national and international laws regulating the electronic flow of data across

international network
A network that links systems between countries.

international boundaries, often called *transborder data flow*. Some countries have strict laws limiting the use of telecommunications and databases, making normal business transactions such as payroll costly, slow, or even impossible.

Boehringer Ingelheim GmbH is a huge pharmaceutical manufacturer with $7.6 billion in revenues and 32,000 employees in 60 nations. The Ingelheim, Germany–based company is using an international network to consolidate data from its widespread operations and financial applications and present key financial information on a daily, weekly, or monthly basis. "I want to be told where I stand and where we are heading," says Boehringer's chief financial officer, Holger Huels. "I like to [be able to] see negative trends and counter them as fast as possible." In addition, Boehringer is now able to close its books for most of its divisions just two hours after the close of business at the end of each month.[8]

Communications Software and Protocols

Communications software provides a number of important functions in a network. Most communications software packages provide error checking and message formatting. In some cases, when there is a problem, the software can indicate what is wrong and suggest possible solutions. Communications software can also maintain a log listing all jobs and communications that have taken place over a specified period of time. In addition, data security and privacy techniques are built in to most packages.

The **network operating system (NOS)** controls the computer systems and devices on a network and allows them to communicate with each other. When network equipment (such as printers, plotters, and disk drives) is required, the NOS makes sure that these resources are correctly used.

With **network management software**, a manager can monitor the use of individual computers and shared hardware (such as printers), scan for viruses, and ensure compliance with software licenses. Some of the many benefits of network management software include fewer hours spent on routine tasks (such as installing new software), faster response to problems, and greater overall network control.

Communications protocols are rules and standards that make communications possible. A number of communications protocols are used by companies and organizations of all sizes. Just as standards are important in building computer and database systems, established protocols help ensure communications among computers of different types and from different manufacturers. Several common protocols are summarized in Table 4.4. Of special interest are some of the emerging protocols to support wireless transmission and wireless LANs, including Bluetooth and 802.11b (Wi-Fi). Wi-Fi allows faster transmissions than Bluetooth and can support connections across longer distances—up to 300 feet compared with just 30 feet for Bluetooth. However, Bluetooth chips are less expensive to make and consume less power than Wi-Fi chips. They are also easier to build in to small devices that run on batteries, such as cell phones and palmtop computers. In offering public Wi-Fi access at their restaurants, McDonald's, Schlotsky's Deli, and Starbucks hope that people will not only work, check e-mail, or chat online but also stay longer and spend more money or select them over a competitor.

In addition to using common communications protocols, various hardware devices are used to enable the high-speed switching of messages from one network to another. These devices are summarized in Table 4.5.

Now that you've learned the basics of telecommunications and networks, the following section discusses a network of global proportions—the Internet.

communications software
The software that provides a number of important functions in a network, such as error checking and data security.

network operating system (NOS)
The systems software that controls the computer systems and devices on a network and allows them to communicate with each other.

network management software
The software that enables a manager on a networked desktop to monitor the use of individual computers and shared hardware (like printers), scan for viruses, and ensure compliance with software licenses.

communications protocol
A standard set of rules that control a telecommunications connection.

Protocol	Description
Open Systems Interconnection (OSI)	A protocol that divides data communications functions into seven distinct layers to simplify the development, operation, and maintenance of complex telecommunications networks
Transmission Control Protocol/Internet Protocol (TCP/IP)	The primary telecommunications protocol of the Internet, developed in the 1970s
Systems Network Architecture (SNA)	The communications protocol used with IBM and IBM-compatible computers
IEEE 802.3 Ethernet	The telecommunications protocol often used with local area networks that ensures compatibility among devices so that many people can attach to a common communications media to share network facilities and resources
Frame relay	A packet-switching protocol for cost-efficient data transmission of traffic between LANs and end points in a WAN
Asynchronous Transfer Mode (ATM)	A switching technology that organizes data into 53-byte cells and can support transmission speeds of up to 10 Gbps
FireWire	The Apple Computer standard for connecting peripheral devices (printers, scanners, cameras, etc.) to the personal computer
Bluetooth	The telecommunications specification that describes how cellular phones, computers, faxes, personal digital assistants, printers, and other electronic devices can be connected wirelessly over short distances (up to 30 feet or so)
IEEE 802.11b (Wi-Fi)	A protocol used to establish a wireless LAN
IEEE 802.11g	A faster version of the Wi-Fi protocol, enabling data transmission at up to 54 Gbps
IEEE 802.16 (WiMax)	A protocol designed to support wireless MANs and be compatible with European standards
IEEE 802.20 (Mobile Broadband Wireless Access, MBWA)	A protocol that operates with existing cellular towers and promises the same coverage area as a mobile phone system with the speed of a Wi-Fi system

Table 4.4

Common Communications Protocols

Device	Function
Bridge	A device used to connect two or more networks that use the same communications protocol
Switch	A telecommunications device that routes incoming data from any one of many input ports to a specific output port that takes the data toward its intended destination
Router	A device or software in a computer that determines the next network point to which a data packet should be forwarded toward its destination
Hub	A place of convergence where data arrives from one or more directions and is forwarded out in one or more directions
Gateway	A network point that acts as an entrance to another network

Table 4.5

Network Switching Devices

USE AND FUNCTIONING OF THE INTERNET

Internet
A collection of interconnected networks, all freely exchanging information.

The Internet is the world's largest computer network. Actually, the **Internet** is a collection of interconnected networks, all freely exchanging information (see Figure 4.9). Research firms, colleges, and universities have long been part of the Internet, and now businesses, high schools, elementary schools, and other organizations are joining up as well. Nobody knows exactly how big the Internet is because it is a collection of separately run, smaller computer networks. There is no single place where all the connections are registered.

Figure 4.9

Routing Messages over the Internet

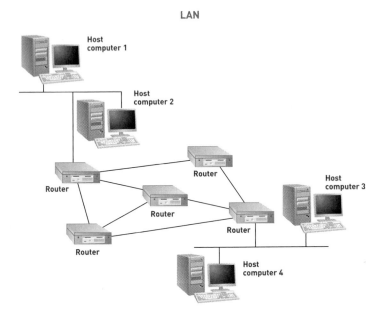

The ancestor of the Internet was **ARPANET**, a project started by the U.S. Department of Defense (DoD) in 1969. ARPANET was both an experiment in reliable networking and a means to link DoD and military research contractors, including a large number of universities doing military-funded research. (*ARPA* stands for the Advanced Research Projects Agency, the branch of the Defense Department in charge of awarding grant money. The agency is now known as DARPA—the added *D* is for *Defense*.) ARPANET was highly successful, and every university in the country wanted to sign up. This wildfire growth made it difficult to manage ARPANET, particularly the large and rapidly growing number of university sites on it. So, ARPANET was broken into two networks: MILNET, which included all military sites, and a new, smaller ARPANET, which included all the nonmilitary sites. The two networks remained connected, however, through use of the **Internet Protocol (IP)**, which enabled traffic to be routed from one network to another as needed. All the networks connected to the Internet speak IP, so they all can exchange messages.

ARPANET
A project started by the U.S. Department of Defense (DoD) in 1969 as both an experiment in reliable networking and a means to link DoD and military research contractors, including a large number of universities doing military-funded research.

Internet Protocol (IP)
The communications standard that enables traffic to be routed from one network to another as needed.

The Internet is increasingly going wireless. In addition to land-based systems, the Internet is also available at sea and in the air. It is now possible to go online while aboard a cruise ship. Some cruise lines are installing Internet appliances that allow crew members to access the Internet to get news and send and receive e-mails. Some airline companies are offering Internet service on their flights. Truck drivers use wireless Internet access from the cab of their trucks to check weather and shipments and to send and receive e-mail to and from family and friends.[9] Physicians use wireless Internet access to speed ordering prescriptions and recording patient treatments.[10] With this more accurate record keeping, the approach also has the potential to reduce errors and harmful interactions of different drugs prescribed to patients.

How the Internet Works

The Internet transmits data from one computer (called a *host*) to another (see Figure 4.9). If the receiving computer is on a network to which the first computer is directly connected, it can send the message directly. If the receiving computer is not on a network to which the sending computer is connected, the sending computer relays the message to another computer that can forward it. The message may be sent through a router to reach the forwarding computer. The forwarding host, which presumably is attached to at least one other network, in turn delivers the message directly if it can or passes it to yet another forwarding host. It is quite common for a message to pass through a dozen or more forwarders on its way from one part of the Internet to another.

The various networks that are linked to form the Internet work pretty much the same way—they pass data around in chunks called *packets*, each of which carries the addresses of its sender and its receiver. The set of conventions used to pass packets from one host to another is known as the Internet Protocol (IP). The best known is the **Transmission Control Protocol (TCP)**, which operates at the Transport layer. TCP is so widely used as the transport-layer protocol that many people refer to TCP/IP, the combination of TCP and IP used by most Internet applications. Adhering to the same technical standards allows the more than 100,000 individual computer networks owned by governments, universities, nonprofit groups, and companies to constitute the Internet. As soon as a network following these standards links to a **backbone**—one of the Internet's high-speed, long-distance communications links—it becomes part of the worldwide Internet community.

Each computer on the Internet has an assigned address called its **Uniform Resource Locator, or URL**, to identify it to other hosts. The URL gives those who provide information over the Internet a standard way to designate where Internet elements such as servers, documents, and newsgroups can be found. For example, take a look at the URL for Course Technology, *http://www.course.com*.

The "http" specifies the access method and tells your software to access this particular file using the Hypertext Transport Protocol. This is the primary method for interacting with the Internet.

The "www" part of the address signifies that the address is associated with the World Wide Web service discussed later. The "course.com" part of the address is the domain name that identifies the Internet host site. Domain names must adhere to strict rules. They always have at least two parts separated by dots (periods). For some Internet addresses, the rightmost part of the domain name is the country code (au for Australia, ca for Canada, dk for Denmark, fr for France, jp for Japan, etc.). Many Internet addresses have a code denoting affiliation categories (Table 4.6 contains a few popular categories). The leftmost part of the domain name identifies the host network or host provider, which might be the name of a university or business.

Transmission Control Protocol (TCP)
A widely used transport-layer protocol that is used in combination with IP by most Internet applications.

backbone
One of the Internet's high-speed, long-distance communications links.

Uniform Resource Locator (URL)
An assigned address on the Internet for each computer.

Affiliation ID	Affiliation
com	business organizations
edu	educational sites
gov	government sites
net	networking organizations
org	organizations

Table 4.6

U.S. Top-Level Domain Affiliations

There are hundreds of thousands of registered domain names. Some people, called *cybersquatters*, have registered domain names in the hope of selling the names to corporations or people at a later date. The domain name Business.com, for example, sold for $7.5 million. But some companies are fighting back, suing people who register domain names in hopes of trying to sell them to companies. Today, the Internet Corporation for Assigned Names and Numbers has the authority to resolve domain name disputes. Under new rules, if an address is found to be "confusingly similar" to a registered trademark, the owner of the domain name has no legitimate interest in the name. The rule was designed in part to prevent cybersquatters.

Accessing the Internet

There are numerous ways to connect to the Internet (see Figure 4.10), but Internet access is not distributed evenly throughout the world. See the "Ethical and Societal Issues" feature for a discussion of the challenges of global Internet access. Which access method is chosen is determined by the size and capability of the organization or individual.

Figure 4.10

Several Ways to Access the Internet

There are several ways to access the Internet, including using a LAN server, dialing in to the Internet using SLIP or PPP, or using an online service with Internet access.

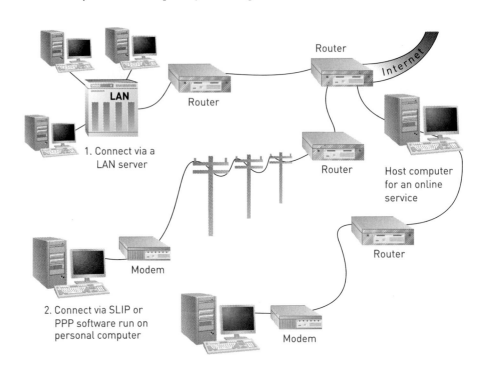

1. Connect via a LAN server

2. Connect via SLIP or PPP software run on personal computer

3. Connect via an online service

Connect via a LAN Server

This approach requires the user to install on his or her PC a network adapter card and Open Datalink Interface (ODI) or Network Driver Interface Specification (NDIS) packet drivers. These drivers allow multiple transport protocols to run on one network card simultaneously. LAN servers are typically connected to the Internet at 56 Kbps or faster. In addition, the higher cost of this service can be shared among several dozen LAN users to allow a reasonable cost per user. Additional costs associated with a LAN connection to the Internet include the cost of the software mentioned at the beginning of this section.

Connect via SLIP/PPP

Serial Line Internet Protocol (SLIP)

A communications protocol that transmits packets over telephone lines.

Point-to-Point Protocol (PPP)

A communications protocol that transmits packets over telephone lines.

This approach requires a modem and the TCP/IP protocol software plus **Serial Line Internet Protocol (SLIP)** or **Point-to-Point Protocol (PPP)** software. SLIP and PPP are two communications protocols that transmit packets over telephone lines, allowing dial-up access to the Internet. If you are running Windows, you will also need Winsock. Users must also have an Internet service provider that lets them dial into a SLIP/PPP server. SLIP/PPP accounts can be purchased for $30 a month or less from regional providers. With all this in place, a modem is used to call in to the SLIP/PPP server. After the connection is made, you are on the Internet and can access any of its resources. The costs include the cost of the modem and software, plus the service provider's charges for access to the SLIP/PPP server. The speed of this Internet connection is limited to the slower of your computer's modem and the speed of the modem of the SLIP/PPP server to which you connect.

Entrepreneurs Work to Lessen the Global Divide

Despite the global spread of information and communications technologies, large parts of the world remain technologically disconnected. The United States has more computers than the rest of the world combined, and when assessed by region, Internet use is dominated by North Americans. To develop a truly global economy, those who enjoy the benefits of technology need to help those less fortunate get connected—and they are!

Many entrepreneurs and business leaders are investing heavily to bring technology to remote corners of the world. In 2000, Martin Varsavsky, an Argentinean telecommunications entrepreneur based in Spain, established the Varsavsky Foundation, an organization committed to bringing tools of learning, from basic bricks-and-mortar classrooms to online cyberlibraries, to children of all ages throughout the world. The foundation further aims to fully integrate technology and the Internet into learning.

Perhaps the most famous entrepreneur turned philanthropist is Microsoft's Bill Gates. The Bill and Melinda Gates Foundation has provided funds to bring technologies to remote areas of the world, such as the snow-swept terrain of Canada's Northwest Territories. There they have set up a system of cyberlibraries in six remote communities to connect rural residents to the world, providing Internet access, training, and software.

A number of organizations have been established to work with business leaders to help others. The World Economic Forum is an independent international organization committed to improving the state of the world. The forum allows the world's leaders to collaborate to address global issues and promote global citizenship among its corporate members. In 2000, the World Economic Forum launched the Global Digital Divide Initiative (GDDI) to develop public and private partnerships to bridge the gap between those who have information and communication technology (ICT) access, skills, and resources and those who do not.

The Jordan Education Initiative, a 2003 GDDI program, brought together leaders from the information technology and telecom industries, such as Dell, HP, IBM, Intel, Microsoft, Siemens, Skillsoft, and Sun Microsystems. They worked with Jordanian authorities to improve education in the kingdom. Ninety-six "Discovery Schools" are serving as a test bed for how ICT can benefit schools and their pupils. Though focused on the advancement of learning in Jordan, the plan also stimulates the growth of the local information technology industry through infrastructure improvements and e-content development.

The digital divide occurs on many levels. While the World Economic Forum is working at the global level, the Aspira Association works on the national level in the United States. Aspira is a confederation of independent organizations throughout the United States and Puerto Rico. Aspira plans to use a $1.7 million grant to develop a national model for training Hispanic parents and students in information technology at more than 40 community technology centers across the country.

Several Latin American governments are realizing that technology may be a lifeline for their people in hard economic times. In Venezuela, the government has set up 243 "infocentres," offering free access to the World Wide Web in libraries, museums, city halls, and the offices of nongovernmental organizations, and it hopes to open 100 more. In 2002, the Mexican government launched the E-Mexico program, with the aim of installing 3,200 "digital community centers" in schools, community centers, city halls, libraries, and health clinics in rural villages and towns.

Despite these efforts, many still need assistance. Some cultures prefer to do without modern technological conveniences, which contradict religious beliefs and cultural values. In creating global relationships, business leaders must be sensitive to the beliefs and values of their international partners. Global business efforts and IS design should accommodate the needs of all and act as a bridge between cultures.

Critical Thinking Questions

1. How do businesses such as Microsoft benefit from their philanthropic investments?
2. What advantages do Internet-connected cultures have over those without information and communications technology?

What Would You Do?

Your global business has finally reached what many would consider a comfortable level of success. After reinvesting in the company and fulfilling all other financial obligations, your chief financial officer is looking for suggestions on how to invest $200,000. Some of your competitors have made big splashes in the media with what you've previously considered to be extravagant donations to charities. You wonder whether this might be an option for some or all of your remaining profits.

3. Out of all the needs in the world, what philanthropic venture would most interest you?
4. Would the type of industry and locations in which you work have an impact on how you invest?

SOURCES: The Digital Divide Network, *www.digitaldividenetwork.org*, accessed March 12, 2004; Mugo Macharia, "ASPIRA Brings Digital Opportunity to the Latino Community," *Digital Divide Network, www.digital dividenetwork.org*, accessed March 12, 2004; Humberto Márquez, "Telecentres to Narrow Digital Divide," *Terra Viva Online*, December 12, 2003, *www.ipsnews.net/focus/tv_society/viewstory.asp?idn=74*; "Jordan Education Initiative to Roll Out e-Learning across the Kingdom and Beyond," *World Economic Forum* press release, June 21, 2003, *www.weforum.com*; "Wired in the Wilderness: Providing Opportunities to NorthWest Territories," *Bill and Melinda Gates Foundation Story Gallery, www.gatesfoundation.org/Story Gallery/*, accessed March 12, 2004; The Varsavsky Foundation Web site, *www.varsavskyfoundation.org/*, accessed March 12, 2004.

Connect via an Online Service

This approach requires nothing more than what is required to connect to any of the online information services, such as a modem, standard communications software, and an online information service account. Increasingly, online services are offering DSL, satellite, and cable connection to the Internet, which provide faster speeds. There is normally a fixed monthly cost for basic services, including e-mail. Additional fees usually apply for DSL, satellite, or cable access, although these costs are falling. The online information services provide a wide range of services, including e-mail and the World Wide Web. America Online and Microsoft Network are examples of such services.

Other Ways to Connect

In addition to computers, many other devices can be connected to the Internet, including cell phones, PDAs, and home appliances. These devices also require specific protocols and approaches to connect. For example, *wireless application protocol (WAP)* is used to connect cell phones and other devices to the Internet.

Internet Service Providers

An **Internet service provider (ISP)** is any company that provides individuals and organizations with access to the Internet. ISPs do not offer the extended informational services offered by commercial online services such as America Online or EarthLink. There are literally thousands of Internet service providers, ranging from universities making unused communications line capacity available to students and faculty to major communications giants such as AT&T and MCI. To use this type of connection, you must have an account with the service provider and software that allows a direct link via TCP/IP.

Internet service provider (ISP)
Any company that provides individuals or organizations with access to the Internet.

To use an ISP such as MSN, you must have an account with the service provider and software that allows a direct link via TCP/IP.

In choosing an Internet service provider, users consider these important criteria: cost, reliability, security, availability of enhanced features, and the service provider's general reputation. Reliability is critical because if your connection to the ISP fails, it interrupts your communication with customers and suppliers. Among the value-added services ISPs provide are electronic commerce, networks to connect employees, networks to connect with business partners, host computers to establish your own Web site, Web transaction processing, network security and administration, and integration services. Many corporate IS managers welcome the chance to turn to ISPs for this wide range of services because they do not have the in-house expertise and cannot afford the time to develop such services from scratch. In addition, when organizations go with an ISP-hosted network, they can also tap the ISP's national infrastructure at minimum cost. That is important when a company has offices spread across the country.

In most cases, ISPs charge a monthly fee that can range from $15 to $30 for unlimited Internet connection through a standard modem. The fee normally includes e-mail. Some ISPs, however, are experimenting with low-fee or no-fee Internet access. But there are strings attached to the no-fee offers in most cases. Some free ISPs require that customers provide detailed demographic and personal information. In other cases, customers must put up with extra advertising. For example, a *pop-up ad* is a window that is displayed when someone visits a Web site. It pops up and advertises a product or service. Some e-commerce retailers have posted ads that resemble computer-warning messages and have been sued for deceptive advertising.[11] A *banner ad* appears as a banner or advertising window that you can ignore or click to go to the advertiser's Web site. Table 4.7 identifies several corporate Internet service providers.

Internet Service Provider	Web Address
AT&T's WorldNet Service	www.att.net
BellSouth	www.bellsouth.com
EarthLink	www.earthlink.net
Sprint	www.sprint.com

Table 4.7

A Representative List of Internet Service Providers

Many ISPs and online services offer broadband Internet access through digital subscriber lines (DSLs), cable, or satellite transmission. Most broadband users pay $50 or less per month for unlimited service. Broadband use has spread globally; more than 70 percent of households in South Korea have broadband access, while about 35 percent of Canadian households have broadband connections.[12] A study by the Pew & American Life Project reports that more than 40 percent of U.S. Web households have broadband connections.[13] Most other countries offer broadband, including Italy (Telecom Italia), Britain (BT Group), Switzerland (Swisscom AG), France (France Telecom), and many others.[14] Some businesses and universities use the very fast T1 lines to connect to the Internet.

THE WORLD WIDE WEB

The World Wide Web was developed by Tim Berners-Lee at CERN, the European Organization for Nuclear Research in Geneva. He originally conceived of it as an internal document-management system. This server can be located at *http://public.web.cern.ch/public.* From this modest beginning, the **World Wide Web** (the Web, WWW, or W3) has grown to a collection of tens of thousands of independently owned computers that work together as one in an Internet service. These computers, called *Web servers*, are scattered all over the world and contain every imaginable type of data. Thanks to the high-speed Internet circuits connecting them and some clever cross-indexing software, users are able to jump from one Web computer to another effortlessly—creating the illusion of using one big computer. Because of its ability to handle multimedia objects, including linking multimedia objects distributed on Web servers around the world, the Web has become the most popular means of information access on the Internet today.

The Web is a menu-based system that uses the client/server model. It organizes Internet resources throughout the world into a series of menu pages, or screens, that appear on your computer. Each Web server maintains pointers, or links, to data on the Internet and can retrieve that data. Web *plug-ins* can help provide additional features to standard Web sites. Macromedia's Flash and RealPlayer are examples of Web plug-ins.

World Wide Web (WWW, or W3)
A collection of tens of thousands of independently owned computers that work together as one in an Internet service.

home page

A cover page for a Web site that has graphics, titles, and text.

hypermedia

The tools that connect the data on Web pages, allowing users to access topics in whatever order they want.

Hypertext Markup Language (HTML)

The standard page description language for Web pages.

HTML tags

The codes that let the Web browser know how to format text–as a heading, as a list, or as body text–and whether images, sound, and other elements should be inserted.

Data can exist on the Web as ASCII characters, word processing files, audio files, graphic and video images, or any other sort of data that can be stored in a computer file. A Web site is like a magazine, with a cover page called a **home page** that has color graphics, titles, and text. All the highlighted type (sometimes underlined) is hypertext, which links the on-screen page to other documents or Web sites. **Hypermedia** connects the data on pages, allowing users to access topics in whatever order they want. As opposed to a regular document, which you read linearly, hypermedia documents are more flexible, letting you explore related documents at your own pace and navigate in any direction. For example, if a document mentions the Egyptian pharaohs, you can choose to see a picture of the pyramids, jump into a description of the building of the pyramids, and then jump back to the original document. Hypertext links are maintained using URLs.

A *Web portal* is an entry point or doorway to the Internet. Web portals include AOL, MSN, Yahoo!, and others. For example, some people use Yahoo.com as their Web portal, which means they have set Yahoo! as their starting point. When they enter the Internet, the Yahoo! Web site appears. You can use Yahoo! to search the Internet, send e-mail, get directions for a trip, buy products and services, find the address and phone number of friends or relatives, and more.

Hypertext Markup Language (HTML) is the standard page description language for Web pages. One way to think about HTML is as a set of highlighter pens in different colors that you use to mark up plain text to make it a Web page—red for the headings, yellow for bold, and so on. The **HTML tags** let the browser know how to format the text: as a heading, as a list, or as body text. HTML also tells whether images, sound, and other elements should be inserted. Users mark up a page by placing HTML tags before and after a word or words. Figure 4.11 shows a simple document and its corresponding HTML tags.

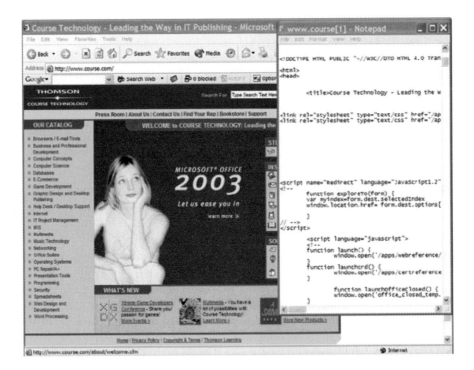

Figure 4.11

Sample Hypertext Markup Language

Shown at the left on the screen is a document, and at the right are the corresponding HTML tags.

A number of newer Web standards are gaining in popularity, including Extensible Markup Language (XML), Extensible Hypertext Markup Language (XHTML), Cascading Style Sheets (CSS), Dynamic HTML (DHMTL), and Wireless Markup Language (WML), which can display Web pages on small screens, such as smart phones and PDAs. XHTML is a combination of XML and HTML that has been approved by the World Wide Web Consortium (W3C).

Web Browsers

A **Web browser** creates a unique, hypermedia-based menu on your computer screen that provides a graphical interface to the Web. The menu consists of graphics, titles, and text with hypertext links. The hypermedia menu links you to Internet resources, including text documents, graphics, sound files, and newsgroup servers. As you choose an item or resource, or move from one document to another, you might be jumping between computers on the Internet without knowing it, while the Web handles all the connections. The beauty of Web browsers and the Web is that they make surfing the Internet fun. Just clicking with a mouse on a highlighted word or graphic whisks you effortlessly to computers halfway around the world. Most browsers offer basic features, such as support for backgrounds and tables, the ability to view a Web page's HTML source code, and a way to create hot lists of your favorite sites. Web browsers enable net surfers to view more complex graphics and 3-D models, as well as audio and video material. Microsoft Internet Explorer and Netscape are examples of Web browsers.

Web browser
The software that creates a unique, hypermedia-based menu on a computer screen, providing a graphical interface to the Web.

Search Engines

Looking for information on the Web is a little like browsing in a library—without the card catalog, it is extremely difficult to find information. Web search tools—called **search engines**—take the place of the card catalog. Most search engines, such as Yahoo.com and Google.com, are free. They make money by charging advertisers to put ad banners on their search engines. Companies often pay a search engine for a *sponsored link*, which is usually displayed at the top of the list of links for an Internet search.

search engine
A Web search tool.

AltaVista is a popular search engine on the Web.

The Web is a huge place, and it gets bigger with each passing day, so even the largest search engines do not index all Internet pages. Even if you do find a search site that suits you, your query might still miss the mark. So when searching the Web, you might want to try more than one search engine to expand the total number of potential Web sites of interest. In addition, searches can use words, such as AND and OR to refine the search. Searches can also use filters, such as displaying Web sites in English only. Filters limit searches to a language, certain file formats, a range of dates, and more. Some search engines also have a subject directory that allows people to get information on various industries and organizations.

Web Programming Languages

Java is an object-oriented programming language from Sun Microsystems based on the C++ programming language, which allows small programs—called *applets*—to be embedded within an HTML document. When the user clicks on the appropriate part of the HTML page to retrieve it from a Web server, the applet is automatically downloaded onto the client workstation, where it begins executing. This reduces the need for computer owners to install huge programs anytime they need a new function. Unlike other programs, Java software can run on any type of computer. Today, Java is being used on computers, cell phones, and a variety of other devices.[15] Cell phones with Java, for example, can give you a weather forecast, allow you to play games, get maps, find the nearest bathroom in some cities, and find which subway to take in London, England.

Java is used by programmers to make Web pages come alive, adding splashy graphics, animation, and real-time updates. Java-enabled Web pages are more interesting than plain Web pages. A financial services company, for example, can use Java to develop a Web-based financial system. The Web server that delivers the Java applet to the Web client is not capable of determining what kind of hardware or software environment the client is running on, and the developer who creates the Java applet does not want to worry about whether it will work correctly on Windows, UNIX, and Mac OS. Java is, thus, often described as a "cross-platform" programming language.

In addition to Java, companies use a variety of other programming languages and tools to develop Web sites. JavaScript, VBScript, and ActiveX (used with Internet Explorer) are Internet languages used to develop Web pages and perform important functions, such as accepting user input. *Hypertext Preprocessor*, or *PHP*, is an open-source programming language. Code or instructions from PHP can be embedded directly into HTML code. PHP's ability to run on different operating systems and database management systems, along with being an open-source language, has made it popular with many Web developers.

Developing Web Content

The art of Web design involves getting around the technical limitations of the Web and using a set of tools to make appealing designs. A number of products make developing and maintaining Web content easier. In short, these products can greatly simplify the creation of a Web page. Microsoft, for example, has introduced a Web development platform called .NET. The .NET platform allows different programming languages to be used and executed. It also includes a rich library of programming code to help build XML Web applications. After a Web site has been constructed, a *content management system (CMS)* can keep the Web site running smoothly. CMS consists of both software and support. Companies that provide CMS can charge from $15,000 to more than $500,000 annually, depending on the complexity of the Web site being maintained and the services being performed. Leading CMS vendors include BroadVision, Focumentum, EBT, FileNet, Open Market, and Vignette. Many of these products are popular with a newer approach to developing and maintaining Web content called *Web services*, discussed next.

Web Services

Web services consist of standards and tools that streamline and simplify communication among Web sites, promising to revolutionize the way we develop and use the Web for business and personal purposes. "Web Services is a tsunami of technology evolution," says Andre Mendes, chief technology officer at Public Broadcasting Service (PBS).[16] Mitsubishi Motors of North America uses Web services to link about 700 automotive dealers on the Internet. The Wells Fargo bank uses Web services to process electronic payments for its 100 largest corporate customers.[17] According to Steve Ellis, executive vice president of Wells Fargo, "Using Web services reduces [account setup] times by 30 to 50 percent for each new customer we add." Web services can also be used to develop new systems to send and receive secure messages between healthcare facilities, doctors, and patients while maintaining patient privacy.

Java
An object-oriented programming language from Sun Microsystems based on C++ that allows small programs (applets) to be embedded within an HTML document.

Web services
The standards and tools that streamline and simplify communication among Web sites for business and personal purposes.

Business Uses of the Web

Businesses everywhere are making large investments to present themselves and their products on the Web, to serve existing customers, and to find new ones. The Web can also cut costs. Aviall, an aviation parts company, invested more than $3 million in a new Web site, Aviall.com. The new Internet site cut ordering costs from $9 per order to $.39 per order.[18] Increasingly, small businesses are using the Internet to boost revenues and decrease costs.[19] As more and more people gain access to the Internet and World Wide Web, their functions are changing drastically. Several of these applications are discussed in the following section.

barnesandnoble.com enables users to buy books, music, and videos online.

INTERNET AND TELECOMMUNICATIONS SERVICES

A wide variety of services are available to help individuals and organizations tap the power of the Internet. Table 4.8 highlights a few of the most common services, which are discussed in the following sections.

E-Mail and Instant Messaging

E-mail has changed the way people communicate. It improves the efficiency of communications by reducing interruptions from the telephone and unscheduled personal contacts. Also, messages can be distributed to multiple recipients easily and quickly without the inconvenience and delay of scheduling meetings. Because past messages can be saved, they can be reviewed later. And because messages are received at a time convenient to the recipient, the recipient has time to respond more clearly and to the point. Many people have two or more e-mail addresses, including free e-mail services.[20]

For large organizations whose operations span a country or the world, e-mail allows people to work around time zone changes. Some users of e-mail estimate that they eliminate two hours of verbal communications for every hour of e-mail use.

The federal government and some states are now proposing legislation to block unwanted e-mail.[21] California, for example, has proposed legislation that would prevent companies from sending unsolicited, commercial e-mail to California residents.[22] The U.S. Congress passed a federal law, called Controlling the Assault of Non Solicited Pornography and

Table 4.8

Summary of Internet and Web Applications

Service	Description
E-mail	Enables you to send text, binary files, sound, and images to others.
Instant messaging	Allows two or more people to communicate instantly on the Internet.
Career information and job searches	Enables you to get up-to-date information on careers and actual jobs.
Web logs (blogs)	Allows people to create and use a Web site to write about their observations, experiences, and feelings on a wide range of topics.
Chat rooms	Enables two or more people to have online text conversations in real time.
Internet phone	Enables you to communicate with others around the world by linking Internet and traditional phone service.
Internet video conferencing	Supports simultaneous voice and visual communications.
Content streaming	Enables you to transfer multimedia files over the Internet so that the data stream of voice and pictures plays more or less continuously.
Shopping on the Web	Allows people to purchase products and services on the Internet.
Web auctions	Lets people bid on products and services.
Music, radio, and video on the Internet	Lets users play or download music, radio, and video.
Additional Internet services	Provides a variety of other services to individuals and companies.

Marketing Act (CAN SPAM), to reduce spam sent by companies in the United States.[23] Unfortunately, legitimate e-mail can get lost. An informal survey of about 10,000 individuals by a columnist for *Information Week* revealed that up to 40 percent of legitimate e-mails are not getting to their proper destinations.[24]

instant messaging
A method that allows two or more individuals to communicate online using the Internet.

Instant messaging is online, real-time communication between two or more people who are connected to the Internet. With instant messaging, two or more screens open up. Each screen displays what one person is typing. Because the typing is displayed on the screen in real time, it is like talking to someone using the keyboard.

A number of companies offer instant messaging, including America Online, Yahoo!, and Microsoft. America Online is one of the leaders in instant messaging, with about 40 million users of its Instant Messenger and about 50 million people using its client program ICQ. In addition to being able to type messages on a keyboard and have the information instantly displayed on the other person's screen, some instant messaging programs are allowing voice communication or connection to cell phones. A wireless service provider announced that it has developed a technology that can detect when a person's cell phone is turned on. With this technology, someone on the Internet can use instant messaging to communicate with someone on a cell phone anywhere in the world. Apple is experimenting with adding audio and video to its instant messaging service, called iChat.[25] The iSight camera can be used to transfer visual images through instant messaging.

Instant messaging services often use a *buddy list* that alerts people when their friends are also online. This feature makes instant messaging even more useful. Instant messaging is so popular that it helps Internet service providers and online services draw new customers and keep old ones.

Internet Cell Phones and Handheld Computers

Increasingly, cell phones, handheld computers, and other devices are being connected to the Internet. Some cell phones, for example, can be connected to the Internet to allow people to search for information, buy products, and chat with business associates and friends. A sales manager for a computer company can use her cell phone to check her company's Internet site to see whether there are enough desktop computers in inventory to fill a large order for an important customer. Using Short Message Service, people can send brief text messages of up to 160 characters between two or more cell phone users.

The service is often called *texting*. Some cell phones also come equipped with digital cameras, FM radios, video games, and small color screens to watch TV. Using multimedia messaging service (MMS), people can send pictures, video, and audio over cell phones to other cell phones or Internet sites. An insurance investigator can use MMS to send photos of a car accident to a central office to process an insurance claim. Of course, cell phones can also be used to send e-mail messages to others. Legislation passed in fall 2003 allows cell phone users to keep their phone numbers when switching to another cell phone company.

In addition to cell phones, handheld computers and other devices can also be connected to the Internet using phone lines or wireless connections, such as Wi-Fi. After connecting, these devices have full access to the Internet and all its applications, which are discussed in this chapter and throughout the book. Managers use handheld computers, such as the Black-Berry or Treo handheld computer, and the Internet to check business e-mail when they are out of the office; sales representatives use them to demonstrate products to customers, check product availability and pricing, and upload customer orders.

Career Information and Job Searching

The Internet is an excellent source of job-related information. People looking for their first job or seeking information about new job opportunities can find a wealth of information. Search engines, such as *www.google.com* and *www.yahoo.com*, can be a good starting point for searching for specific companies or industries. You can use a directory on Yahoo's home page, for example, to explore various industries and careers. Most medium or large companies have Internet sites that list open positions, salaries, benefits, and people to contact for further information. IBM's Web site, *www.ibm.com*, has a link to "Jobs at IBM." When you click on this link, you can get information on jobs with IBM around the world. Some Internet sites specialize in certain careers or industries. The site *www.directmarketingcareers.com* lists direct marketing jobs and careers, and the site *www.einsurancejobs.com* lists insurance jobs. Some sites can help people develop a good résumé and find a good job. They can also help you develop an effective cover letter for a résumé, prepare for a job interview, negotiate a good employment contract, and more. In addition, several Internet sites specialize in helping people get job information and even apply for jobs online, including *www.monster.com* and *www.careerbuilder.com*. You must be careful when applying for jobs online, however. Some bogus companies or Web sites will steal your identity by asking for personal information. People eager to get a job often give their Social Security number, birth date, and other personal information. The result can be no job, large bills on your credit card, and ruined credit.

Web Log (Blog)

A **Web log**, also called a **blog**, is a Web site that people can create and use to write about their observations, experiences, and feelings on a wide range of topics.[26] A *blogger* is a person who creates a blog, while *blogging* refers to the process of placing entries on a blog site.[27] A blog is like a journal. When people post information to a blog, it is placed at the top of the blog. Previous entries on the blog are pushed down. Blogs can be used by anyone or any organization to publish and share information. See the "Information Systems @ Work" feature to learn about practical uses of blogs.

Web log (blog)
A Web site that people can create and use to write about their observations, experiences, and feelings on a wide range of topics.

Inscene Embassy Changes Marketing Focus with Blogs

The Web log, or blog, has become a popular method of Internet communication—so popular, in fact, that many believe it has the potential to change the nature of the Web. Blogs empower regular Webizens (citizens of the Web) to publish their thoughts without any formal knowledge of HTML. Some forms of blogs are similar to usenet newsgroups designed for the Web. Most blog Web sites are centered around a particular topic and can be either one person's ongoing writing or assorted postings from a group of writers or the general public.

Blogs began as underground commentaries published by and for "Internet insiders"—those who spend a substantial portion of their lives online. These original blogs took the form of Web sites centered on the often radical viewpoints of intellectuals and malcontents. Within a few years, though, blogs were discovered by the general public through postings on Web sites that catered to special-interest groups. Blogs eventually made their way into mainstream culture when they were mentioned on network TV, in magazines, and on the radio.

The power of the blog caught the attention of the business community, especially marketers. Businesses are seriously examining the potential of blogs to gather and harness public opinion regarding their products. One of the boldest experiments in product-based blogs has been launched by Inscene Embassy. Inscene is a German fashion label that appeals to the teen market, with fashions that are casual, radical, and cross-cultural. Inscene has drafted young people in several cities around the world—including Tokyo, New York, London, and Berlin—to maintain blogs as cultural ambassadors for the brand.

The blogs create a sense of community among Inscene customers, who use the blog to share their thoughts on fashion and the latest Inscene merchandise. The blog provides Inscene with invaluable information on the way its target market thinks. The paid young people, who act as moderators for the blogs, guide the conversation and filter what is posted to keep the blogs clean of profanity and abuse.

Meg Hourihan, coauthor of the book *We Blog: Publishing Online with Weblogs*, sees a powerful commercial future for blogging. "Commercial Web sites aren't inherently better than personal ones, but they have business models and budgets. They have target audiences that can benefit from the type of focused content produced by bloggers. When a blogger is being paid to maintain a Weblog, he is able to do so full-time, with all his attention focused on the topic of choice."

Blogs are transforming the Web from a medium for only those with HTML knowledge and Web server access to a form of communication and expression for anyone with an Internet connection. Blogs have a unique power and appeal for those who become involved with them. Enthusiasm is contagious, and the ideas and energy generated by discussion on blogs, if harnessed, could be very useful to commercial enterprises.

Discussion Questions

1. How has the use of blogs changed the way marketers collect marketing data at Inscene Embassy?
2. Besides being a collection point for customer opinion, how do the blogs assist Inscene Embassy in selling clothes?

Critical Thinking Questions

3. If you were head of marketing at Inscene, what instructions would you give your paid bloggers to assist in building your product's reputation and maximizing the amount of valuable information you can collect from the blogs?
4. What dangers does Inscene Embassy expose itself to in supporting a public forum centered around its products?

SOURCES: James Lewin, "Learning from Blogs," *ITWorld*, December 17, 2003, *www.itworld.com*; Inscene Embassy Web site, *www.inscene-embassy.de/*, accessed March 20, 2004; Jonathan V. Last, "World Wide Dean," *The Daily Standard*, January 9, 2004, *www.lexis-nexis.com*.

Several Internet sites specialize in helping people get job information and even apply for jobs online.

Blogs exist in a wide variety of topics and areas. For example, the Western States Information Network (WSIN) developed a blog to allow local fire departments, water departments, and similar organizations to share information online.[28] E-mails between departments are posted on the blog. The blog provides a central location where all e-mails can be posted and read. Blogs are used by journalists, people in disaster areas, soldiers in the field, and people who just want to express themselves.[29] Venture capitalists can use *www.ventureblog.com* to investigate Internet or dot.com companies.[30] Blog sites, such as *www.blogger.com* and *www.globeofblogs.com* can include information and tools to help people create and use Web logs. Blogs are easy to set up. You can go to a blog service provider, such as *www.livejournal.com*, create a username and password, select a theme, choose a URL, follow any other instructions, and start making your first entry.

Chat Rooms

A **chat room** is a facility that enables two or more people to engage in interactive "conversations" over the Internet. When you participate in a chat room, dozens of people from around the world might be participating. Multiperson chats are usually organized around specific topics, and participants often adopt nicknames to maintain anonymity. One form of chat room, Internet Relay Chat (IRC), requires participants to type their conversation rather than speak. Voice chat is also an option, but you must have a microphone, sound card and speakers, fast modem or broadband, and voice-chat software compatible with the other participants'.

chat room
A facility that enables two or more people to engage in interactive "conversations" over the Internet.

Internet Phone and Videoconferencing Services

Internet phone service enables you to communicate with others around the world. This service is relatively inexpensive and can make sense for international calls. With some services, it is possible to make a call from someone using the Internet to someone using a standard phone. Cost is often a big factor in use of Internet phones—a call can be as low as 1 cent per minute for calls within the United States. Low rates are also available for calling outside the United States. In addition, voice mail and fax capabilities are available. Some cable TV companies, for example, are offering cable TV, phone service, and caller ID for under $40 a month.[31]

Using *voice-over-IP (VoIP)* technology, network managers can route phone calls and fax transmissions over the same network they use for data—which means no more separate phone bills.[32] Gateways installed at both ends of the communications link convert voice to IP packets and back. With the advent of widespread, low-cost Internet telephony services, traditional long-distance providers are being pushed to either respond in kind or trim their own long-distance rates. The school system in Appleton, Wisconsin, for example, installed a phone system based on the Internet and saved about 30 percent a year in telecommunications costs.[33]

What is especially interesting about VoIP is the way voice is being merged with video and data communications over the Web or a company's data network. In the long run, it's not the cost savings that will boost the market, it's the multimedia capabilities it gives us and the smart call-management capabilities. Travel agents could use voice and video over the Internet to discuss travel plans; Web merchants could use it to show merchandise and take orders; and customers could show suppliers problems with their products.

Internet videoconferencing, which supports both voice and visual communications, is another important Internet application. Microsoft NetMeeting, a utility within Windows, is an inexpensive and easy way for people to meet and communicate on the Web. The Internet can also be used to broadcast sales seminars using presentation software and videoconferencing equipment. These Internet presentations are often called *Webcasts* or *Webinars*. Hardware and software are needed to support videoconferencing. The key here is a video codec to convert visual images into a stream of digital bits and translate back again. The ideal video product will support multipoint conferencing, in which multiple users appear simultaneously on multiple screens.

Content Streaming

content streaming

A method for transferring multimedia files over the Internet so that the data stream of voice and pictures plays more or less continuously without a break, or very few of them; enables users to browse large files in real time.

Content streaming is a method for transferring multimedia files over the Internet so that the data stream of voice and pictures plays more or less continuously, without a break, or with very few of them. It also enables users to browse large files in real time. For example, rather than wait the half-hour it might take for an entire 5-MB video clip to download before they can play it, users can begin viewing a streamed video as it is being received.

Shopping on the Web

Shopping on the Web for books, clothes, cars, medications, and even medical advice can be convenient and easy. Some Internet shoppers are loyal to a few familiar Internet sites. This is good news for well-established and popular Internet sites. To add to their other conveniences, many Web sites offer free shipping and free pickup for returned items that don't fit or don't meet a customer's needs.[34] Table 4.9 shows the number of unique visitors to several Internet shopping sites.[35]

Table 4.9

Unique Visitors to Shopping Sites

(Source: Data from Mylene Mangalinsan, "Yahoo Hopes New Service Turns Searchers into Shoppers," *The Wall Street Journal*, September 23, 2003, p. B1.)

Company	Millions of Unique Visitors
eBay	42.4
Amazon	26.1
Yahoo! Shopping	15.1
DealTime	11.9
Wal-Mart Stores	9.2
Target	7.6
AOL Shopping	7.5
Bizrate.com	7.4
Sears	5.3
MSN Shopping	4.9

Increasingly, people are using bots to help them search for information or shop on the Internet. A **bot** is a software tool that searches the Web for information, products, or prices. A bot, short for *robot*, can find the best prices or features from multiple Web sites. Hotbot.com is an example of an Internet bot.

Web Auctions

A **Web auction** is a way to match people and companies who want to sell products and services with people and companies who want to buy products and services. Web auction sites are a place where businesses are growing their markets or reaching customers in a very low cost-per-transaction basis. Web auctions are transforming the customer-supplier relationship.

In addition to typical products and services, Internet auction sites excel at offering unique and hard-to-find items. Finding these items without an Internet auction site is often difficult, time-consuming, and expensive. The business-to-business application of auction sites is expected to continue. Almost anything you may want to buy or sell can be found on auction sites. One of the most popular auction sites is eBay, which often has millions of auctions occurring at the same time. Traditional companies are even starting their own auction sites.

Although auction Web sites are excellent for matching buyers and sellers, there can be problems with their use. Auction sites on the Web are not always able to determine whether products and services listed by people and companies are legitimate. In one case, a person posted items he didn't own on an auction site and stole $120,000 from auction users.[36] The person who stole the money, who is now in jail, said, "I never knew how easy it was to manipulate people. It was like taking candy from a baby." In addition, some Web sites have had illegal or questionable items offered. Many Web sites have an aggressive fraud investigation system to prevent and help prosecute fraudulent use of their sites. Even with these potential problems, the use of Web auction sites is expected to continue to grow rapidly.

Music, Radio, and Video on the Internet

Music, radio, and video are hot growth areas on the Internet. Audio and video programs can be played on the Internet, or files can be downloaded for later use. Using music players and music formats such as MP3, it is possible to download music from the Internet and listen to it anywhere using small, portable music players.

The Recording Industry Association of America (RIAA) won a legal battle against Napster, preventing the company from allowing free copies of music to be shared over the Internet. More recently, the RIAA has been filing lawsuits against individuals who download music illegally.[37] But a number of companies now offer music downloads for minimal fees, usually under a dollar per song. The Internet is also being used to form music collaborations that would be difficult or impossible otherwise, allowing musicians to record music from long distance.

Radio broadcasts are now available on the Internet. Entire audio books can also be downloaded for later listening, using devices such as the Audible Mobile Player. This technology is similar to the popular books-on-tape media, except you don't need a cassette tape or a tape player.

Other Internet Services and Applications

Other Internet services are constantly emerging. For example, the Internet facilitates distance learning, which has dramatically increased in the last several years. Many colleges and universities now allow students to take courses without ever visiting campus. In fact, you may be taking this course online. Businesses are also taking advantage of distance learning through the Internet. Video cameras can be attached to computers and connected to the Internet.[38] Internet cameras can be used to conduct job interviews, hold group meetings with people around the world, monitor young children at daycare centers, check rental properties and second homes from a distance, and more. People can use the Internet to connect with friends or others with similar interests.[39] Internet sites, such as

bot
A software tool that searches the Web for information, products, prices, and so forth.

Web auction
An Internet site that matches people who want to sell products and services with people who want to purchase those products and services.

ZeroDegrees.com, Tribe.net, and Spoke.com, are examples of social networking Internet sites. Today, manufacturers of sound systems are putting Internet and network capabilities into their devices.[40] For example, high-end Marantz video projectors and McIntosh stereo amplifiers are able to download music from the Internet. Other devices for home use allow people to view photos and see movies throughout a house using wireless networks and the Internet.[41] Gateway and Apex Digital, for example, produce devices that use wireless networks and the Internet to transfer audio and video files from a PC to a TV and back.

The following section discusses the use of corporate intranets and extranets, special applications of Internet technology that are widely used today.

INTRANETS AND EXTRANETS

intranet
An internal corporate network built using Internet and World Wide Web standards and products; used by employees to gain access to corporate information.

An **intranet** is an internal corporate network built using Internet and World Wide Web standards and products. Employees of an organization use it to gain access to corporate information. After getting their feet wet with public Web sites that promote company products and services, corporations are seizing the Web as a swift way to streamline—even transform—their organizations. These private networks use the infrastructure and standards of the Internet and the World Wide Web. A big advantage of using an intranet is that many people are already familiar with Internet technology, so they need little training to make effective use of their corporate intranet.

An intranet is an inexpensive, yet powerful alternative to other forms of internal communication, including conventional computer setups. One of an intranet's most obvious virtues is its ability to slash the need for paper. Because Web browsers run on any type of computer, the same electronic information can be viewed by any employee. That means that all sorts of documents (such as internal phone books, procedure manuals, training manuals, and requisition forms) can be inexpensively converted to electronic form on the Web and be constantly updated. An intranet provides employees with an easy and intuitive approach to accessing information that was previously difficult to obtain. For example, it is an ideal solution to providing information to a mobile sales force that needs access to rapidly changing information. Intranets can also do something far more important. The Bank of New Zealand allows its 16 business units to develop and manage content for its intranet.[42] According to a company spokesperson, "They've never had the ability to have that sense of ownership of the content that we can now give to the business units." The Scottish Water Utility used the same approach with its intranet, which replaced three legacy systems.[43] "Our intranet ownership was held by the internal communications team, and we wanted to get it out into the business to provide a tool for people to share communications across the organization," says Linda Fay, the digital media team leader for Scottish Water.

extranet
A network based on Web technologies that links selected resources of a company's intranet with its customers, suppliers, or other business partners.

A rapidly growing number of companies offer limited network access to selected customers and suppliers. Such networks are referred to as *extranets*, which connect people who are external to the company. An **extranet** is a network that links selected resources of the intranet of a company with its customers, suppliers, or other business partners. Again, an extranet is built around Web technologies.

Security and performance concerns are different for an extranet than for a Web site or network-based intranet. User authentication and privacy are critical on an extranet so that information is protected. Obviously, performance must be good to provide quick response to customers and suppliers. Table 4.10 summarizes the differences between users of the Internet, intranets, and extranets.

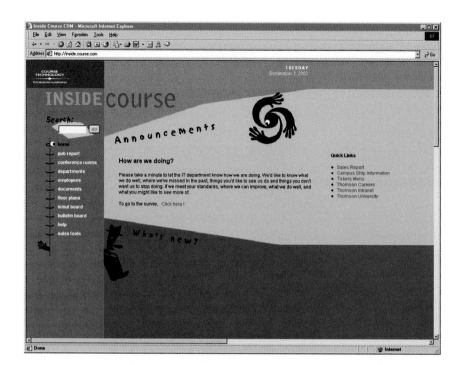

An intranet is an internal corporate network used by employees to gain access to company information.

Type	Users	Need for User ID and Password
Internet	Anyone	No
Intranet	Employees and managers	Yes
Extranet	Business partners	Yes

Table 4.10

Summary of Internet, Intranet, and Extranet Users

Secure intranet and extranet access applications usually require the use of a virtual private network (VPN). A **virtual private network** (**VPN**) is a secure connection between two points across the Internet. VPNs transfer information by encapsulating traffic in IP packets and sending the packets over the Internet, a practice called **tunneling**. Most VPNs are built and run by Internet service providers. Companies that use a VPN from an Internet service provider have essentially outsourced their networks to save money on wide area network equipment and personnel. In using a VPN, a user sends data from his or her personal computer to the company's firewall (discussed later), which also converts the data into a coded form that cannot be easily read by an interceptor. The coded data is then sent via an access line to the company's Internet service provider. From there, the data is transmitted through tunnels across the Internet to the recipient's Internet service provider and then over an access line to the receiving company's firewall, where it is decoded and sent to the receiver's personal computer (see Figure 4.12).

virtual private network (VPN)
A secure connection between two points across the Internet.

tunneling
The process by which VPNs transfer information by encapsulating traffic in IP packets over the Internet.

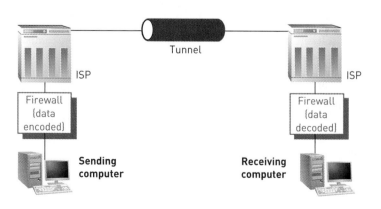

Figure 4.12

Virtual Private Network

NET ISSUES

The topics raised in this chapter apply not only to the Internet and intranets, but also to LANs, private WANS, and every type of network. Control, access, hardware, and security issues affect all networks, so it is important to mention some of these issues.

Management Issues

Although the Internet is a huge, global network, it is managed at the local level; no centralized governing body controls the Internet. Although the U.S. federal government provided much of the early direction and funding for the Internet, the government does not own or manage it. The Internet Society and the Internet Activities Board (IAB) are the closest the Internet has to a centralized governing body. These societies were formed to foster the continued growth of the Internet. The IAB oversees a number of task forces and committees that deal with Internet issues. One of the main functions of the IAB is to manage the network protocols used by the Internet, including TCP/IP. Some universities and government agencies are investigating how the Internet can be controlled to prevent sensitive information and pornographic material from being placed on it.

Service and Speed Issues

Service and speed issues on the Internet are a function of the increasing volume of traffic and more sophisticated Web sites. The growth in Internet traffic continues to be phenomenal. Traffic volume on company intranets is growing even faster than the Internet. Companies setting up an Internet or intranet Web site often underestimate the amount of computing power and communications capacity they need to serve all the "hits" (requests for pages) they get from Web cruisers. Web server computers can be overwhelmed with thousands of hits per hour. In addition, Web sites are becoming more sophisticated with video and audio clips and other features that require faster Internet speeds.

Privacy, Fraud, and Security

As the use of the Internet grows, privacy, fraud, and security issues become even more important. People and companies are reluctant to embrace the Internet unless these issues are successfully addressed.

Privacy

cookie
A text file that an Internet company can place on the hard disk of a computer system to track user movements.

From a consumer perspective, the protection of individual privacy is essential. Yet, many people use the Internet without realizing that their privacy might be in jeopardy. Spyware can hijack your browser, generate pop-up ads, and report your activities to someone else over the Internet. *Spyware* consists of hidden files and information trackers that install themselves secretly when you visit some Internet sites. Many Internet sites use cookies to gather information about people who visit their sites. A **cookie** is a text file that an Internet company can place on the hard disk of a computer system. These text files keep track of visits to the site and the actions people take. To help prevent this potential problem, some companies are developing software to prevent these files from being placed on computer systems. CookieCop, for example, allows Internet users to accept or reject cookies by Internet site. Microsoft Internet Explorer 6 also has the ability to screen Web sites according to their privacy policy. Using the Platform for Privacy Preferences (P3P), Internet Explorer 6 can summarize the privacy policy for Web sites and prevent information from being transmitted from your computer to a Web site that doesn't meet certain criteria. In addition, preferences in Internet browsers can be set to restrict the use of cookies. Many browsers also allow users to easily delete cookies.

Fraud

Internet fraud is another important issue. Some people have received false messages that seem to be from their Internet service providers asking them to update their personal information, including Social Security numbers and credit card information. But instead of going to the Internet service provider, the information is captured and used by online thieves. This type of Internet activity, often called *phishing*, is becoming more prevalent and dangerous. The possibility of Internet fraud has prevented many people from using the Internet. Local, state, and federal agencies are actively pursuing Internet fraud. As law enforcement agencies crack down on fraud, public confidence in using the Internet should increase.

Security with Encryption and Firewalls

When it comes to security on the Internet, it is essential to remember two things. First, there is no such thing as absolute security. Second, plenty of clever people consider it great sport to try to breach any security measures—the better your security, the greater the challenge to them.

From a corporate strategy perspective, security of data is essential. Such approaches as cryptography can help. **Cryptography** is the process of converting a message into a secret code and changing the encoded message back to regular text. The original conversion is called **encryption**. The unencoded message is called *plaintext*. The encoded message is called *ciphertext*. Decryption converts ciphertext back into plaintext (see Figure 4.13). For much of the Cold War era, cryptography was the province of military and intelligence agencies; uncrackable codes were reserved for people with security clearance only.

cryptography
The process of converting a message into a secret code and changing the encoded message back to regular text.

encryption
The conversion of a message into a secret code.

Figure 4.13

Cryptography is the process of converting a message into a secret code and changing the encoded message back into regular text.

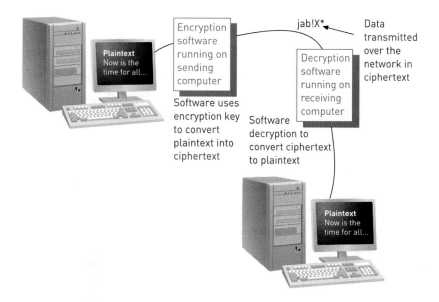

U.S. banks and brokerage houses use the federal government's Data Encryption Standard (DES) algorithm to protect the integrity and confidentiality of fund transfers totaling trillions of dollars a day worldwide. Organizations encrypt the words and videos of their teleconferencing sessions. Individuals encode their electronic mail. And researchers use encryption to hide information about new discoveries from prying eyes.

Encryption is not just for keeping secrets. It can also be used to verify who sent a message and to tell whether the message was tampered with en route. A **digital signature** is a technique used to meet these critical needs for processing online financial transactions. Digital signatures involve a complicated technique that combines the public-key encryption method with a "hashing" algorithm that prevents reconstructing the original message. The hashing algorithm provides further encoding by using rules to convert one set of characters to another set (e.g., the letter *s* is converted to a *v*, *2* is converted to *7*, etc.). Thus, encryption also can prevent electronic fraud by authenticating senders' identities with digital signatures. A digital ID, for example, can be purchased by an individual. When the individual requests sensitive information from a Web site, the digital ID is sent to the site to confirm that individual's

digital signature
The encryption technique used to verify the identity of a message sender for processing online financial transactions.

firewall

A device that sits between an internal network and the Internet, limiting access into and out of a network based on access policies.

identity. A server certificate authenticates its site to users so they can be confident the Web site is safe. Many Web sites display a padlock icon at the bottom of an Internet screen to indicate that the site is encrypted.

The most popular method of preventing unauthorized access to corporate computer data is to construct what is known as a firewall between company computers and the Internet. An Internet **firewall** is a device that sits between your internal network and the Internet. Its purpose is to limit access into and out of your network based on your organization's access policy. A firewall can be anything from a set of filtering rules set up on the router to an elaborate application gateway consisting of one or more specially configured computers that control access.

Firewalls permit desired services on the outside, such as e-mail, to pass. In addition, most firewalls allow access to the Web from inside protected networks. But firewalls deny other, unwanted access. For example, you may be able to log on to systems on the Internet, but users on remote systems cannot log on to your local system because the firewall prevents it.

The U.S. Computer Emergency Readiness Team (*www.us-cert.gov*) responds to virus attacks, network intrusions, and other threats. It also provides security training and consulting services to individual agencies. Vulnerability and threat information is shared with the public. The *www.cert.org* Web site by Carnegie Mellon also contains information about Internet security.

The U.S. Computer Emergency Readiness Team assists civilian government agencies with computer security incidents.

SUMMARY

Principle

Effective communications are essential to organizational success.

Telecommunications refers to the electronic transmission of signals for communications, including telephone, radio, and television.

The elements of a telecommunications system start with a sending unit, such as a person, a computer system, a terminal, or another device, that originates the message. The sending unit transmits a signal to a telecommunications device. The telecommunications device performs a number of functions, which can include converting the signal into a different form or from one type to another. A telecommunications device is a hardware component that allows electronic communication to occur or to occur more efficiently. The telecommunications device then sends the signal through a medium. A telecommunications medium is anything that carries an electronic signal and interfaces between a sending device and a receiving device. The signal is received by another telecommunications device that is connected to the receiving computer. The process can then be reversed.

A communications channel is the transmission medium that carries a message from the source to its receivers. Telecommunications media fall into two broad categories: guided transmission media, in which communications signals are guided along a solid medium, and wireless media, in which the communications signal is sent over airwaves. Guided transmission media include twisted-pair wire cable, coaxial cable, and fiber-optic cable. Wireless media types include microwave, cellular, and infrared.

There are many types of telecommunications carriers. A local exchange carrier is a public telephone company in the United States that provides service to homes and businesses within its defined geographical area, called its *local access and transport area,* through what is called a local *loop*—typically a pair of copper wires, also called the "last mile." The local exchange office can switch calls locally or to long-distance carrier phone offices.

Competitive local exchange carriers compete with the local exchange carriers. These competitors include wireless service providers, satellite TV service providers, cable TV companies, and even power companies.

The three established long-distance providers include AT&T, Sprint, and MCI. Local exchange carriers have been able to compete with them to provide long-distance services due to the Telecommunications Act of 1996.

When an organization needs to use two or more computer systems, one of three basic processing strategies can be followed: centralized, decentralized, or distributed. With centralized processing, all processing occurs in a single location or facility. With decentralized processing, processing devices are placed at various remote locations and do not communicate with each other. With distributed processing, computers are placed at remote locations but are connected to each other via telecommunications devices.

Three commonly used approaches for connecting computers include terminal-to-host, file server, and client/server. With terminal-to-host connections, the application and database reside on the same host computer, and the user interacts with the application and data using a "dumb" terminal. All computations, data access and formatting, and data display are done by an application that runs on the host computer.

In the file server approach, the application and database reside on the same host computer, called the *file server.* The database management system runs on the end user's personal computer or workstation. The file server sends the user the entire file that contains the data requested, including a lot of data the user does not want or need.

A client/server system is a network that connects a user's computer (a client) to one or more host computers (servers). A client is often a PC that requests services from the server, shares processing tasks with the server, and displays the results.

The physical distance between nodes on the network and the communications and services provided by the network determine whether it is a personal area network (PAN), local area network (LAN), metropolitan area network (MAN), or wide area network (WAN). A PAN supports devices within a range of 33 feet or so. A network that connects computer systems and devices within the same geographic area, such as an office building or factory, is a LAN. A MAN connects users and their computers in a geographical area larger than a LAN but smaller than a WAN. WANs tie together large geographic regions, including communications between countries.

Communications software performs important functions such as error checking and message formatting. A network operating system (NOS) controls the computer systems and devices on a network. Network management software enables a manager to monitor the use of individual computers and shared hardware, scan for viruses, and ensure compliance with software licenses.

When people on one network want to communicate with people on another network, they need a common communications protocol and various network devices to do so. A communications protocol is a standard set of rules that control a telecommunications connection, such as OSI, TCP/IP, SNA, Ethernet, frame relay, FireWire, Bluetooth, Wi-Fi, IEEE 802.11g, WiMax, and Mobile Broadband Wireless Access. Various devices are also used in telecommunications. A bridge connects two or more networks that use the same communications protocol. A switch routes incoming data from one of many input ports to a specific output port toward its intended destination. A router determines the next network point to which a data packet should be forwarded. A hub is a place of convergence where data arrives from one or more directions and is forwarded out in one or more other directions. A gateway is a network point that acts as an entrance to another network.

Principle

The Internet is like many other technologies—it provides a wide range of services, some of which are effective and practical for use today, others are still evolving, and still others will fade away from lack of use.

The Internet started with ARPANET, a project started by the U.S. Department of Defense (DoD). Today, the Internet is the world's largest computer network composed of interconnected networks. The Internet transmits data from one computer (called a *host*) to another. The set of conventions used to pass packets from one host to another is known as the Internet Protocol (IP). Many other protocols are used in conjunction with IP. The best known is the Transmission Control Protocol (TCP). Each computer on the Internet has an assigned address to identify it from other hosts, called its *Uniform Resource Locator (URL)*. There are several ways to connect to the Internet: via a LAN whose server is an Internet host, via SLIP or PPP, and via an online service that provides Internet access.

An Internet service provider (ISP) is any company that provides individuals or organizations with access to the Internet. Among the value-added services ISPs provide are electronic commerce, networks to connect employees, networks to connect business partners, Web-site hosting, Web transaction processing, network security and administration, and integration services.

Principle

Originally developed as a document-management system, the World Wide Web is a menu-based system that is easy to use for personal and business applications.

The Web is a collection of independently owned computers that work together as one. High-speed Internet circuits connect these computers, and cross-indexing software enables users to jump from one Web computer to another effortlessly.

A Web site is like a magazine, with a cover page called a *home page* that has graphics, titles, and black and highlighted text. Web pages are loosely analogous to chapters in a book. Hypertext links are maintained using URLs, a standard way of coding the locations of the Hypertext Markup Language (HTML) documents. In addition to HTML, a number of newer Web standards are gaining in popularity, including Extensible Markup Language (XML), Extensible Hypertext Markup Language (XHTML), Cascading Style Sheets (CSS), Dynamic HTML (DHMTL), and Wireless Markup Language (WML).

A Web browser reads HTML and creates a unique, hypermedia-based menu on your computer screen that provides a graphical interface to the Web. Internet Explorer and Netscape are examples of Web browsers. A search engine helps find information on the Internet. Popular search engines include Yahoo! and Google.

Java is an object-oriented programming language from Sun Microsystems based on C++ that allows small programs—applets—to be embedded within an HTML document. When the user clicks on the appropriate part of the HTML page to retrieve it from a Web server, the applet is downloaded onto the client workstation, where it begins executing.

Internet and Web applications include e-mail and instant messaging; Internet cell phones and handheld computers; career information and job searching; Web logs (blogs); chat rooms; Internet phone and videoconferencing services; content streaming; shopping on the Web; Web auctions; music, radio, and video; and distance learning.

An intranet is an internal corporate network built using Internet and World Wide Web standards and products. It is used by the employees of an organization to gain access to corporate information.

An extranet is a network that links selected resources of the intranet of a company with its customers, suppliers, or other business partners. Security and performance concerns are different for an extranet than for a Web site or network-based intranet. User authentication and privacy and peformance are critical on an extranet.

Principle

Because the Internet and the World Wide Web are becoming more universally used and accepted for business, management, service and speed, privacy, and security issues must continually be addressed and resolved.

Management issues and service and speed affect all networks. No centralized governing body controls the Internet. The growth in Internet traffic continues to be phenomenal, and companies setting up an Internet or intranet Web site often underestimate the amount of computing power and communications capacity they need. Privacy, fraud, and security issues must continually be addressed and resolved. Cryptography techniques and firewalls help combat information thieves and provide as much security as possible.

CHAPTER 4: SELF ASSESSMENT TEST

Effective communications are essential to organizational success.

1. A _____ carries an electronic signal between sender and receiver.

   a. modem
   b. communications protocol
   c. transmission media
   d. telecommunications

2. A(An) _____ provides service to homes and businesses within a defined geographical area.

3. Two broad categories of transmission media are:

   a. guided and wireless
   b. shielded and unshielded
   c. twisted and untwisted
   d. infrared and microwave

4. Switching equipment that routes phone calls and messages within a building and connects internal phone lines to a few phone company lines is called a PBX. True or False?

5. A(An) _____ is a network that supports the interconnection of information technology devices within a range of 33 feet or so.

The Internet is like many other new technologies—it provides a wide range of services, some of which are effective and practical for use today, others are still evolving, and still others will fade away from lack of use.

6. The _____ was the ancestor of the Internet. It was developed by the U.S. Department of Defense.

7. On the Internet, what enables traffic to flow from one network to another?

   a. Internet Protocol
   b. ARPANET
   c. Uniform Resource Locator
   d. LAN server

Originally developed as a document-management system, the World Wide Web is a menu-based system that is easy to use for personal and business applications.

8. A Web log, also called a *blog*, is an online Web site that people can create and use to write about their observations, experiences, and feelings on a wide range of topics. True or False?

9. What allows two or more people to engage in online, interactive "conversation" over the Internet?

   a. content streaming
   b. chat rooms
   c. e-mail
   d. hypermedia

10. _____ can be used to route phone calls over networks and the Internet.

11. What is the standard page description language for Web pages?

    a. Home Page Language
    b. Hypermedia Language
    c. Java
    d. Hypertext Markup Language (HTML)

12. A(An) _____ is a network based on Web technology that links customers, suppliers, and others to the company.

Because the Internet and the World Wide Web are becoming more universally used and accepted for business, management, service and speed, privacy, and security issues must continually be addressed and resolved.

13. What sits between an internal network or computer and the Internet to prevent unauthorized access to a computer system?

    a. digital signature
    b. firewall
    c. extranet
    d. intranet

14. A(An) _____ is a text file that a Web site places on the hard drive of a computer to keep track of visits to the site and the actions people take while at that site.

CHAPTER 4: SELF-ASSESSMENT TEST ANSWERS

(1) c (2) local exchange carrier (3) a (4) True (5) personal area network (6) ARPANET (7) a (8) True (9) b (10) Voice over IP (VoIP) (11) d (12) extranet (13) b (14) cookie

REVIEW QUESTIONS

1. Describe the elements and steps involved in the telecommunications process.
2. Identify three types of guided telecommunications media.
3. Define the term *computer network*.
4. How is client/server computing different from file server computing?
5. What role do the bridge, router, gateway, and switch play in a network?
6. Describe a local area network.
7. What is a metropolitan area network?
8. Explain the naming conventions used to identify Internet host computers.
9. What is a domain name?
10. Briefly describe three different ways to connect to the Internet. What are the advantages and disadvantages of each approach?
11. What is an Internet service provider? What services do they provide?
12. What is an Internet chat room?
13. What is content streaming?
14. Briefly describe a Web auction.
15. What is the Web? Is it another network like the Internet or a service that runs on the Internet?
16. What is a URL? How is it used?
17. What is HTML? How is it used?
18. What is a Web browser? How is it different from a Web search engine?
19. What is an intranet? Provide three examples of the use of an intranet.
20. What is an extranet? How is it different from an intranet?
21. What is cryptography?
22. What are firewalls? How are they used?

DISCUSSION QUESTIONS

1. Why is an organization that employs centralized processing likely to have a different management decision-making philosophy than an organization that employs distributed processing?
2. Briefly discuss the pros and cons of e-mail versus voice mail. Under what circumstances would you use one and not the other?
3. What issues would you expect to encounter in establishing an international network for a large, multinational company?
4. Identify at least seven communications protocols. Why do you think that there are so many protocols? Will the number of protocols increase or shrink over time?
5. Instant messaging is being widely used today. Describe how this technology could be used in a business setting. Are there any drawbacks or limitations to using instant messaging in a business setting?
6. Briefly describe how the Internet phone service operates. Discuss the potential impact that this service could have on traditional telephone services and carriers.
7. The U.S. federal government is against the export of strong cryptography software. Discuss why this might be so. What are some of the pros and cons of this policy?
8. Identify three companies with which you are familiar that are using the Web to conduct business. Describe their use of the Web.
9. What is voice over IP (VoIP), and how could it be used in a business setting?
10. Getting music, radio, and video programs from the Internet is easier than in the past, but some companies are still worried that people will illegally obtain copies of this programming without paying the artists and producers royalties. If you were an artist or producer, what would you do?
11. How could you use the Internet if you were a traveling salesperson?
12. Briefly summarize the differences in how the Internet, a company intranet, and an extranet are accessed and used.

PROBLEM-SOLVING EXERCISES

1. Do research on the Web to find several popular Web auction sites. After researching these sites, use a word processor to write a report on the advantages and potential problems of using a Web auction site to purchase a product or service.

Also discuss the advantages and potential problems of selling a product or service on a Web auction site. How could you prevent scams on an auction Web site?

 2. Develop a brief proposal for creating a business Web site. How could you use Web services to make creating and maintaining the Web site easier and less expensive?

Develop a simple spreadsheet to analyze the income you need to cover your Web site and other business expenses.

 3. You are a manager of a small company with a new Web site. How would you avoid privacy, fraud, and security problems? Develop a brief report describing what you would do. Using a graphics program, diagram how you would protect your Internet site from outside hackers.

TEAM ACTIVITIES

 1. Form a team to identify the public locations (airport, public library, Starbucks, etc.) in your area where the capability of wireless LAN connections are available. Visit at least two locations and write a brief paragraph discussing your experience at each location trying to connect to the Internet.

2. With your teammates, identify a company that is making effective use of a company extranet. Find out all you can about its extranet. Try to speak with one or more of the customers or suppliers who use the extranet and ask what benefits it provides from their perspective.

WEB EXERCISES

1. Go online and identify three companies that provide firewall software. Do a comparison of these options and state which one you think is the best and why.

2. The Internet can be a powerful source of information about various industries and organizations. Locate several industry or organization Web sites. Which one is the best? Why?

CAREER EXERCISES

1. There are a number of online job-search companies, including Monster.com. Investigate one or more of these companies and research the positions available in the telecommunications industry, including the Internet. You

may be asked to summarize your findings for your class in a written or verbal report.

2. Use the Internet to explore the starting salaries, benefits, and job descriptions for three career areas.

VIDEO QUESTIONS

Watch the video clip **Online Storage** and answer these questions:

1. What are the advantages of storing files using services such as *briefcase.yahoo.com*?

2. Do you think that Internet storage may become our primary form of storage in the future, over storing our files on local hard drives? What will it take, in terms of network performance, storage capacity, and price to make this a reality?

CASE STUDIES

Case One

Eastman Chemical Revamps with Web Services

Eastman Chemical Company is the world's largest producer of polyester plastics for packaging; a leading supplier of raw materials for paints, inks, and graphic arts; and a marketer of more than 1,200 chemicals, fibers, and plastics products. The company is, by all measures, successful, with recent annual sales revenue of $5.8 billion. Eastman has approximately 15,000 employees in more than 30 countries and manufacturing sites strategically located in 17 different countries.

Eastman maintains a set of application servers that provide information services to employees worldwide. The servers contain a variety of operating systems (OSs): a large IBM server runs a proprietary OS, and others run Microsoft Windows 2000 and Windows NT OSs. Recently, Eastman Chemical decided to revamp its information systems. It had been using packaged applications and wanted to develop its own custom applications to meet all the needs of its staff. Eastman IS specialists turned to Web service technologies as the ideal solution for connecting differing computer platforms over the Internet.

Web services provide standardized interfaces between diverse computer systems so that they can automatically make and fulfill data requests over the Internet. Before Eastman's IS crew could begin the development process, they first needed to dissect the ways their current systems were being used. To do that, they studied the company's application servers, assessing what the applications did, stripping off the user interfaces and exposing the application functions as services, says Carroll Pleasant, an associate analyst in Eastman's emerging digital technologies group.

After the group defined the "services" that the current system provided, they interviewed the users to learn how the system could be improved. "The users will be the ones deciding what the business processes will be, rather than having the applications determine the business process for them," explained Pleasant.

In developing the new Web services–based systems, the team ran into many challenges. Although Web services technology has been around for a few years, the technology is still in its infancy and is just beginning to affect businesses. Although many businesses are using isolated Web service applications for special tasks, not many have been brave enough to base an entire corporate information system on the

technology. Much of the work Pleasant and his team are conducting is considered pioneering work. "It's going to take a long, long time for everything to switch over to Web services and a service-oriented architecture," Pleasant says. "We see the movement going on with almost all of our vendors. We're confident they're going this route. But it takes time to get there."

In the meantime, Eastman employees from around the world are enjoying the first of the Web services rollouts. The new applications include rich graphical user interfaces that "deploy like a Web application," Pleasant says. "The user just clicks on a shortcut and points to the application running on the server." From any Eastman location in the world, the Web service communicates with the server and quickly delivers the requested information.

Web services are destined to change the landscape of the Internet. Services that support a wide variety of needs are already passing their data back and forth over the Internet in high volumes. Applications as sophisticated as storage management and customer relationship management and as simple as stock quotations are employing Web services. As Web services proliferate, some people are becoming concerned that the services' demands on bandwidth may overcome the ability of the Internet to support them.

Discussion Questions

1. What advantages do Web services promise Eastman Chemical Company over its previous information systems?
2. What risks are Pleasant and his crew taking by being the first to deploy the new technology? What benefits might they enjoy?

Critical Thinking Questions

3. How do Web services meet the unique challenges, such as those that Eastman faces, of running applications on the Web?
4. How does the "services" approach of Web services make sense for achieving high-quality information systems?

SOURCES: Carol Sliwa, "Enterprises Take Early Lead in Web Services Integration Projects," *Computerworld*, June 16, 2003, *www.computerworld.com*; Eastman Chemical Company profile, *www.eastman.com/About_Eastman/ The_Company/Company_Profile.asp*, accessed March 19, 2004; W3C Web Services Activity Web site, *www.w3.org/2002/ws/*, accessed March 19, 2004.

Case Two

Best Western: First to Provide Free Internet

Internet access has become a necessity for today's business traveler. Those in the hospitality industry have picked up on this fact and are working to profit from it. Many well-known hotel chains are providing Internet hookups in rooms and profiting from their hourly usage rates. But Best Western has taken a unique and different approach.

Best Western is using the value of high-speed Internet access to gain a competitive advantage by offering it to customers as a standard, free service. Best Western realizes that Internet access is more popular with customers than even local phone service—particularly because most business travelers carry their own cell phones that accommodate both local and long-distance calls. High-speed Internet access provides business travelers with valued access to e-mail, the Web, and corporate VPNs.

Best Western believes that travelers are tired of paying extra to access their e-mail and will appreciate this free, no-hassle service. Research seems to support this belief. A recent survey by Yesawich, Pepperdine, Brown & Russell showed that 65 percent of business travelers agree that free access to the Internet from their room is extremely or very desirable. Many hotel chains charge $10 per day for the service, and customers suffer not only from the expense but also from the inconvenience of an "activation procedure" that connects them to the Internet and automatically updates the customer's billing record in the database.

Best Western plans to offer free access in all of its 2,300 hotels in the United States, Canada, and the Caribbean. Best Western is the first major chain to offer this service free to its customers. According to Tom Higgins, CEO of Best Western, the company is responding to popular demand. "It's the number one amenity requested by virtually everyone, especially businesspeople." Best Western's initial goal is to wire at least 15 percent of the rooms in each hotel. It also plans to create wireless access in each hotel's public areas.

The rest of the hotel industry is watching. Best Western's move will no doubt influence the other large chains to follow suit. In the near future, business travelers will be assured of Internet access wherever they may roam. Free high-speed Internet access, combined with increased use of VPN technologies, will affect the way businesspeople work. Employers are more likely to send employees on the road, and employees will be less tethered to their offices because the office will virtually travel with each employee.

The hotel industry is just one of several industries working to keep people online. The transportation industry—trains, planes, and cruise lines—is offering high-speed Internet access, and restaurants, coffee shops, airports, and most places where people congregate are providing wireless high-speed access. Although most are charging for the service, any might decide to follow Best Western's lead and offer the service for free—to get a leg up on the competition.

Discussion Questions

1. Do you think Best Western's free high-speed Internet service will win over customers? Why?
2. If Best Western's plan fails and does not improve its customer base, what will Best Western have lost in the effort? How might it recoup its losses?

Critical Thinking Questions

3. What expenses are incurred when a business provides high-speed Internet access to its customers?
4. Considering the expenses of providing the service, and assuming that half the customers use the service, how much profit is made from the service at $10 per day per customer? How might Best Western make up its loss of the $10-per-day service charge?

SOURCES: James Lewin, "Will This Western Cause a Stampede?" *ITWorld*, January 28, 2004, *www.itworld.com*; "Best Western Plans Industry's Largest High Speed Internet Rollout," *PR Newswire*, November 3, 2003, *www.lexis-nexis.com*; Best Western Web site, *www.bestwestern.com*, accessed March 20, 2004.

NOTES

Sources for the opening vignette: Stacy Williams, "Leading Pasta Maker Increases Productivity, Reduces Costs with Cisco WLAN," *Cisco Newsroom*, February 27, 2004, *http://newsroom.cisco.com/dlls/2004/ts_022704.html*; Barilla Group Web site, *www.barillagroup.com/index.htm*, accessed March 9, 2004; "IP Telephony/VOIP" Cisco Web site, *www.cisco.com/en/US/tech/tk652/tk701/tech_protocol_family_home.html*, accessed March 9, 2004.

1. Mearian, Lucas, "Wells Fargo Upgrades Online Apps," *Computerworld*, February 16, 2004, accessed at *www.computerworld.com*.
2. Ho, David, "High Speed Internet Access Over Power Lines Looks Promising, Government Says," *The San Diego Union-Tribune*, January 15, 2003, accessed at *www.signonsandiego.printthis.*

clickability.com.
3. Backover, Andrew, and McCarthy, Michael, "Cable Firms Wired about Offering Net Phone Calls," *USA Today*, December 10, 2003, page 7B.
4. Ewalt, David M., "Cellular Data Services Promise," *InformationWeek*, January 12, 2004, accessed at *www.informationweek.com*.
5. Marlin, Steven, "The 24-Hour Supply Chain," *InformationWeek*, January 26, 2004, accessed at *ww.informationweek.com*.
6. Sullivan, Laurie, "Tire Maker Looks for Low-Cost Efficiencies," *InformationWeek*, February 23, 2004, accessed at *www.informationweek.com*.
7. Hamblen, Matt, "Las Vegas Schools Mix IP, Digital Communications," *Computerworld*, December 8, 2003, accessed at *www.*

computerworld.com.

8. Songini, Marc L., "Case Study: Boehringer Cures Slow Reporting," *Computerworld*, July 23, 2003, accessed at *www.computerworld.com.*

9. Fillion, Roger, "Wireless Internet Access Is Popping Up in Unexpected Places," *Rocky Mountain News*, January 19, 2004, p. 1B.

10. Brand, Rachel, "Wireless to Wellness," *Rocky Mountain News*, February 2, 2004, p. 1B.

11. Morrissey, Brian, "DoubleClick Hit with Deceptive Ad Suit," *InternetNews.com*, July 22, 2003.

12. Rosenbush, Steve, "How The U.S. Lags," *Business Week*, March 1, 2004, p. 39.

13. Mullaney, Timothy, "At Last, The Web Hits 100 MPH," *Business Week*, June 23, 2003, p. 80.

14. Pringle, David, "Europe's Broadband Battle Spans Borders," *The Wall Street Journal*, January 15, 2004, p. B4.

15. Jesdanum, Anick, "Phones Juiced for Java," *Rocky Mountain News*, July 14, 2003, p. 6B.

16. Violino, Bob, "Waves of Change," *Computerworld*, May 19, 2003, p. 28.

17. Gralla, Preston, "Web Services in Action," *Computerworld*, May 19, 2003, p. 36.

18. Alexander, Steve, "Best in Class: Web Site Adds Inventory Control and Forecasting," *Computerworld*, February 24, 2003, p.45.

19. Pflughoeft, Kurt et al., "Multiple Conceptualization of Small Business Web Use and Benefit," *Decision Sciences*, Summer 2003, p. 467.

20. Mossberg, Walter, "Checking Out E-Mail for Cheapskates," *The Wall Street Journal*, August 27, 2003, p. D1.

21. Tynan, Daniel, "Uncle Sam Vs. Spam," *PC World*, August 2003, p. 123.

22. Mangalindan, Mylene, "California Gets Serious About Spam," *The Wall Street Journal*, September 24, 2003, p. A3.

23. Metz, Cade, "Can E-Mail Survive," *PC Magazine*, February 17, 2004, p. 65.

24. Langa, Fred, "E-Mail—Hideously Unreliable," *InformationWeek*, January 12, 2004.

25. Mossberg, Walter, "Apple Out to Raise Bar for Audio, Video," *Rocky Mountain News*, August 18, 2003, p. 5B.

26. Goldsborough, Reid, "Blogs," *Arrivals*, August-September 2003, p. 42.

27. Staff, "Wacky Questions," *Rocky Mountain News*, August 5, 2003, p. 2D.

28. Verton, Dan, "Blogs Play a Role in Homeland Security," *Computerworld*, May 12, 2003, p. 12.

29. Ernst, Warren, "Building Blogs," *PC Magazine*, June 30, 2003, p. 58.

30. Dreier, Troy, "Blog On," *PC Magazine*, September 2, 2003, p. 154.

31. Prince, Marcelo, "Dialing for Dollars," *The Wall Street Journal*, May 19, 2003, p. R9.

32. Park, Andrew, "Net Phones Start Ringing Up Customers," *Business Week*, December 29, 2003, p. 45.

33. Staff, "Clear Signal for Internet Phones," *Business Week*, June 23, 2003, p. 84.

34. Spencer, Jane, "I Ordered That?" *The Wall Street Journal*, September 4, 2003, p. D1.

35. Mangalinsan, Mylene, "Yahoo Hopes New Service Turns Searchers into Shoppers," *The Wall Street Journal*, September 23, 2003, p. B1.

36. Warner, Melanie, "eBay's Worst Nightmare," *Fortune*, May 26, 2003, p. 89.

37. Gomes, Lee, "RIAA Takes Off the Gloves," *The Wall Street Journal*, September 15, 2003, p. B1.

38. Brandt, Andrew, "Network Cameras for All," *PC World*, February 2004, p. 60.

39. Kandra, Anne, "Can You Profit from Online Networking?" *PC World*, February 2004, p. 45.

40. Schoenberger, Chana, "Everyone Is Trying to Sell Digital Convergence to Homeowners," *Forbes*, February 2, 2004, p. 90.

41. Clark, Don, "Gadgets Grow Up," *The Wall Street Journal*, January 8, 2004, p. D1.

42. Wood, Richard, "Intranet in Staff Hands," *New Zealand Infotech Weekly*, February 9, 2004, p. 3.

43. Knights, Miya, "Intranet Consolidates Utility's Legacy Systems," *Computing*, January 29, 2004, p. 11.

Australian Film Studio Benefits from Broadband Networks...and Location

Sigi Goode
The Australian National University

It's been a hectic 24 hours at Fire Is a Liquid Films, a small Australian film-production studio. The firm's 18 staff members have just finished rendering scenes from three feature films, pitched ideas for two TV commercials, and delivered two milestone [progress] updates to production clients. Thanks to the power of broadband networking, the employees have accomplished most of this work without leaving their small studio. Of his company's tight deadlines and advanced technology, Toni Brasting, Managing Director, explains, "We play to our strengths in terms of location and technology.

"Our work ranges from cleaning up footage to improve colour or contrast, up to full-blown CGI modelling and rendering. We work with TV commercials right up to feature film houses, and from one or two frames up to entire scenes. We can capitalise on our geographic location. With most feature films being shot in the U.S. or Europe, [clients] can send us their digital footage before they go to bed, and we can work on the scene during our daytime. By the time they wake up, the processed footage is ready and waiting for them. We rely heavily on our broadband network infrastructure to move digital data around. Depending on the quality and duration of the footage, the upload process can take anywhere from a few minutes for a brief scene to a few hours for an entire feature at DVD quality. Most of our clients prefer to deal with individual scenes in this way because it allows them to shuffle their work flow priorities around.

"Most films and commercials used to be shot on 35mm analogue film, which can take time to process. The trend is moving more and more towards digital video because it can be easily edited and manipulated. With 35mm, if you wanted to bid on work with a film studio, it helped if you were fairly close by so [that] you could have access to the film reels for digitising purposes. With the advent of digital video and broadband networking, we can bid on projects anywhere in the world, from the *Matrix* to *28 Days Later*. The actors can get a better feel for each scene, and the director can place each scene in greater context: their vision comes to life quicker.

"For editing, we mostly use Final Cut Pro on our Macintosh systems—the Macs are easy to use, and they look good, too (which impresses clients when they tour the studio). They run Mac OS X, which was originally based on BSD Unix. Occasionally, we need to develop a new technique for a project, so it's useful to have access to software source code. Mac OS X gives us good support for our own open-source Linux tools.

"Another technique we use is laptop imaging. Clients like the feeling that we're taking care of them and that they are our focus. When one of our teams goes to meet a client for a pitch or a milestone update, they can pull a laptop off the shelf and build a customised hard-drive image right from their desktop with Norton Ghost or one of our home-brewed imaging applications. The image software 'bakes' a hard-drive image on the fly, which includes the OS, editing software, and relevant film footage. If necessary, we can also include correspondence and storyboards—we can have that client's entire relationship history right there in the meeting. We're also about to start phasing in tablet PCs with scribe pens so that we can work on potential storyboards during the meeting itself.

"The other important part of our processing platform is our render farm. It's basically a room full of identical Linux-based PCs, which work on a problem in parallel. We use commodity PC hardware and high-speed networking to allow each machine to work on the problem at once. It means we can get supercomputer performance at a fraction of the cost. We use our desktop machines to edit in draft resolution (which is of inferior quality but much faster to render or preview). When we're happy with the cut, we 'rush it' to the render farm to process the final piece of footage. We also offer render farm leasing services, so other film companies can send us their raw model files, use the render farm over the network, and then download their completed footage. We then invoice them for processing time—they never have to leave their offices."

Discussion Questions

1. How might open-source software benefit a small business? What problems might a small business encounter when using open-source software?
2. The studio profiled in this case processes tasks during a client firm's downtime. What other tasks could be accomplished more inexpensively or more efficiently by using providers in other geographic locations?
3. What other types of firms could also benefit from a render farm? What firms do you know of that actively make use of this technology?

Critical Thinking Questions

4. Are modern firms too reliant on electronic networking? If so, what can be done about this drawback? What paper-based systems could the firm use if its electronic systems fail?
5. How might wireless networking improve the operation of systems such as this firm's render farm?

Note: All names have been changed at the request of the interviewee.

Virtual Learning Environment Provides Instruction Flexibility

Vida Bayley

Coventry University, United Kingdom

One of the major components of Coventry University's teaching and learning strategy has been the introduction and use of a Virtual Learning Environment (VLE) utilising the WebCT (Web Course Tools) product platform. The Joint Information Systems Committee (JISC) in the UK has set out a definition of a Virtual Learning Environment and states that it refers to the components that support online interactions of various kinds taking place between learners and tutors.

A central aim of the university was to enhance face-to-face teaching and to offer greater flexibility to both teaching staff and students. It created a central support unit— CHED (Centre for Higher Education Development)—to introduce and develop new initiatives in teaching and learning across the university. In addition, a task force of academic staff was established to lead and facilitate these initiatives in collaboration with CHED. Through these developments and with the support of the university's Computing Services unit, WebCT was implemented as a core instrument in achieving the university's educational vision. The WebCT application allows course material and resources to be placed on the Web and provides staff and students with a set of tools that can be customised to course and module requirements. In brief, it also contains the following functions, which JISC considers to be essential features of a VLE:

- controlled access to curriculum with mapping to elements that can be individually assessed and recorded
- tracking mechanisms for student activities and performance
- support of online learning through content development
- communication and feedback mechanisms between various participants in the learning process
- incorporation of links to other internal/external systems

In line with the aims and goals of the VLE strategy, Coventry Business School has been active for a number of years in the development of curricula and modules to create innovative learning communities and to engage students in the learning process. One of the most successful programmes the Business School developed was the B.A. in Business Enterprise (B.A.B.E.) degree. The Business Enterprise course was a learning programme designed to develop graduates who would be able to proactively manage business-related tasks, solve business problems, and influence the business environment in which they work. A central element of the programme was that it would enable students to lay the foundation for lifelong learning, in which students take responsibility for continual learning and personal development through a variety of media. The incorporation of WebCT into the B.A.B.E. programme marked a significant departure from established methods of teaching, learning, and assessment in the Business School.

Although the school had previously used a variety of computer-based activities within the learning environment, their use was limited to individual modules and in many cases even restricted. In accounting and information technology courses, software products such as EQL tutorials (EQL International Ltd is a developer of e-learning and computer-based assessment solutions) were used as supplementary teaching material and for online assessment or self-test exercises. The WinEcon package (PC-based introductory economics software) had been developed with Bristol University through a consortium of eight UK economics departments and was used by the Economics subject group in course teaching. In other subject areas, the teaching staff provided students with Web URLs related to particular reading or assessment topics, which could be easily accessed via the Internet. In contrast, the Business Enterprise programme was designed for delivery via the World Wide Web, so module delivery, programme management, and staff/student communication were all built around WebCT.

This programme was the first to be offered by the university via the World Wide Web. And it was significant because it was not designed solely for distance learning but rather to make use of the range of tools offered by the WebCT platform for teaching and learning both within a classroom setting and off campus via the Net. WebCT tools were used to structure and to provide Web-based access to course and module materials, and communications facilities were adapted to dovetail with existing classroom-based teaching and learning. Of particular importance was the consideration by the teaching team of how the Web software features could add value to the traditional teaching/learning modes and methods that they would be supporting or replacing. This evaluation in turn spurred individual lecturers to examine their teaching practice, stimulated ideas about the applicability of Web-based teaching, and encouraged staff to develop their knowledge of and skills related to the university's intranet and the Internet. Once these opportunities were identified and placed within the context of the module and course objectives, lecturers then customised the WebCT features and integrated them within the overall teaching and learning framework. The benefit to students is flexibility: Course and module content is accessible from outside the university and outside of teaching hours. Course materials either can be placed within WebCT as HTML documents (so they can be edited directly online) or can be uploaded as individual applications, such as PowerPoint slides, Word documents, and Excel spreadsheets.

Effective communication and feedback are important aspects of a successful learning environment. One way of facilitating peer group communication and communication between students and staff was to use the discussion forum/bulletin board facilities provided by the software. The B.A.B.E. programme used this feature as the primary mode of delivering information to the student community. All course- and module-related announcements, such as class rescheduling, changes to assessment deadlines, and meeting arrangements, for example, were made via the main bulletin board. The flexibility that the software provides for creating discussion areas allowed lecturers to construct arenas for students to work individually on specific projects yet be involved as group members in discussions centred on common themes. Being asynchronous, these discussion groups allowed students to become involved in an ongoing exchange of ideas, reflect on their own and others' contributions, and thereby facilitate their own learning. In addition to the discussion forum, WebCT incorporates a chat facility that can enable a small group of students to participate in online, real-time conversations. An additional advantage of this type of synchronous communication is that both students and lecturers could develop group management and e-moderation skills. In addition, the Business School linked the Web e-mail facilities available in WebCT to the e-mail provided to all students by Coventry University, allowing messages to be transmitted within modules. This meant that messages were module specific and did not get lost in the larger system—especially in modules with large numbers of students.

The student presentation area of WebCT was used to allow students to create their own Web pages, which could be displayed and shared, thus providing them with an opportunity to gain new skills and an understanding of an important part of the processes involved in managing a business enterprise.

The Business School also uses STile, a portal that was developed to provide students with a range of resources that are available on the university's intranet. The portal contains an easy-to-use link to WebCT.

In the course of the development of the B.A.B.E. programme, it became apparent that for the success of new IT innovations, such as Web-based teaching, the project staff must identify and involve key stakeholders, particularly when a radical restructuring of teaching and learning methods is needed. These stakeholders include administrative personnel, teaching professionals, and also those who are at the very centre of learning activity—the student body. The university benefited greatly by actively involving students and staff in the design, use, and evaluation of their educational environment. The experience of the B.A.B.E. venture highlighted the fact that that in order to transform the students' learning experience, all the stakeholders needed to understand how to construct a fit between the needs and expectations of the learner, the skills and pedagogical tools of the practitioners, and the tools available within the VLE. The rewards have been great. The establishment of a VLE, with WebCT as its centre, has generated a new excitement in the approach to university teaching, stimulating reflection, reinvention, and transformation and creating a "journey of learning" for both students and teaching staff.

Although the B.A.B.E. degree programme has now been modified under the current course structure, many of its innovative features have been transferred and embedded into teaching and learning processes across the whole of the undergraduate business programme.

In addition, WebCT has enabled the Business School to introduce and develop with business partners new, forward-looking projects, such as the development and delivery of the Post Graduate Certificate, Post Graduate Diploma, and MA in Communications Management via the Cable and Wireless Virtual Academy. It has also played a significant role in the establishment of learning communities and development of work-based learning projects with public-sector organisations such as the National Health Service, local authorities, and private-sector enterprises.

Discussion Questions

1. Why do you think that a discussion forum would motivate you to use the Internet to access learning resources?
2. What do you think the use of e-mail facilities would contribute to your learning?

Critical Thinking Questions

3. Discuss the advantages that a VLE with Internet access could offer students in developing their learning. Are there any disadvantages?
4. Identify and discuss the different types of contributions that various stakeholders could make to a student's learning process.

Business Information Systems

Chapter 5 Electronic Commerce and Transaction Processing Systems

Chapter 6 Information and Decision Support Systems

Chapter 7 Specialized Information Systems: Artificial Intelligence, Expert Systems, Virtual Reality, and Other Systems

CHAPTER · 5 ·

Electronic Commerce and Transaction Processing Systems

PRINCIPLES	LEARNING OBJECTIVES
▪ E-commerce is a new way of conducting business, and as with any other new application of technology, it presents both opportunities for improvement and potential problems.	▪ Identify several advantages of e-commerce. ▪ Identify some of the major challenges that companies must overcome to succeed in e-commerce and m-commerce. ▪ Describe some of the current uses and potential benefits of m-commerce. ▪ Identify several e-commerce applications.
▪ E-commerce requires the careful planning and integration of a number of technology infrastructure components.	▪ Outline the key components of technology infrastructure that must be in place for e-commerce to succeed. ▪ Discuss the key features of the electronic payment systems needed to support e-commerce.
▪ An organization's transaction processing system (TPS) must support the routine, day-to-day activities that occur in the normal course of business and help a company add value to its products and services.	▪ Identify the basic activities and business objectives common to all TPSs. ▪ Explain some key control and management issues associated with TPSs. ▪ Identify the challenges multinational corporations must face in planning, building, and operating their TPSs.
▪ Implementation of an enterprise resource planning (ERP) system enables a company to achieve many benefits by creating a highly integrated set of systems.	▪ Discuss the advantages and disadvantages associated with the implementation of an ERP system.

INFORMATION SYSTEMS IN THE GLOBAL ECONOMY

SPRINT, UNITED STATES; AND SONY, JAPAN

Handheld Entertainment

The birth of telecommunications has revolutionized the manner in which we conduct business. Early telecommunications networks allowed businesses to carry out long-distance transactions for the first time. Similarly, the development and growth of the Internet supported widespread electronic commerce between businesses and consumers. Today, the explosion in the mobile communications market, along with the merger of voice and data networks, has brought e-commerce to consumers over their cell phones. Buying products and services has never been easier. Whenever and wherever the impulse may strike, a purchase is only a button-push away. The products that are most likely to succeed in this market are music and entertainment.

Sony has joined with Sprint to provide music products over its cellular network. Currently, the companies are focusing on a variety of ring tunes or ringers. Sprint is offering downloadable ringers created by the Sony Music Mobile Products Group over its PCS Vision network. Ringers range from animated polyphonic tones, to actual clips from popular songs, to specialized sound and voice recordings. For example, you could download a clip of your favorite rap or rhythm and blues artist or have Bugs Bunny notify you of an incoming call. The sound is accompanied by animated graphics on the phone's color digital display. Sprint customers apparently appreciate this customization; in a recent year, Sprint sold more than 20 million ringers and screen savers to a network of 3.2 million Sprint PCS Vision customers.

The companies are not stopping with ringers. Soon you may be able to use your cell phone to download and store full songs and then listen to them with headphones, or you could transfer them to your computer or entertainment center. Nokia is designing an application it calls "visual radio." Visual radio allows handsets to receive FM radio signals and matches the audio content with related pictures, graphics, and other content. The application provides a convenient means for listeners to purchase and download music they hear on the radio over their cell phone. "What we're bringing to the table with visual radio is impulse buying," says Reidar Wasenius, a senior project manager with Nokia's multimedia group. "You happen to hear something in a certain mood, and the radio station offers you the purchase opportunity. You do it there and then." Sprint is working on a similar technology that includes music videos and sports video clips.

Partnerships between record companies and telecommunications companies are multiplying as the industries reconfigure to take advantage of this new and booming market. Music from BMG artists such as Britney Spears, Maroon5, Three Days Grace, the Strokes, Kenny Chesney, and Pink are available as ring tunes for $2.50 each.

If music is the forerunner in mobile e-commerce, then mobile games are next in line. Major electronic gaming providers such as SEGA, Namco, and THQ have partnered with Sprint to offer dozens of games that can be downloaded and played on the tiny cell phone display.

Security is another issue with wireless communications. Technologies provided by companies such as BREW and Qpass allow for safe and secure transactions between the customer and the service provider. Using a secure wallet application, customers can save and encrypt their personal and financial information on their cell phone, making purchasing as easy as entering a password.

Electronic commerce has evolved to support every type of transaction, from multimillion-dollar deals between corporate giants to micropayment transactions between people on the street. It is anticipated that by 2010 only 17 percent of transactions will be made with cash, 22 percent with checks, and 61 percent with electronic funds transfers, debit cards, and credit cards. With so much of our economy relying on e-commerce, the technology industry has stepped up its efforts to provide the necessary security and stability to support safe and efficient transactions.

As you read this chapter, consider the following:

- What opportunities and benefits has e-commerce provided businesses and consumers?
- What effect do transaction processing systems have on the efficiency of an organization and its reputation with its customers?

Why Learn About Electronic Commerce and Transaction Processing Systems?

Electronic commerce and transaction processing systems have transformed many areas of our lives and careers. One fundamental change has been the manner in which companies interact with their suppliers, customers, government agencies, and other business partners. As a result, most organizations today have or are considering setting up business on the Internet and sophisticated transaction processing systems. To be successful, all members of the organization need to participate in that effort. As a sales or marketing manager, you will be expected to help define your firm's e-commerce business model. Customer service employees can expect to participate in the development and operation of their firm's Web site. Other employees who work directly with customers—whether in sales, customer service, or marketing—require high-quality transaction processing systems to provide good customer service. Such workers may use a transaction processing system to check the inventory status of ordered items, view the production planning schedule to tell the customer when the item will be in stock, or enter data into the shipment planning system to schedule delivery to the customer.

Clearly, as an employee in today's organization, you must understand what role e-commerce and transaction processing systems can play, how to capitalize on their many opportunities, and how to avoid their pitfalls.

This chapter begins by providing a brief overview of the exciting world of e-commerce and defines its various components.

AN INTRODUCTION TO ELECTRONIC COMMERCE

business-to-consumer (B2C) e-commerce
A form of e-commerce in which customers deal directly with the organization, avoiding any intermediaries.

business-to-business (B2B) e-commerce
A form of e-commerce in which the participants are organizations.

Early e-commerce news profiled start-up companies that used Internet technology to compete with the traditional players in an industry. For example, Amazon.com challenged well-established booksellers Walden Books and Barnes and Noble. Like Amazon, the companies discussed in the opening vignette provide an example of **business-to-consumer (B2C) e-commerce**, in which customers deal directly with an organization and avoid any intermediaries. Other types of e-commerce are **business-to-business (B2B) e-commerce**, in which the participants are organizations, and **consumer-to-consumer (C2C) e-commerce**, which involves consumers selling directly to other consumers. General Electric Aircraft Engine is the world's leading producer of large and small jet engines for commercial and military aircraft. It operates a B2B Web site that allows users of its engines to order spare parts and maintenance supplies. eBay is an example of a C2C e-commerce site; customers buy and sell items directly to each other through the site.

Aside from the major categories of e-commerce, companies are also using Internet technologies to enhance their current operations, such as inventory control and distribution. But whatever model is used, successful implementation of e-business requires significant changes to existing business processes and substantial investment in IS technology.

Over the past few years, we have learned a lot about the practical limitations of e-commerce. It has become painfully clear that before companies can achieve profits, they must understand their business, their consumers, and the constraints of e-commerce. Although it once seemed so, selling consumer goods online in a virtual storefront may not always be a great way to compete. And inventing a new use for cutting-edge technology isn't necessarily enough to guarantee a successful business.

Still, e-commerce is not dead; it is maturing and evolving, with the focus currently shifted from B2C to B2B. E-commerce is a useful tool for connecting business partners in a virtual supply chain to cut resupply times and reduce costs. More than 80 percent of U.S. companies have experimented with some form of online procurement, although most are channeling less than 10 percent of their total procurement online, according to data from Forrester Research and the Institute for Supply Management.[1] In contrast, according to the U.S. Department of Commerce, B2C e-commerce sales accounted for barely 2 percent of all retail sales. Yet, for 2003, B2C e-commerce sales rose by 26.3 percent—to an estimated at $54.9 billion.[2] And the trend is spreading globally. Forrester Research predicts that online shopping revenue in European countries will rise from $40 billion in 2004 to $167 billion by 2009.[3]

Businesses and individuals use e-commerce to reduce transaction costs, speed the flow of goods and information, improve the level of customer service, and enable close coordination among manufacturers, suppliers, and customers. E-commerce also enables consumers and companies to gain access to worldwide markets. Dallas-based Aviall expanded into a worldwide provider of supply chain management services for hundreds of aviation parts manufacturers and airlines. Its clients use Aviall's Web site for ordering, inventory control, and demand forecasting. To move its services online, the firm spent $40 million to develop a Web site and install several supporting information systems, including online purchasing, sales force automation, order entry, financial management, inventory control and warehouse management, product allocation, and purchasing forecasting systems. The Web site generates $60 million of the company's $800 million in annual revenue.[4]

Business processes that are strong candidates for conversion to e-commerce are those that are paper based and time-consuming and those that can make business more convenient for customers. Thus, some of the first business processes that companies converted to an e-commerce model were those related to buying and selling. For example, after Cisco Systems, the maker of Internet routers and other telecommunications equipment, put its procurement operation online in 1998, the company reported that it halved cycle times and saved an additional $170 million in material and labor costs. Similarly, Charles Schwab & Co. slashed transaction costs by more than half by shifting brokerage transactions from traditional channels such as retail and phone centers to the Internet.

The E-Commerce Supply Chain

All business organizations contain a number of value-added processes. The supply chain management process is a key value chain that, for most companies, offers tremendous business opportunities if converted to e-commerce. **Supply chain management** is composed of three subprocesses: demand planning to anticipate market demand, supply planning to allocate the right amount of enterprise resources to meet demand, and demand fulfillment to fulfill demand quickly and efficiently (see Figure 5.1). The objective of demand planning is to understand customers' buying patterns and develop aggregate, collaborative long-term, intermediate-term, and short-term forecasts of customer demand. Supply planning includes strategic planning, inventory planning, distribution planning, procurement planning, transportation planning, and supply allocation. The goal of demand fulfillment is to provide fast, accurate, and reliable delivery for customer orders. Demand fulfillment includes order capturing, customer verification, order promising, backlog management, and order fulfillment.

Figure 5.1

Supply Chain Management

Conversion to e-commerce supply chain management provides businesses an opportunity to (1) increase revenues or decrease costs by eliminating time-consuming and labor-intensive steps throughout the order and delivery process, (2) improve customer satisfaction by enabling customers to view detailed information about delivery dates and order status, and (3) reduce inventory including raw materials, safety stocks, and finished goods. Achieving this goal requires integrating all subprocesses that exchange information and move goods between suppliers and customers, including manufacturers, distributors, retailers, and any other enterprise within the extended supply chain.

Business-to-Business (B2B) E-Commerce

Although the business-to-consumer (B2C) market grabs more of the news headlines, the business-to-business (B2B) market is considerably larger and is growing much more rapidly. Business-to-business e-commerce offers enormous opportunities. It allows manufacturers to buy at a low cost worldwide, and it offers enterprises the chance to sell to a global market right from the start. Moreover, e-commerce offers great promise for developing countries, helping them to enter the prosperous global marketplace, and hence helping reduce the gap between rich and poor countries.

Some companies, such as those in the automotive and aerospace industries, have been conducting e-commerce for decades through the use of **electronic data interchange** (EDI), an intercompany, application-to-application communication of data in standard format, permitting the recipient to perform the functions of a standard business transaction, such as processing purchase orders. EDI uses network systems and follows standards and procedures that allow output from one system to be processed directly as input to other systems, without human intervention. This technology eliminates the need for paper documents and substantially cuts down on costly errors. Orders and inquiries are transmitted from the customer's computer to the manufacturer's computer. The manufacturer's computer can then determine when new supplies are needed and can automatically place orders by connecting with the supplier's computer.

Because of the costs involved in buying new technology, the EDI capabilities of most small businesses are nonexistent or extremely limited. A few major retailers and manufacturers enlisted the help of third parties to bring smaller firms into their EDI supply chain. For example, SPS Commerce specializes in hooking up small businesses like American Outdoor Products (25 employees) to big supply chains like Recreational Equipment Incorporated (REI). SPS built an Internet-based application that translates EDI ordering and shipping requirements so that workers can access them through a Web browser on their PC.

Many companies have now gone beyond simple EDI-based applications to launch e-commerce initiatives with suppliers, customers, and employees to address business needs in new areas.

electronic data interchange (EDI)
An intercompany, application-to-application communication of data in standard format for business transactions.

Business-to-Consumer (B2C) E-Commerce

E-commerce for consumers is gaining broad acceptance, although some shoppers are not yet convinced that it is worthwhile to connect to the Internet, search for shopping sites, wait for the images to download, try to figure out the ordering process, and then worry about whether their credit card numbers will be stolen by a hacker. But attitudes are changing, and an increasing number of shoppers are benefiting from the convenience of e-commerce. In time-strapped households, consumers are asking themselves, "Why waste time fighting crowds in shopping malls when from the comfort of home I can shop online anytime and have the goods delivered directly?" These shoppers have found that many goods and services are cheaper when purchased via the Web—for example, stocks, books, newspapers, airline tickets, and hotel rooms. They can also get information about automobiles, cruises, loans, insurance, and home prices to cut better deals. More than just a tool for placing orders, the Internet is emerging as a paradise for comparison shoppers. Internet shoppers can, for example, unleash shopping bots or access sites such as Google, Excite, or Yahoo! to browse the Internet and obtain lists of items, prices, and merchants.

In March 2004, Yahoo! announced plans to buy European comparison-shopping site Kelkoo. France-based Kelkoo reaches about 10 percent of all European Internet users and counts more than 2,500 individual merchants among its paying customers. The site enables shoppers to compare prices on 3 million products in 25 categories, such as books, electronics, mobile phones, movies, music, and travel services. Yahoo!'s purchase of Kelkoo highlights the convergence of Web searches and online shopping, which will help Yahoo! achieve its goal to create the most comprehensive and best user experience on the Web. The purchase will also give Yahoo! another way to help marketers reach consumers by leveraging the Web's global reach. Yahoo! can now say to its biggest advertisers, companies such as Coca-Cola and General Motors, that they can reach global audiences through its portal.[5]

Many manufacturers and retailers have outsourced the physical logistics of delivering merchandise to cybershoppers—the storing, packing, shipping, and tracking of products. To provide this service, DHL, Federal Express, United Parcel Service, and other delivery firms have developed software tools and interfaces that directly link customers' ordering, manufacturing, and inventory systems with their own system of highly automated warehouses, call centers, and worldwide shipping network. The goal is to make the transfer of all information and inventory—from the manufacturer to the delivery firm to the consumer—fast and simple.

For example, when a customer orders a printer at the Hewlett-Packard Web site, that order actually goes to FedEx, which stocks all the products that HP sells online at a dedicated e-distribution facility in Memphis, a major FedEx shipping hub. FedEx ships the order, which triggers an e-mail notification to the customer that the printer is on its way and an inventory notice to HP that the FedEx warehouse now has one fewer printer in stock (see Figure 5.2). For product returns, HP enters return information into its own system, which is linked to FedEx. This signals a FedEx courier to pick up the unwanted item at the customer's house or business. Customers don't need to fill out shipping labels or package the item. Instead, the FedEx couriers use information transmitted over the Internet to a computer in their trucks to print a label from a portable printer attached to their belts. FedEx has control of the return, and HP can monitor its progress from start to finish.

Consumer-to-Consumer (C2C) E-Commerce

Consumer-to-consumer (C2C) e-commerce involves consumers selling directly to other consumers. Often this exchange is done through Web auction sites such as eBay, which enabled people to sell in excess of $8 billion in merchandise in one three-month period in 2004 to other consumers by auctioning the items off to the highest bidder.[6] The growth of C2C is responsible for reducing the use of the classified pages of a newspaper to advertise and sell personal items.

Product and Information Flow for HP Printers Ordered over the Web

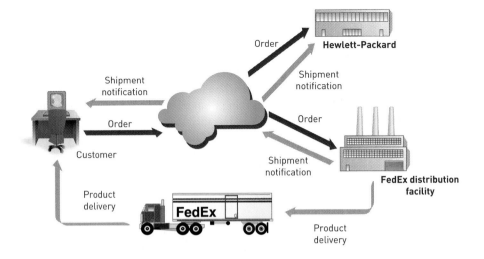

Mobile Commerce

mobile commerce (m-commerce)

The use of wireless devices such as PDAs and cell phones to place orders and conduct business.

wireless application protocol (WAP)

A standard set of specifications for Internet applications that run on handheld, wireless devices.

Mobile commerce (m-commerce) relies on the use of wireless devices, such as personal digital assistants, cell phones, and smart phones, to place orders and conduct business. Handset manufacturers such as Ericsson, Motorola, Nokia, and Qualcomm are working with communications carriers such as Cingular and Sprint to develop such wireless devices and their related technology. To ensure user-friendliness, the interface between the wireless device and its user must make purchasing an item nearly as easy as it is to purchase it on a PC. In addition, network speed must improve so that users do not become frustrated. Security is also a major concern, with two major issues: the security of the transmission itself and the trust that the transaction is being made with the intended party. Encryption can provide secure transmissions. Digital certificates, discussed later in this chapter, can ensure that transactions are made between the intended parties. To address the limitations of wireless devices (small screen size, limited input capabilities, less processing speed and bandwidth than desktop computers), the industry has undertaken a standardization effort for their Internet communications. The **wireless application protocol (WAP)**, is a standard set of specifications for Internet applications that run on handheld, wireless devices. WAP is a key technology underlying m-commerce.

Because m-commerce devices usually have a single user, they are ideal for accessing personal information and receiving targeted messages for a particular consumer. Through m-commerce, companies can reach individual consumers to establish one-to-one marketing relationships and permit communication to occur whenever it is convenient—in short, anytime and anywhere. Here are just a few examples of potential m-commerce applications:

- Banking customers can use their wireless, handheld devices to access their accounts and pay their bills.
- Clients of brokerage firms can view stock prices and company research as well as conduct trades to fit their schedules.
- Information services such as financial news, sports information, and traffic updates can be delivered to individuals whenever they want.
- On-the-move retail consumers can place and pay for orders instantaneously.
- Telecommunications service users can view service changes, pay bills, and customize their services.
- Retailers and service providers can send potential customers advertising, promotions, or coupons to entice them to try their services as they move past their place of business.

Albertson's, the $36 billion food and drug retailer, is using m-commerce technology to improve customer service. Shoppers at some of the retailer's stores can scan the bar codes of their items with handheld devices to record purchases, tally costs, receive special offers, and check out and pay. The handheld devices run off in-store wireless networks. At checkout time, customers can elect to use an "Express Pay Station," where they scan an "end-of-trip" bar code at the station and then automatically download the contents on the scanner into the register.[7]

Global E-Commerce and M-Commerce

The use of the Internet is growing rapidly in markets throughout Europe, Asia, and Latin America. So, e-commerce sites are broadening their focus from North American consumers. In Europe, the Nordic countries are the most wired, and 48 percent of Sweden's online population has ordered something in the past three months. In Asia, China represents a huge market opportunity for companies around the globe with a population exceeding 1.2 billion and an estimated 50 to 100 million Internet users. Japan has an estimated 100 million Internet users.[8] South Korea is a leader in terms of access to broadband technology, with more than 80 percent of total households having access to high-speed Internet services at less than $30 per month. Such easy accessibility encourages almost every South Korean to go online. In India, low personal computer ownership and Internet usage, an immature telecommunications infrastructure, lack of security, and high access costs slow the growth of e-commerce. In Latin America, Brazil and Mexico, the economic giants of Latin America, are leading the way in terms of developing e-commerce businesses and the requisite infrastructure, legal, and regulatory foundations necessary to support them. The entry into a global marketplace also comes with many challenges, as illustrated by the "Ethical and Societal Issues" special-interest feature.

The market for m-commerce in North America is expected to mature much later than in Western Europe and Japan for several reasons. In North America, responsibility for network infrastructure is fragmented among many providers, consumer payments are usually done by credit card, and most Americans are unfamiliar with mobile data services. In most Western European countries, communicating via wireless devices is common, and consumers are much more willing to use m-commerce. Japanese consumers are generally enthusiastic about new technology and are much more likely to use mobile technologies for making purchases.

E-COMMERCE APPLICATIONS

E-commerce is being applied to the retail and wholesale, manufacturing, marketing, investment and finance, and auction industries. Here are some current uses in those areas.

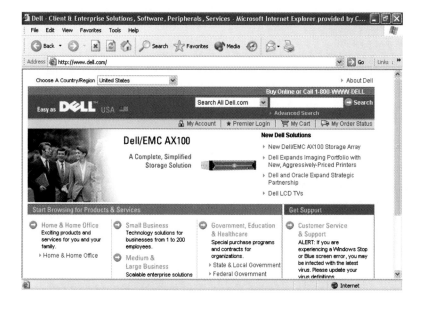

Dell sells its products through the Dell.com Web site.

ETHICAL AND SOCIETAL ISSUES

Canadian Prescription Drug Web Sites

The globalization of e-commerce presents businesses with challenges above and beyond cultural adaptations. Along with convenience, it has brought fierce competition for goods and services. Online pharmacies may be the most dramatic example of this fact.

As with many industries, pharmaceutical companies price their products differently for different countries. Prices for prescription medications are determined by a country's economic well-being, its medical insurance structure, the amount of demand for medications, competition in the market, and other factors. The result is that Americans pay more for prescription drugs than consumers in most other countries. This fact became obvious to the public when Canadian pharmacies began selling prescription drugs on the Web.

Web sites such as *www.CanadaPharmacy.com*, *www.Canada Meds.com*, and many others showed Americans that their northern neighbors pay only 30 to 50 percent what is charged in the United States for the same brand-name drugs. Naturally, many U.S. citizens began purchasing their drugs online from Canadian pharmacies. The federal government took a stand against buying prescriptions from foreign countries, citing the lack of regulation as a potential health risk for customers. The FDA Web site states, "It is illegal for anyone, including a foreign pharmacy, to ship prescription drugs that are not approved by the FDA into the U.S. even though the drug may be legal to sell in that pharmacy's country."

While the FDA discouraged the practice, increasing numbers of U.S. citizens began buying drugs online. New U.S.-based online businesses have also arisen, such as *www.BurlingtonDrug Club.com*, which act as intermediaries between U.S. citizens and Canadian pharmacies. Paid a percentage of sales by the Canadian pharmacies, these businesses assist Americans with finding the best price from the most reliable and trustworthy pharmacies.

State governments, frustrated by the federal government's inability to provide their citizens with affordable prescription drugs, have begun researching ways to sidestep federal laws and obtain Canadian drugs for their state workers and senior citizens. More than 25 states are now exploring the mass purchase of Canadian drugs.

Pharmaceutical companies are taking action to stop Canadian pharmacies from selling to anyone outside that country's borders. Such giants as Pfizer, GlaxoSmithKline, and Eli Lilly have systematically begun drug rationing in Canada. Canadians complain that the rationing barely leaves enough medicine to meet their own demand. Pfizer also cut off delivery to two Canadian prescription drug wholesalers because they were sending medicine back to the United States. Despite these efforts, U.S. prescriptions from Canada are growing by 18 percent annually and should top $1 billion in 2004.

Many in favor of global prescription medicine pricing are making serious accusations against the FDA. "We are here to remind the FDA they don't work for the big drug companies or their lobbyists," said Illinois Democratic Governor Rod Blagojevich. "They work for you and me."

The online prescription medicine business offers lessons on many levels and is indicative of the issues that we must contend with while building global e-commerce markets. How the U.S. government and the pharmaceutical companies resolve this issue will set a precedent for many other such cases that are sure to arise in the future.

Critical Thinking Questions

1. How effective can national laws be in controlling online purchases from other countries? What action could the federal government take if it wanted to put an end to re-imported drugs?
2. What solution would you recommend that might satisfy the pharmaceutical companies, the FDA, and the states and their citizens?

What Would You Do?

As the promotional manager for LA Artists, Inc., you are responsible for building an audience for 26 Los Angeles performers. One of the bands you represent has a huge local following, and you think it has the potential to make it really big. The band's Web site has helped build a national following, with huge sales in 20 states, which led to the band's recent national tour. LA Artists has gradually increased the price of CDs sold on Web site to make the most of the band's recent popularity. The next step is to begin promoting the band in other countries. To build an international audience, you will need to entice those who have not yet heard the band with discounted prices on CDs. You have decided that the $12 CDs sold online in the United States will be marked down to $5 for foreign sales.

3. Is it fair that U.S. citizens should be charged more than those outside the country to purchase the band's CDs?
4. How can you implement this e-commerce system so that international customers visiting the band's Web site will be charged less for CDs without offending U.S. fans? How can you be sure that U.S. citizens won't take advantage of the discounted prices?

SOURCES: Francis R. Carroll, "Rationing of Canadian Drugs Is Enough to Make You Sick," *Worcester Telegram & Gazette, Inc.*, March 29, 2004, *www.lexis-nexis.com*; "Prescription Drug Importation: Company That Helps Purchase Canadian Drugs Opens in Oklahoma," *Health & Medicine Week*, February 16, 2004, *www.lexis-nexis.com*; Todd Richmond, "FDA Blasts Wisconsin Drug Web site," *The Associated Press*, March 18, 2004; "State of Wisconsin, Prescription Drug Resource Center" Web site, accessed April 17, 2004, *drugsavings.wi.gov*; FDA Web site, accessed April 17, 2004, *www.fda.gov*.

Retail and Wholesale

E-commerce is being used extensively in retailing and wholesaling. **Electronic retailing**, sometimes called *e-tailing*, is the direct sale of products or services by businesses to consumers through electronic storefronts, which are typically designed around the familiar electronic catalog and shopping cart model. Companies such as Office Depot, Wal-Mart, and many others have used the same model to sell wholesale to employees of corporations. There are tens of thousands of electronic retail Web sites—selling literally everything from soup to nuts. In addition, cybermalls are another means to support retail shopping. A **cybermall** is a single Web site that offers many products and services at one Internet location—similar to a regular shopping mall. An Internet cybermall pulls multiple buyers and sellers into one virtual place, easily reachable through a Web browser. For example, PC Mall is a hardware, software, and consumer electronics retailer that sells items for the home, garden, and office décor; patriotic merchandise; gifts, collectibles, toys, and games; electronics; travel accessories; business supplies; sporting goods; tools and home hardware; health and beauty products; jewelry; and more.[9]

A key sector of wholesale e-commerce is spending on manufacturing, repair, and operations (MRO) goods and services—from simple office supplies to mission-critical equipment, such as the motors, pumps, compressors, and instruments that keep manufacturing facilities up and running smoothly. MRO purchases often approach 40 percent of a manufacturing company's total revenues, but the purchasing system can be haphazard, without automated controls. In addition to these external purchase costs, companies face significant internal costs resulting from outdated and cumbersome MRO management processes. For example, studies show that a high percentage of manufacturing downtime often results from not having the right part at the right time in the right place. The result is lost productivity and capacity. E-commerce software for plant operations provides powerful comparative searching capabilities to enable managers to identify functionally equivalent items, helping them spot opportunities to combine purchases for cost savings. Comparing various suppliers, coupled with consolidating more spending with fewer suppliers, leads to decreased costs. In addition, automated workflows are typically based on industry best practices, which can streamline processes.

Manufacturing

One approach taken by many manufacturers to raise profitability and improve customer service is to move their supply chain operations onto the Internet. Here they can form an **electronic exchange** to join with competitors and suppliers alike, using computers and Web sites to buy and sell goods, trade market information, and run back-office operations, such as inventory control, as shown in Figure 5.3. With such an exchange, the business center is not a physical building but a network-based location where business interactions occur. This approach has greatly speeded up the movement of raw materials and finished products among all members of the business community, thus reducing the amount of inventory that must be maintained. It has also led to a much more competitive marketplace and lower prices. Private exchanges are owned and operated by a single company. The owner uses the exchange to trade exclusively with established business partners. Public exchanges are owned and operated by industry groups. They provide services and a common technology platform to their members and are open, usually for a fee, to any company that wants to use them.

One example of a successful exchange is the WorldWide Retail Exchange (WWRE) founded in 2000 by 17 international retailers to enable participants in the food, general merchandise, textile/home, and drugstore sectors to simplify and automate supply chain processes. Current membership consists of 64 retail industry leaders from around the world with a combined revenue of more than $900 billion. Over its first four years of operation, the WWRE saved its members more than $1 billion on maintenance, repair, and operating equipment and private-label goods.[10] Nick Parnaby, global director of member development at the WorldWide Retail Exchange, says that the typical member saves 13 percent, but that number rises to 20 percent if member companies "pool their spend" with one another to improve economies of scale.[11]

electronic retailing (e-tailing)
The direct sale from business to consumer through electronic storefronts, typically designed around an electronic catalog and shopping cart model.

cybermall
A single Web site that offers many products and services at one Internet location.

electronic exchange
An electronic forum where manufacturers, suppliers, and competitors buy and sell goods, trade market information, and run back-office operations.

Figure 5.3

Model of an Electronic Exchange

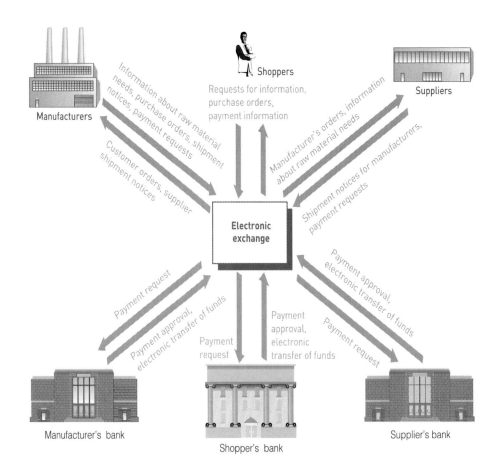

At the turn of the twenty-first century, Internet and brick-and-mortar companies set up more than 1,000 online marketplaces in 70 industries. Many of these exchanges are no longer operating because they failed to bring their members true business benefits. Today there are about 200 exchanges in existence. Table 5.1 provides a partial list of some of the more successful electronic exchanges.

Table 5.1

Some Successful Electronic Exchanges

Source: Adapted from Steve Ulfelder, "B2B Survivors," *Computerworld*, February 2, 2004, accessed at *www.computerworld.com*.

Exchange	Industry
Covisint	Automotive
Elemica	Chemical
Exostar	Defense
E2Open Inc.	Technology
Global Healthcare Exchange	Healthcare
Pantellos Group	Utility
Trade-Ranger	Energy
Transora	Consumer goods
UCCnet	Consumer goods
WorldWide Retail Exchange	Retail

The members of successful exchanges have benefited greatly from their services. Cinergy Corp., an energy supplier in the Midwest, routes 50 percent of its purchase orders through the Pantellos exchange and has dramatically reduced the errors in its order processing. Overall, members of Pantellos have saved an aggregate $315 million since the exchange was founded. The Dow Chemical Company saved thousands of hours of effort and hopes to reduce inventories by as much as 50 percent through use of the Elemica exchange.[12]

Several strategic and competitive issues are associated with the use of exchanges. Many companies distrust their corporate rivals and fear they may lose trade secrets through participation in such exchanges. Suppliers worry that the online marketplaces and their auctions will drive down the prices of goods and favor buyers. Suppliers also can spend a great deal of money in the setup to participate in multiple exchanges. For example, more than a dozen new exchanges have appeared in the oil industry, and the printing industry is up to more than 20 online marketplaces. Until a clear winner emerges in particular industries, suppliers are more or less forced to sign on to several or all of them. Yet another issue is potential government scrutiny of exchange participants—anytime competitors get together to share information, it raises questions of collusion or antitrust behavior.

Many companies that already use the Internet for their private exchanges have no desire to share their expertise with competitors. At Wal-Mart, the world's number-one retail chain, executives turned down several invitations to join exchanges in the retail and consumer goods industries. Wal-Mart is pleased with its in-house exchange, Retail Link, which connects the company to 7,000 worldwide suppliers that sell everything from toothpaste to furniture.

Marketing

The nature of the Web allows firms to gather much more information about customer behavior and preferences than they could using other marketing approaches. Marketing organizations can measure many online activities as customers and potential customers gather information and make their purchase decisions. Analysis of this data is complicated because of the Web's interactivity and because each visitor voluntarily provides or refuses to provide personal data such as name, address, e-mail address, telephone number, and demographic data. Internet advertisers use the data they gather to identify specific portions of their markets and target them with tailored advertising messages. This practice, called **market segmentation**, divides the pool of potential customers into subgroups, which are usually defined in terms of demographic characteristics such as age, gender, marital status, income level, and geographic location.

market segmentation
The identification of specific markets to target them with advertising messages.

comScore Networks is a global information provider for large companies seeking information on consumer behavior to boost their marketing, sales, and trading strategies.

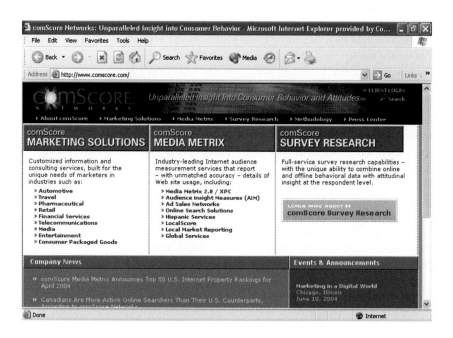

Technology-enabled relationship management is a new twist on establishing direct customer relationships made possible when firms promote and sell on the Web. **Technology-enabled relationship management** occurs when a firm obtains detailed information about a customer's behavior, preferences, needs, and buying patterns and uses that information to set prices, negotiate terms, tailor promotions, add product features, and otherwise customize its entire relationship with that customer.

DoubleClick is a leading global Internet advertising company that leverages technology and media expertise to help advertisers use the power of the Web to build relationships with customers. The DoubleClick Network is its flagship product, a collection of high-traffic and well-recognized sites on the Web including MSN, Sports Illustrated, Continental Airlines, the Washington Post, CBS, and more than 1,500 others. This network of sites is coupled with DoubleClick's proprietary DART targeting technology, which allows advertisers to target their best prospects based on the most precise profiling criteria available. DoubleClick then places a company's ad in front of those best prospects. DART powers more than 60 billion ads per month and is trusted by top advertising agencies. Comprehensive online reporting lets advertisers know how their campaign is performing and what type of users are seeing and clicking on their ads. This high-level targeting and real-time reporting provide speed and efficiency not available in any other medium. The system is also designed to track advertising transactions, such as impressions and clicks, to summarize these transactions in the form of reports and to compute DoubleClick Network member compensation.

Investment and Finance

The Internet has revolutionized the world of investment and finance. Perhaps the changes have been so great because this industry had so many built-in inefficiencies and so much opportunity for improvement.

Online Stock Trading

Before the World Wide Web, if you wanted to invest in stocks, you called your broker and asked what looked promising. He'd tell you about two or three companies and then would try to sell you shares of a stock or perhaps a mutual fund. The sales commission was well over $100 for the stock (depending on the price of the stock and the number of shares purchased) or as much as an 8 percent sales charge on the mutual fund. If you wanted information about the company before you invested, you would have to wait two or three days for a one-page Standard and Poor's stock report providing summary information and a chart of the stock price for the past two years to arrive in the mail. Once you purchased or sold the stock, it would take two days to get an order confirmation in the mail, detailing what you paid or received for the stock.

The brokerage business adapted to the Internet faster than any other arm of finance. To make a trade, all you need to do is log on to the Web site of your online broker, and with a few keystrokes and a few clicks of your mouse to identify the stock and number of shares involved in the transaction, you can buy and sell securities in seconds. In addition, an overwhelming amount of free information is available to online investors—from the latest Securities & Exchange filings to the rumors spread in chat rooms. See Table 5.2 for a short list of the more valuable sites.

One indispensable tool of the online investor is a portfolio tracker. This tool allows you to enter information about the securities you own—ticker symbol, number of shares, price paid, and date purchased—at a tracker Web site. You can then access the tracker site to see how your stocks are doing. (There is typically a 15- to 20-minute delay between the price displayed at the site and the price at which the stock is actually being sold.) In addition to reporting the current value of your portfolio, most sites provide access to news, charts, company profiles, and analyst ratings on each of your stocks. You can also program many of the trackers to watch for certain events (e.g., stock price change of more than +/− 3 percent in a single day). When one of the events you specified occurs, an "alert" symbol is posted next to the affected stock. Table 5.3 lists a number of the more popular tracker Web sites.

Table 5.2

Web Sites Useful to Investors

Name of Site	URL	Description
The Street Dot Com	www.thestreet.com	Provides information, advice, and recommendations based on popular radio/TV personality Jim Cramer
Quote.com	http://finance.lycos.com	Enables you to get stock quotes and price charts and access message boards for selected stocks
Kiplinger	www.kiplinger.com	Provides personal financial advice
Elite Trader	www.elitetrader.com	Virtual gathering place for day traders with bulletin boards and chat rooms
DRIP Advisor	www.dripadvisor.com	Covers the basics of dividend reinvestment programs (DRIPs), what companies offer DRIPs, and how to start a DRIP
EDGAR Online	www.edgar-online.com	Provides access to company filings with the Securities and Exchange Commission (SEC)
Federal Filings Online	www.fedfil.com	Dow Jones directory of documents filed with the federal government, including bankruptcy proceedings, initial public offering (IPO) filings, SEC reports, and court cases

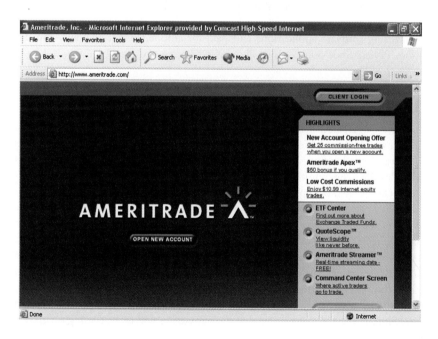

Ameritrade.com is an online brokerage site that offers information, tools, and account-management services for investors.

Online Banking

Online banking customers can check balances of their savings, checking, and loan accounts; transfer money among accounts; and pay their bills. These customers enjoy the convenience of not writing checks in longhand, think they have a better knowledge of their current balances, and appreciate the reduction of expenditures on envelopes and stamps. All of the nation's major banks and many of the smaller banks enable their customers to pay bills online. The number of Americans who pay at least some of their bills online is estimated to exceed 35 million.

Table 5.3

Popular Stock Tracker Web Sites

Name of the Web Stock Tracker Site	URL
MSN MoneyCentral	http://moneycentral.msn.com/investor
Quicken.com	www.quicken.com
The Motley Fool	www.fool.com
Yahoo!	http://quote.yahoo.com
Morningstar	www.morningstar.com

electronic bill presentment
A method of billing whereby the biller posts an image of your statement on the Internet and alerts you by e-mail that your bill has arrived.

Here's how electronic bill payment works. You first set up a list of frequent payees, along with their addresses and a code describing the type of payment, such as "home mortgage." Then, when you go online to pay your bills, you simply enter the code or name assigned to the check recipient, the amount of the check, and the date you want it paid. In many cases, the bank still prints and mails a check, so you have to time your online transactions to allow for bank processing and mail delays. But most bill-paying programs allow you to schedule recurring payments for every week, month, or quarter, which you might want to do for your auto loan or health insurance bill.

The next advance in online bill paying is **electronic bill presentment**, which eliminates all paper, right down to the bill itself. With this process, the biller posts an image of your statement on the Internet and alerts you by e-mail that your bill has arrived. You then direct your bank to pay it. Some vendors such as Edocs Inc. and Avolent Inc. offer electronic bill presentment and payment software that users must purchase and install on their computers. Other companies, such as Xign Corp. and Metavante Corp., offer electronic bill presentment and payment as a hosted service to subscribers. After deciding whether to purchase electronic bill presentment as software or as a service, customers must choose how to link the system to their existing software applications and data for automated bill payment and dispute resolution. Payless Shoe Source, Inc., chose Xign's service to distribute invoices to about 45,000 vendors.[13]

Auctions

The Internet has created many new options for C2C e-commerce, including electronic auctions, in which geographically dispersed buyers and sellers can come together. A special type of auction called *bidding* allows a prospective buyer to place only one bid for an item or a service. Priceline.com's initial business model enabled consumers to achieve significant savings by naming their own price for goods and services. Priceline.com took these consumer offers and then presented them to sellers, who filled as much of the demand as they wished at price points determined by the buyers.

eBay selected Pitney Bowes to provide its postage service, which gives its customers easy access to U.S. Postal Service (USPS) shipping services. With this new tool, eBay users can purchase postage online, pay for it using their PayPal accounts, and print their shipping label from their computer—all from the eBay Web site. Once the label is purchased, both the buyer and the seller will be able to track the delivery status of the package online from the eBay site. Simplifying the shipping part of the transaction will make trading on eBay faster and easier.[14]

Now that we've examined some of the applications of e-commerce, let's look at some of the technical issues related to information systems and technology that make it possible.

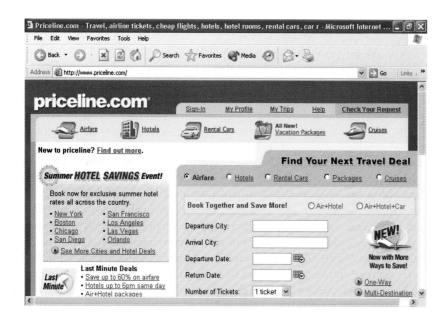

E-COMMERCE TECHNOLOGY, INFRASTRUCTURE, AND DEVELOPMENT

For e-commerce to succeed, a complete set of hardware, software, and network components must be chosen carefully and integrated to support a large volume of transactions with customers, suppliers, and other business partners worldwide. Online consumers frequently complain that poor Web site performance (e.g., slow response time and "lost" orders) drives them to abandon some e-commerce sites in favor of those with better, more reliable performance. This section provides a brief overview of the key technology infrastructure components (see Figure 5.4).

Figure 5.4

Key E-Commerce Technical Components

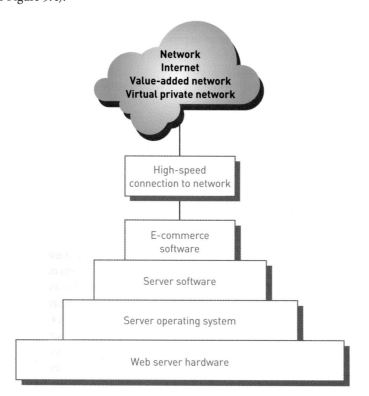

Hardware

A Web server complete with the appropriate software is key to successful e-commerce. The amount of storage capacity and computing power required of the Web server depends primarily on two things—the software that must run on the server and the volume of e-commerce transactions that must be processed. Although business managers and information systems staff can define the software to be used, it is difficult for them to estimate how much traffic the site will generate. As a result, the most successful e-commerce solutions are designed to be highly scalable so that they can be upgraded to meet unexpected user traffic.

Many companies decide that a third-party Web service provider is the best way to meet their initial e-commerce needs. A Web service rents out space on its computer system and provides a high-speed connection to the Internet, which minimizes the initial setup costs for e-commerce. The service provider can also provide personnel trained to operate, troubleshoot, and manage the Web server. Other companies decide to take full responsibility for acquiring, operating, and supporting their own Web server hardware and software, but this approach requires considerable up-front capital and a set of skilled and trained individuals. Whichever approach is taken, there must be adequate hardware backup to avoid a major business disruption in case of a failure of the primary Web server.

Software

Each e-commerce Web server must have software to perform a number of fundamental services, including security and identification authentication, retrieval and sending of Web pages, and Web page construction. The two most popular Web server software packages are Apache HTTP Server and Microsoft Internet Information Server.

Web site development tools include features such as an HTML/visual Web page editor (e.g., Microsoft's FrontPage, NetStudio's NetStudio, SoftQuad's HoTMetaL Pro), software development kits that include sample code and code development instructions for languages such as Java or Visual Basic, and Web page upload support to move Web pages from a development PC to the Web site. Which tools are bundled with the Web server software depends on which Web server software you select.

Web page construction software uses Web editors to produce Web pages—either static or dynamic. *Static Web pages* always contain the same information—for example, a page that provides text about the history of the company or a photo of corporate headquarters. *Dynamic Web pages* contain variable information and are built in response to a specific Web visitor's request. For example, if a Web site visitor inquires about the availability of a certain product by entering a product identification number, the Web server will search the product inventory database and generate a dynamic Web page based on the current product information it found, thus fulfilling the visitor's request. This same request by another visitor later in the day may yield different results due to ongoing changes in product inventory. A server that handles dynamic content must be able to access information from a variety of databases. The use of open database connectivity enables the Web server to assemble information from different database management systems, such as SQL Server, Oracle, and Informix.

Once you have located or built a host server, including the hardware, operating system, and Web server software, you can begin to investigate and install e-commerce software. There are three core tasks that **e-commerce software** must support: catalog management, product configuration, and shopping cart facilities.

Catalog management software combines different product data formats into a standard format for uniform viewing, aggregating, and integrating catalog data into a central repository for easy access, retrieval, and updating of pricing and availability changes. The data required to support large catalogs is almost always stored in a database on a computer that is separate from, but accessible to, the e-commerce server machine. The effort to build and maintain online catalogs can be substantial.

Corporate Express sells furniture, paper, computer supplies, and office equipment. It maintains catalogs tailored to each customer's format, terminology, and buying practices. If certain customers don't buy office furniture from Corporate Express, office furniture is blocked from their version of the catalog. Corporate Express also maintains a list of items

Web site development tools
Tools used to develop a Web site, including HTML or visual Web page editor, software development kits, and Web page upload support.

Web page construction software
Software that uses Web editors and extensions to produce both static and dynamic Web pages.

e-commerce software
Software that supports catalog management, product configuration, and shopping cart facilities.

catalog management software
Software that automates the process of creating a real-time interactive catalog and delivering customized content to a user's screen.

that each customer orders frequently, as well as the special terms and prices that the customer has negotiated. Such attention to customization is greatly appreciated by customers because it makes their ordering process easier. Corporate buyers also appreciate the fact that their employees can only purchase prearranged items so that "maverick" buying is eliminated.[15]

Customers need help when an item they are purchasing has many components and options. **Product configuration software** tools assist B2B salespeople with matching their company's products to customer needs. Buyers use the new Web-based product configuration software to build the product they need online with little or no help from salespeople. For example, Dell customers use product configuration software to build the computer of their dreams. Use of such software can expand into the service arena as well, with consumer loans and financial services to help people decide what sort of loan or insurance is best for them.

Today many e-commerce sites use an **electronic shopping cart** to track the items selected for purchase, allowing shoppers to view what is in their cart, add new items to it, or remove items from it, as shown in Figure 5.5. To order an item, the shopper simply clicks that item. All the details about it—including its price, product number, and other identifying information—are stored automatically. If the shopper later decides to remove one or more items from the cart, he or she can do so by viewing the cart's contents and removing any unwanted items. When the shopper is ready to pay for the items, he or she clicks a button (usually labeled "proceed to checkout") and begins a purchase transaction. Clicking the "Checkout" button displays another screen that usually asks the shopper to fill out billing, shipping, and payment method information and to confirm the order.

product configuration software
Software used by buyers to build the product they need online.

electronic shopping cart
A model commonly used by many e-commerce sites to track the items selected for purchase, allowing shoppers to view what is in their cart, add new items to it, and remove items from it.

Figure 5.5

Electronic Shopping Cart

An electronic shopping cart (or bag) allows online shoppers to view their selections and add or remove items.

Electronic Payment Systems

Electronic payment systems are a key component of the e-commerce infrastructure. Check writing is giving way to electronic payments, with checks representing a decreasing percent of non-cash payments. Current e-commerce technology relies on user identification and encryption to safeguard business transactions. Actual payments are made in a variety of ways, including electronic cash; electronic wallets; and credit, charge, debit, and smart cards.

Authentication technologies are used by organizations to confirm the identity of a user requesting access to information or assets. A **digital certificate** is an attachment to an e-mail message or data embedded in a Web site that verifies the identity of a sender or Web site.

digital certificate
An attachment to an e-mail message or data embedded in a Web page that verifies the identity of a sender or a Web site.

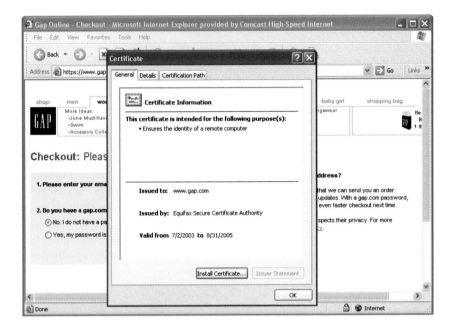

A security system assures customers that information they provide, such as credit card numbers, to retailers cannot be viewed by anyone else on the Web.

One indicator of the security associated with a Web site is visible on screen. Look at the bottom left corner of your browser before sending your credit card number to an e-commerce vendor. If you use Netscape Navigator, make sure you see a solid key in a small blue rectangle. If you use Microsoft Internet Explorer, the words "Secure Web site" appear near a little gold lock. If you're worried about how secure a secure connection is, visit the Netcraft.com site. At this site you can type in any Web site address and determine the equipment being used for secure transactions. One more tip: To ensure security, you should always use the newest browser available. The newer the browser, the better the security.

Electronic Cash

Electronic cash is an amount of money that is computerized, stored, and used as cash for e-commerce transactions. A consumer must open an account with a bank to obtain electronic cash. Whenever the consumer wants to withdraw electronic cash to make a purchase, he or she accesses the bank via the Internet and presents proof of identity—typically a digital certificate. After the bank verifies the consumer's identity, it issues the consumer the requested amount of electronic cash and deducts the same amount from the consumer's account. The electronic cash is stored in the consumer's electronic wallet on his or her computer's hard drive, or on a smart card (both are discussed later).

Consumers can spend their electronic cash when they locate e-commerce sites that accept electronic cash for payment. The consumer sends electronic cash to the merchant for the specified cost of the goods or services. The merchant validates the electronic cash to be certain it is not forged and belongs to the customer. Once the goods or services are shipped to the consumer, the merchant presents the electronic cash to the issuing bank for deposit. The bank then credits the merchant's account for the transaction amount, minus a small service charge.

electronic cash
An amount of money that is computerized, stored, and used as cash for e-commerce transactions.

Electronic Wallets

Online shoppers quickly tire of repeatedly entering their shipment and payment information each time they make a purchase. An **electronic wallet** holds credit card information, electronic cash, owner identification, and address information. It provides this information at an e-commerce site's checkout counter. When consumers click on items to purchase, they can then click on their electronic wallet to order the item, thus making online shopping much faster and easier.

electronic wallet
A computerized stored value that holds credit card information, electronic cash, owner identification, and address information.

Credit, Charge, Debit, and Smart Cards

Online shoppers use credit and charge cards for the majority of their Internet purchases. A credit card, such as Visa or MasterCard, has a preset spending limit based on the user's credit limit, and each month the user can pay off a portion of the amount owed or the entire credit card balance. Interest is charged on the unpaid amount. A charge card, such as American Express, carries no preset spending limit, and the entire amount charged to the card is due at the end of the billing period. Charge cards do not involve lines of credit and do not accumulate interest charges.

Debit cards look like credit cards or automated teller machine (ATM) cards, but they operate like cash or a personal check. While a credit card is a way to "buy now, pay later," a debit card is a way to "buy now, pay now." Debit cards allow you to spend only what is in your bank account. It is a quick transaction between the merchant and your personal bank account. When you use a debit card, your money is quickly deducted from your checking or savings account. Credit, charge, and debit cards currently store limited information about you on a magnetic stripe. This information is read each time the card is swiped to make a purchase. All credit card customers are protected by law from paying any more than $50 for fraudulent transactions. At Visa, online purchases account for the highest amount of purchase fraud—24 cents for every $100 spent, compared with 6 cents for every $100 overall. Indeed, the risk of bogus credit card transactions has slowed the growth of e-commerce by exposing merchants to substantial losses and making online shoppers nervous. Based on the results of its annual survey of e-commerce crime, security company CyberSource estimates online crooks will make away with $1.6 billion of $94 billion (about 1.7 percent) of annual U.S. business-to-consumer e-commerce revenue.[16]

MasterCard estimates that 7 percent of its $922 billion in annual card purchases are now taking place on the Web. As a result, the firm now requires that high-volume merchants and payment processors conduct quarterly assessments of their Web sites, or MasterCard will stop doing business with them. Both host and network-based software offerings provide such assessments. Host-based tools reside on servers and desktops and use software agents that can log on to a host to report their findings to a management console. Network-based tools scan remotely in ways similar to a hacker probe. The agents check user access logs and the identity of those who accessed financial data on the server. However, if there is no coordinated staff approach, network administrators might think that the network is under attack because of the scans.[17]

The **smart card** is a credit card–sized device with an embedded microchip to provide electronic memory and processing capability. Smart cards can be used for a variety of purposes, including storing a user's financial facts, health insurance data, credit card numbers, and network identification codes and passwords. They can also store monetary values for spending.

Smart cards are better protected from misuse than conventional credit, charge, and debit cards because the smart card information is encrypted. Conventional credit, charge, and debit cards clearly show your account number on the face of the card. The card number, along with a forged signature, is all that a thief needs to purchase items and charge them against your card. A smart card makes credit theft practically impossible because a key to unlock the encrypted information is required, and there is no external number that a thief can identify and no physical signature a thief can forge.

Smart cards have been around for more than a decade and are widely used in Europe, Australia, and Japan, but they have not caught on in the United States. Use has been limited because there are so few smart card readers to record payments, and U.S. banking regulations have slowed smart card marketing and acceptance as well. American Express launched its Blue card smart card in 1999. You can use a smart card reader that attaches to your PC monitor to make online purchases with your American Express card. You must visit the American Express Web site to get an electronic wallet to store your credit card information and shipping address. When you want to buy something online, you go to the checkout screen of a Web merchant, swipe your Blue card through the reader, type in a password, and you're done. The digital wallet automatically tells the vendor your credit card number, its expiration date, and your shipping information.

smart card
A credit card–sized device with an embedded microchip to provide electronic memory and processing capability.

As we pointed out earlier, e-commerce applications often provide customers the ability to order and pay for products and request service and information. As such, these applications form the basis for a class of information systems called *transaction processing systems*, which we discuss next.

FedEx adds value to its service by providing timely and accurate on-line data on the exact location of a package.

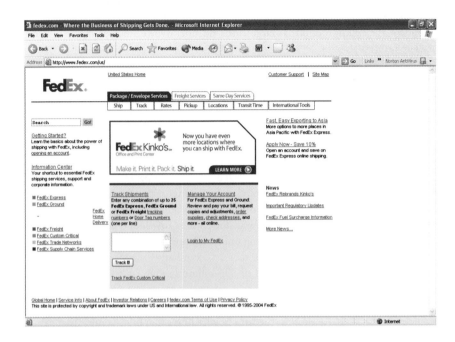

AN OVERVIEW OF TRANSACTION PROCESSING SYSTEMS

Transaction processing was one of the first business processes to be computerized, and without information systems, recording and processing business transactions would consume huge amounts of an organization's resources. The transaction processing system (TPS) also provides employees involved in other business processes—the management information system/decision support system (MIS/DSS) and the special-purpose information systems—with data to help them achieve their goals. A TPS serves as the foundation for the other systems (see Figure 5.6). Transaction processing systems perform routine operations such as sales ordering and billing, often performing the same operations daily or weekly. The amount of support for decision making that a TPS directly provides managers and workers is low.

Figure 5.6

TPS, MIS/DSS, and Special Information Systems in Perspective

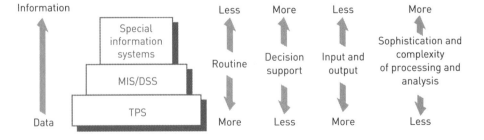

These systems require a large amount of input data and produce a large amount of output without requiring sophisticated or complex processing. As we move from transaction processing to management information/decision support, and special-purpose information systems, we see less routine, more decision support, less input and output, and more sophisticated and complex processing and analysis. But the increase in sophistication and complexity in moving from transaction processing does not mean that it is less important to a business. In most cases, all these systems start as a result of one or more business transactions.

Every organization has TPSs, which process the detailed data necessary to update records about the fundamental business operations of the organization. These systems include order entry, inventory control, payroll, accounts payable, accounts receivable, and the general ledger, to name just a few. The input to these systems includes basic business transactions such as customer orders, purchase orders, receipts, time cards, invoices, and customer payments. The result of processing business transactions is that the organization's records are updated to reflect the status of the operation at the time of the last processed transaction. Automated TPSs consist of all the components of a computer-based information system (CBIS), including databases, telecommunications, people, procedures, software, and hardware devices used to process transactions. The processing activities include data collection, data editing, data correction, data manipulation, data storage, and document production.

The U.S. Bureau of Customs and Border Protection (part of the U.S. Department of Homeland Security) has the busiest TPS in the world. At peak workloads, the bureau's system processes an incredible 51,448 transactions per second. The bureau's TPS uses an Advantage CA-Datacom database from Computer Associates and runs on IBM eServer zSeries hardware.[18]

Traditional Transaction Processing Methods and Objectives

When computerized TPSs first evolved, only one method of processing was available. All transactions were collected in groups, called *batches*, and processed together. With **batch processing systems**, business transactions are accumulated over a period of time and prepared for processing as a single unit or batch (see Figure 5.7a). The time period during which transactions are accumulated is whatever length of time is needed to meet the needs of the users of that system. For example, it may be important to process invoices and customer payments for the accounts receivable system daily. On the other hand, the payroll system may receive time cards and process them biweekly to create checks and update employee earnings records as well as to distribute labor costs. The essential characteristic of a batch processing system is that there is some delay between the occurrence of the event and the eventual processing of the related transaction to update the organization's records.

Today's computer technology allows another processing method, called *online, real-time*, or **online transaction processing (OLTP)**. With this form of data processing, each transaction is processed immediately, without the delay of accumulating transactions into a batch (see Figure 5.7b). As soon as the input is available, a computer program performs the necessary processing and updates the records affected by that single transaction. Consequently, at any time, the data in an online system always reflects the current status. When you make an airline reservation, for instance, the transaction is processed, and all databases, such as seat occupancy and accounts receivable, are updated immediately. This type of processing is absolutely essential for businesses that require data quickly and update it often, such as airlines, ticket agencies, and stock investment firms. Many companies have found that OLTP helps them provide faster, more efficient service—one way to add value to their activities in the eyes of the customer. Increasingly, companies are using the Internet to perform many OLTP functions.

Even though the technology exists to run TPS applications using online processing, it is not done for all applications. For many applications, batch processing is more appropriate and cost-effective. Payroll transactions and billing are typically done via batch processing. Specific goals of the organization define the method of transaction processing best suited for the various applications of the company. Now that we have seen the big picture of how organizations integrate their various TPS functions into a comprehensive system, let's take a closer look at the basic activities that all TPSs perform.

batch processing system
Method of computerized processing in which business transactions are accumulated over a period of time and prepared for processing as a single unit or batch.

online transaction processing (OLTP)
Computerized processing in which each transaction is processed immediately, without the delay of accumulating transactions into a batch.

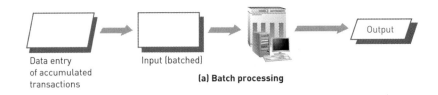

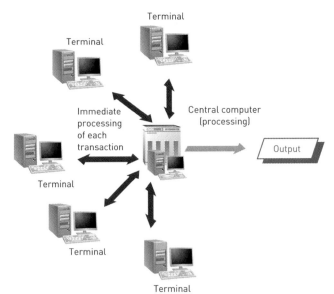

Figure 5.7

Batch Versus Online Transaction Processing

(a) Batch processing inputs and processes data in groups. (b) In online processing, transactions are completed as they occur.

(a) Batch processing

(b) Online transaction processing

Transaction Processing Activities

All TPSs perform a common set of basic data-processing activities. TPSs capture and process data that describes fundamental business transactions. This data is used to update databases and to produce a variety of reports that people both within and outside the enterprise use. The business data goes through a **transaction processing cycle** that includes data collection, data editing, data correction, data manipulation, data storage, and document production (see Figure 5.8).

transaction processing cycle
The process of data collection, data editing, data correction, data manipulation, data storage, and document production.

Figure 5.8

Data Processing Activities Common to TPSs

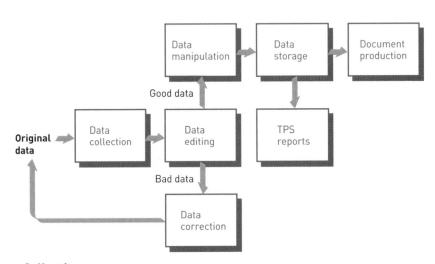

Data Collection

The process of capturing and gathering all data necessary to complete transactions is called **data collection**. In some cases it can be done manually, such as by collecting hand-written sales orders or changes to inventory. In other cases, data collection is automated via special input devices such as scanners, point-of-sale devices, and terminals.

data collection
The process of capturing and gathering all data necessary to complete transactions.

Data collection begins with a transaction (e.g., taking a customer order) and results in the origination of data that is input to the TPS. Data should be captured at its source, and it should be recorded accurately, in a timely fashion, with minimal manual effort, and in a form that can be directly entered into the computer rather than keying the data from a document. This approach is called *source data automation*. An example of source data automation is the use of automated devices at a retail store to speed the checkout process—either UPC codes read by a scanner or RFID signals picked up when the items approach the checkout stand. The use of both UPC bar codes and RFID tags is quicker and more accurate than having a clerk enter codes manually at the cash register. The product ID for each item is determined automatically, and its price is found in the item database. The point-of-sale TPS uses the price data to determine the customer's bill. The store's inventory and purchase databases record the number of units of an item purchased, the date, the time, and the price. The inventory database generates a management report notifying the store manager to reorder items that have fallen below the reorder quantity. The detailed purchases database can be used by the store or sold to marketing research firms or manufacturers for detailed sales analysis (see Figure 5.9).

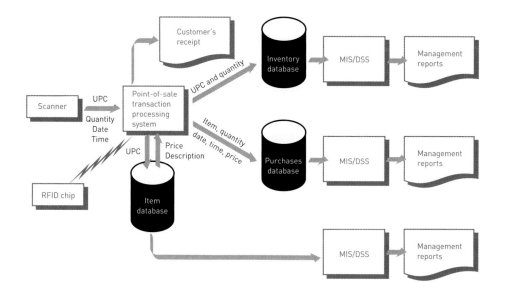

Figure 5.9

Point-of-Sale TPS

The purchase of items at the checkout stand updates a store's inventory database and its database of purchases.

Data Editing

An important step in processing transaction data is to perform **data editing** for validity and completeness to detect any problems. For example, quantity and cost data must be numeric and names must be alphabetic; otherwise, the data is not valid. Often the codes associated with an individual transaction are edited against a database containing valid codes. If any code entered (or scanned) is not present in the database, the transaction is rejected.

data editing
The process of checking data for validity and completeness.

Data Correction

It is not enough simply to reject invalid data. The system should also provide error messages that alert those responsible for the data editing function. Error messages must specify what problem is occurring so that corrections can be made. A **data correction** involves reentering miskeyed or misscanned data that was found during data editing. For example, a UPC that is scanned must be in a master table of valid UPCs. If the code is misread or does not exist in the table, the checkout clerk is given an instruction to rescan the item or key in the information manually.

data correction
The process of reentering miskeyed or misscanned data that was found during data editing.

Data Manipulation

Another major activity of a TPS is **data manipulation**, the process of performing calculations and other data transformations related to business transactions. Data manipulation can include classifying data, sorting data into categories, performing calculations, summarizing results, and storing data in the organization's database for further processing. In a payroll

data manipulation
The process of performing calculations and other data transformations related to business transactions.

TPS, for example, data manipulation includes multiplying an employee's hours worked by the hourly pay rate. Overtime calculations, federal and state tax withholdings, and deductions are also performed.

Data Storage

data storage
The process of updating one or more databases with new transactions.

Data storage involves updating one or more databases with new transactions. Once the update process is complete, this data can be further processed and manipulated by other systems so that it is available for management decision making. Thus, although transaction databases can be considered a by-product of transaction processing, they have a pronounced effect on nearly all other information systems and decision-making processes in an organization.

Document Production and Reports

document production
The process of generating output records and reports.

TPSs produce important business documents. **Document production** involves generating output records and reports. These documents may be hard-copy paper reports or displays on computer screens (sometimes referred to as *soft copy*). Paychecks, for example, are hard-copy documents produced by a payroll TPS, while an outstanding balance report for invoices might be a soft-copy report displayed by an accounts receivable TPS. Often, results from one TPS are passed downstream as input to other systems. For example, the results of updating the inventory database are used to create the stock exception report (a type of management report) of items whose inventory level is below the reorder point.

In addition to major documents such as checks and invoices, most TPSs provide other useful management information and decision support, such as printed or on-screen reports that help managers and employees perform various activities. A report showing current inventory is one example; another might be a document listing items ordered from a supplier to help a receiving clerk check the order for completeness when it arrives. A TPS can also produce reports required by local, state, and federal agencies, such as statements of tax withholding and quarterly income statements.

Basic TPS Applications

TPSs perform several critical functions for an organization's day-to-day activities. Two of the most basic are order processing systems and purchasing and accounting systems.

Order Processing Systems

Because transaction processing systems were first built to handle the give and take between customers and suppliers, we can gain an understanding of how they work by examining several common TPSs that support order processing (see Table 5.4).

Table 5.4

Systems That Support Order Processing

Order Processing	Purchasing	Accounting
• Order entry	• Inventory control (raw materials, packing materials, spare parts, and supplies)	• Budget
• Sales configuration		• Accounts receivable
• Shipment planning	• Purchase order processing	• Payroll
• Inventory control (finished product)	• Receiving	• Asset management
• Invoicing and billing	• Accounts payable	• General ledger
• Customer interaction		
• Routing and scheduling		

order processing systems
Systems that process order entry, sales configuration, shipment planning, shipment execution, inventory control, invoicing, customer relationship management, and routing and scheduling.

Order processing systems include order entry, sales configuration, shipment planning, shipment execution, inventory control, invoicing, customer relationship management, and routing and scheduling. The business processes supported by these systems are so critical to the operation of an enterprise that the order processing systems are sometimes referred to as the "lifeblood of the organization." Figure 5.10 is a system-level flowchart that shows the various systems and the information that flows between them. A rectangle represents a system, a line represents the flow of information from one system to another, and a circle represents any entity outside the system—in this case, the customer. What is key to note here is how these TPSs work together as an integrated whole to support major business processes.

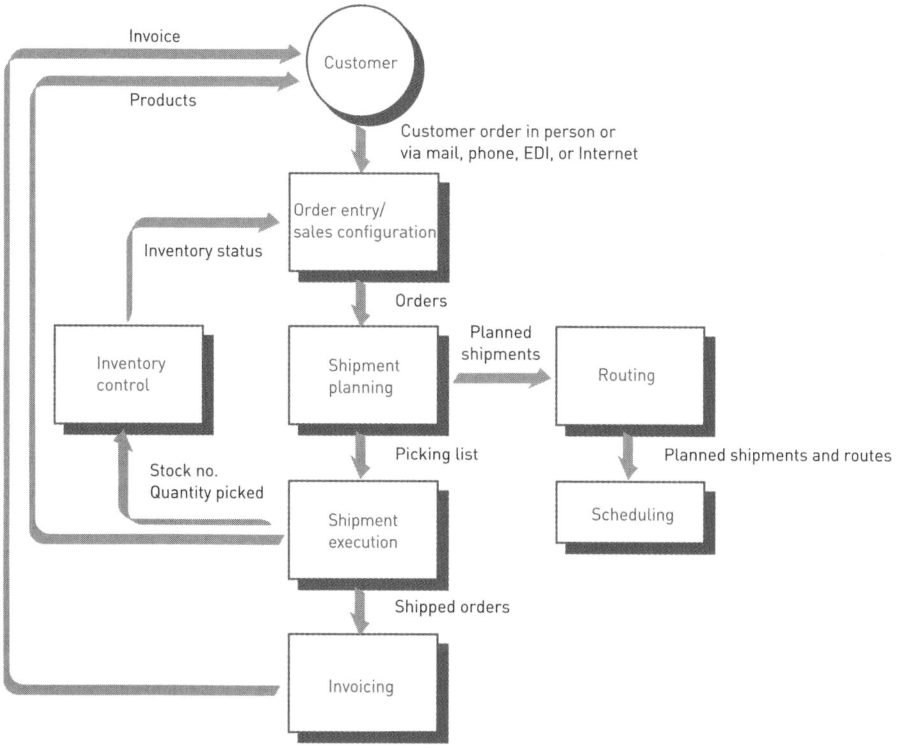

Figure 5.10

Order Processing Systems

Purchasing and Accounting Systems

Transaction processing systems support many areas of the business. For example, the purchasing TPSs include inventory control, purchase order processing, receiving, and accounts payable. This integrated set of systems enables an organization to plan, manage, track, and pay for its purchases of raw materials, parts, and services. The accounting TPSs include budget, accounts receivable, payroll, asset management, and general ledger systems. This integrated set of systems enables an organization to plan, manage, track, and control its cash flow and revenue.

Figure 5.11 shows the total integration of a firm's TPSs.

TPS CONTROL AND MANAGEMENT ISSUES

Transaction processing systems are the backbone of any organization's information systems. They capture facts about the fundamental business operations of the organization—facts without which orders cannot be shipped, customers cannot be invoiced, and employees and suppliers cannot be paid. In addition, the data captured by the TPSs flows downstream to other systems in the organization. Like any structure, an organization's information systems are only as good as the foundation on which they are built. In fact, most organizations would grind to a screeching halt if their TPSs failed.

Business Continuity Planning

Business continuity planning identifies the business processes that must be restored first in the event of a disaster to get the business's operations restarted with minimum disruption; it also specifies the actions that must be taken and by whom to restore operations. Order processing and shipping are examples of business processes that must be resumed as quickly as possible. Disasters can be natural emergencies such as a flood, a fire, or an earthquake or interruptions in business processes such as labor unrest, terrorist activity, hacker attack, or erasure of an important file. Key actions include safe evacuation of all employees, assessment

business continuity planning
Identification of the business processes that must be restored first in the event of a disaster and specification of what actions should be taken and who should take them to restore operations.

of the disaster's impact, relocation to alternate work spaces, backup and recovery of important electronic and manual business records, and use of alternate equipment.

Figure 5.11

Integration of a Firm's TPSs

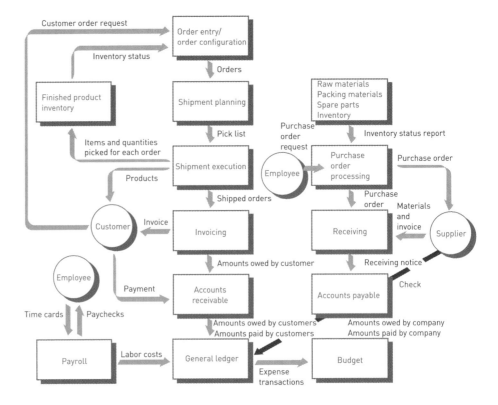

One of the first steps of business continuity planning is to identify potential threats or problems, such as natural disasters, employee misuse of personal computers, and poor internal control procedures. Business continuity planning also involves disaster preparedness. Business managers should occasionally hold an unannounced "test disaster"—similar to a fire drill—to ensure that the disaster plan is effective.

Companies vary widely in the thoroughness and effectiveness of their business continuity planning, and some have a harder time resuming business than others. However difficult the process, companies cannot afford to be unprepared for operational outages. Recently, disasters such as the Northeast power blackout and the California wildfires wreaked havoc with U.S. businesses. Blackout losses to workers and investors were estimated to be $4.2 billion by Anderson Economic Group, while Safeco Corporation estimates business wildfire claims at $3 million.[19]

Companies like Iron Mountain provide a secure, off-site environment for records storage. In the event of a disaster, vital data can be recovered.

(Source: Geostock/Getty Images.)

Transaction Processing System Audit

The accounting scandals of 2001 and 2002 spurred corporate board members, lawmakers, regulators, and stockholders to pressure corporate executives to produce accurate financial reports and do so in a timely fashion. In July 2002, the Sarbanes-Oxley Act was enacted, which set deadlines for public companies to implement procedures to ensure their audit committees could document financial data, validate earnings reports, and verify the accuracy of information. In response to the new requirements, business managers are demanding that their financial systems provide real-time data feeds and expenditure and sales updates so they can ensure their numbers are correct. Unfortunately, some organizations were unable to produce accurate reports. New World Pasta is the leading U.S. maker of dry pasta, with such brands as American Beauty, Creamette, Ideal, Ronzoni, and San Giorgio. In 2004, the firm filed for Chapter 11 bankruptcy protection, blaming two years of accounting problems caused by faulty TPSs.[20]

The Sarbanes-Oxley Act focused attention on the security of data and systems. In addition, regulators are interpreting the Financial Services Modernization Act (Gramm-Leach-Bliley) as requiring systems security for financial service providers, including specific standards to protect customer privacy. In the healthcare industry, the Health Insurance Portability and Accountability Act (HIPAA) defines regulations covering healthcare providers to ensure that their patient data is adequately protected.[21]

Clearly, the CIO must act to prevent the kind of accounting irregularities or loss of data privacy that can get their companies into trouble and erase investor confidence. One key step is to conduct a **transaction processing system audit** that attempts to answer four basic questions:

- Does the system meet the business need for which it was implemented?
- What procedures and controls have been established?
- Are these procedures and controls being used properly?
- Are the information systems and procedures producing accurate and honest reports?

In addition to these four basic auditing questions, other areas are typically investigated during an audit. These areas include the distribution of output documents and reports, the training and education associated with existing and new systems, and the time necessary to perform various tasks and to resolve problems and bottlenecks in the system. General areas of improvement are also investigated and reported during the audit.

In establishing the integrity of the computer programs and software, an audit trail must be established. The **audit trail** allows the auditor to trace any output from the computer system back to the source documents. With many of the real-time and time-sharing systems available today, it is extremely difficult to follow an audit trail. In many cases, no record of system inputs exists; thus, the audit trail is destroyed. In such cases, the auditor must investigate the actual processing in addition to the inputs and outputs of the various programs. In an attempt to safeguard the privacy of medical records, the federal Health Insurance Portability and Accountability Act requires that healthcare organizations, healthcare providers, and insurance companies establish an audit trail for each patient record. As an individual's medical record moves, each application it touches must imprint it with an identifier that cites every person who handled it and for what purpose.

While the CIO must take an active role in ensuring the integrity of financial reporting systems, that's not the same as guaranteeing the material accuracy of financial statements. At some of the companies under scrutiny, not even the most conscientious and informed CIO could have uncovered the financial irregularities that were occurring.

International Issues

Businesses are increasingly operating across country borders or around the globe, and these operations complicate the functioning of an organization's TPS. Multinational corporations using TPSs must address many issues and complexities in planning, building, and operating their TPSs. Different languages and cultures, disparities in IS infrastructure, varying laws and customs rules, and multiple currencies are among the challenges of linking all the business partners, customers, and subsidiaries of a multinational company.

transaction processing system audit
An examination of the TPS to answer whether the system meets the business need for which it was implemented, what procedures and controls have been established, whether these procedures and controls are being used properly, and whether the information systems and procedures are producing accurate and honest reports.

audit trail
Documentation that allows the auditor to trace any output from the computer system back to the source documents.

ENTERPRISE RESOURCE PLANNING

Flexibility and quick response are hallmarks of business competitiveness. Access to information at the earliest possible time can help businesses serve customers better, raise quality standards, and assess market conditions. Enterprise resource planning (ERP) is a key factor in instant access. Although some think that ERP systems are only for extremely large companies, this is not the case. Small and midsized companies, which generally include those ranging from $50 million to $500 million in annual revenue, represent the greatest growth for ERP companies. ERP systems are commonly used in manufacturing companies, colleges and universities, professional service organizations, retailers, and healthcare organizations.[22] A few leading vendors of ERP systems are listed in Table 5.5.

Table 5.5

Some ERP Software Vendors

Software Vendor	Name of Software
Oracle	Oracle Manufacturing
SAP America	SAP R/3
Ross Systems	iRenaissance
QAD	MFG/Pro
Lawson Software	ERP Solutions

An Overview of Enterprise Resource Planning

The key to ERP is real-time monitoring of business functions, which permits timely analysis of key issues such as quality, availability, customer satisfaction, performance, and profitability. Financial and planning systems automatically receive information from manufacturing and distribution. When something happens on the manufacturing line that affects a business situation—for example, packing material inventory drops to a certain level, which affects the ability to deliver an order to a customer—a message is triggered for the appropriate person in purchasing. The key steps in running a manufacturing organization using an ERP system are outlined here.

NetERP software from NetSuite provides tightly integrated, comprehensive ERP solutions for businesses, giving them access to real-time business intelligence, thus enabling better decision making.

(Source: Courtesy of NetSuite Inc.)

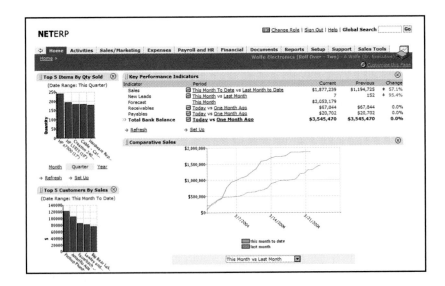

Develop Demand Forecast

For a manufacturing organization, the traditional planning process begins with the preparation of a long-term demand forecast. This forecast is prepared up to 18 months in advance and attempts to predict the weekly amount of each product to be purchased over this time period. The development of the forecast may require special software modules, historical data related to shipments, and discussions with members of sales, manufacturing, and finance organizations. Some organizations require years to implement an accurate, reliable demand forecasting process.

Deduct Demand Forecast from Inventory

As the forecasted demand is deducted from existing inventory to project future inventory levels, the system will display points at which additional finished products need to be produced.

Determine What Is Needed for Production

A bill of materials (BOM, a sort of "recipe" of parts and ingredients needed) for each item to be produced is used to translate finished product requirements into detailed lists of requirements for raw materials and packaging materials, which will be required to make and ship each finished product item.

Check Inventory for Needed Raw Materials

The forecasted needs for raw materials and packaging materials are subtracted from the existing inventory of these items. Again, the system eventually displays and triggers purchases of additional items required to meet production needs.

Schedule Production

The ERP production planning module uses the demand forecast and finished product inventory data to determine the week-by-week production schedules.

Assess Need for Additional Production Resources

The production schedule may reveal interesting insights, such as the need to build additional manufacturing capacity, hire additional workers, or develop new suppliers to provide sufficient raw materials. These new requirements can be input into the purchasing system and human resource modules of the ERP system and be used by managers in that area to develop future plans.

Financial Forecasting

All the generated data can be fed into the financial module of the ERP system to prepare a profit and loss forecast statement to assess the firm's future profitability. This profit forecast in turn can be used to help establish new budget limits for the upcoming year.

Lean manufacturing is a new manufacturing model some companies are considering to avoid problems with the demand-forecast approach just described. Lean manufacturing aims to improve efficiency, eliminate product backlogs, and synchronize production to *actual* customer demand rather than long-term forecasts. The major ERP vendors are now offering lean manufacturing solutions to help companies build products to actual demand, rather than forecasts. The solutions include supplier self-service portals that enable customers to log on to a Web site and enter their planned orders for weeks in advance. This enables the manufacturer to produce to actual demand rather than a long-range forecast of dubious accuracy.[23]

Implementation of a lean manufacturing ERP system from Ross Systems has enabled Michael Angelo's Gourmet Foods to dramatically reduce the time it takes to develop, test, and distribute new products. The maker of specialty frozen and refrigerated products such as lasagna and calzones used to spend six months to bring new products to market. Now the company can do so in fewer than 90 days. Time to market is a critical success factor for a specialty-foods maker because retailers deal with finicky consumers who change their minds frequently. With the new ERP system, Michael Angelo's is able to operate using a just-in-time production process. Its freshly made ingredients are received and products are cooked, packaged, and shipped to retail customers within 24 hours. Nearly 85 percent of Michael Angelo's orders are scheduled via EDI. The receipt of orders initiates the whole process. When an order is received, inventory is checked in real time, so the company knows whether it has ingredients on hand to fill that order. Receipt of an order also triggers a series of data exchanges with the other applications in the ERP system, alerting buyers what to purchase and production managers what to schedule.[24] In addition to manufacturing and finance, ERP systems can also support the human resource, sales, and distribution functions. This integration breaks through traditional corporate boundaries and can dramatically affect the entire organization.

Advantages and Disadvantages of ERP

Increased global competition, new needs of executives for control over the total cost and product flow through their enterprises, and ever-more-numerous customer interactions are driving the demand for enterprise-wide access to real-time information. ERP offers integrated software from a single vendor to help meet those needs. ERP vendors have also developed specialized systems for specific applications and market segments. The "Information Systems @ Work" feature describes the benefits one lending company realized by adopting an ERP system.

The primary benefits of implementing ERP include elimination of inefficient or outdated systems, easing adoption of improved work processes, improving access to data for operational decision making, and technology standardization. Disadvantages include the considerable time and expense in getting an ERP system up and running, compatibility problems with other systems, and the risk inherent in having only one vendor. Let's look at these pros and cons more closely.

Elimination of Costly, Inflexible Legacy Systems

Adoption of an ERP system enables an organization to eliminate dozens or even hundreds of separate systems and replace them with a single, integrated set of applications for the entire enterprise. In many cases, these systems are decades old, the original developers are long gone, and the systems are poorly documented. As a result, the systems are extremely difficult to fix when they break, and adapting them to meet new business needs takes too long. They become an anchor around the organization that keeps it from moving ahead and remaining competitive. An ERP system helps match the capabilities of an organization's information systems to its business needs—even as these needs evolve.

Improvement of Work Processes

Competition requires companies to structure their business processes to be as effective and customer oriented as possible. ERP vendors do considerable research to define the best business processes. They gather requirements of leading companies within the same industry and combine them with findings from research institutions and consultants. The individual application modules included in the ERP system are then designed to support these **best practices**, the most efficient and effective ways to complete a business process. Thus, implementation of an ERP system ensures good work processes based on best practices. For example, for managing customer payments, the ERP system's finance module can be configured to reflect the most efficient practices of leading companies in an industry. This increased efficiency ensures that everyday business operations follow the optimal chain of activities, with all users supplied the information and tools they need to complete each step.

Increase in Access to Data for Operational Decision Making

ERP systems operate via an integrated database, using essentially one set of data to support all business functions. So, decisions on optimal sourcing or cost accounting, for instance, can be run across the enterprise from the start, rather than looking at separate operating units and then trying to coordinate that information manually or reconciling data with another application. The result is an organization that looks seamless, not only to the outside world but also to the decision makers who are deploying resources within the organization.

The data is integrated to provide excellent support for operational decision making and allows companies to provide greater customer service and support, strengthen customer and supplier relationships, and generate new business opportunities. For example, once a salesperson makes a new sale, the business data captured during the sale is distributed to related transactions for the financial, sales, distribution, and manufacturing business functions in other departments.

best practices
The most efficient and effective ways to complete a business process.

ERP Consolidation Opens Doors for College Lender

Academic Management Services (AMS), a 500-employee academic loan provider, is a recent winner of *CIO* Magazine's Enterprise Value Award. The company developed an enterprise system that so improved its work processes that it moved into new markets, increased in value, and eventually was acquired by educational lending giant Sallie Mae. To learn from this Cinderella story, let's start at the beginning.

AMS loan counselors work with clients over the phone to determine an academic loan program that best matches their needs and overall financial picture. The counselor's primary responsibility is to determine which of six categories of loans best suits the client's needs. In AMS's previous system, each loan classification had its own online system, which counselors used to determine clients' eligibility. During a phone interview, the counselor would have to navigate between several interfaces, which was not only difficult for the counselor but also inconvenient for the customers. AMS realized that if it integrated the processes and systems into a single ERP platform, it could save counselors time and streamline operations. The ERP system the firm created was named ICE—Integrated Counseling and Enrollment.

Early ERP systems were designed primarily for manufacturing and production planning. In the mid-1990s ERP solutions expanded to include ordering systems, financial and accounting systems, asset management, and human resource management systems. By the late 1990s, ERP solutions were again broadened to include systems such as AMS's, allowing them to consolidate information across their organizations.

One of the advantages of consolidating systems into an ERP system is a reduction of data redundancy. Streamlining databases translates to significant savings in time and expense by not having to cross-reference or maintain multiple databases and by being able to respond to customer inquiries quicker. "Many companies have pockets of information relative to their customers," explains AMS CIO John Mariano. "The more that's integrated and the more business intelligence you can gain from that, the more you are able to deliver value across the company."

The consolidation at AMS greatly increased the efficiency of the loan counselors because they were able to describe, recommend, and approve the loan package that best matched the families' needs. AMS's loan agents now handle 1 million outbound and 300,000 inbound calls annually—90 percent more calls than they did prior to ICE. This increase in call volume results from each counselor's increased efficiency and the system's ability to accommodate more counselors. A family's complete financial profile is entered into ICE

for processing. The software does much of the work in deciding which loan option is the best fit, presenting recommendations to the counselors for consideration. Loan counselors can authorize loans within the system, often while the potential customer is on the phone. The ability to approve a loan in this manner can reduce the waiting period for families and helps AMS loan counselors handle more calls.

Consolidating disparate systems also helped AMS's information systems budgets by consolidating and shutting down redundant systems. In implementing the ICE project, AMS saved on licensing and maintenance fees in both its systems and processes. "At one time, we had four separate [database installations] of Oracle," says CIO Mariano. "Now we've consolidated to one. It has cut our IT expenses by approximately 40 percent."

The ICE system also helped AMS enter a new line of business— consolidation loans, in which borrowers combine numerous education loans into a single, lower-interest loan. These types of loans now account for nearly 25 percent of AMS's revenue. "They've created a new market ability, created more reliable partnerships with universities and did it all for $311,500," says Paul Gaffney, executive vice president and CIO of Staples, a member of the Enterprise Value Awards judging panel. "They're the poster child for IT value."

Discussion Questions

1. How have AMS loan counselor's activities changed since the implementation of the ICE system? How has the balance shifted between time spent with computer systems and time spent with customers?
2. How has the ICE system allowed AMS loan counselors to be more effective?

Critical Thinking Questions

3. Consolidating systems into a centralized ERP system is a common activity for many businesses lately. What, do you think, has caused business systems to become so divided in years past?
4. What is the advantage of hiring large firms like SAP to design an ERP, rather than hiring smaller vendors who specialize in subsystems?

SOURCES: Lafe Low, "They Got It Together," *CIO Magazine*, February 15, 2004, *www.cio.com;* Tuition Pay Web site, accessed May 23, 2004, at *https://secure.tuitionpay.com;* Sallie Mae Web site, accessed May 23, 2004, at *www.salliemae.com/about/abt_ams.html.*

Upgrade of Technology Infrastructure

An ERP system provides an organization with the opportunity to upgrade and simplify the information technology it employs. In implementing ERP, a company must determine which hardware, operating systems, and databases it wants to use. Centralizing and formalizing these decisions enables the organization to eliminate the hodgepodge of multiple hardware platforms, operating systems, and databases it is currently using—most likely from a variety of vendors. Standardization on fewer technologies and vendors reduces ongoing maintenance and support costs as well as the training load for those who must support the infrastructure.

Expense and Time in Implementation

Getting the full benefits of ERP is not simple or automatic. Although ERP offers many strategic advantages by streamlining a company's TPSs, ERP is time-consuming and is difficult and expensive to implement. Some companies have spent years and tens of millions of dollars implementing ERP systems. A survey by the Meta Group on leading ERP vendors' work for 200 user companies in 12 industries revealed that ERP investments were made over three to five years. On average, 25 percent of the cost was spent on software, 40 percent on professional services, and 25 percent on internal staff. The software cost amounted to 1 to 3 percent of the company's annual revenue, with smaller companies spending a greater percentage. The implementation time for these projects was approximately 20 months. Another seven months were needed before the benefits were realized.[25]

Difficulty Implementing Change

In some cases, a company has to radically change how it operates to conform to the ERP's work processes—its best practices. These changes can be so drastic to long-time employees that they retire or quit rather than go through the change. This exodus can leave a firm short of experienced workers.

Difficulty Integrating with Other Systems

Most companies have other systems that must be integrated with the ERP system. These systems can include financial analysis programs, Internet operations, and other applications. Many companies have experienced difficulties making these other systems operate with their ERP system. Other companies need additional software to create these links. In October 2003 Goodyear Tire & Rubber Co. disclosed that it would have to restate its financial results back to 1998 because of financial errors resulting from a faulty implementation of ERP software and a set of older applications that are used for intercompany billing. The company said that it expects to lower the net income it reported during the restatement period by as much as $100 million. The ERP system was installed in 1999 and runs Goodyear's core accounting functions.[26]

Risks in Using One Vendor

The high cost to switch to another vendor's ERP system makes it extremely unlikely that a firm will do so. So, once a company has adopted an ERP system, the vendor knows that it has a "captive audience" and has less incentive to listen and respond to customer issues. The high cost to switch also creates a high level of risk—in the event the ERP vendor allows its product to become outdated or goes out of business. Picking an ERP system involves not just choosing the best software product but also choosing the right long-term business partner.

Risk of Implementation Failure

Implementing an ERP system is extremely challenging and requires tremendous amounts of resources, the best IS people, and plenty of management support. Failed ERP installations often result from implementation problems rather than shortcomings in the software itself. And when there are problems with an ERP implementation, it can be expensive. In 2004, Ohio's attorney general filed a lawsuit against an ERP vendor seeking $510 million in damages stemming from an allegedly faulty installation of the company's ERP and student administration applications at Cleveland State University. Cleveland State was the first school to install a full set of the ERP vendor's student administration applications. But after it began using the software in 1998, university officials blamed the technology for problems in processing financial aid, enrolling transfer students, and recording grades. The problems led to more than $5 million in lost revenue because of an inability to track and collect receivables, plus additional unexpected expenditures to purchase a second mainframe and server with an Oracle database.[27]

Obviously, firms that decide to implement ERP systems should do so with careful planning and a clear idea of gains they can make competitively. The positives can then significantly outweigh the negatives.

SUMMARY

Principle

E-commerce is a new way of conducting business, and as with any other new application of technology, it presents both opportunities for improvement and potential problems.

Businesses and individuals use e-commerce to reduce transaction costs, speed the flow of goods and information, improve the level of customer service, and enable the close coordination of actions among manufacturers, suppliers, and customers. E-commerce also enables consumers, companies, and developing countries to gain access to worldwide markets.

Business-to-business (B2B) e-commerce allows organizations to buy and sell at a low cost worldwide. B2B e-commerce is currently the largest type of e-commerce. By using business-to-consumer (B2C) e-commerce, a provider of consumer services can eliminate the middlemen between it and its end consumer, sometimes squeezing costs and inefficiencies out of the supply chain and leading to higher profits and lower prices.

Consumer-to-consumer (C2C) e-commerce involves consumers selling directly to other consumers. Online auctions are the chief method of C2C e-commerce.

Supply chain management is composed of three subprocesses: demand planning to anticipate market demand, supply planning to allocate the right amount of enterprise resources to meet demand, and demand fulfillment to fulfill demand quickly and efficiently. Conversion to e-commerce supply chain management provides businesses an opportunity to (1) increase revenues or decrease costs by eliminating time-consuming and labor-intensive steps throughout the order and delivery process, (2) improve customer satisfaction by enabling customers to view detailed information about delivery dates and order status, and (3) reduce inven- tory including raw materials, safety stocks, and finished goods. But all subprocesses that exchange information and move goods between suppliers and customers must be upgraded.

The use of the Internet is growing rapidly in markets throughout Europe, Asia, and Latin America. Companies that want to succeed on the Web cannot ignore the growth in non-U.S. markets.

Many manufacturers and retailers have outsourced the physical logistics of delivering merchandise to cybershoppers. The goal is to make the transfer of all information and inventory—from the manufacturer to the delivery firm to the consumer—fast and simple.

Mobile commerce is the use of wireless devices such as PDAs, cell phones, and smart phones to facilitate the sale of goods or services—anytime, anywhere. The wireless application protocol (WAP) enables development of m-commerce software for wire-less devices.

Electronic retailing (e-tailing) is the direct sale from a business to consumers through electronic storefronts designed around an electronic catalog and shopping cart model. A cybermall is a single Web site that offers many products and services at one Internet location.

Manufacturers are joining electronic exchanges, where they can join with competitors and suppliers to use computers and Web sites to buy and sell goods, trade market information, and run back-office operations such as inventory control. They are also using e-commerce to improve the selling process by moving customer queries about product availability and prices online.

The Web allows firms to gather much more information about customer behavior and preferences than they could using other marketing approaches. This new technology has greatly enhanced the practice of market segmentation and enabled companies to establish closer relationships with their customers. The Internet has also revolutionized the world of investment and finance, especially online stock trading and online banking. In addition, the Internet has created many options for electronic auctions, where geographically dispersed buyers and sellers can come together.

Principle

E-commerce requires the careful planning and integration of a number of technology infrastructure components.

A number of infrastructure components must be chosen and integrated to support a large volume of transactions with customers, suppliers, and other business partners worldwide. These components include Web server hardware, Web site development tools, Web page construction software, catalog management software, product configuration software, and electronic shopping carts.

Organizations use authentication technology to confirm the identify of a user requesting access to information or assets. A digital certificate is an attachment to an e-mail message or data embedded in a Web page that verifies the identity of a sender or a Web site. Actual payments for e-commerce are made in a variety of ways, including electronic cash, electronic wallets, and credit, charge, debit, and smart cards.

Principle

An organization's transaction processing system (TPS) must support the routine, day-to-day activities that occur in the normal course of business and help a company add value to its products and services.

Transaction processing systems (TPSs) are at the heart of most information systems in businesses today. TPSs consist of all the components of a CBIS, including databases, networks, people, procedures, software, and hardware devices to process transactions. All TPSs perform data collection, which involves the capture of source data to complete a set of transactions; data editing, which checks for data validity and completeness; data correction, which involves providing feedback of a potential problem and enabling users to change the data; data manipulation, which is the performance of calculations, sorting, categorizing, summarizing, and storing data for further processing; data storage, which involves placing transaction data into one or more databases; and document production, which involves outputting records and reports.

The primary methods of TPSs include batch and online transaction processing.

TPS applications are seen throughout an organization. The order processing systems include order entry, sales configuration, shipment planning, shipment execution, inventory control, invoicing, customer relationship management, and routing and scheduling. The purchasing transaction processing systems include inventory control, purchase order processing, accounts payable, and receiving. The accounting systems include the budget, accounts receivable, payroll, asset management, and general ledger.

Because of the importance of a TPS to the ongoing operation of an organization, a business resumption plan that anticipates and minimizes the effects of disasters is mandatory. Business continuity planning focuses primarily on two issues: maintaining the integrity of corporate information and keeping the information system running until normal operations can be resumed.

The accounting scandals of 2001 and 2002 spurred corporate board members, lawmakers, regulators, and stockholders to pressure corporate executives to produce accurate financial reports and do so in a timely fashion. The Sarbanes-Oxley Act is directed at achieving this goal. In addition, many organizations conduct a transaction processing system audit to assess their systems.

Numerous complications arise that multinational corporations must address in planning, building, and operating their TPSs. These challenges include dealing with different languages and cultures, disparities in IS infrastructure, varying laws and customs rules, and multiple currencies.

Principle

Implementation of an enterprise resource planning (ERP) system enables a company to achieve many benefits by creating a highly integrated set of systems.

Enterprise resource planning (ERP) software is a set of integrated programs that manage a company's vital business operations for an entire multisite, global organization. It must be able to support multiple legal entities, multiple languages, and multiple currencies. Although the scope of an ERP system may vary from vendor to vendor, most ERP systems provide integrated software to support manufacturing and finance. In addition to these core business processes, some ERP systems are capable of supporting business functions such as human resources, sales, and distribution.

Implementation of an ERP system can provide many advantages, including elimination of costly, inflexible legacy systems; providing improved work processes; providing access to data for operational decision making; and creating the opportunity to upgrade technology infrastructure. Some of the disadvantages associated with an ERP system are that they are time-consuming, difficult, and expensive to implement.

CHAPTER 5: SELF-ASSESSMENT TEST

E-commerce is a new way of conducting business, and as with any other new application of technology, it presents both opportunities for improvement and potential problems.

1. Successful implementation of e-commerce requires _____ and _____.

    a.   changes to existing business processes; substantial investment in IS technology

    b.   conversion to XML software standards; Java programming scripts

    c.   implementation of tight security standards; Web site personalization

    d.   market segmentation; Web site globalization

2. Amazon.com is an example of what form of e-commerce?

   a. A2B
   b. B2B
   c. B2C
   d. C2C

E-commerce requires the careful planning and integration of a number of technology infrastructure components.

3. Which of the following is NOT one of three subprocesses of supply chain management?

   a. demand planning to anticipate market demand
   b. supply planning to allocate the right amount of enterprise resources to meet demand
   c. demand fulfillment to fulfill demand quickly and efficiently
   d. promotion to increase customer demand

4. The use of electronic exchanges continues to grow rapidly. True or False?

5. _____ management software combines different product data formats into a standard format for uniform viewing, aggregating, and integrating catalog data into a central repository for easy access, retrieval, and updating of pricing and availability changes.

6. A smart card makes credit theft practically impossible because a key to unlock the encrypted information is required, and there is no external number that a thief can identify and no physical signature a thief can forge. True or False?

An organization's transaction processing system (TPS) must support the routine, day-to-day activities that occur in the normal course of business and help a company add value to its products and services.

7. Which of the following sets of characteristics are usually associated with transaction processing systems?

   a. sophisticated and complex processing and analysis
   b. process large amounts of data and produce large amounts of output
   c. batch processing only
   d. produce exception reports and support drill-down analysis

8. A form of TPS where business transactions are accumulated over a period of time and prepared for processing as a single unit is called _____.

9. Which of the following statements is TRUE?

   a. Business continuity focuses on the actions that must be taken to restore computer operations and services in the event of a disaster.
   b. Business continuity planning identifies the business processes that must be restored first in the event of a disaster to get the business's operations restarted with minimum disruption; it also specifies the actions that must be taken and by whom to restore operations.
   c. Companies vary widely in the thoroughness and effectiveness of their business continuity planning.
   d. all of the above

10. This act was passed in July 2002 and set deadlines for public companies to implement procedures that ensure their audit committees can document underlying financial data to validate earnings reports and meet demands for accuracy.

    a. Gramm-Leach-Bliley Act
    b. Sarbanes-Oxley Act
    c. HIPAA
    d. none of these

Implementation of an enterprise resource planning (ERP) system enables a company to achieve many benefits by creating a highly integrated set of systems.

11. Which of the following is a primary benefit of implementing an ERP system?

    a. elimination of inefficient systems
    b. easing adoption of improved work processes
    c. improving access to data for operational decision making
    d. all of the above

12. Because it is so critical to the operation of an organization, most companies are able to implement an ERP system without major difficulty. True or False?

CHAPTER 5: SELF-ASSESSMENT TEST ANSWERS

(1) a (2) c (3) d (4) False (5) Catalog (6) True (7) b (8) batch processing (9) d (10) b (11) d (12) False

REVIEW QUESTIONS

1. Define the term *e-commerce*. Identify and briefly describe three different forms of e-commerce. Which form is the largest in terms of dollar volume?

2. What is electronic data interchange? What industries have been leaders in the use of EDI?

3. What is supply chain management?

4. What benefits can a firm achieve through conversion to an e-commerce supply chain management system?

5. Define *m-commerce*. Which forms of e-commerce can it support?

6. What are some of the special limitations that complicate the use of handheld devices for m-commerce?

7. What are some of the issues associated with the use of electronic exchanges?

8. What is the wireless application protocol?

9. Briefly explain the differences among smart, credit, charge, and debit cards.

10. What is technology-enabled relationship management?

11. Identify the key elements of technology infrastructure required to successfully implement e-commerce within an organization.

12. What basic transaction processing activities are performed by all transaction processing systems?

13. Distinguish between a batch processing system and an online processing system.

14. Identify four complications that multinational corporations must address in planning, building, and operating their TPSs.

15. What is the difference between data editing and data correction?

16. A business continuity plan focuses on what two issues?

DISCUSSION QUESTIONS

1. Why are many manufacturers and retailers outsourcing the physical logistics of delivering merchandise to shoppers? What advantages does such a strategy offer? Are there any potential issues or disadvantages?

2. Wal-Mart, the world's number-one retail chain, has turned down several invitations to join exchanges in the retail and consumer goods industries. Is this good or bad for the overall U.S. economy? Why?

3. Identify and briefly describe three potential m-commerce applications.

4. Discuss the use of e-commerce to improve spending on manufacturing, repair, and operations (MRO) goods and services.

5. Discuss the pros and cons of e-commerce companies capturing data about you as you visit their sites.

6. What is an electronic exchange? How successful have they been? What are some of the benefits and issues associated with their use?

7. Discuss the difference between electronic bill payment and electronic bill presentment. What are the benefits and some issues associated with each?

8. Assume that you are the owner of a small grocery store. Describe the importance of capturing complete, accurate transactions of customer purchases.

9. Imagine that you are the new IS manager for a *Fortune* 1000 company. An internal audit has revealed a number of problems with your firm's existing accounting systems. Prepare a brief outline of a talk you will make to senior company managers outlining the results of the audit and your next steps.

10. What is the advantage of implementing ERP as an integrated solution to link multiple business processes? What are some of the challenges and potential problems?

11. You are the key user of the firm's inventory-control system and have been asked to perform an IS audit of this system. Outline the steps you would take to complete the audit.

12. Identify five or six key elements or pieces of information that should be included in a firm's business continuity plan. What steps would you define your firm's business continuity plan?

PROBLEM-SOLVING EXERCISES

1. As a team, develop a set of criteria you would use to evaluate various business-to-consumer Web sites on the basis of ease of use, protection of consumer data, security of payment process, etc. Develop a simple spreadsheet containing these criteria. Evaluate five different popular Web sites using the criteria you developed.

2. Do research to learn more about the use of WAP and other specifications being developed to support m-commerce. Briefly describe the specifications you uncover. Who is behind the development of these standards? Which standards seem to be gaining the broadest acceptance? Prepare a one- to two-page report for your instructor.

3. Assume that you are starting an online grocery store (for nonperishable items only) that will allow users to "browse" your aisles electronically via the Web and make their purchase selections. Once an order is complete, the items are pulled from a warehouse, packed, and shipped overnight. Using a graphics program, draw a diagram that shows the different ways you will interact with your customers. Use a word processing program to develop a list of key facts you would like to capture about each customer and about each contact with a customer.

TEAM ACTIVITIES

1. As a team, choose an idea for an e-commerce Web site—products or services you would provide. Develop an implementation plan that outlines the steps you need to take and the decisions you must make to set up the Web site and make it operational.

2. Assume that your team has formed a consulting firm to perform an external audit of a firm's accounting information systems. Develop a list of at least ten questions you would ask as part of your audit. What specific inputs and outputs from various systems would you want to see? Visit a company and perform the audit based on these questions or role play the scenario if a live visit is impossible.

WEB EXERCISES

1. Find a Web site that provides investment portfolio tracking. As a team, select five stocks to make an imaginary purchase and allocate $100,000 among them. Check back in one week and determine the change in value in your investment. Make a list of the various features and analyses that are available at this Web site. Compare the change in the value of your portfolio with that of other teams. Compare the various features and analyses available at your Web site with the sites used by other teams.

2. Using the Internet, identify several companies that have implemented an ERP system in the last two years. Classify the implementations as success, partial success, or failure. What is your basis for making this classification?

CAREER EXERCISES

1. For your chosen career field, describe how you might use or be involved with e-commerce. If you have not chosen a career yet, answer this question for someone in marketing, finance, or human resources.

2. ERP software vendors need business systems analysts that understand both information systems and business processes. Make a list of six or more specific qualifications needed to be a strong business systems analyst.

VIDEO QUESTIONS

Watch the video clip **Find the Best Deals Online** and answer these questions:

1. What services do Web sites like Epinions.com offer consumers that are unavailable through traditional brick-and-mortar mall shopping?

2. What scam did the reporter warn about regarding e-commerce vendor reviews and ratings?

CASE STUDIES

Case One

RealEstate.com: Buying and Selling Homes Online

The real estate market might be considered the holy grail of B2C e-commerce. Many respectable companies have searched for it, but none have returned with the prize. Consider Cendant, one of the foremost providers of travel and real estate services in the world. In 1997, it published a Web site that automated the time-consuming process of closing a home sale, eliminating huge amounts of paperwork. That business failed. It then produced Move.com, a one-stop Web site with home listings and tools for home buyers and sellers. That idea also failed. More recently, Cendant tried to create a discount online real estate broker called Blue Edge. That, too, failed. The reason for these failures seemed to stem from the customer's unwillingness to trust important transactions to automated systems rather than competent human brokers. Cendant's frustrations with online real estate were shared by many others seeking to win the grail.

The dot-com boom spawned dozens of Internet-based brokers and other cyber real estate ideas. "Microsoft, Yahoo!, everyone and their brother poured a lot of money into real estate," said Brad Inman, a longtime publisher of real estate trade news. "They come in starry-eyed and say these Realtors are idiots. They spend a lot of money. And they realize they can't unlock the role of the Realtor. So they get out."

After ten years of failed attempts at this market, most have given up, figuring that the role of the Real Estate broker simply cannot be automated. Some people, however, find this $1.5 trillion market irresistible and continue to look for new e-commerce services that will attract home buyers. Barry Diller, CEO of InterActiveCorp, is just such a person. InterActiveCorp is the world's leading multibrand interactive commerce company. InterActiveCorp owns Expedia.com, Hotels.com, the HSN channel, Ticketmaster, Match.com, Hotwire.com, LendingTree.com, and other companies.

On the subject of online real estate, Diller commented, "If you think about real estate, with one million brokers and trillions of dollars in commerce, and information that now is able to be accessed by individuals with great ease, it is a perfect place for us to play. Right now, online adoption is tiny, and over time it will grow."

Studies show that the Web does play a role in the real estate market. Three-quarters of home buyers turn to the Internet to look at listings and learn about neighborhoods. Then they turn to realtors to visit homes and negotiate their deals. Diller wants to build a business that links Web shoppers with brokers. He bought RealEstate.com to accomplish his goal. RealEstate.com is run under the popular LendingTree.com umbrella, but unlike LendingTree, the site focuses exclusively on real estate: finding a home, finding a realtor, and finding a mortgage company.

You might wonder how a business can profit from connecting customers with realtors. The answer is that the service is provided at the expense of the realtor. While RealEstate.com isn't out to replace the realtor, it does take a substantial cut from the realtor's commission: one third to be exact. A portion of the percentage of the commission that RealEstate.com earns is given back to the customer in the form of gift cards for The Home Depot. For example, if you purchase a $220,000 home through a realtor you met through RealEstate.com, you would get a $1,000 gift card for shopping at The Home Depot.

Cutting into realtor's commissions, which have fallen over the last decade from 6 percent to 5.1 percent, is making RealEstate.com some enemies in the industry. No one has more at stake from real estate commissions than Cendant, which controls more than 25 percent of the brokerage business through its Century 21, Coldwell Banker, and ERA franchises. Cendant has brought charges against InterActiveCorp, owner of RealEstate.com, for false advertising. Cendant is also pushing the National Association of Realtors to adopt rules that would prevent some listings from being used on Web sites that earn most of their money from referrals. Without access to all listings in an area, RealEstate.com would lose credibility and customers.

Is InterActiveCorp worried? With the growing success of RealEstate.com, there are bound to be many legal battles to win their piece of the grail. But with its transactions exceeding $1.7 billion annually, and annual earnings of more than $16 million in fees from brokers, RealEstate.com seems to be providing a service that interests home buyers. It appears that InterActiveCorp has finally achieved what so many others have attempted but failed at: delivering a profitable e-commerce real estate service.

Discussion Questions

1. Why is it so much more difficult to sell homes over the Web than other products such as air travel and hotel rooms?
2. Besides homes, what other products are difficult to sell on the Web? Why?

Critical Thinking Questions

3. Why do people seem to distrust automated systems when it comes to real estate transactions?
4. Do you think that the National Association of Realtors should be allowed to withhold listings from companies like RealEstate.com? Why or why not?

SOURCES: Saul Hansell, "Point. Click. En Garde!" *New York Times*, April 18, 2004, *www.nytimes.com*; "LendingTree Completes Acquisition of Real Estate.com," *PR Newswire*, December 23, 2003, *www.lexisnexis.com*; Real Estate.com Web site, accessed April 22, 2004; InterActiveCorp Web site, accessed April 22, 2004, *www.iac.com*; Cendant Web site, accessed April 22, 2004, *www.cendant.com*.

Case Two

Point and Pay in Japan with KDDI

An important part of a TPS is its ability to handle a variety of payment mechanisms. Cash, check, credit card, or debit card are traditional methods of payment that most TPSs support. Recent trends indicate that society is moving away from paper currency in favor of electronic funds transfers. In Chapter 2 you read about new technologies that scan paper checks and translate them into electronic funds transfers. Other technologies are being developed and tested to eliminate both paper and plastic forms of currency, allowing transactions to take place anywhere, anytime.

Japanese citizens are testing a new use for their cell phone handsets. A trial service provided by Japanese cellular carrier KDDI Corp. and supported by several Japanese credit card issuers allows customers to point their cell phone handset at a cash register and push a button to pay. The technology is based on a recently released standard called Infrared for Financial Messaging (IrFM). IrFM employs the long-established Infrared Data Association (IrDA) Standard used in TV remote controls to transfer data wirelessly between devices at close range. While KDDI is providing the hardware for the trial, Harex InfoTech is providing the software.

The new point-and-pay technology includes numerous applications. A typical transaction involves having the customer work through prompts and menus on the cell phone screen to select a credit card, enter a four-digit PIN, and send the credit card information to a receiver. An Express Pay service can be used for micropayments—low-cost payments at vending machines, the subway, or bus. Express Pay allows the customer to pay by simply pressing a button on the side of the handset without having to navigate through the multistep menu routine.

Besides conducting business transactions, IrFM technology can be used for security clearance or for electronic attendance taking. Imagine pointing and clicking a cell phone as you enter the classroom for lecture!

Japan has more than 8 million users of infrared-enabled mobile phones. Low-cost infrared receivers connected to existing TPSs, including cash registers, point-of-sale terminals, and vending machines, communicate with IrFM handsets. Industry watchers expect the Japanese to embrace the technology quickly, and new commercial rollouts are planned in the near future. Harex is also testing point-and-pay technology in South Korea and at the University of Southern California's Marshall School of Business.

Discussion Questions

1. What conveniences are provided to the customer and vendor through IrFM?
2. Why has Japan been selected to test this new technology?

Critical Thinking Questions

3. What security concerns surround wireless trans- actions such as those enabled by IrFM?
4. What precautions need to be built into the IrFM system to guard against the theft of people's cell phones and consequently their identity?

SOURCES: Jørgen Sundgot, "Point-and-pay with IrFM," *InfoSynch World*, January 9, 2003, *www.infosyncworld.com/news/n/2868.html*; "Infrared Mobile Phone Payment System, Export to Japan," *PR Newswire*, April 14, 2003, *www.lexisnexis.com*; Harex Infotech Web site, accessed May 23, 2004, at *www.mzoop.com*.

NOTES

Sources for the opening vignette: "Is That a Radio in Your Pocket?" *PR Newswire*, January 8, 2004, *www.lexis-nexis.com*; "Sprint and Sony Music Entertainment Announce Broad Strategic Partnership to Distribute Mobile Entertainment Content," *Sprint News Releases*, June 30, 2003, *www.sprint.com*; Brian Garrity, "Wireless Deals Focus on Ring Tunes," *Billboard*, April 3, 2004, *www.lexis-nexis.com*, "Sony Ericsson and Handango Announce Easy Over-the-Air Download to SmartPhones," *Hugin*, October 20, 2003, *www.lexis-nexis.com*; Rebecca Harper, "The New Currency: Kiss Your Cash Good-bye," *Wired*, April 2004, p. 59.

1. Chabow, Eric, "E-Commerce Continues to Grow Very Nicely," *InformationWeek*, May 1, 2003, accessed at *www.informationweek.com*.
2. Rosencrance, Linda, "Commerce Department: E-Commerce Sales Up Sharply in Q4," *Computerworld*, February 23, 2004, accessed at *www.computerworld.com*.
3. Regan, Keith, "Yahoo Pays $575M for E-Shopping Site," *E-Commerce Times*, March 26, 2004, accessed at *www.ecommercetimes.com*.
4. Alexander, Steve, "Web Site Adds Inventory Control and Forecasting," *Computerworld*, February 24, 2003, accessed at *www.pcomputerworld.com*.
5. Regan, Keith, "Yahoo Pays $575M for E-Shopping Site," *E-Commerce Times*, March 26, 2004, accessed at *www.ecommercetimes.com*.
6. "Update: eBay Profit Nearly Doubles, Outlook Raised," *Reuters*, April 21, 2004, accessed at *http://news.moneycentral.msn.com*.
7. Bacheldor, Beth, "Albertson's Technology Brings Handhelds to Customers," *InformationWeek*, April 8, 2004, accessed at *www.informationweek.com*.

8. Lyman, Jay, "Internet Users in China Number Nearly 80 Million," *E-Commerce Times*, January 15, 2004, accessed at *www.ecommercetimes.com*.

9. "Holiday Prep: PC Mall Fights Online Fraud with Three Tiered Defense," *Network World Fusion*, December 16, 2003, accessed at *www.nwfusion.com*.

10. World Wide Retail Exchange Web site, accessed at *www.worldwideretailexchange.org/cs/en/about_wwre/overview.htm* on April 14, 2004.

11. Ulfelder, Steve, "B2B Exchange Survivors," *Computerworld*, February 2, 2004, accessed at *www.computerworld.com*.

12. Ulfelder, Steve, "B2B Survivors," *Computerworld*, February 2, 2004, accessed at *www.computerworld.com*.

13. Scheier, Robert L., "The Price of E-Payment," *Computerworld*, May 26, 2003, accessed at *www.computerworld.com*.

14. Rosencrance, Linda, "Ebay Selects Pitney Bowes," *Computerworld*, March 9, 2004 accessed at *www.computerworld.com*.

15. Anthes, Gary H., "B2B: Corporate Express Goes Direct," *Computerworld*, September 1, 2003, accessed at *www.computerworld.com*.

16. "Holiday Prep: PC Mall Fights Online Fraud with Three Tiered Defense," *Network World Fusion*, December 16, 2003, accessed at *www.nwfusion.com*.

17. Messmer, Ellen, "Feeling Vulnerable? Try Assessment Tools," *Network World Fusion*, April 5, 2004, accessed at *www.nwfusion.com*.

18. Whiting, Rick, "The World's Hardest-Working Databases," *InformationWeek*, February 13, 2004, accessed at *www.informationweek.com*.

19. D'Anton, Helen, "Companies Get Ready for the Unexpected," *InformationWeek*, January 19, 2004, accessed at *www.informationweek.com*.

20. "New World Pasta Files for Ch 11," *USA Today*, May 11, 2004, page B1.

21. Hunker, Jeffrey, "New Security Imperative: Demonstrating Results," *Information Week*, April 12, 2004, accessed at *www.informationweek.com*.

22. Bacheldor, Beth, "Midtier ERP Customers Command Attention," *Information Week*, April 26, 2004, accessed at *www.informationweek.com*.

23. Bacheldor, Beth, "PeopleSoft Aims at Lean Manufacturing," *Information Week*, March 19, 2004, accessed at *www.informationweek.com*.

24. Bacheldor, Beth, "Faster Food via ERP," *InformationWeek*, April 21, 2004, accessed at *www.informationweek.com*.

25. Fox, Pimm, "The Art of ERP Done Right," *Computerworld*, May 19, 2003, accessed at *www.computerworld.com*.

26. Songini, Marc L., "Goodyear Hits $100M Bump with ERP System," *Computerworld*, November 3, 2003, accessed at *www.computerworld.com*.

27. Songini, Marc L., "University Hits PeopleSoft with $510 M Lawsuit," *Computerworld*, March 26, 2004, accessed at *www.computerworld.com*.

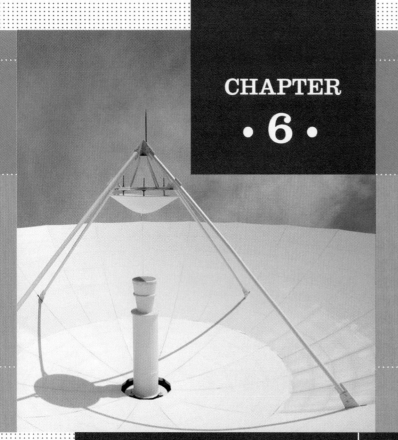

CHAPTER
· 6 ·

Information and Decision Support Systems

PRINCIPLES	LEARNING OBJECTIVES
▪ Good decision-making and problem-solving skills are the key to developing effective information and decision support systems.	▪ Define the stages of decision making. ▪ Discuss the importance of implementation and monitoring in problem solving.
▪ The management information system (MIS) must provide the right information to the right person in the right fashion at the right time.	▪ Explain the uses of MISs and describe their inputs and outputs. ▪ Discuss information systems in the functional areas of business organizations.
▪ Decision support systems (DSSs) are used when the problems are unstructured.	▪ List and discuss important characteristics of DSSs that give them the potential to be effective management support tools. ▪ Identify and describe the basic components of a DSS.
▪ Specialized support systems, such as group support systems (GSSs) and executive support systems (ESSs), use the overall approach of a DSS in situations such as group and executive decision making.	▪ State the goals of a GSS and identify the characteristics that distinguish it from a DSS. ▪ Identify the fundamental uses of an ESS and list the characteristics of such a system.

INFORMATION SYSTEMS IN THE GLOBAL ECONOMY

KIKU-MASAMUNE SAKE BREWING CO., LTD, JAPAN

Ancient Company Profits from State-of-the-Art Business Intelligence System

The Kano family began brewing sake in the small town of Mikage, Kobe, Japan, in 1659. The sake wine that they brewed, named *Kiku-Masamune*—translated as chrysanthemum sake—became a staple in Japan. Over the centuries, it has played an important role in Japanese culture and religion. In 1877, Kiku-Masamune became a global commodity when the company began exporting to England. Today, Kiku-Masamune sake is exported to many countries, and until recently, it enjoyed a monopoly position in the sake market.

In the mid-1990s, U.S. sales for Kiku-Masamune skyrocketed when sushi became the latest dining rage. With sushi bars opening nearly everywhere, the sake once described as "moonlight steeped in spring rain" enjoyed unprecedented sales. The company took advantage of its competitive position and began educating U.S. consumers through sake seminars in Los Angeles and New York City. The attention drawn to sake and its leading producer, Kiku-Masamune, did not go unnoticed by some ambitious entrepreneurs.

Competition in the sake industry increased both within Japan and around the world. Also, the Japanese were beginning to favor other alcoholic beverages imported from the West. With competition from an estimated 1,700 Japanese sake breweries and a dwindling client base, Kiku-Masamune was losing business both at home and abroad. The company knew that it could no longer rely solely on tradition and quality. It needed to learn more about the market—and its customers. But Kiku-Masamune had one big advantage over newcomers: its experience. The key to using its expertise lay in the millions of records it had collected over its long business history.

The company's paper-based record store was huge, with files on over 1.9 million vendors and 300,000 expense records, both of which were increasing by 100,000 records each year. Kiku-Masamune was in dire need of an electronic system that could manage these records and provide up-to-the-minute reports on the condition of the company, its suppliers, and customers. As Keiji Fukuda, systems manager in the company's accounting department observes, "Of course, if we'd been prepared to spend all our time trying to unearth crucial data, we could have continued indefinitely with our paper-based operations. But what if somebody needed to look at a document submitted several years ago? We simply didn't have the tools to meet those kinds of demands."

Kiku-Masamune contracted with a U.S. system developer to design an information and decision support system customized for the company's needs. "Before implementing the system, it was like we were only seeing pieces of the puzzle but never the big picture. Now we're able to identify where changes are needed in our contracts and drill down to the information we need to use our sales resources most effectively," says Yoshihiko Handa, sales and marketing director at Kiku-Masamune.

The new system produces scheduled reports annually, quarterly, monthly, weekly, and daily that provide information on key indicators such as performance by retailer and profits by product and customer. Kiku-Masamune is able to use this information to determine which of its customers require attention and to expand its customer base. The new system also provides the company with information to better manage its manufacturing process. Kiku-Masamune can provide customers with "just in time" delivery of sake, which saves both Kiku-Masamune and its customers the expense of storing large inventories.

The system has been successful in providing Kiku-Masamune with a strong competitive advantage in the sake market. The company is working to train all 500+ employees on the new information and decision support system. The ultimate goal is to arm all employees with information to make the right decisions quickly and effectively, resulting in more profitable and efficient operations overall.

Business intelligence (BI) solutions such as the one designed for Kiku-Masamune are in increasing demand. In spite of a recent tough economic climate, the BI market experienced strong growth. Research company IDC estimates that the BI market will reach $7.5 billion in 2006. Business intelligence development projects show consistently high returns on investment and are increasingly recognized as key for business success.

BI systems are one of many types of information and decision support systems that are implemented to provide up-to-the-second information for decision making at all levels within a business. Information and decision support systems support decisions in all functional units within an organization: financial, manufacturing, marketing, human resource, and others.

Kiku-Masamune uses its new information and decision support system to price its premium sake astutely, to guide decision making, and to help optimize sales opportunities. Effective information gathering and information sharing are clearly at the center of the company's new-found calm amid the storm of competition.

As you read this chapter, consider the following:

- How can management information systems guide a company to achieve a competitive advantage within an industry?
- What characteristics of an information and decision support system make it valuable?

Why Learn About Information and Decision Support Systems?

You have seen throughout this book how information systems can make you more efficient and streamline manual systems. The true potential of information systems, however, is in helping you and your coworkers make more informed decisions. This chapter shows you how to slash costs, increase profits, and uncover new opportunities for your company. A loan committee at a bank or credit union can use a group support system to help them determine who should receive loans. Transportation coordinators can use management information reports to find the least expensive way to ship products to market and to solve bottlenecks. Store managers can use decision support systems to help them decide what and how much inventory to order to meet customer needs and increase profits. An entrepreneur who owns and operates a temporary storage company can use vacancy reports to help him or her determine what price to charge for new storage units that are being built. Everyone wants to be a better problem solver and decision maker. This chapter shows you how information systems can help. We begin with an overview of decision making and problem solving.

As seen in the opening vignette, information and decision support are the lifeblood of today's organizations. Thanks to information and decision support systems, managers and employees can obtain useful information in real time. As you saw in Chapter 5, the TPS captures a wealth of data. When this data is filtered and manipulated, it can provide powerful support for managers and employees. The ultimate goal of management information and decision support systems is to help managers and executives at all levels make better decisions and solve important problems. The result can be increased revenues, reduced costs, and the realization of corporate goals. We begin by investigating decision making and problem solving.

DECISION MAKING AND PROBLEM SOLVING

Every organization needs effective decision making. In most cases, strategic planning and the overall goals of the organization set the course for decision making, helping employees and business units achieve their objectives and goals. Often, information systems also assist with strategic planning and problem solving. Good decision analysis, for example, can contribute millions or even billions of dollars to a large chemical or photographic company's profits.

Decision Making as a Component of Problem Solving

In business, one of the highest compliments you can receive is to be recognized by your colleagues and peers as a "real problem solver." Problem solving is a critical activity for any business organization. After a problem has been identified, the problem-solving process begins with decision making. A well-known model developed by Herbert Simon divides the **decision-making phase** of the problem-solving process into three stages: intelligence, design, and choice. This model was later incorporated by George Huber into an expanded model of the entire problem-solving process (see Figure 6.1).

decision-making phase
The first part of problem solving, including three stages: intelligence, design, and choice.

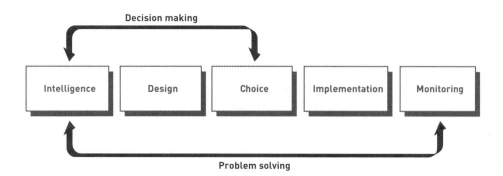

The first stage in the problem-solving process is the **intelligence stage**. During this stage, potential problems or opportunities are identified and defined. Information is gathered that relates to the cause and scope of the problem. During the intelligence stage, resource and environmental constraints are investigated. For example, exploring the possibilities of shipping tropical fruit from a farm in Hawaii to stores in Michigan would be done during the intelligence stage. The perishability of the fruit and the maximum price consumers in Michigan are willing to pay for the fruit are problem constraints. Aspects of the problem environment that must be considered in this case include federal and state regulations regarding the shipment of food products.

In the **design stage**, alternative solutions to the problem are developed. In addition, the feasibility of these alternatives is evaluated. In our tropical fruit example, the alternative methods of shipment, including the transportation times and costs associated with each, would be considered. During this stage, the problem solver might determine that shipment by freighter to California and then by truck to Michigan is not feasible because the fruit would spoil.

The final stage of the decision-making phase, the **choice stage**, requires selecting a course of action. In our tropical fruit example, the Hawaiian farm might select the method of shipping by air to Michigan as its solution. The choice stage would then conclude with selection of the actual air carrier. As you will see later, various factors influence choice; the apparently easy act of choosing is not as simple as it might first appear.

Figure 6.1

How Decision Making Relates to Problem Solving

The three stages of decision making—intelligence, design, and choice—are augmented by implementation and monitoring to result in problem solving.

intelligence stage
The first stage of decision making, in which potential problems or opportunities are identified and defined.

design stage
The second stage of decision making, in which alternative solutions to the problem are developed.

choice stage
The third stage of decision making, which requires selecting a course of action.

problem solving

A process that goes beyond decision making to include the implementation stage.

implementation stage

A stage of problem solving in which a solution is put into effect.

monitoring stage

The final stage of the problem-solving process, in which decision makers evaluate the implementation.

Problem solving includes and goes beyond decision making. It also includes the **implementation stage**, when the solution is put into effect. For example, if the Hawaiian farmer's decision is to ship the tropical fruit to Michigan as air freight using a specific air freight company, implementation involves informing the farming staff of the new activity, getting the fruit to the airport, and actually shipping the product to Michigan.

The final stage of the problem-solving process is the **monitoring stage**. In this stage, decision makers evaluate the implementation to determine whether the anticipated results were achieved and to modify the process in light of new information. Monitoring can involve feedback and adjustment. For example, after the first shipment of fruit, the Hawaiian farmer might learn that the flight of the chosen air freight firm routinely makes a stopover in Phoenix, Arizona, where the plane sits on the runway for a number of hours while loading additional cargo. If this unforeseen fluctuation in temperature and humidity adversely affects the fruit, the farmer might have to readjust his solution to include a new air freight firm that does not make such a stopover, or perhaps he would consider a change in fruit packaging.

Programmed Versus Nonprogrammed Decisions

programmed decisions

The decisions made using a rule, procedure, or quantitative method.

In the choice stage, various factors influence the decision maker's selection of a solution. One such factor is whether the decision can be programmed. **Programmed decisions** are made using a rule, procedure, or quantitative method. For example, to say that inventory should be ordered when inventory levels drop to 100 units is to adhere to a rule. Programmed decisions are easy to computerize using traditional information systems. It is simple, for example, to program a computer to order more inventory when inventory levels for a certain item reach 100 units or fewer. Most of the processes automated through transaction processing systems share this characteristic: The relationships between system elements are fixed by rules, procedures, or numerical relationships. Management information systems are also used to solve programmed decisions by providing reports on problems that are routine and in which the relationships are well defined (structured problems).

Ordering more inventory when inventory levels drop to specified levels is an example of a programmed decision.

(Source: Courtesy of Symbol Technologies.)

nonprogrammed decisions

The decisions that deal with unusual or exceptional situations.

Nonprogrammed decisions, however, deal with unusual or exceptional situations. In many cases, these decisions are difficult to quantify. Determining the appropriate training program for a new employee, deciding whether to start a new type of product line, and weighing the benefits and drawbacks of installing a new pollution control system are examples. Each of these decisions contains many unique characteristics for which the application of rules or procedures is not so obvious. Today, decision support systems are used to solve a variety of nonprogrammed decisions, in which the problem is not routine and rules and relationships are not well defined (unstructured or ill-structured problems).

Optimization, Satisficing, and Heuristic Approaches

In general, computerized decision support systems can either optimize or satisfice. An **optimization model** finds the best solution, usually the one that will best help the organization meet its goals. For example, an optimization model can find the appropriate number of products an organization should produce to meet a profit goal, given certain conditions and assumptions. Optimization models utilize problem constraints. A limit on the number of available work hours in a manufacturing facility is an example of a problem constraint. Some spreadsheet programs, such as Microsoft Excel, have optimizing features (see Figure 6.2). An appliance manufacturer, for example, can use an optimization program to help it reduce the time and cost of manufacturing appliances and increase profits by millions of dollars. The Scheduling Appointments at Trade Events (SATE) software package is an optimization program that schedules appointments between buyers and sellers at trade shows and meetings. The optimization software also allows decision makers to explore various alternatives. The software has been used at the Australian Tourism Exchange, the largest travel fair in the Southern Hemisphere.[1] Schindler, the world's largest escalator company, used an optimization technique to plan maintenance programs.[2] One important decision that Schindler and many other companies face is how much preventive maintenance to perform. Good preventive maintenance can save a company from making costly emergency repairs, which in Schindler's case involves sending maintenance teams and equipment to office buildings and retail stores to repair out-of-service escalators. However, too much preventive maintenance wastes valuable resources. Using a quantitative optimization program, Schindler was able to save about $1 million annually in total maintenance costs.

> **optimization model**
> A process to find the best solution, usually the one that will best help the organization meet its goals.

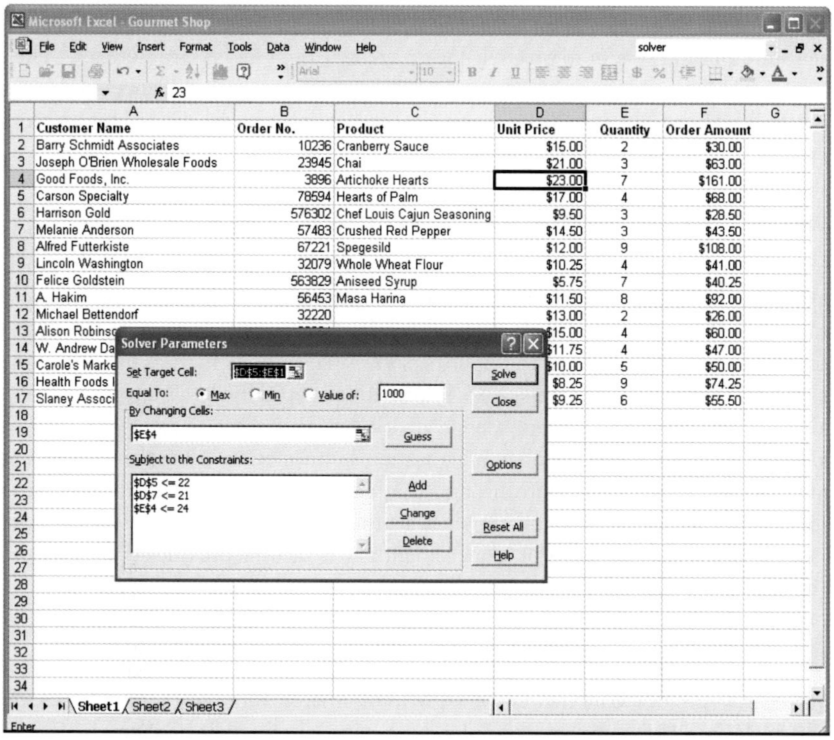

Figure 6.2

Some spreadsheet programs, such as Excel, have optimizing routines. This figure shows Solver, which can find an optimal solution given certain constraints.

A **satisficing model** is one that finds a good—but not necessarily the best—problem solution. Satisficing is usually used because modeling the problem properly to get an optimal decision would be too difficult, complex, or costly. Satisficing normally does not look at all possible solutions but only at those likely to give good results. Consider a decision to select a location for a new plant. To find the optimal (best) location, you would have to consider all cities in the United States or the world. A satisficing approach would be to consider only five or ten cities that might satisfy the company's requirements. Limiting the options may not result in the best decision, but it will likely result in a good decision, without spending

> **satisficing model**
> A model that will find a good—but not necessarily the best—problem solution.

heuristics
The commonly accepted guidelines
or procedures that usually find a
good solution.

the time and effort to investigate all cities. Satisficing is a good alternative modeling method because it is sometimes too expensive to analyze every alternative to get the best solution.

Heuristics, often referred to as "rules of thumb"—commonly accepted guidelines or procedures that usually find a good solution—are very often used in decision making. A heuristic that baseball team managers use is to place batters most likely to get on base at the top of the lineup, followed by the power hitters who'll drive them in to score. An example of a heuristic used in business is to order four months' supply of inventory for a particular item when the inventory level drops to 20 units or fewer; even though this heuristic might not minimize total inventory costs, it might be a very good rule of thumb to avoid stockouts without too much excess inventory. Trend Micro, a provider of antivirus software, has developed an antispam product that is based on heuristics.[3] The software examines e-mails to find those most likely to be spam.

In addition to using problem-solving models, decision makers also use information systems to improve the efficiency and quality of their decisions. One such system is a management information system, discussed next.

AN OVERVIEW OF MANAGEMENT INFORMATION SYSTEMS

Management information systems (MISs) can often give companies and other organizations a competitive advantage by providing the right information to the right people in the right format and at the right time. The U.S. military coalition in Iraq, for example, used a management information system to precisely locate enemy troops and their resources.[4] The MIS used global positioning systems (GPSs) for satellite mapping and surveillance and frequently updated information obtained from troops in the field to get and display information on enemy troops and U.S. troops and equipment.

Management Information Systems in Perspective

The primary purpose of an MIS is to help an organization achieve its goals by providing managers with insight into the regular operations of the organization so that they can control, organize, and plan more effectively and efficiently. One important role of the MIS is to provide the right information to the right person in the right fashion at the right time. In short, an MIS provides managers with information, typically in reports, that supports effective decision making and provides feedback on daily operations. Figure 6.3 shows the role of MISs within the flow of an organization's information. Note that business transactions can enter the organization through traditional methods or via the Internet or an extranet connecting customers and suppliers to the firm's transaction processing systems. The use of MISs spans all levels of management. That is, they provide support to and are used by employees throughout the organization.

Inputs to a Management Information System

As shown in Figure 6.3, data that enters an MIS originates from both internal and external sources, including the company's supply chain, first discussed in Chapter 1. The most significant internal data sources for an MIS are the organization's various TPSs and ERP systems and related databases. As discussed in Chapter 3, companies also use data warehouses and data marts to store valuable business information. Business intelligence, also discussed in Chapter 3, can be used to turn a database into powerful information throughout the organization. Other internal data comes from specific functional areas throughout the firm.

External sources of data can include customers, suppliers, competitors, and stockholders, whose data is not already captured by the TPS, as well as other sources, such as the Internet. In addition, many companies have implemented extranets to link with selected suppliers and other business partners to exchange data and information.

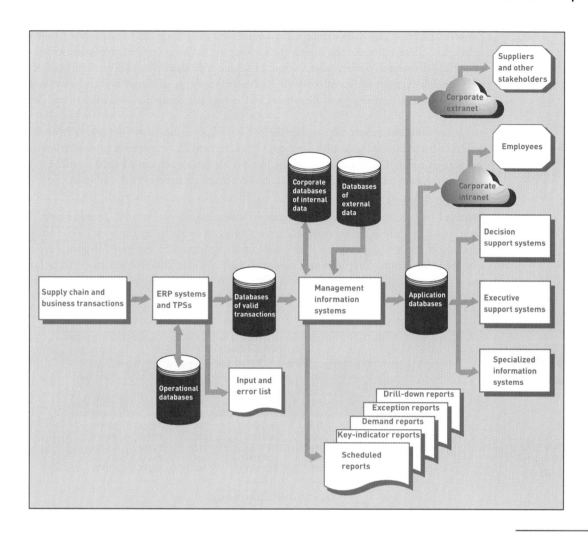

Outputs of a Management Information System

The output of most management information systems is a collection of reports that are distributed to managers. Kodak, for example, uses an MIS to send important sales information to its sales reps.[5] According to James Sanford, senior manager of sales communication and strategy, "Like a lot of companies, Kodak was good at collecting data but not very good at sharing and updating that data." Management reports can come from various company databases through data mining, first introduced in Chapter 3. Data mining allows a company to sift through a vast amount of data stored in databases, data warehouses, and data marts to produce a variety of reports, including scheduled reports, key-indicator reports, demand reports, exception reports, and drill-down reports (see Figure 6.4).

Scheduled Reports

Scheduled reports are produced periodically, or on a schedule, such as daily, weekly, or monthly. For example, a production manager could use a weekly summary report that lists total payroll costs to monitor and control labor and job costs. A manufacturing report generated once a day to monitor the production of a new item is another example of a scheduled report. Other scheduled reports can help managers control customer credit, the performance of sales representatives, inventory levels, and more. Boehringer Ingelheim, a large German drug company with more than $7 billion in revenues and thousands of employees in 60 countries, uses a variety of reports to allow it to respond rapidly to changing market conditions. According to the company's chief financial officer (CFO), "I want to be told where I stand and where we are heading. I like to be able to see negative trends and counter them as fast as possible." The company uses Cognos, Inc.'s Impromptu MIS to develop scheduled reports on costs for its various operations. Managers can drill down into more levels of detail to individual transactions if they want.[6] Xcel Energy has developed an online reporting system for its customers in Colorado to help them control their electricity bills.[7] The program, called InfoSmart, allows customers to analyze their energy use. "This is one instance where knowledge truly does equal power," says Debbie Mukherjee, product portfolio manager at Xcel. The scheduled reports allow customers to determine the energy efficiency of their homes and major appliances. It also shows them how their energy use compares with other homes in their area.

A **key-indicator report** summarizes the previous day's critical activities and is typically available at the beginning of each workday. These reports can summarize inventory levels, production activity, sales volume, and the like. Key-indicator reports are used by managers and executives to take quick, corrective action on significant aspects of the business.

Demand Reports

Demand reports are developed to give certain information upon request. In other words, these reports are produced on demand. For example, an executive might want to know the production status of a particular item—a demand report can be generated to give the requested information. Suppliers and customers can also use demand reports. FedEx, for example, provides demand reports on its Web site to allow its customers to track packages from their source to their final destination. Students in Idaho and other states can go to the

key-indicator report

A summary of the previous day's critical activities; typically available at the beginning of each workday.

demand reports

The reports that are developed to give certain information at a person's request.

Figure 6.4

Reports Generated by an MIS

The five types of reports are (a) scheduled, (b) key indicator, (c) demand, (d) exception, and (e–h) drill down.

(Source: George W. Reynolds, *Information Systems for Managers*, Third Edition. St. Paul, MN: West Publishing Co., 1995.)

(a) Scheduled Report

Daily Sales Detail Report

Prepared: 08/10/06

Order #	Customer ID	Salesperson ID	Planned Ship Date	Quantity	Item #	Amount
P12453	C89321	CAR	08/12/06	144	P1234	$3,214
P12453	C89321	CAR	08/12/06	288	P3214	$5,660
P12454	C03214	GWA	08/13/06	12	P4902	$1,224
P12455	C52313	SAK	08/12/06	24	P4012	$2,448
P12456	C34123	JMW	08/13/06	144	P3214	$720
....	....	....		....	....	....

(b) Key-Indicator Report

Daily Sales Key-Indicator Report

	This Month	Last Month	Last Year
Total Orders Month to Date	$1,808	$1,694	$1,914
Forecasted Sales for the Month	$2,406	$2,224	$2,608

(c) Demand Report

Daily Sales by Salesperson Summary Report

Prepared: 08/10/06

Salesperson ID	Amount
CAR	$42,345
GWA	$38,950
SAK	$22,100
JWN	$12,350
....	....
....	....

(d) Exception Report

Daily Sales Exception Report—Orders Over $10,000

Prepared: 08/10/06

Order #	Customer ID	Salesperson ID	Planned Ship Date	Quantity	Item #	Amount
P12345	C89321	GWA	08/12/06	576	P1234	$12,856
P22153	C00453	CAR	08/12/06	288	P2314	$28,800
P23023	C32832	JMN	08/11/06	144	P2323	$14,400
....	....	....	....	....	....	....
....	....	....	....	....	....	....

Figure 6.4 (cont.)

Reports Generated by an MIS

(e) First-Level Drill-Down Report

Earnings by Quarter (Millions)			
	Actual	Forecast	Variance
2nd Qtr. 2007	$12.6	$11.8	6.8%
1st Qtr. 2007	$10.8	$10.7	0.9%
4th Qtr. 2006	$14.3	$14.5	-1.4%
3rd Qtr. 2006	$12.8	$13.3	-3.8%

(f) Second-Level Drill-Down Report

Sales and Expenses (Millions)			
Qtr: 2nd Qtr. 2007	Actual	Forecast	Variance
Gross Sales	$110.9	$108.3	2.4%
Expenses	$ 98.3	$ 96.5	1.9%
Profit	12.6	$ 11.8	6.8%

(g) Third-Level Drill-Down Report

Sales by Division (Millions)			
Qtr: 2nd Qtr. 2007	Actual	Forecast	Variance
Beauty Care	$ 34.5	$ 33.9	1.8%
Health Care	$ 30.0	$ 28.0	7.1%
Soap	$ 22.8	$ 23.0	-0.9%
Snacks	$ 12.1	$ 12.5	-3.2%
Electronics	$ 11.5	$ 10.9	5.5%
Total	$110.9	$108.3	2.4%

(h) Fourth-Level Drill-Down Report

Sales by Product Category (Millions)			
Qtr: 2nd Qtr. 2007 Division: Health Care	Actual	Forecast	Variance
Toothpaste	$12.4	$10.5	18.1%
Mouthwash	$ 8.6	$ 8.8	-2.3%
Over-the-Counter Drugs	$ 5.8	$ 5.3	9.4%
Skin Care Products	$ 3.2	$ 3.4	-5.9%
Total	$30.0	$28.0	7.1%

Rate-MyProfessor Internet site to get ratings of faculty members.[8] The Laurel Pub Company, a bar and pub chain with more than 630 outlets in England, uses demand reports to generate important sales data when requested.[9] The company expects to save about £500,000 over a five-year period. Other examples of demand reports include reports requested by executives to show the hours worked by a particular employee, total sales to date for a product, and so on.

Exception Reports

Exception reports are reports that are automatically produced when a situation is unusual or requires management action. For example, a manager might set a parameter that generates a report of all inventory items with fewer than the equivalent of five days of sales on hand. This unusual situation requires prompt action to avoid running out of stock on the item. The exception report generated by this parameter would contain only items with fewer than five days of sales in inventory. Detroit-based financial services provider Comerica uses exception reports to assemble a list of customer inquiries that have been open a period of time without some progress or closure.[10] The company wants to improve its customer service. According to the senior vice president of the company, "We'll be able to see who's doing real well, and reward those folks. And we'll be able to see who's not engaged yet and help them." Exception reports are also used to help fight terrorism.[11] The Matchmaker System scans airline passenger lists and displays an exception report of passengers that could be a threat so that authorities can remove the suspected passengers before the plane takes off. The system was developed in England as part of its Defence Evaluation Research Agency.

Drill-Down Reports

Drill-down reports provide increasingly detailed data about a situation. Through the use of drill-down reports, analysts are able to see data at a high level first (such as sales for the entire company), then at a more detailed level (such as the sales for one department of the company), and then a very detailed level (such as sales for one sales representative).

exception reports
The reports that are automatically produced when a situation is unusual or requires management action.

drill-down reports
The reports that provide increasingly detailed data about a situation.

Developing Effective Reports

Management information system reports can help managers develop better plans, make better decisions, and obtain greater control over the operations of the firm, but in practice, the types of reports can overlap. For example, a manager can demand an exception report or set trigger points to be warned about items contained in a key-indicator report. In addition, some software packages can be used to produce, gather, and distribute reports from different computer systems. Certain guidelines should be followed in designing and developing reports to yield the best results. Table 6.1 explains these guidelines.

Guidelines	Reason
Tailor each report to user needs.	The unique needs of the manager or executive should be considered, requiring user involvement and input.
Spend time and effort producing only reports that are useful.	Once instituted, many reports continue to be generated even though no one uses them anymore.
Pay attention to report content and layout.	Prominently display the information that is most desired. Do not clutter the report with unnecessary data. Use commonly accepted words and phrases. Managers can work more efficiently if they can easily find desired information.
Use management by exception reporting.	Some reports should be produced only when there is a problem to be solved or an action that should be taken.
Set parameters carefully.	Low parameters may result in too many reports; high parameters mean valuable information could be overlooked.
Produce all reports in a timely fashion.	Outdated reports are of little or no value.
Periodically review reports.	Review reports at least once a year to make sure all reports are still needed. Review report content and layout. Determine whether additional reports are needed.

FUNCTIONAL ASPECTS OF THE MIS

Most organizations are structured along functional lines or areas. This functional structure is usually apparent from an organization chart, which typically shows vice presidents under the president. Some of the traditional functional areas are finance, manufacturing, marketing, human resource, and other specialized information systems. The MIS can be divided along those functional lines to produce reports tailored to individual functions (see Figure 6.5).

Financial Management Information Systems

financial MIS
An information system that provides financial information to all financial managers within an organization.

A **financial management information system (financial MIS)** provides financial information not only for executives but also for a broader set of people who need to make better decisions on a daily basis. Financial MISs are used to streamline reports of transactions. The 200-year-old New York Stock Exchange is investigating its electronic reports of trading.[12] The new system should cut costs and be more efficient.[13] Other financial MISs attempt to detect stock-market fraud and abuse.[14] The Financial Services Authority in England is spending about £4 million to identify stock-market abuse and fraud with a new system called Surveillance

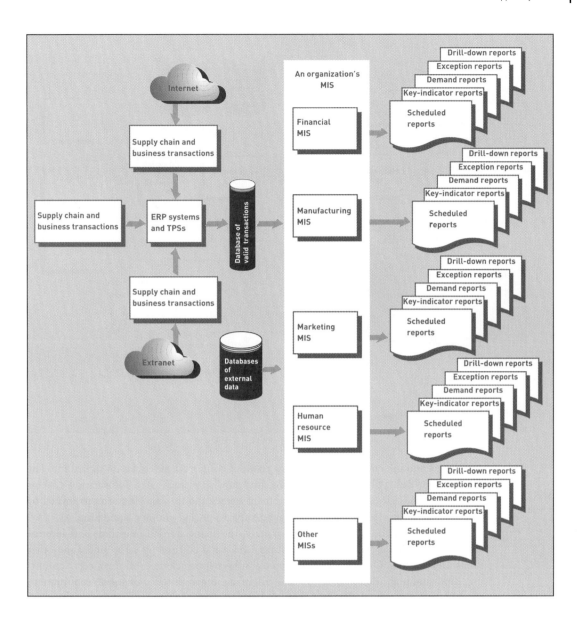

Figure 6.5

The MIS is an integrated collection of functional information systems, each supporting particular functional areas.

and Automated Business Reporting Engine (SABRE). The system should be fully operational by 2006. According to one SABRE spokesperson, "SABRE is a simple database, and we use a number of predetermined and flexible reports to extract information from the system." Most financial MISs perform the following functions:

- Integrate financial and operational information from multiple sources, including the Internet, into a single system.
- Provide easy access to data for both financial and nonfinancial users, often through use of the corporate intranet to access corporate Web pages of financial data and information.
- Make financial data immediately available to shorten analysis turnaround time.
- Enable analysis of financial data along multiple dimensions—time, geography, product, plant, and customer.
- Analyze historical and current financial activity.
- Monitor and control the use of funds over time.

Figure 6.6 shows typical inputs, function-specific subsystems, and outputs of a financial MIS, including profit and loss, auditing, and uses and management of funds.

In addition to providing information for internal control and management, financial MISs often are required to provide information to outside individuals and groups, including stockholders and federal agencies. Public companies are required to disclose their financial results to stockholders and the public. The federal government also requires financial

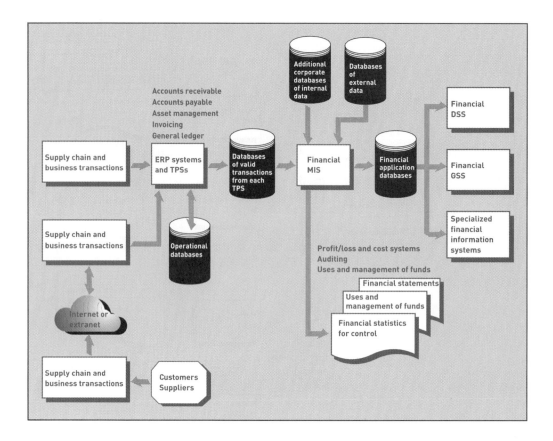

Figure 6.6

Overview of a Financial MIS

statements and information systems. As a result of antiterrorism legislation signed into law by President Bush, financial service firms must now implement new information systems designed to make it easier for law enforcement agencies to find and freeze assets owned by suspected terrorists.[15] The legislation also attempts to uncover money laundering.

Depending on the organization and its needs, the financial MIS can include both internal and external systems that assist in acquiring, using, and controlling cash, funds, and other financial resources. Other important financial subsystems include profit/loss, cost accounting, and auditing. *Auditing* is the analysis of an organization's financial condition to determine whether financial statements and reports produced by the financial MIS are accurate. Each subsystem interacts with the TPS in a specialized way and has information outputs that assist financial managers in making better decisions. These outputs include profit/loss and cost systems reports, internal and external auditing reports, and uses and management of funds reports. With all of their advantages, these subsystems are not foolproof, however. Some critics point to the Enron bankruptcy fiasco of the early 2000s, which resulted in many employees and investors losing huge sums of money, as an example of an external audit not showing the true picture of the company.

Manufacturing Management Information Systems

More than any other functional area, manufacturing has been revolutionized by advances in technology. As a result, many manufacturing operations have been dramatically improved over the last decade. Also, with the emphasis on greater quality and productivity, having an efficient and effective manufacturing process is becoming even more critical. The use of computerized systems is emphasized at all levels of manufacturing—from the shop floor to the executive suite. The use of the Internet has also streamlined all aspects of manufacturing. Figure 6.7 gives an overview of some of the manufacturing MIS inputs, subsystems, and outputs.

The manufacturing MIS subsystems and outputs monitor and control the flow of materials, products, and services through the organization. As raw materials are converted to finished goods, the manufacturing MIS monitors the process at almost every stage. European

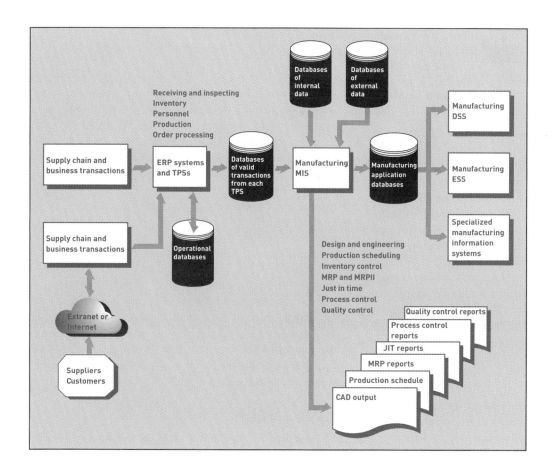

airplane manufacturer Airbus, for example, is using a manufacturing MIS to monitor and control its suppliers and parts to reduce costs.[16] New technology could make this process easier. Using specialized computer chips and tiny radio transmitters, companies can monitor materials and products through the entire manufacturing process. Procter & Gamble, Gillette, Wal-Mart, and Target have funded research into this new manufacturing MIS. Car manufacturers, which convert raw steel, plastic, and other materials into a finished automobile, also monitor their manufacturing processes. Auto manufacturers add thousands of dollars of value to the raw materials they use in assembling a car. If the manufacturing MIS also lets them provide additional service, such as customized paint colors on any of their models, it has added further value for customers. In doing so, the MIS helps provide the company the edge that can differentiate it from competitors. The success of an organization can depend on the manufacturing function.

Marketing Management Information Systems

A **marketing MIS** supports managerial activities in product development, distribution, pricing decisions, promotional effectiveness, and sales forecasting. Act!, for example, is a marketing automation tool to manage sales contacts, client e-mail systems, and group sales meetings.[17] Marketing functions are increasingly being performed on the Internet. A number of companies are developing Internet marketplaces to advertise and sell products.[18] The amount spent on online advertising is worth billions of dollars annually. Software can measure how many customers see the advertising.[19] According to a senior manager of CDW, a direct marketer of hardware and software products, "A satisfied customer is one who sees you as meeting expectations. A loyal customer, on the other hand, wants to do business with you again and will recommend you to others." CDW uses a software product called SmartLoyalty to analyze customer loyalty. Read the "Information Systems @ Work" special-interest feature to see how one golf equipment manufacturer uses a marketing MIS.

Figure 6.7

Overview of a Manufacturing MIS

marketing MIS
An information system that supports managerial activities in product development, distribution, pricing decisions, promotional effectiveness, and sales forecasting.

A "TaylorMade" Information and Decision Support System

For some people, being an ambassador for the best-selling golf club manufacturer would be a dream job. Traveling to all the best country clubs, playing one round of golf after another with the pros while they try out your clubs, and retiring to the clubhouse for dining and more golf chat might sound like heaven on earth. In reality, it is not as glamorous as it sounds. The sales associates for Taylor-Made, makers of the number-one driver on the PGA tour, spend 60 percent of their time taking inventory in golf shops and nearly all the rest of their time traveling to visit the next customer.

Until recently, TaylorMade sales associates also spent a considerable amount of time shuffling papers and entering data into an antiquated information system. They maintained records on each contact and customer so that they could review inventory records, purchasing history, and other notes prior to customer visits to determine what products to bring along on the trip. With hundreds of customers to manage, the amount of information became unmanageable.

TaylorMade's management team was struggling to keep up with the data as well. It had a group of "order management system programmers," who produced business reports for managers. The group was constantly buried under backlogged report requests. Managers were afraid to request reports because of the group's get-in-line attitude. Clearly, it was time for an IS overhaul.

TaylorMade hired a systems development team to create an information management system to provide the sales associates and managers with the information they required to make informed decisions. Scheduled key-indicator reports provide important information for guiding sales strategies. Managers in the manufacturing division rely on exception reports to let them know which types of golf clubs are running low in the warehouse. The sales force uses demand reports to find out who their best customers are and on what areas of the market they should focus. All TaylorMade employees are able to access reports from the new system through a convenient Web-based self-service interface. They no longer need to wait in line for the "order management system programmers" to fulfill report requests. Better yet, the sales associates use the bar-code scanners on their handheld PCs to quickly take inventory at country clubs; information is automatically uploaded to the home office's database.

The system is available to employees over the corporate network or the Web. No matter where they are in the world, sales associates can log on instantly and access information by region, territory, or representative. A sales representative might need to know how left-handed golf clubs are selling on the West Coast versus the East or whether a harsh winter increased sales of golf jackets. "The biggest advantage will be to visually and actively manage all parameters and key metrics of our business," said IS Director Tom Collard. "The system will help us in that we won't be stockpiling inventory three months in advance. You are actively getting signals from the sales force on what the demand truly is."

TaylorMade's new system has inspired the company to invest in other information systems. It has recently installed a new customer relationship management system that has allowed it to automate much of its customer service, reduce its staff, and improve customer satisfaction. By implementing new and effective information and decision support systems, TaylorMade has extended its reputation of high quality beyond its products into every aspect of its business.

Discussion Questions

1. How has TaylorMade increased the efficiency of its sales associates?
2. How have information systems increased TaylorMade's ability to gain a competitive advantage in the golf equipment market?

Critical Thinking Questions

3. How do you think the handheld power wielded by Taylor-Made sales associates affects their level of self-esteem and job satisfaction?
4. How do you think the information system has affected relationships between the customer and sales associate now that the associate can field any question with the push of a button?

SOURCES: "KANA Honored with *CRM Magazine* 2004 Service Leader Award for Web Self-Service," *Business Wire*, March 2, 2004; Business Objects Customers in the Spotlight Web site, accessed April 26, 2004, *www.businessobjects.com/customers/spotlight/taylormade.asp*; Kana ROI Success Stories Web site, accessed April 26, 2004, *www.kana.com*; TaylorMade Web site, accessed April 26, 2004, *www.taylormadegolf.com*.

Customer relationship management (CRM) programs, available from some ERP vendors, help a company manage all aspects of customer encounters.[20] After installing a CRM system, U.S. Steel was able to increase its cash flow by millions of dollars.[21] CRM software can help a company collect customer data, contact customers, educate customers about new products, and sell products to customers through an Internet site. An airline, for example, can use a CRM system to notify customers about flight changes. Other airlines have also benefited from CRM and use Web sites to allow customers to look up possible wait times. Yet, not all CRM systems and marketing sites on the Internet are successful. Customization and ongoing maintenance of a CRM system can also be expensive. Figure 6.8 shows the inputs, subsystems, and outputs of a typical marketing MIS.

Figure 6.8

Overview of a Marketing MIS

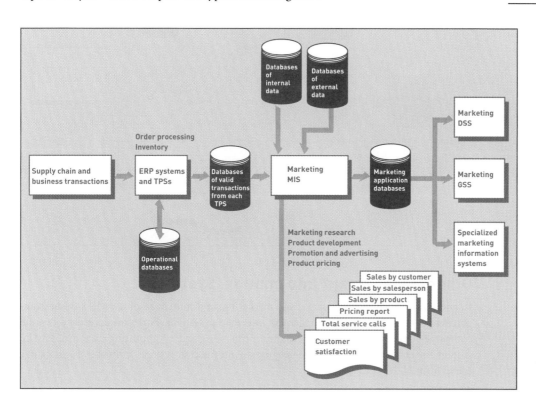

Human Resource Management Information Systems

A **human resource MIS**, also called the *personnel MIS*, is concerned with activities related to employees and potential employees of the organization. Because the personnel function relates to all other functional areas in the business, the human resource MIS plays a valuable role in ensuring organizational success. Some of the activities performed by this important MIS include workforce analysis and planning, hiring, training, job and task assignment, and many other personnel-related issues. An effective human resource MIS allows a company to keep personnel costs at a minimum while serving the required business processes needed to achieve corporate goals. Figure 6.9 shows some of the inputs, subsystems, and outputs of the human resource MIS.

Human resource subsystems and outputs range from the determination of human resource needs and hiring through retirement and outplacement. Most medium and large organizations have computer systems to assist with human resource planning, hiring, training and skills inventory, and wage and salary administration. Outputs of the human resource MIS include reports such as human resource planning reports, job application review profiles, skills inventory reports, and salary surveys.

human resource MIS
An information system that is concerned with activities related to employees and potential employees of an organization; also called a *personnel MIS*.

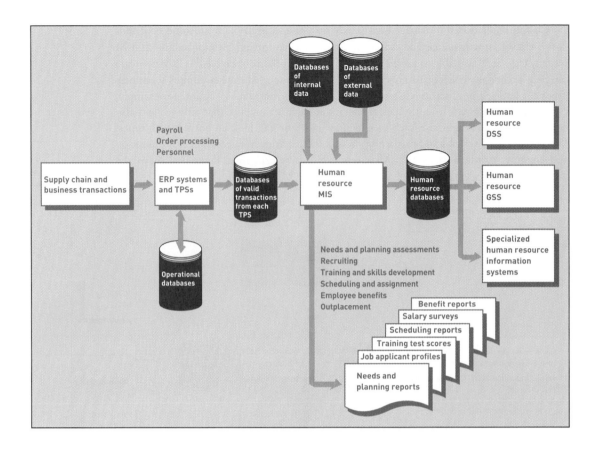

Figure 6.9

**Overview of a Human
Resource MIS**

Other Management Information Systems

In addition to finance, manufacturing, marketing, and human resource MISs, some companies have other functional management information systems. For example, most successful companies have well-developed accounting functions and a supporting accounting MIS. Also, many companies make use of geographic information systems for presenting data in a useful form.

Accounting MISs

accounting MIS
An information system that provides aggregate information on accounts payable, accounts receivable, payroll, and many other applications.

In some cases, accounting works closely with financial management. An **accounting MIS** performs a number of important activities, providing aggregate information on accounts payable, accounts receivable, payroll, and many other applications. The organization's TPS captures accounting data, which is also used by most other functional information systems.

Some smaller companies hire outside accounting firms to assist them with their accounting functions. These outside companies produce reports for the firm using raw accounting data. In addition, many excellent integrated accounting programs are available for personal computers in small companies. Depending on the needs of the small organization and its personnel's computer experience, using these computerized accounting systems can be a very cost-effective approach to managing information.

Geographic Information Systems

**geographic information system
(GIS)**
A computer system capable of assembling, storing, manipulating, and displaying geographic information, that is, data identified according to their locations.

Increasingly, managers want to see data presented in graphical form. A **geographic information system (GIS)** is a computer system capable of assembling, storing, manipulating, and displaying geographically referenced information, that is, data identified according to their locations. A GIS enables users to pair maps or map outlines with tabular data to describe aspects of a particular geographic region. For example, sales managers might want to plot total sales for each county in the states they serve. Using a GIS, they can specify that each county be shaded to indicate the relative amount of sales—no shading or light shading represents no or little sales, and deeper shading represents more sales. Although many software products have seen declining revenues, the use of GIS software is increasing by more than 10 percent per year on average.[22] Edens & Avant, a $2.3 billion real estate investment firm, uses a GIS from Environmental Systems Research Institute to increase profits and plan for potential disasters at the shopping centers its owns and manages.[23] According to David Beitz, geographic and marketing information systems manager for the company, "The cost is tremendous if you build a shopping center somewhere and a major tenant leaves. You're going to lose a lot of money."

You saw earlier in this chapter that management information systems (MISs) provide useful summary reports to help solve structured and semistructured business problems. Decision support systems (DSSs) offer the potential to assist in solving both semistructured and unstructured problems.

AN OVERVIEW OF DECISION SUPPORT SYSTEMS

A DSS is an organized collection of people, procedures, software, databases, and devices used to support problem-specific decision making and problem solving. The focus of a DSS is on decision-making effectiveness when faced with unstructured or semistructured business problems. As with a TPS and an MIS, a DSS should be designed, developed, and used to help an organization achieve its goals and objectives. Decision support systems offer the potential to generate higher profits, lower costs, and better products and services. For example, healthcare organizations use DSSs to improve patient care and reduce costs.[24] Decision support systems can also monitor and improve patient care. One diabetes Web site developed a DSS to provide customized treatment plans and reports.[25] The Web site uses powerful DSS software and grid computing, where the idle capacity of a network of computers is harnessed.

Capabilities of a Decision Support System

Developers of decision support systems strive to make them more flexible than management information systems and to give them the potential to assist decision makers in a variety of situations. Table 6.2 lists a few DSS applications. DSSs can assist with all or most problem-solving phases, decision frequencies, and different degrees of problem structure. DSS approaches can also help at all levels of the decision-making process. This section investigates these DSS capabilities. A single DSS might provide only a few of these capabilities, depending on its uses and scope.

Table 6.2

Selected DSS Applications

Company or Application	Description
ING Direct	The financial services company uses a DSS to summarize the bank's financial performance. The bank needed a measurement and tracking mechanism to determine how successful it was and to make modifications to plans in real time.
Cinergy Corporation	The electric utility developed a DSS to reduce lead time and effort required to make decisions in purchasing coal.
U.S. Army	It developed a DSS to help recruit, train, and educate enlisted forces. The DSS uses a simulation that incorporates what-if features.
National Audubon Society	It developed a DSS called Energy Plan (EPLAN) to analyze the impact of U.S. energy policy on the environment.
Hewlett-Packard	The computer company developed a DSS called Quality Decision Management to help improve the quality of its products and services.
Virginia	The state of Virginia developed the Transportation Evacuation Decision Support System (TEDSS) to determine the best way to evacuate people in case of a nuclear disaster at its nuclear power plants.

Support for Problem-Solving Phases

The objective of most decision support systems is to assist decision makers with the phases of problem solving. As previously discussed, these phases include intelligence, design, choice, implementation, and monitoring. A specific DSS might support only one or a few phases. During the implementation phase of a DSS, the state of California ordered 19 drug companies to distribute about $150 million of scarce drugs to about 150 California hospitals and clinics.[26] A DSS was developed to distribute about 90 percent of the drugs required by the court decision. This approach was able to distribute more than 125 different drugs in 20 drug categories as required by law. By supporting all types of decision-making approaches, a DSS gives the decision maker a great deal of flexibility in getting computer support for decision-making activities.

Support for Different Decision Frequencies

ad hoc DSS

A DSS concerned with situations or decisions that come up only a few times during the life of the organization.

institutional DSS

A DSS that handles situations or decisions that occur more than once, usually several times a year or more. An institutional DSS is used repeatedly and refined over the years.

Decisions can range on a continuum from one-of-a-kind to repetitive decisions. One-of-a-kind decisions are typically handled by an ad hoc DSS. An **ad hoc DSS** is concerned with situations or decisions that come up only a few times during the life of the organization; in small businesses, they might happen only once. For example, a company might be faced with a decision on whether to build a new manufacturing facility in another area of the country. Repetitive decisions are addressed by an institutional DSS. An **institutional DSS** handles situations or decisions that occur more than once, usually several times a year or more. An institutional DSS is used repeatedly and refined over the years. Examples of institutional DSSs include systems that support portfolio and investment decisions and production scheduling. These decisions might require decision support numerous times during the year. For example, DSSs are used to solve computer-related problems that can occur multiple times through the day.[27] With this approach, the DSS monitors computer systems second by second for problems and takes action to prevent problems, such as slowdowns and crashes, and to recover from them when they occur. One IBM engineer believes that this approach, called *autonomic computing*, is the key to the future of computing. Between these two extremes are decisions managers make several times but not regularly or routinely.

Support for Different Problem Structures

highly structured problems

Problems that are straightforward and require known facts and relationships.

semistructured or unstructured problems

More complex problems in which the relationships among the data are not always clear, the data may be in a variety of formats, and the data is often difficult to manipulate or obtain.

As discussed previously, decisions can range from highly structured and programmed to unstructured and nonprogrammed. **Highly structured problems** are straightforward, requiring known facts and relationships. **Semistructured** or **unstructured problems**, on the other hand, are more complex. The relationships among the data are not always clear, the data may be in a variety of formats, and the data is often difficult to manipulate or obtain. In addition, the decision maker might not know the information requirements of the decision in advance. For example, a DSS has been used to support sophisticated and unstructured investment analysis and make substantial profits for traders and investors.[28] Some DSS trading software is programmed to place buy and sell orders automatically without a trader manually entering a trade, based on parameters set by the trader.

Support for Various Decision-Making Levels

Decision support systems can help managers at different levels within the organization. Operational managers can get assistance with daily and routine decision making. Tactical decision makers can be supported with analysis tools to ensure proper planning and control. At the strategic level, DSSs can help managers by providing analysis for long-term decisions requiring both internal and external information (see Figure 6.10).

Figure 6.10

Decision-Making Level

Strategic-level managers are involved with long-term decisions, which are often made infrequently. Operational managers are involved with decisions that are made more frequently.

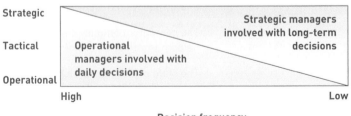

A Comparison of DSS and MIS

A DSS differs from an MIS in numerous ways, including the type of problems solved; the support given to users; the decision emphasis and approach; and the type, speed, output, and development of the system used. Table 6.3 lists brief descriptions of these differences.

Table 6.3

Comparison of DSSs and MISs

Factor	DSS	MIS
Problem Type	A DSS is good at handling unstructured problems that cannot be easily programmed.	An MIS is normally used only with more structured problems.
Users	A DSS supports individuals, small groups, and the entire organization. In the short run, users typically have more control over a DSS.	An MIS supports primarily the organization. In the short run, users have less control over an MIS.
Support	A DSS supports all aspects and phases of decision making; it does not replace the decision maker—people still make the decisions.	This is not true of all MIS systems—some make automatic decisions and replace the decision maker.
Emphasis	A DSS emphasizes actual decisions and decision-making styles.	An MIS usually emphasizes information only.
Approach	A DSS is a direct support system that provides interactive reports on computer screens.	An MIS is typically an indirect support system that uses regularly produced reports.
System	The computer equipment that provides decision support is usually online (directly connected to the computer system) and related to real time (providing immediate results). Computer terminals and display screens are examples—these devices can provide immediate information and answers to questions.	An MIS, using printed reports that may be delivered to managers once a week, may not provide immediate results.
Speed	Because a DSS is flexible and can be implemented by users, it usually takes less time to develop and is better able to respond to user requests.	An MIS's response time is usually longer.
Output	DSS reports are usually screen oriented, with the ability to generate reports on a printer.	An MIS, however, typically is oriented toward printed reports and documents.
Development	DSS users are usually more directly involved in its development. User involvement usually means better systems that provide superior support. For all systems, user involvement is the most important factor for the development of a successful system.	An MIS is frequently several years old and often was developed for people who are no longer performing the work supported by the MIS.

COMPONENTS OF A DECISION SUPPORT SYSTEM

At the core of a DSS are a database and a model base. In addition, a typical DSS contains a **dialogue manager**, which allows decision makers to easily access and manipulate the DSS and to use common business terms and phrases. Finally, access to the Internet, networks, and other computer-based systems permits the DSS to tie into other powerful systems, including the TPS or function-specific subsystems. Internet software agents, for example, can be used in creating powerful decision support systems. Figure 6.11 shows a conceptual model of a DSS.

dialogue manager
The user interface that allows decision makers to easily access and manipulate the DSS and to use common business terms and phrases.

The Database

The database management system allows managers and decision makers to perform *qualitative analysis* on the company's vast stores of data in databases, data warehouses, and data marts, discussed in Chapter 3. A *data-driven DSS* primarily performs qualitative analysis based on the company's databases. Data-driven DSSs tap into vast stores of information contained in the corporate database, retrieving information on inventory, sales, personnel, production, finance, accounting, and other areas. Data mining and business intelligence, introduced in Chapter 3, are often used in a data-driven DSS. Airline companies, for example, use a data-driven DSS to help them identify customers for round-trip flights between major cities. The data-driven DSS can be used to search a data warehouse to contact thousands of

Conceptual Model of a DSS

DSS components include a model base; database; external database access; access to the Internet and corporate intranet, networks, and other computer systems; and a dialogue manager.

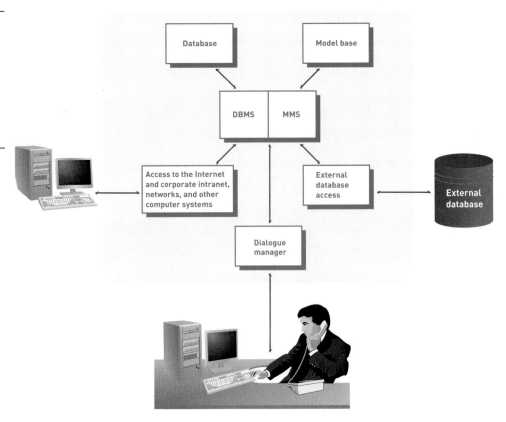

customers who might be interested in an inexpensive flight. A casino can use a data-driven DSS to search large databases to get detailed information on patrons. It can tell how much they spend a day on gambling, and more. TUI, a travel company in Europe, uses a data-driven DSS to make better decisions to help reduce costs and increase efficiency.[29] According to the company's financial director, "We will make significant cost savings. It will allow us to become more efficient and provide us with one version of the truth."

A database management system can also connect to external databases such as the Internet, libraries, government databases, and more. The combination of internal and external database access can give key decision makers a better understanding of the company and its environment.

The Model Base

model base
A part of a DSS that provides decision makers access to a variety of models and assists them in decision making.

The **model base** allows managers and decision makers to perform *quantitative analysis* on both internal and external data. A *model-driven DSS* primarily performs mathematical or quantitative analysis. The model base gives decision makers access to a variety of models so that they can explore different scenarios and see their effects. Ultimately, it assists them in the decision-making process. For example, Bayer, the large drug company, had to choose whether to research a gene that could dispose certain people to asthma.[30] The company developed a DSS model that saved it six months of analysis time. At the choice stage, Bayer decided to explore which genes might cause asthma. According to the head of the research and development department of a competing company, "This is revolutionary. You can change genes with the stroke of a keypad. We've never had the tools like these, to look at the complexity of disease." In the broadest sense, a model-driven DSS takes a pile of data and turns it into useful information. In the sports world, AC Milan, a football (soccer) league champion, uses a model-driven DSS to reduce injuries by 90 percent.[31] The software models help players determine their diets, training schedule, and mental attitude. According to the coach, skeptical athletes changed their minds about the program once it was in place: "When they [the players] realized that it would make them more healthy and prolong their playing careers, they became much more accepting."

Model management software (MMS) is often used to coordinate the use of models in a DSS, including financial, statistical analysis, graphical, and project management models. Depending on the needs of the decision maker, one or more of these models can be used (see Table 6.4).

model management software (MMS)
The software that coordinates the use of models in a DSS.

Model Type	Description	Software That Can Be Used
Financial	Provides cash flow, internal rate of return, and other investment analysis	Spreadsheet, such as Excel
Statistical	Provides summary statistics, trend projections, hypothesis testing, and more	Statistical program, such SPSS or SAS
Graphical	Assists decision makers in designing, developing, and using graphic displays of data and information	Graphics programs, such as PowerPoint
Project Management	Handles and coordinates large projects; also used to identify critical activities and tasks that could delay or jeopardize an entire project if they are not completed in a timely and cost-effective fashion	Project management software, such as Project

The Dialogue Manager

The dialogue manager allows users to interact with the DSS to obtain information. It assists with all aspects of communications between the user and the hardware and software that constitute the DSS. In a practical sense, to most DSS users, the dialogue manager is the DSS. Upper-level decision makers are often less interested in where the information came from or how it was gathered than that the information is both understandable and accessible.

Table 6.4

DSSs often use financial, statistical, graphical, and project management models.

GROUP SUPPORT SYSTEMS

The DSS approach has resulted in better decision making for all levels of individual users. However, many DSS approaches and techniques are not suitable for a group decision-making environment. Although not all workers and managers are involved in committee meetings and group decision-making sessions, some tactical and strategic-level managers can spend more than half their decision-making time in a group setting. Such managers need assistance with group decision making. A **group support system (GSS)**, also called a *group decision support system* and a *computerized collaborative work system*, consists of most of the elements in a DSS, plus software to provide effective support in group decision-making settings (see Figure 6.12). Lowe & Partners Worldwide, a global advertising company, uses a GSS called *swarming technology* to link experts in many diverse areas to make important advertising decisions.[32] The company electronically links account executives from Hong Kong, England, India, the United States, and other areas with experts in various industries to help craft advertising programs. According to a company executive, "We're discovering resources we didn't know existed."

group support system (GSS)
The software application that consists of most of the elements in a DSS, plus software to provide effective support in group decision-making settings; also called *group decision support system* or *computerized collaborative work system*.

Characteristics of a GSS That Enhance Decision Making

It is often said that two heads are better than one. When it comes to decision making, a GSS's unique characteristics have the potential to result in better decisions. Developers of these systems try to build on the advantages of individual support systems while adding new approaches unique to group decision making. For example, some GSSs can allow the exchange of information and expertise among people without personal meetings or direct face-to-face interaction. The following are some characteristics that can improve and enhance decision making.

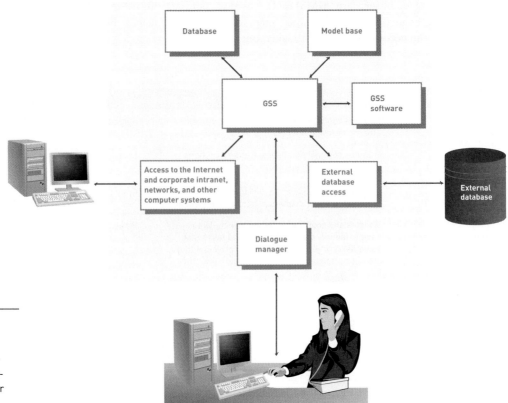

Figure 6.12

Configuration of a GSS

A GSS contains most of the elements found in a DSS, plus software to facilitate group member communications.

Special Design

The GSS approach acknowledges that special procedures, devices, and approaches are needed in group decision-making settings. These procedures must foster creative thinking, effective communications, and good group decision-making techniques.

Ease of Use

Like an individual DSS, a GSS must be easy to learn and use. Systems that are complex and hard to operate will seldom be used. Many groups have less tolerance than do individual decision makers for poorly developed systems.

Flexibility

Two or more decision makers working on the same problem might have different decision-making styles and preferences. Each manager makes decisions in a unique way, in part because of different experiences and cognitive styles. An effective GSS not only has to support the different approaches that managers use to make decisions but also must find a means to integrate their different perspectives into a common view of the task at hand.

Decision-Making Support

A GSS can support different decision-making approaches, including the **delphi approach**, in which group decision makers are geographically dispersed throughout the country or the world. This approach encourages diversity among group members and fosters creativity and original thinking in decision making. Another approach, called **brainstorming**, which often consists of members offering ideas "off the top of their heads," fosters creativity and free thinking. The **group consensus approach** forces members in the group to reach a unanimous decision. The Shuttle Project Engineering Office at the Kennedy Space Center has used the consensus-ranking organizational-support system (CROSS) to evaluate space projects in a group setting. The group consensus approach analyzes the benefits of various projects and their probabilities of success. CROSS is used to evaluate and prioritize advanced space projects.[33]

delphi approach

A decision-making approach in which group decision makers are geographically dispersed; this approach encourages diversity among group members and fosters creativity and original thinking in decision making.

brainstorming

A decision-making approach that often consists of members offering ideas "off the top of their heads."

group consensus approach

A decision-making approach that forces members in the group to reach a unanimous decision.

With the **nominal group technique**, each decision maker can participate; this technique encourages feedback from individual group members, and the final decision is made by voting, similar to a system for electing public officials.

Anonymous Input

Many GSSs allow anonymous input, in which the person giving the input is not known to other group members. For example, some organizations use a GSS to help rank the performance of managers. Anonymous input allows the group decision makers to concentrate on the merits of the input without considering who gave it. In other words, input given by a top-level manager is given the same consideration as input from employees or other members of the group. Some studies have shown that groups using anonymous input can make better decisions and have superior results compared with groups that do not use anonymous input. Anonymous input, however, can result in flaming, in which an unknown team member posts insults or even obscenities on the GSS.

Reduction of Negative Group Behavior

One key characteristic of any GSS is the ability to suppress or eliminate group behavior that is counterproductive or harmful to effective decision making. In some group settings, dominant individuals can take over the discussion, which can prevent other members of the group from presenting creative alternatives. In other cases, one or two group members can sidetrack or subvert the group into areas that are nonproductive and do not help solve the problem at hand. Other times, members of a group might assume they have made the right decision without examining alternatives—a phenomenon called *groupthink*. If group sessions are poorly planned and executed, the result can be a tremendous waste of time. Today, many GSS designers are developing software and hardware systems to reduce these types of problems. Procedures for effectively planning and managing group meetings can be incorporated into the GSS approach. A trained meeting facilitator is often employed to help lead the group decision-making process and to avoid groupthink.

Parallel Communication

With traditional group meetings, people must take turns addressing various issues. One person normally talks at a time. With a GSS, every group member can address issues or make comments at the same time by entering them into a PC or workstation. These comments and issues are displayed on every group member's PC or workstation immediately. Parallel communication can speed meeting times and result in better decisions.

Automated Record Keeping

Most GSSs can keep detailed records of a meeting automatically. Each comment that is entered into a group member's PC or workstation can be anonymously recorded. In some cases, literally hundreds of comments can be stored for future review and analysis. In addition, most GSS packages have automatic voting and ranking features. After group members vote, the GSS records each vote and makes the appropriate rankings.

GSS Software

GSS software, often called *groupware* or *workgroup software*, helps with joint workgroup scheduling, communication, and management.[34] One popular package, Lotus Notes, can capture, store, manipulate, and distribute memos and communications that are developed during group projects. It can also incorporate knowledge management, discussed in Chapter 3, into the Lotus Notes Package. Some companies standardize on messaging and collaboration software, such as Lotus Notes. Microsoft's NetMeeting product supports application sharing in multiparty calls. Microsoft Exchange is another example of groupware. This software allows users to set up electronic bulletin boards, schedule group meetings, and use e-mail in a group setting. NetDocuments Enterprise can be used for Web collaboration. The groupware is intended for legal, accounting, and real estate businesses. A Breakout Session feature allows two people to take a copy of a document to a shared folder for joint revision and work. The software also permits digital signatures and the ability to download and work on shared documents on handheld computers. Other GSS software packages include Collabra Share, OpenMind, and TeamWare. All of these tools can aid in group decision making.

nominal group technique
A decision-making approach that encourages feedback from individual group members, and the final decision is made by voting, similar to the way public officials are elected.

In addition to stand-alone products, GSS software is increasingly being incorporated into existing software packages. Today, some transaction processing and enterprise resource planning packages include collaboration software. Some ERP producers (see Chapter 5), for example, have developed groupware to facilitate collaboration and to allow users to integrate applications from other vendors into the ERP system of programs. Today, groupware can interact with wireless devices. Research In Motion, the maker of Blackberry software, offers mobile communications, access to group information, meeting schedules, and other services that can be directly tied to groupware and servers.[35] In addition to groupware, GSSs use a number of tools discussed previously, including the following:

- E-mail and instant messaging (IM)
- Videoconferencing
- Group scheduling
- Project management
- Document sharing

GSS software allows work teams to collaborate and reach better decisions—even if they work across town, in another region, or on the other side of the globe.

(Source: © Mark Richards/Photo Edit.)

GSS Alternatives

Group support systems can take on a number of network configurations, depending on the needs of the group, the decision to be supported, and the geographic location of group members. The frequency of GSS use and the location of the decision makers are two important factors (see Figure 6.13).

Figure 6.13

GSS Alternatives

The decision room may be the best alternative for group members who are located physically close together and who need to make infrequent decisions as a group. By the same token, group members who are situated at distant locations and who frequently make decisions together may require a wide area decision network to accomplish their goals.

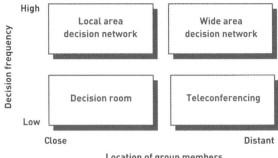

decision room

A room that supports decision making, with the decision makers in the same building, combining face-to-face verbal interaction with technology to make the meeting more effective and efficient.

The Decision Room

The **decision room** is ideal for situations in which decision makers are located in the same building or geographic area and the decision makers are occasional users of the GSS approach. In these cases, one or more decision rooms or facilities can be set up to accommodate the GSS approach. Groups, such as marketing research teams, production management groups, financial control teams, or quality-control committees, can use the decision rooms when needed. The decision room alternative combines face-to-face verbal interaction with technology-aided formalization to make the meeting more effective and efficient. Decision rooms, however, can be expensive to set up and operate. A typical decision room is shown in Figure 6.14.

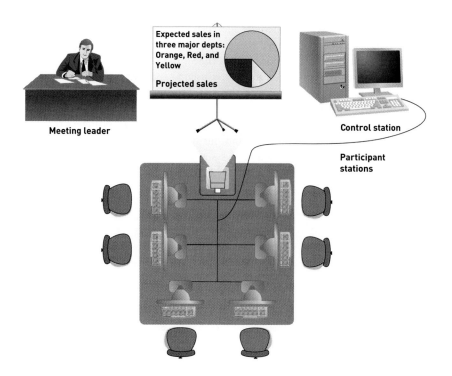

Meeting leader

Expected sales in three major depts: Orange, Red, and Yellow

Projected sales

Control station

Participant stations

Figure 6.14

The GSS Decision Room

For group members who are in the same location, the decision room is an optimal GSS alternative. This approach can use both face-to-face and computer-mediated communication. By using networked computers and computer devices, such as project screens and printers, the meeting leader can pose questions to the group, instantly collect their feedback, and, with the help of the governing software loaded on the control station, process this feedback into meaningful information to aid in the decision-making process.

The Local Area Decision Network

The local area decision network can be used when group members are located in the same building or geographic area and under conditions in which group decision making is frequent. In these cases, the technology and equipment of the GSS approach is placed directly into the offices of the group members. Usually this is accomplished via a local area network (LAN).

The Teleconferencing Alternative

Teleconferencing is used when the decision frequency is low and the location of group members is distant. These distant and occasional group meetings can tie together multiple GSS decision-making rooms across the country or around the world. Using long-distance communications technology, these decision rooms are electronically connected in teleconferences and videoconferences. This alternative can offer a high degree of flexibility. The GSS decision rooms can be used locally in a group setting or globally when decision makers are located throughout the world. GSS decision rooms are often connected through the Internet.

The Wide Area Decision Network

The wide area decision network is used when the decision frequency is high and the location of group members is distant. In this case, the decision makers require frequent or constant use of the GSS approach. Decision makers located throughout the country or the world must be linked electronically through a wide area network (WAN). The group facilitator and all group members are geographically dispersed. In some cases, the model base and database are also geographically dispersed. This GSS alternative allows people to work in **virtual workgroups**, in which teams of people located around the world can work on common problems.

The Internet is increasingly being used to support wide area decision networks. As discussed in Chapter 4, a number of technologies, including videoconferencing, instant messaging, chat rooms, and telecommuting, can be used to assist the GSS process. In addition, many specialized wide area decision networks make use of the Internet for group decision making and problem solving.

virtual workgroups
Teams of people who are located around the world working on common problems.

EXECUTIVE SUPPORT SYSTEMS

executive support system (ESS)

A specialized DSS that includes all hardware, software, data, procedures, and people used to assist senior-level executives within the organization.

Because top-level executives often require specialized support when making strategic decisions, many companies have developed systems to assist executive decision making. This type of system, called an **executive support system (ESS)**, is a specialized DSS that includes all hardware, software, data, procedures, and people used to assist senior-level executives within the organization. In some cases, an ESS, also called an *executive information system (EIS)*, supports decision making of members of the board of directors, who are responsible to stockholders. These top-level decision-making strata are shown in Figure 6.15. See the "Ethical and Societal Issues" feature to find out how these top executives are becoming involved in security for valuable information systems.

Figure 6.15

The Layers of Executive Decision Making

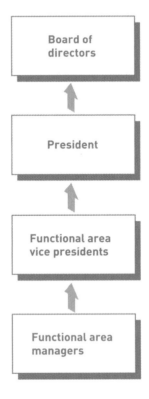

An ESS can also be used by individuals farther down in the organizational structure. Once targeted at the top-level executive decision makers, ESSs are now marketed to—and used by—employees at other levels in the organization. In the traditional view, ESSs give top executives a means of tracking critical success factors. Today, all levels of the organization share information from the same databases. However, for our discussion, we will assume ESSs remain in the upper management levels, where they indicate important corporate issues, indicate new directions the company may take, and help executives monitor the company's progress.

Executive Support Systems in Perspective

An ESS is a special type of DSS, and like a DSS, an ESS is designed to support higher-level decision making in the organization. The two systems are, however, different in important ways. DSSs provide a variety of modeling and analysis tools to enable users to thoroughly analyze problems—that is, they allow users to *answer* questions. ESSs present structured information about aspects of the organization that executives consider important—in other words, they allow executives to *ask* the right questions.

CEOs "Called to Action" Regarding Information Security

The U.S. government is pushing CEOs to take responsibility for the nation's critical information infrastructure. "In this era of increased cyber attacks and information security breaches, it is essential that all organizations give information security the focus it requires," said Amit Yoran, Director of the National Cyber Security Division, Information Analysis and Infrastructure Protection (IAIP) within the U.S. Department of Homeland Security. "[By] addressing these cyber and information security concerns, the private sector will not only strengthen its own security, but help protect the homeland as well."

As noted in a report titled *Information Security Governance: A Call to Action*, a group of CEOs observed that information security is too often treated as a technical issue and passed along to the CIO and technical department to handle. But information systems are so important that top executives and boards of directors must become involved and make security an integral part of core business operations. The report was authored by the Corporate Governance Task Force of the National Cyber Security Partnership. This CEO-led task force is responsible for identifying cybersecurity roles and responsibilities within the corporate management structure.

Part of this effort involves developing best practices and metrics that bring accountability to three key elements of a cybersecurity system: people, processes, and technology. The underlying philosophy is that without cooperation from individual companies and organizations, it is impossible to secure the U.S.'s information infrastructure. Will private corporations cooperate? "There's a significant amount of compliance already," said task force cochairman Arthur Coviello, CEO and president of RSA Security, Inc. "It's hard to imagine that any CEO could not take this as a significant responsibility."

To coax companies to comply, the Department of Homeland Security is establishing an awards program for companies that meet or exceed the security guidelines. Organizations are asked to state on their Web sites that they intend to use the tools developed by the Corporate Governance Task Force to assess their performance and report the results to their boards of directors. Some believe that if there is not widespread voluntary support of the task force's recommendations, the government might need to apply added incentives. Orson Swindle, a member of the Federal Trade Commission, stated that if the task force report fails to get the attention of CEOs and boards of directors, "I have no doubt that some sort of regulation will be passed."

Critical Thinking Questions

1. How might organizations secure sensitive information from cybercriminals? Consider the security concerns and possible solutions regarding data stored in databases and also the important information that is transferred over networks and often printed in reports.
2. How could the government get 100 percent cooperation from private corporations? Is such a goal attainable? If attained, will it guarantee national security? Why or why not?

What Would You Do?

As CEO of Boston Chemical, you have been called on to audit your information systems, identify areas of security concerns, and address the concerns with remedies. The initial audit indicates that much of your information infrastructure is outdated and unable to provide security that measures up to today's standards. An overhaul will cost tens of thousands of dollars and require significant downtime for the company—neither of which the company can afford. In short, it is impossible for you to comply with the recommendations of the task force and stay in business. However, you understand the importance of complying, especially because your company deals in dangerous chemicals.

3. Considering that there are no current laws governing compliance, would it be best to simply ignore the task force recommendation? What other options might there be?
4. What might the government do for businesses such as yours that are unable to comply due to financial constraints?

SOURCES: "Corporate Governance Task Force of the National Cyber Security Partnership Releases Industry Framework," *Entrust*, April 12, 2004, *www.entrust.com*; Dan Verton, "CEOs Urged to Take Control of Cybersecurity," *Computerworld*, April 12, 2004, *www.computerworld.com*; National Cyber Security Partnership (NCSP) Web site, accessed April 25, 2004, *www.cyberpartnership.org*.

The following are general characteristics of ESSs:

- *Are tailored to individual executives.* ESSs are typically tailored to individual executives; DSSs are not tailored to particular users. An ESS is an interactive, hands-on tool that allows an executive to focus, filter, and organize data and information.
- *Are easy to use.* A top-level executive's most critical resource can be his or her time. Thus, an ESS must be easy to learn and use and not overly complex.
- *Have drill-down abilities.* An ESS allows executives to drill down into the company to determine how certain data was produced. Drill down allows an executive to get more detailed information if needed.
- *Support the need for external data.* The data needed to make effective top-level decisions is often external—information from competitors, the federal government, trade associations and journals, consultants, and so on. An effective ESS is able to extract data useful to the decision maker from a wide variety of sources, including the Internet and other electronic publishing sources such as legal and public business information from LexisNexis.
- *Can help with situations that have a high degree of uncertainty.* There is a high degree of uncertainty with most executive decisions. Handling these unknown situations using modeling and other ESS procedures helps top-level managers measure the amount of risk in a decision.
- *Have a future orientation.* Executive decisions are future oriented, meaning that decisions will have a broad impact for years or decades. The information sources to support future-oriented decision making are usually informal—from organizing golf partners to tying together members of social clubs or civic organizations.
- *Are linked with value-added business processes.* Like other information systems, executive support systems are linked with executive decision making about value-added business processes. For instance, executive support systems can be used by car-rental companies to analyze trends.

Capabilities of Executive Support Systems

The responsibility given to top-level executives and decision makers brings unique problems and pressures to their jobs. The following is a discussion of some of the characteristics of executive decision making that are supported through the ESS approach. As you will note, most of these characteristics are related to an organization's overall profitability and direction. An effective ESS should have the capability to support executive decisions with many of these capabilities, such as strategic planning and organizing, crisis management, and more.

Support for Defining an Overall Vision
One of the key roles of senior executives is to provide a broad vision for the entire organization. This vision includes the organization's major product lines and services, the types of businesses it supports today and in the future, and its overriding goals.

Support for Strategic Planning
ESSs also support strategic planning. **Strategic planning** involves determining long-term objectives by analyzing the strengths and weaknesses of the organization, predicting future trends, and projecting the development of new product lines. It also involves planning the acquisition of new equipment, analyzing merger possibilities, and making difficult decisions concerning downsizing and the sale of assets if required by unfavorable economic conditions.

Support for Strategic Organizing and Staffing
Top-level executives are concerned with organizational structure. For example, decisions concerning the creation of new departments or downsizing the labor force are made by top-level managers. Overall direction for staffing decisions and effective communication with labor unions are also major decision areas for top-level executives. ESSs can be employed to help analyze the impact of staffing decisions, potential pay raises, changes in employee benefits, and new work rules.

strategic planning
The process of determining long-term objectives by analyzing the strengths and weaknesses of the organization, predicting future trends, and projecting the development of new product lines.

Support for Strategic Control

Another type of executive decision relates to strategic control, which involves monitoring and managing the overall operation of the organization. Effective ESS approaches can help top-level managers make the most of their existing resources and control all aspects of the organization.

Support for Crisis Management

Even with careful strategic planning, a crisis can occur. Major disasters, including hurricanes, tornadoes, floods, earthquakes, fires, and terrorist activities, can totally shut down major parts of the organization. Handling these emergencies is another responsibility for top-level executives. In many cases, strategic emergency plans can be put into place with the help of an ESS. These contingency plans help organizations recover quickly if an emergency or crisis occurs.

Decision making is a vital part of managing businesses strategically. Information and decision support, group support, and executive support systems help employees by tapping existing databases and providing them with current, accurate information. The increasing integration of all business information systems—from TPSs to MISs to DSSs—can help organizations monitor their competitive environment and make better-informed decisions. Organizations can also use specialized business information systems, discussed in the next chapter, to achieve their goals.

SUMMARY

Principle

Good decision-making and problem-solving skills are the key to developing effective information and decision support systems.

Every organization needs effective decision making and problem solving to reach its objectives and goals. Problem solving begins with decision making. A well-known model developed by Herbert Simon divides the decision-making phase of the problem-solving process into three stages: intelligence, in which potential problems or opportunities are identified and defined; design, in which alternative solutions are developed and explored; and choice, in which the best course of action is selected and later evaluated. Decision making is a component of problem solving. Problem solving also includes implementation, during which a solution is put into effect, and monitoring and modification, if needed.

Decisions can be programmed or nonprogrammed. Programmed decisions are made using a rule, a procedure, or a quantitative method, such as ordering more inventory at certain points. A nonprogrammed decision deals with unusual or exceptional situations, such as determining the best training program for a new employee. Decision makers can use different approaches: optimization (finding the best solution), satisficing (finding a good solution when a problem is too complex to use optimization), or heuristics (using a "rule of thumb" or commonly used guideline to find a good decision).

Principle

The management information system (MIS) must provide the right information to the right person in the right fashion at the right time.

A management information system is an integrated collection of people, procedures, databases, and devices that provide managers and decision makers with information to help achieve organizational goals. An MIS can provide managers with insight into the regular operations of the organization so that they can control, organize, and plan more effectively and efficiently. The primary difference between the reports generated by the TPS and those generated by the MIS is that MIS reports support managerial decision making at the higher levels of management.

Data that enters the MIS originates from both internal sources (an organization's TPS, ERP system, data warehouse, or data mart) and external sources (the Internet, extranets, customers, suppliers, competitors, and stockholders). The output of most MISs is a collection of reports that are distributed to managers: scheduled reports, key-indicator reports, demand reports, exception reports, and drill-down reports. Most MISs are organized along the functional lines of an organization and include financial, manufacturing, marketing, human resource, and other specialized systems, such as accounting and geographic information systems. The primary output of these functional MISs are summary reports that assist in managerial decision making.

Principle

Decision support systems (DSSs) are used when the problems are unstructured.

A decision support system (DSS) is an organized collection of people, procedures, software, databases, and devices working to support managerial decision making. DSSs provide assistance through all phases and different frequencies of the problem-solving process. An ad hoc DSS addresses unique, infrequent decision situations; an institutional DSS handles routine decisions. Highly structured problems, semistructured problems, and unstructured problems can be supported with a DSS. A DSS can also support different managerial levels, including strategic, tactical, and operational-level managers. A common database is often the link that ties together a company's TPS, MIS, and DSS.

The components of a DSS are the database, the model base, the dialogue manager, and a link to external databases, the Internet, the corporate intranet, extranets, networks, data warehouses, and data marts. A data-driven DSS primarily performs qualitative analysis based on the company's databases. Data mining is often used in a data-driven DSS. The model base contains the models used by the decision maker, such as financial, statistical, graphical, and project management models. A model-driven DSS primarily performs mathematical or quantitative analysis. Model management software (MMS) is often used to coordinate the use of models in a DSS. The dialogue manager provides a dialogue management facility to assist in communications between the system and the user. Access to other computer-based systems permits the DSS to tie into other powerful systems, including the TPS or function-specific subsystems.

Principle

Specialized support systems, such as group support systems (GSSs) and executive support systems (ESSs), use the overall approach of a DSS in situations such as group and executive decision making.

A group support system (GSS), also called a *group decision support system* or *computerized collaborative work system*, consists of most of the elements in a DSS plus software to provide effective support in group decision-making settings. GSSs are typically easy to learn and use and can offer specific or general decision-making support. GSS software, also called *groupware*, is specially designed to help generate lists of decision alternatives and perform data analysis, letting people work on joint documents and files over a network. The frequency of GSS use and the location of the decision makers influence the GSS alternative chosen. The decision room supports users in a single location that meet infrequently. Local area networks can be used when group members are located in the same geographic area and users meet regularly. Teleconferencing is used when decision frequency is low and the location of group members is distant. A wide area network is used when the decision frequency is high and the location of group members is distant.

Executive support systems (ESSs) are specialized decision support systems designed to meet the needs of senior management. They indicate issues of importance to the organization, indicate new directions the company might take, and help executives monitor the company's progress. ESSs are typically easy to use, offer a wide range of computer resources, and handle a variety of internal and external data. In addition, the ESS performs sophisticated data analysis, offers a high degree of specialization, and provides flexibility and comprehensive communications abilities. An ESS also supports individual decision-making styles. Some of the major decision-making areas that can be supported through an ESS are providing an overall vision, strategic planning and organizing, strategic control, and crisis management.

CHAPTER 6: SELF-ASSESSMENT TEST

Good decision-making and problem-solving skills are the key to developing effective information and decision support systems.

1. The first stage of the decision-making process is the _____.

    a. initiation phase
    b. intelligence phase
    c. design phase
    d. choice phase

2. Problem solving is one of the phases of decision making. True or False?
3. The final stage of problem solving is _____.
4. A decision that inventory should be ordered when inventory levels drop to 500 units is an example of a(an) _____.

    a. synchronous decision
    b. asynchronous decision
    c. nonprogrammed decision
    d. programmed decision

5. A(An) _____ model will find the best solution, usually the one that will best help the organization meet its goals.

6. A satisficing model is one that will find a good problem solution but not necessarily the best problem solution. True or False?

The management information system (MIS) must provide the right information to the right person in the right fashion at the right time.

7. What summarizes the previous day's critical activities and is typically available at the beginning of each workday?

    a. key-indicator report
    b. demand report
    c. exception report
    d. database report

8. The uses and management of funds is a subsystem of the _____.

    a. marketing MIS
    b. financial MIS
    c. manufacturing MIS
    d. auditing MIS

9. Another name for the _____ MIS is the personnel MIS because it is concerned with activities related to employees and potential employees of the organization.

Decision support systems (DSSs) are used when the problems are unstructured.

10. The focus of a decision support system is on decision-making effectiveness when faced with unstructured or semistructured business problems. True or False?

11. One-of-a-kind decisions are typically handled by a(an) _____ DSS.

12. What component of a decision support system allows decision makers to easily access and manipulate the DSS and to use common business terms and phrases?

    a. the knowledge base
    b. the model base
    c. the dialogue manager
    d. the expert system

Specialized support systems, such as group support systems (GSSs) and executive support systems (ESSs), use the overall approach of a DSS in situations such as group and executive decision making.

13. In a GSS, what approach or technique uses voting to make the final decision in a group setting?

    a. group consensus
    b. group think
    c. nominal group technique
    d. delphi

14. A type of software that helps with joint workgroup scheduling, communication, and management is called _____.

15. The local area decision network is the ideal GSS alternative for situations in which decision makers are located in the same building or geographic area and the decision makers are occasional users of the GSS approach. True or False?

16. A(An) _____ supports the actions of members of the board of directors, who are responsible to stockholders.

CHAPTER 6: SELF-ASSESSMENT TEST ANSWERS

(1) b (2) False (3) monitoring (4) d (5) optimization (6) True (7) a (8) b (9) human resource (10) True (11) ad hoc (12) c (13) c (14) groupware or workgroup software (15) False (16) executive information system (EIS)

REVIEW QUESTIONS

1. What are the stages of problem solving?
2. What is the difference between decision making and problem solving?
3. What is a satisficing model? Describe a situation when it should be used.
4. Give an example of an optimization model.
5. What are the basic kinds of reports produced by an MIS?
6. What guidelines should be followed in developing reports for management information systems?
7. What are the functions performed by a financial MIS?
8. Describe the functions of a manufacturing MIS.
9. What is a human resource MIS? What are its outputs?
10. List and describe some special types of MISs.

11. What is a geographic information system?
12. Describe the difference between a structured and an unstructured problem and give an example of each.
13. Define *decision support system*. What are its characteristics?
14. Describe the difference between a data-driven and a model-driven DSS.
15. What are the components of a decision support system?
16. State the objective of a group support system (GSS) and identify three characteristics that distinguish it from a DSS.
17. Identify three group decision-making approaches often supported by a GSS.
18. What is an executive support system? Identify three fundamental uses for such a system.

DISCUSSION QUESTIONS

1. Select an important problem you had to solve during the last two years. Describe how you used the decision-making and problem-solving steps discussed in this chapter to solve the problem.
2. What is the relationship between an organization's ERP and TPS systems and its MISs? What is the primary role of MISs?
3. How can MISs be used to support the objectives of the business organization?
4. Describe a financial MIS for a *Fortune* 1,000 manufacturer of food products. What are the primary inputs and outputs? What activities does it need to perform?
5. How can a strong financial MIS provide strategic benefits to a firm?
6. Why is auditing so important in a financial MIS? Give an example of an audit that failed to disclose the true nature of the financial position of a firm. What was the result?
7. You have been hired to develop an MIS and a DSS for a manufacturing company. Describe what information you would include in printed reports and what information you would provide using a screen-based decision support system.
8. Pick a company and research its human resource MIS. Describe how its system works. What improvements could be made to its human resource MIS?
9. You have been hired to develop a DSS for a car company, such as Ford or GM. Describe how you would use both data-driven and model-driven DSSs.
10. Imagine that you are the chief financial officer (CFO) for a service organization. You are concerned with the integrity of the firm's financial data. What steps might you take to ascertain the extent of any problems?
11. What functions do DSSs support in business organizations? How does a DSS differ from a TPS and an MIS?
12. How is decision making in a group environment different from individual decision making, and why are information systems that assist in the group environment different? What are the advantages and disadvantages of making decisions as a group?
13. You have been hired to develop group support software. Describe the features you would include in your new GSS.
14. The use of ESSs should not be limited to the executives of the company. Do you agree or disagree? Why?
15. Imagine that you are the vice president of manufacturing for a *Fortune* 1,000 manufacturing company. Describe the features and capabilities of your ideal ESS.

PROBLEM-SOLVING EXERCISES

1. You have been asked to select GSS software to help your company make better decisions to market a new product. Using the Internet, find and investigate three types of software that your company could use for collaborative decision making. Use your word processor to describe what you found and the advantages and disadvantages of each GSS software package.
2. Review the summarized consolidated statement of income for the manufacturing company shown in the 2nd column. Use graphics software to prepare a set of bar charts that shows data for this year compared with the data for last year.

   a. Operating revenues increase by 3.5% while operating expenses increase 2.5%.
   b. Other income and expenses decrease to $13,000.
   c. Interest and other charges increase to $265,000.

Operating Results (in millions)

Operating Revenues	$2,924,177
Operating Expenses (including taxes)	2,483,687
Operating Income	440,490
Other Income and Expenses	13,497
Income before Interest and Other Charges	453,987
Interest and Other Charges	262,845
Net Income	191,142
Average Common Shares Outstanding	147,426
Earnings per share	1.30

If you were a financial analyst tracking this company, what detailed data might you need to perform a more complete analysis? Write a brief memo summarizing your data needs.

3. As the head buyer for a major supermarket chain, you are constantly being asked by manufacturers and distributors to stock their new products. More than 50 new items are introduced each week. Many times, these products are launched with national advertising campaigns and special promotional allowances to both retailers. To add new products, the amount of shelf space allocated to existing products must be reduced, or items must be eliminated altogether.

Develop a marketing MIS that you can use to estimate the change in profits from adding or deleting an item from inventory. Your analysis should include input such as estimated weekly sales in units, shelf space allocated to stock an item (measured in units), total cost per unit, and sales price per unit. Your analysis should calculate total annual profit by item and then sort the rows in descending order based on total annual profit.

TEAM ACTIVITIES

1. Have your team members select two industries. Using only the Internet, have your team work together to make a group decision to develop a list of the 10 best companies in each industry. Your team might be asked to prepare a report on your decision and how difficult it was to use only the Internet for communications.
2. Have your team make a group decision about how to solve the most frustrating aspect of college or university life. Appoint one or two members of the team to disrupt the meeting with negative group behavior. After the meeting, have your team describe how to prevent this negative group behavior. What GSS software features would you suggest to prevent the negative group behavior your team observed?
3. Imagine that you and your team have decided to develop an ESS software product to support senior executives in the music recording industry. What are some of the key decisions these executives must make? Make a list of the capabilities that such a system must provide to be useful. Identify at least six sources of external information that will be useful to its users.

WEB EXERCISES

1. Most companies typically have a number of functional MISs, such as finance. Find the site of two finance companies, such as a bank or a brokerage company. Compare these sites. Which one do you prefer? How could these sites be improved? (Hint: If you are having trouble, try Yahoo!. It should have a listing for "Business and Economy" on its home page. From there, you can go to "companies" and then "finance." There will be several menu choices from there.) You might be asked to develop a report or send an e-mail message to your instructor about what you found.
2. Use the Internet to explore two or more software packages that can be used to make group decisions easier. Summarize your findings in a report.
3. Software, such as the Excel spreadsheet, is often used to find an optimal solution to maximize profits or minimize costs. Search the Internet using Yahoo!, Google, or another search engine to find other software packages that offer optimization features. Write a report describing one or two of the optimization software packages. What are some of the features of the package?

CAREER EXERCISES

1. What decisions are critical for success in a career that interests you? What specific types of reports could help you make better decisions on the job? Give three specific examples.
2. How often do you think you will be involved with group decision making for your career of choice? How can groupware be used to help your team make better decisions? What specific groupware would you like to use on the job?

VIDEO QUESTIONS

Watch the video clip **Army's Virtual World** and answer these questions.

1. The "war on terror" has ushered us into a new form of battle in which individual soldiers rather than generals and platoon leaders are often responsible for making important decisions in the field. Describe the new system presented in this video and how it is being used to train soldiers to quickly make correct decisions.

2. How might this method of using virtual simulations to train decision makers be applied to more typical business-related decision support systems.

CASE STUDIES

Case One

Sensory Systems Provide Better-Tasting Products for Kraft

Have you ever wondered what makes Oreo cookies so delicious? The crunch of the cookie balances perfectly with the cool, semisoft icing. The bittersweet dark chocolate flavor is offset by the sweet, creamy center in perfect harmony. For those who enjoy extremes, the cookie can be twisted apart, allowing the icing to be scraped off and consumed by itself for a sugary rush followed by a mild cookie chaser. For some, Oreos may be the perfect food product.

Who is responsible for this heavenly confection? How is it that every Oreo you eat is as equally delicious as the last one or the one you ate several years ago? As you might have guessed, information and decision support systems play a large role in the production of this perfect cookie, as well as all of the other Kraft products that you enjoy.

Kraft produces hundreds of popular food products, such as Oreos, Jell-O, Post breakfast cereals, Ritz crackers, Tombstone pizza, and Grey Poupon mustard, to name just a few. Kraft products cover the full range of flavors, textures, and eating experiences. Kraft brands hold the top position in 21 of the 25 top food categories in the United States and 21 of the top 25 country categories internationally.

Kraft designs and tests its food sensations in a large lab facility in Glenview, Illinois. The product developers and sensory technologists who work there have food development and production down to a science. To ensure consistent flavor and appearance, Kraft tests its foods throughout the manufacturing process and assigns numerical measurements that quantify the flavor, color, aroma, and other attributes of each product. Concepts that most people express with words such as "chewy," "sweet," "crunchy," and "creamy" are assigned precise definitions and numerical scales to standardize product information.

The characteristics of an Oreo cookie, and all other Kraft products, are stored as a series of numbers and attributes in SENECA, the Sensory and Experimental Collection Application. SENECA assists Kraft in getting more out of its sensory data than ever before.

"SENECA has a database of information that includes everything you might be interested in, in terms of sensory tests," says Beth Knapp, lead systems developer at Kraft. Data stored in SENECA is collected through any of three diverse testing methods employed by Kraft:

- Discrimination testing compares two or more products.
- Descriptive testing employs professional sensory panelists to evaluate all aspects of a product from texture to taste.
- Consumer testing measures the personal responses and opinions of the general public.

Kraft's SENECA application takes the collected data from all of these tests and makes it available for analysis and reuse. The system builds models, histories, and trends based on consumer testing and then evaluates product changes based on discrimination and descriptive testing. Reports are generated to inform Kraft senior managers of products that are well received or perhaps falling out of favor. The system also allows them to drill down into the information to discover the qualities of the product that might be responsible. Based on this information, Kraft is able to capitalize on the qualities that people find most favorable.

SENECA is a powerful tool that provides Kraft with insight into its product line development. Kraft also employs a second information system to guide its manufacturing process. The process variation reduction (PVR) system ensures consistent flavor and appearance for every Kraft product—to make sure that every Ritz cracker tastes as good as the last. "Variation reduction also creates higher average quality," says Knapp, "which means happier customers and more repeat sales."

Kraft's PVR application boosts production, allowing the company to produce more salable cookies, crackers, and other food items per hour. The PVR application evaluates every manufacturing procedure, from recipe instructions to cookie dough shapes and sizes. Reports identify steps in the production process that create excess variation and helps executives improve processes for those areas.

What's more, the PVR system has been deployed over the Internet so that all of Kraft's manufacturing units can take advantage of it. According to Knapp, the PVR process will help Kraft increase revenues and reduce costs. "Process variation reduction has the potential to generate significant cost savings for each facility implementing these procedures," she says.

Ultimately, the SENECA and PVR applications will help ensure consistent, high-quality products for Kraft customers. "To make the ideal cookie, it has to have good shelf life and an excellent flavor," says Keith Eberhardt, a statistician at Kraft. "And if you keep it closer to the ideal, then the product that reaches customers will be even better."

Discussion Questions

1. Why is it useful for food manufacturers to represent the attributes of food products—taste, color, texture, aroma—using a numeric system?

2. After a winning product has been designed, why is it important to mass-produce the product consistently every time?

Critical Thinking Questions

3. Can the development and manufacturing methods used at Kraft be applied to some other industrial product with equal success? Provide an example and rationale.

4. Could some products benefit from inconsistencies in production? Name some. Any food items? Why?

SOURCES: SAS Success Stories Web site, accessed April 26, 2004, *www.sas.com/success/kraft.html*; Kraft Web site, accessed April 26, 2004, *www.kraft.com*.

Case Two

Sanoma Magazines Follow Key Performance Indicators to Success

Sanoma Magazines ranks as the fourth largest European magazine publisher, with approximately 230 magazines published in ten countries. It is the market leader in consumer magazines in Finland, Belgium, the Czech Republic, Hungary, and the Netherlands. Its net sales totaled €1028.4 million in 2003, and it employs about 3,900 people. Sanoma Magazines is headquartered in Amsterdam.

Sonoma Magazine's Belgium division found itself facing some tough competition. It set about finding a more efficient system for reporting and analysis to lead to faster, better-informed decision making. The company's two business units, advertising and reader sales, each had its own production center and set of software applications. Because the two systems worked from separate data sources, management found it difficult, if not impossible, to track and understand information between departments and across the entire organization.

Yves Gilbert, senior systems designer at Sanoma Magazines Belgium described it this way: "We had requests for reports coming and going in all directions, and our 30-person IT staff had to handle them all. This not only created bottlenecks with IT and frustration for everyone but also led to reporting discrepancies and an overall lack of confidence in our reporting process." An information system that provides inaccurate information is more dangerous than no information system at all. And one that employees don't trust is useless.

The IT management team at Sanoma Magazines Belgium set out to correct the problem by asking its executives and analysts what exactly it needed an information and decision support system to provide in order to get the broader, big-picture view of the business that they desired. "Sanoma needed to define lists of key performance indicators (KPIs) that would help managers see and assess changes over time, understand the economic climate, foresee events, make and justify decisions based on accurate metrics, and analyze the impact of actions," explains Gilbert.

Sanoma Magazines Belgium chose BusinessObjects business intelligence software to build its solution. The software applications that are included in the solution are the following:

- BusinessObjects Application Foundation and BusinessObjects Analytics, an integrated suite of enterprise analytic applications, to assess the market more accurately and gather information on reader and advertiser relations
- BusinessObjects Data Integrator to transfer corporate data sources to data marts and operate its data flow

Together, these applications allowed Sanoma Magazines Belgium to centralize corporate data and run high-level analytical functions that make use of key indicators to let managers know the state of the company at any given moment.

All of the units within the organization are finding the BusinessObjects reporting capabilities extremely valuable. The advertising department uses BusinessObjects Analytics to create individualized marketing programs. Employees can then drill down into a global list of KPIs for more information and more detailed reports to increasingly refine their analysis. The sales force uses the system to detect changes in the competitive environment and critical customer accounts and act accordingly. Marketing can analyze the performance of the company's magazines, pricing strategy, and marketing campaigns. They can also monitor all of these indicators over a period of time to make comparisons from one year to the next and take five-year averages.

The human resource department uses BusinessObjects HR Intelligence (Workforce Analytics module) to help manage staff, profiles, completed training programs, and costs. The finance management department uses BusinessObjects

Finance Intelligence (Revenue Cycle Analytics module) to analyze expenses by cost center, as well as monitor payment periods, unpaid bills, cash generated, budgets allocated, and budgets spent.

Today, users analyzing their business activity have a coherent picture of the company and a single version of the truth. Everyone is working from the same base, with the same tool, and using the same method for analyzing business.

Discussion Questions

1. What role do key performance indicators (KPIs) play in the decision-making process? How do they assist managers in taming information overload?
2. What advantage did Sonoma Magazines Belgium gain when it centralized its data source and standardized its systems?

Critical Thinking Questions

3. How did Sonoma Magazines Belgium benefit from adopting an off-the-self solution rather than developing its own information system from scratch?
4. How have information systems, such as the one used at Sonoma Magazines Belgium, affected the magazine industry in terms of competition, workforce composition, and effectiveness?

SOURCES: Business Objects "Customers in the Spotlight" Web site, accessed April 26, 2004, *www.businessobjects.com/customers/spotlight/sanoma.asp*; Sanoma Magazines Web site, accessed April 27, 2004, *www.sanomamagazines.fi.*

NOTES

Sources for the opening vignette: "Sake Producer Provides Old-World Taste and New-World Competitive Edge," *Business Objects Success Stories*, accessed April 25, 2004, *www.businessobjects.com/customers/spotlight/kikumasamune.asp*; Kiku-Musamune Web site, accessed April 25, 2004, *www.kikumasamune.com/*, Business Objects Business Intelligence Web page, accessed April 25, 2004, *www.businessobjects.com/products/bistandardization/default.asp*.

1. Mills, P. et al., "Scheduling Appointments at Trade Events for the Australian Tourist Commission," *Interfaces,* May-June, 2003, p. 12.
2. Blakeley, Fred et al., "Optimizing Periodic Maintenance," *Interfaces*, January, 2003, p. 67.
3. Savage, Marcia, "Trend Micro Debuts Heuristical Antispam Solution," *Asia Computer Weekly*, March 17, 2003.
4. Samuelson, Douglas, "The Netware in Iraq," *OR/MS Today*, June 2003, p. 20.
5. King, Julia, "Sealing the Deal and Collecting Their Due," *Computerworld*, March 31, 2003, p. 52.
6. Songini, Marc, "Boehringer Cures Slow Reporting," *Computerworld*, July 21, 2003, p. 30.
7. Draper, Heather, "Xcel Site Turns Up the Heat on Waste," *Rocky Mountain News*, March 3, 2003, p. 1B.
8. Giegerich, Steve, "Putting Professors to the Test," *Summit Daily News,* February 20, 2003, p. B1.
9. Thomas, Daniel, "Pub Chain Cuts a Better Deal," *Computer Weekly,* January 27, 2004, p. 10.
10. Nelson, Kristi, "Referral System Leads to Connectivity," *Bank System & Technology*, March 1, 2003, p. 16.
11. Warren, Peter, "Software Could Aid Anti-Terrorism Fight," *Computing*, January 15, 2004, p. 1.
12. Kelly, Kate, "Big Board Chief Mulls Changing Trading System," *The Wall Street Journal*, January 30, 2004, p. C1.
13. Kelly, Kate, "A Little Scary," *The Wall Street Journal*, February 2, 2004, p. C1.
14. Huber, Nick, "FSA to Target Stock Market Abuse," *Computer Weekly,* January 20, 2004, p. 14.
15. Thibodeau, P. and Mearian, L. "Terrorism Taxes IT Planning," *Computerworld,* November 5, 2001, p. 1.
16. Hamblen, Matt, "Aircraft Maker Turns to Sourcing Software for New Military Plane," *Computerworld,* December 1, 2003, p. 7.
17. Ellison, Carol, "Unwire Your Sales Force," *PC Magazine,* September 2, 2003, p. 34.
18. Mangalindan, Mylene, "Web Ads on the Rebound," *The Wall Street Journal*, August 25, 2003, p. B1.
19. Leon, Mark, "True-Blue Customers," *Computerworld*, August 11, 2003, p. 37.
20. Gentle, Michael, "CRM: Ready or Not," *Computerworld,* August 18, 2003, p. 40.
21. Breskin, Ira, "Rust Belt CRM," *Computerworld,* December 15, 2003, p. 42.
22. Staff, "GIS Software Market Sees Dynamic Growth," *Electric Light & Power*, February, 2003, p. 6.
23. Mitchell, Robert, "Web Services Put GIS on the Map," *Computerworld*, December 15, 2003, p. 30.
24. Brewin, Bob, "Feds Test Network to Send Alerts," *Computerworld,* March 31, 2003, p. 19.
25. Bulkeley, William, "Diabetes Web Site to Provide Customized Treatment Plans," *The Wall Street Journal*, May 14, 2003, p. D3.
26. Swamnathan, J., "Decision Support for Allocating Scarce Drugs," *Interfaces*, March 2003, p. 1.
27. Grimes, Seth, "Autonomic Computing—Major Vendors Are Applying Decision Support Techniques to Service-Centric Computing," *Intelligent Enterprise*, November 15, 2002, p. 18.
28. Tan, Kopin, "Technology Transforms Options Traders," *The Wall Street Journal*, April 30, 2003, p. C14.
29. Nash, Emma, "Travel Firm Uses BI Tools to Compete," *Computing,* January 29, 2004, p. 11.
30. Barrett, Amy, "Feeding the Pipeline," *Business Week*, May 12, 2003, p. 78.
31. Thomas, Daniel, "Software Improves AC Milan's Game," *Computing,* January 29, 2004, p. 6.
32. Melymuka, Kathleen, "Swarming Technology Helps Widely Dispersed Experts," *Computerworld*, July 28, 2003, p. 35.
33. Tavana, Madjid, "CROSS for Evaluating and Prioritizing Advanced Technology Projects at NASA," *Interfaces*, May-June 2003, p. 40.
34. Schultz, Keith, "Lotus Notes and Domino 6," *New Architect,* January 1, 2003, p. 40.
35. Dornan, Andy, "Mobile E-Mail," *Network Magazine*, February 1, 2003, p. 34.

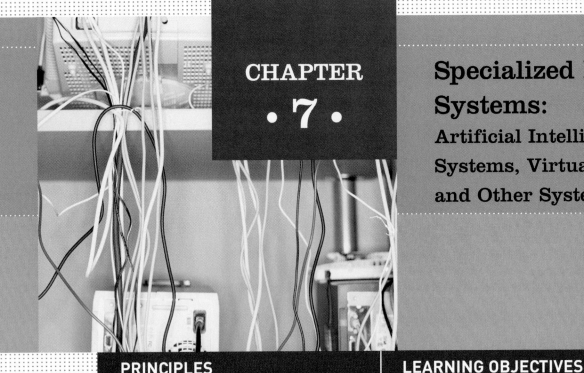

CHAPTER · 7 ·

Specialized Information Systems:
Artificial Intelligence, Expert Systems, Virtual Reality, and Other Systems

PRINCIPLES	LEARNING OBJECTIVES
■ Artificial intelligence systems form a broad and diverse set of systems that can replicate human decision making for certain types of well-defined problems.	■ Define the term *artificial intelligence* and state the objective of developing artificial intelligence systems.
	■ List the characteristics of intelligent behavior and compare the performance of natural and artificial intelligence systems for each of these characteristics.
	■ Identify the major components of the artificial intelligence field and provide one example of each type of system.
■ Expert systems can enable a novice to perform at the level of an expert but must be developed and maintained very carefully.	■ List the characteristics and basic components of expert systems.
	■ Identify at least three factors to consider in evaluating the development of an expert system.
	■ Outline and briefly explain the steps for developing an expert system.
	■ Identify the benefits associated with the use of expert systems.
■ Virtual reality systems have the potential to reshape the interface between people and information technology by offering new ways to communicate information, visualize processes, and express ideas creatively.	■ Define the term *virtual reality* and provide three examples of virtual reality applications.
■ Specialized systems can help organizations and individuals achieve their goals.	■ Discuss examples of specialized systems for organizational and individual use.

INFORMATION SYSTEMS IN THE GLOBAL ECONOMY ➤
AMAZON.COM, UNITED STATES

Amazon Leverages Artificial Intelligence Against Fraud

The amount of information stored in and flowing through many of today's information systems has become so massive that keeping up with it is beyond human capacity. Because computers are ideally suited for quickly handling large amounts of data, many companies are designing computer programs to automate the management and interpretation of data. For these programs to be most effective, they need to be able to think and interpret trends in the data as a human being would—but much, much faster.

For large retailers such as Amazon.com, losses to credit card fraud are substantial. Amazon.com has 35 million customers who access its products through five Web sites adapted for different countries and languages: *www.amazon.com*, *www.amazon.fr*, *www.amazon.co.uk*, *www.amazon.de*, and *www.amazon.co.jp*. Amazon turned to SAS Institute to develop the foundation for its state-of-the-art fraud-detection system.

Fraud-detection techniques are not typically publicized—the less people know about them, the more effective they are. Even Amazon won't fully disclose the details of its artificial intelligence (AI) system. The company is happy to disclose, however, the effectiveness of its new AI fraud-detection system. According to Jaya Kolhatkar, Amazon.com's director of fraud detection, the new system greatly reduced the cases of fraud at Amazon. In the first six months of the system's use, fraud rates were halved.

The system developed for Amazon uses classic AI techniques that simulate human intelligence, such as decision trees and neural networks, to analyze each transaction. The neural networks work within the system to analyze patterns in the data to "learn" which patterns represent fraud—much the same way that human beings learn. The system works on multiple levels. One portion of the system analyzes transactions as they occur, while the credit card number is being approved. Another subsystem crunches data in Amazon's huge transaction database, looking for fraudulent activities in past transactions.

"Fraudsters generally follow similar patterns of behavior," says Kolhatkar. "That makes it easier to detect fraud because you can look for corresponding patterns in transaction and customer data." For example, Amazon knows that fraudsters tend to purchase goods—such as electronics—that they can dispose of easily. Also, they do not ship the goods to the same address that is used for billing, so an order not shipped to the billing address might be an indication of a fraudulent transaction. They also tend to use the fastest possible shipping method. Of course, use of these features does not mean that fraud has definitely occurred, but combined with other indicators, they would be pointers to follow up on.

Amazon's fraud-detection system analyzes the behavioral patterns of fraudsters and builds predictive scores that indicate the likelihood of fraudulent behavior. "We run these scores against the customer database," says Kolhatkar. "We then use SAS to prioritize the results. Obviously, we have to investigate a case of potential fraud very thoroughly before beginning legal action, so we prioritize the results of running the fraud scores and begin with the highest priority cases."

Analyzing large amounts of data to turn up useful and valuable information—such as fraudulent credit card purchases—is just one of the many applications of AI. AI systems are taking over many human tasks that people find either tedious, dangerous, or too big to handle. In short, AI is becoming a valuable extension of human intelligence.

As you read this chapter, consider the following:

- How are AI systems empowering businesses to attain goals previously unattainable?
- What are the strengths and limitations of AI systems, and how do they relate to the strengths and weaknesses of human beings?

Why Learn About Specialized Information Systems?	Specialized information systems are used in almost every industry. If you are a production manager at an automotive company, you might oversee robots that attach windshields to cars or paint body panels. If you are a young stock trader, you might use a special system called *neural networks* to uncover patterns and make millions of dollars trading stocks and stock options. If you are a marketing manager for a PC manufacturer, you might use virtual reality on a Web site to show customers your latest laptop and desktop computers. If you are in the military, you might use computer simulation as a training tool to prepare you for combat. If you work for a petroleum company, you might use an expert system to determine where to drill for oil and gas. Many additional examples of the use of these specialized information systems are discussed throughout this chapter. Learning about these systems will help you discover new ways to use information systems in your day-to-day work.

artificial intelligence (AI)
The ability of computers to mimic or duplicate the functions of the human brain.

At a Dartmouth College conference in 1956, John McCarthy proposed the use of the term **artificial intelligence** (**AI**) to describe computers with the ability to mimic or duplicate the functions of the human brain. Many AI pioneers attended this first conference; a few predicted that computers would be as "smart" as people by the 1960s. The prediction has not yet been realized, but the benefits of artificial intelligence in business and research can be seen today, and research continues.

AN OVERVIEW OF ARTIFICIAL INTELLIGENCE

Science fiction novels and popular movies have featured scenarios of computer systems and intelligent machines taking over the world. Steven Hawking, who is the Lucasian Professor of Mathematics at Cambridge University (a position once held by Isaac Newton) and author of *A Brief History of Time* said, "In contrast with our intellect, computers double their performance every 18 months. So the danger is real that they could develop intelligence and take over the world." Computer systems such as Hal in the classic movie *2001: A Space Odyssey* and the movie *A.I.* are futuristic glimpses of what might be. These accounts are fictional, but we see the real application of many computer systems that use the notion of AI. These systems help to make medical diagnoses, explore for natural resources, determine what is wrong with mechanical devices, and assist in designing and developing other computer systems. This chapter explores the exciting applications of artificial intelligence, expert systems, virtual reality, and some specialized systems to see what the future really might hold.

Artificial Intelligence in Perspective

Artificial intelligence systems include the people, procedures, hardware, software, data, and knowledge needed to develop computer systems and machines that demonstrate characteristics of intelligence. Researchers, scientists, and experts on how humans think are often involved in developing these systems.

The Nature of Intelligence

From the early AI pioneering stage, the research emphasis has been on developing machines with **intelligent behavior**. But machine intelligence is hard to achieve. According to Steve Grand, who developed a robot called Lucy and was given an award for his work in artificial intelligence, "True machine intelligence, let alone consciousness, is a very long way off."[1] Asimo, a robot from Honda, is making progress but has a long way to go. According to Honda's head of the European Research Institute, "Asimo is a marvelous walking machine, a masterpiece of engineering, but the next stage is to enable it to develop the ability to think for itself."[2]

The *Turing Test* attempts to determine whether the responses from a computer with intelligent behavior are indistinguishable from responses from a human. No computer has passed the Turing Test, developed by Alan Turing, a British mathematician. The Loebner Prize offers money and a gold medal for anyone developing a computer that can pass the Turing Test (see *www.loebner.net*). Some of the specific characteristics of intelligent behavior include the ability to do the following:

- *Learn from experience and apply the knowledge acquired from experience.* Being able to learn from past situations and events is a key component of intelligent behavior and is a natural ability of humans, who learn by trial and error. This ability, however, must be carefully programmed into computer systems. Today, researchers are developing systems that have this ability. For instance, computerized AI chess software can learn to improve while playing human competitors.[3] In one match, Garry Kasparov competed against a personal computer with AI software developed in Israel, called Deep Junior.[4] This match in 2003 was a 3-3 tie, but Kasparov picked up something the machine would have no interest in, $700,000. This chapter explores the exciting applications of artificial intelligence and looks at what the future really might hold.[5]

- *Handle complex situations.* Humans are involved in complex situations. World leaders face difficult political decisions regarding terrorism, conflict, global economic conditions, hunger, and poverty. In a business setting, top-level managers and executives are faced with a complex market, challenging competitors, intricate government regulations, and a demanding workforce. Even human experts make mistakes in dealing with these situations. Developing computer systems that can handle perplexing situations requires

artificial intelligence systems
The people, procedures, hardware, software, data, and knowledge needed to develop computer systems and machines that demonstrate the characteristics of intelligence.

intelligent behavior
The ability to learn from experiences and apply knowledge acquired from experience, handle complex situations, solve problems when important information is missing, determine what is important, react quickly and correctly to a new situation, understand visual images, process and manipulate symbols, be creative and imaginative, and use heuristics.

Computers like Deep Junior attempt to learn from past chess moves. While its predecessor, Deep Blue, was a powerful supercomputer, Deep Junior runs on standard computer hardware. Its logic system calculates around 2 to 3 million combinations per second.

(Source: REUTERS/Chip East REUTERS/Landov.)

careful planning and elaborate computer programming. Trying to unlock the mysteries of DNA and the human genome is a complex process.[6] Artificial intelligence can be used to make complex research decisions. According to Eric Lander, director of the Broad Institute at MIT and Harvard, "Computers looked at DNA sequences and decided the next experiment that needed to be done was to close gaps in the genome. The computer knew what experiments to order up, without human input."

- *Solve problems when important information is missing.* The essence of decision making is dealing with uncertainty. Quite often, decisions must be made even when we lack information or have inaccurate information, because obtaining complete information is too costly or impossible. Today, AI systems can make important calculations, comparisons, and decisions even when missing information.

- *Determine what is important.* Knowing what is truly important is the mark of a good decision maker. Developing programs and approaches to allow computer systems and machines to identify important information is not a simple task.

- *React quickly and correctly to a new situation.* A small child, for example, can look over a ledge or a drop-off and know not to venture too close. The child reacts quickly and correctly to a new situation. Computers, on the other hand, do not have this ability without complex programming.

- *Understand visual images.* Interpreting visual images can be extremely difficult, even for sophisticated computers. Moving through a room of chairs, tables, and other objects can be trivial for people but extremely complex for machines, robots, and computers. Such machines require an extension of understanding visual images, called a **perceptive system**. Having a perceptive system allows a machine to approximate the way a human sees, hears, and feels objects. Military robots, for example, use cameras and perceptive systems to conduct reconnaissance missions to detect enemy weapons and soldiers.[7] Detecting and destroying them can save lives. According to Col. Bruce Jette, "I don't have any problems writing to iRobot, saying I'm sorry your robot died, can we get another? That's a lot easier letter to write than to a father or mother." The military and the iRobot company are not releasing information about the PacBot robot.[8] From 50 to 100 robots are being used in Iraq.

- *Process and manipulate symbols.* People see, manipulate, and process symbols every day. Visual images provide a constant stream of information to our brains. By contrast, computers have difficulty handling symbolic processing and reasoning. Although computers excel at numerical calculations, they aren't as good at dealing with symbols and three-dimensional objects. Recent developments in machine-vision hardware and software, however, allow some computers to process and manipulate symbols on a limited basis.

- *Be creative and imaginative.* Throughout history, some people have turned difficult situations into advantages by being creative and imaginative. For instance, when shipped

perceptive system
A system that approximates the way a human sees, hears, and feels objects.

a lot of defective mints with holes in the middle, an enterprising entrepreneur decided to market these new mints as Lifesavers instead of returning them to the manufacturer. Ice cream cones were invented at the St. Louis World's Fair when an imaginative store owner decided to wrap ice cream with a waffle from his grill for portability. Developing new and exciting products and services from an existing (perhaps negative) situation is a human characteristic. Few computers have the ability to be truly imaginative or creative in this way, although software has been developed to enable a computer to write short stories.

- *Use heuristics.* With some decisions, people use heuristics (rules of thumb arising from experience) or even guesses. In searching for a job, you might decide to rank companies you are considering according to profits per employee. Today, some computer systems, given the right programs, obtain good solutions that use approximations instead of trying to search for an optimal solution, which would be technically difficult or too time-consuming.

This list of traits only partially defines intelligence. Unlike virtually every other field of IS research in which the objectives can be clearly defined, the term *intelligence* is a formidable stumbling block. One of the problems in artificial intelligence is arriving at a working definition of real intelligence against which to compare the performance of an artificial intelligence system.

The Difference Between Natural and Artificial Intelligence

Since the term *artificial intelligence* was defined in the 1950s, experts have disagreed about the difference between natural and artificial intelligence. Can computers be programmed to have common sense? Profound differences exist, but they are declining in number (see Table 7.1). One of the driving forces behind AI research is an attempt to understand how humans actually reason and think. It is believed that the ability to create machines that can reason will be possible only once we truly understand our own processes for doing so.

Table 7.1

A Comparison of Natural and Artificial Intelligence

Attributes	Natural Intelligence (Human)	Artificial Intelligence (Machine)
The ability to use sensors (eyes, ears, touch, smell)	High	Low
The ability to be creative and imaginative	High	Low
The ability to learn from experience	High	Low
The ability to be adaptive	High	Low
The ability to afford the cost of acquiring intelligence	High	Low
The ability to use a variety of information sources	High	High
The ability to acquire a large amount of external information	High	High
The ability to make complex calculations	Low	High
The ability to transfer information	Low	High
The ability to make a series of calculations rapidly and accurately	Low	High

The Major Branches of Artificial Intelligence

AI is a broad field that includes several specialty areas, such as expert systems, robotics, vision systems, natural language processing, learning systems, and neural networks (see Figure 7.1). Many of these areas are related; advances in one can occur simultaneously with or result in advances in others.

Figure 7.1

A Conceptual Model of Artificial Intelligence

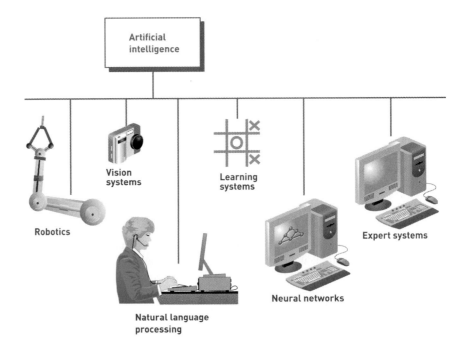

Expert Systems

expert system
Hardware and software that stores knowledge and makes inferences, similar to a human expert.

An **expert system** consists of hardware and software that stores knowledge and makes inferences, similar to a human expert. Because of their many business applications, expert systems are discussed in more detail in the next several sections of the chapter.

Robotics

robotics
Mechanical or computer devices that perform tasks requiring a high degree of precision or that are tedious or hazardous for humans.

Robotics involve developing mechanical or computer devices that can paint cars, make precision welds, and perform other tasks that require a high degree of precision or are tedious or hazardous for humans. The NASA shuttle crash of the early 2000s, for example, has led some people to recommend using robots instead of humans to explore space and perform scientific research.[9] Some robots, like the ER-1, can be used for entertainment.[10] By placing a laptop computer on top, attaching a camera, and installing the necessary software, the ER-1 can maneuver around objects. With an optional gripper arm, the robot can pick up small objects. Researchers hope that Lucy, a robot orangutan, will some day be able to help scientists determine how the brain works.[11] Paro, a harp seal robot, has been used as a therapeutic tool.[12] According to Takanore Shibata, an engineer for Japan's National Institute of Advanced Industrial Science and Technology, "We know that pet therapy helps physically, psychologically, and socially, and Paro does the same thing for people who are unable to care for a live pet." Contemporary robotics combines both high-precision machine capabilities with sophisticated controlling software. The controlling software in robots is what is most important in terms of AI. The processor in an advanced industrial robot today works at about 10 million instructions per second (MIPS)—no smarter than an insect. To achieve anything even approaching human intelligence, the robot processor must achieve 100 trillion operations per second.

Many applications of robotics exist, and research into these unique devices continues. Manufacturers use robots to assemble and paint products. Welding robots have enabled firms to manufacture top-quality products and reduce labor costs while shortening delivery time to their customers. Some robots, such as Honda's Asimo, can shake hands, dance, and even reply to simple questions. The technology used in a robot's legs might offer improvements to people with disabilities. One robot uses technology that is based on the whiskers of a rat to allow the robot to navigate through close spaces.[13] Known as "whiskerbot," the robot uses whisker-like sensors to determine the closeness, size, and texture of objects it touches. Other robots can be used to lay fiber-optic cable or to entertain disabled children while they receive needed therapy and exercise. A surgical robot named da Vinci performed a gall bladder operation on a 16-year-old patient in Denver, Colorado.[14] The surgical robot reduces scars, doesn't introduce bacteria that some doctors do, and can be more precise than a physician. It can also be easier and more comfortable for doctors to operate using a robot than their own hands. Some surgical robots cost more than $1 million and have multiple surgical arms and sophisticated vision systems.[15] Dr. Makimoto, chief of technology at Sony Corporation has predicted that in 50 years, robot soccer players will be able to beat the best human soccer teams.[16] "Some think it's a crazy idea but robot engineers are very serious about this!" says Dr. Makimoto.

Robots can be used in situations that are hazardous or inaccessible to humans. The Rover was a remote-controlled robot used by NASA to explore the surface of Mars.

(Source: Courtesy of NASA.)

Vision Systems

Another area of AI involves vision systems. **Vision systems** include hardware and software that permit computers to capture, store, and manipulate visual images and pictures. The U.S. Justice Department uses vision systems to perform fingerprint analysis, with almost the same level of precision as human experts. The speed with which the system can search through a huge database of fingerprints has brought quick resolution to many long-standing mysteries. Vision systems are also effective at identifying people based on facial features. A California wine bottle manufacturer uses a computerized vision system to inspect wine bottles for flaws.[17] The vision system saves the bottle producer both time and money. The company produces about 2 million wine bottles per day.

vision systems
The hardware and software that permit computers to capture, store, and manipulate visual images and pictures.

Natural Language Processing

Natural language processing allows a computer to understand and react to statements and commands made in a "natural" language, such as English. Restoration Hardware, for example, has developed a Web site that uses natural language processing to allow its customers to quickly find what they want on its site.[18] The natural language processing system corrects spelling mistakes, converts abbreviations into words and commands, and allows people to ask questions in English. In addition to making it easier for customers, Restoration Hardware has seen an increase in revenues as a result of the use of natural language processing.

natural language processing
Processing that allows the computer to understand and react to statements and commands made in a "natural" language, such as English.

There are three levels of voice recognition: command (recognition of dozens to hundreds of words), discrete (recognition of dictated speech with pauses between words), and continuous (recognition of natural speech, such as Naturally Speaking). For example, a natural language processing system can retrieve important information without making the user type in commands or search for key words. With natural language processing, users speak into a microphone connected to a computer and have the computer convert the electrical impulses generated from the voice into text files or program commands. With some simple natural language processors, you say a word into a microphone and type the same word on the keyboard to train the system to recognize the spoken words. The computer then matches the sound with the typed word. With more advanced natural language processors, recording and typing words is not necessary. Upstart Natural Machine is making its Java code for verbal AI available to others. The company hopes that everyone will benefit by making the computer code open and available.

Dragon Systems' Naturally Speaking 7 Essentials uses continuous voice recognition, or natural speech, allowing the user to speak to the computer at a normal pace without pausing between words. The spoken words are transcribed immediately onto the computer screen.

(Source: Courtesy of ScanSoft, Inc.)

Learning Systems

learning systems
A combination of software and hardware that allows the computer to change how it functions or reacts to situations based on feedback it receives.

Another part of AI deals with **learning systems**, a combination of software and hardware that allows a computer to change how it functions or reacts to situations based on feedback it receives. For example, some computerized games have learning abilities. If the computer does not win a game, it remembers not to make the same moves under the same conditions again. Learning systems software requires feedback on the results of actions or decisions. At a minimum, the feedback needs to indicate whether the results are desirable (winning a game) or undesirable (losing a game). The feedback is then used to alter what the system will do in the future.

Neural Networks

neural network
A computer system that can simulate the functioning of a human brain.

An increasingly important aspect of AI involves neural networks. A **neural network** is a computer system that can act like or simulate the functioning of a human brain. The systems use massively parallel processors in an architecture that is based on the human brain's own mesh-like structure. In addition, neural network software can be used to simulate a neural network using standard computers. Neural networks can process many pieces of data at once and learn to recognize patterns. A chemical company, for example, can use neural network software to analyze a large amount of data to control chemical reactors. Fujitsu Laboratories used neural networks to improve motor coordination and movement in its robots.[19] The neural network software gives the robot a smooth walk and allows it to get up if it falls. The "Ethical and Societal Issues" feature explains yet another possible application of neural networks—predicting human behavior.

Some of the specific features of neural networks include the following:

- The ability to retrieve information even if some of the neural nodes fail
- Fast modification of stored data as a result of new information
- The ability to discover relationships and trends in large databases
- The ability to solve complex problems for which all the information is not present

Biologically Inspired Algorithms Fight Terrorists and Guide Businesses

Soon people might be collaborating with intelligent machines to anticipate future human actions. Although scientists have been using computers to forecast predictable natural events such as weather and earthquakes for years, the technology has recently been turned to predicting human behavior. Scientists could use such technology to explore the future of humanity, businesses can use it to anticipate the moves of their competitors and customers, and government and law enforcement can use it to catch the bad guys.

As with many truly revolutionary technologies, funding for this research is coming from the U.S. government, through the Defense Advanced Research Projects Agency (DARPA)—the same organization that produced the Internet. Research in this new intelligence technology is taking place as part of a $54 million program known as Genoa II. The goal of the project is to employ machine intelligence to anticipate future terrorist threats. Researchers hope to make it possible for humans and computers to "think together" in real time to "anticipate and preempt terrorist threats," according to official program documents. "In Genoa II, we are interested in collaboration between people and machines," said Tom Armour, Genoa II program manager at DARPA, "We imagine software agents working with humans...and having different sorts of software agents also collaborating among themselves."

The challenge lies in getting computers to mimic the human brain's ability to reduce complexity. Although computers are good at carrying out rules-based algorithms, such as those required for playing chess, they aren't so good at more complex deciphering, such as finding a word hidden in a picture. Researchers hope to change that through the use of biologically inspired algorithms. "One way to make computers more intelligent and lifelike is to look at living systems and imitate them," says Melanie Mitchell, an associate professor at Oregon Health & Science University's School of Science & Engineering in Portland.

Neural networks are one form of biologically inspired algorithms. They mimic the neurons in the human brain to identify logical patterns in data and to produce very sophisticated learning. Another form of biologically inspired algorithms, genetic algorithms, is inspired by evolution. Here, a computer program evolves a solution to a problem rather than requiring a person to try to engineer one. Also, researchers are beginning to produce security applications that mimic the human immune system, which attacks other sorts of invaders.

Private-sector researchers are studying "cognitive amplifiers," which can enable software to model current situations and predict "plausible futures." They are on the verge of creating practical applications to support cognitive machine intelligence, associative memory, biologically inspired algorithms, and Bayesian inference networks, which are based on a branch of mathematical probability theory that says uncertainty about the world and outcomes of interest can be modeled by combining common sense with evidence observed in the real world.

"Some of the core algorithms we are working with have been around for centuries," says Ron Kolb, director of technology at San Francisco–based Autonomy, Inc., a firm that makes advanced pattern-recognition and knowledge management software. "It's just now that we're finding the practical applications for them." Kolb explains, "We're able to produce an algorithm that says here are the patterns that exist, here are the important patterns that exist, here are the patterns that contextually surround the data, and as new data enters the stream, we're able to build associative relationships to learn more as more data is digested by the system."

Grant Evans, CEO of A4Vision in Cupertino, California, and an expert in cognitive machine intelligence and biologically inspired algorithms, believes that he knows where the future of this research is leading. "Now we're integrating cognitive machine intelligence in the form of video with avatars—3-D digital renderings of real people, that can see and track you."

This is just the kind of talk that makes privacy advocates nervous. They fear that the government and businesses will go too far in collecting social information and infringe on the privacy rights of individuals. Genoa II might be shelved because of its central role in the controversial Terrorism Information Awareness program. Private-sector researchers say that many significant advances are still possible and are, in fact, already happening. Many automotive and aerospace manufacturers have used rudimentary pieces of this technology to save millions of dollars by leveraging developmental expertise across functional areas, says Kolb. "We're no longer looking for information, information is looking for us," he says.

Critical Thinking Questions

1. Why is it that computers might be able to do a better job of predicting human behavior than human beings?
2. What precautions would be wise when developing systems that study human behavior to predict future events?

What Would You Do?

Your company has implemented biologically inspired algorithms to project trends in sales for your product lines. The AI program has indicated that the popularity of one of your best-selling products will begin to decline in the first quarter of next year.

3. How might this prediction influence your treatment of the product line and your business?
4. What might you do to change the future that was predicted by the intelligent machine?

SOURCES: Dan Verton, "Using Computers to Outthink Terrorists," *Computerworld*, September 1, 2003, *www.computerworld.com*, A4Vision Web site, accessed May 31, 2004, *www.a4vision.com*.

Neural networks excel at pattern recognition. For example, neural network computers can be used to read bank check bar codes despite smears or poor-quality printing. Pattern recognition can be used to help prevent terrorism by analyzing and matching images from multiple cameras focusing on people or locations.[20] Many retail stores use Falcon Fraud Manager, a neural network system, to detect possible credit card fraud. Falcon Fraud Manager is used to protect more than 450 million credit card accounts. The Counter Fraud Service in England uses neural network software to detect fraud.[21] Since the late 1990s, the Counter Fraud Service has saved enough money to build three new hospitals, conduct 12,000 heart transplants, or perform 46,000 hip replacements. Fraud has dropped by 30 percent, and the money recovered has increased 700 percent—all helped by the neural network system. Some hospitals use neural networks to determine a patient's likelihood of contracting cancer or other diseases. The speed of genomic research can be increased with software that includes neural network features. Sandia Laboratories has developed a neural network system to give soldiers in a military conflict real-time advice on strategy and tactics.[22]

Neural networks work particularly well when it comes to analyzing detailed trends. Large amusement parks and banks use neural networks to determine staffing needs based on customer traffic—a task that requires precise analysis, down to the half-hour. Increasingly, businesses are firing up neural networks to help them navigate ever-thicker forests of data and make sense of myriad customer traits and buying habits. Computer Associates has developed Neugents, neural intelligence agents that "learn" patterns and behaviors and predict what will happen next. For example, Neugents can be used to track the habits of insurance customers and predict which ones will not renew, say, an automobile policy. They can then suggest to an insurance agent what changes might be made in the policy to persuade the consumer to renew it. The technology also can track individual users at e-commerce sites and their online preferences so that they don't have to input the same information each time they log on—their purchasing history and other data will be recalled each time they access a Web site.

Other Artificial Intelligence Applications

A few other artificial intelligence applications exist in addition to those just discussed. A **genetic algorithm** is an approach to solving large, complex problems in which a number of repeated operations or models change and evolve until the best one emerges. The approach is based on the theory of evolution that requires variation and natural selection. The first step is to change or vary a number of competing solutions to the problem. This can be done by changing the parts of a program or combining different program segments into a new program, similar to the evolution of species, in which the genetic makeup of a plant or animal mutates or changes over time. The second step is to select only the best models or algorithms that continue to evolve. Programs or program segments that are not as good as others are discarded, similar to natural selection or "survival of the fittest," in which only the best species survive and continue to evolve. This process of variation and natural selection continues until the genetic algorithm yields the best possible solution to the original problem. For example, some investment firms use genetic algorithms to help select the best stocks or bonds. Genetic algorithms are also used in computer science and mathematics.[23] Genetic algorithms can help companies determine which orders to accept for maximum profit.[24] The approach helps companies select the orders that will increase profits and take full advantage of the company's production facilities. Genetic algorithms are also being used to make better decisions in developing inputs to neural networks.[25]

An **intelligent agent** (also called an *intelligent robot* or *bot*) consists of programs and a knowledge base used to perform a specific task for a person, a process, or another program. Like a sports agent who searches for the best endorsement deals for a top athlete, an intelligent agent often searches to find the best price, the best schedule, or the best solution to a problem. The programs used by an intelligent agent can search through large amounts of data, while the knowledge base refines the search or accommodates user preferences. Often used to search the vast resources of the Internet, intelligent agents can help people find information on an important topic or the best price for a new digital camera. Intelligent agents can also be used to make travel arrangements, monitor incoming e-mail for viruses or junk mail, and coordinate meetings and schedules of busy executives. In the human resource field, intelligent

genetic algorithm
An approach to solving large, complex problems in which a number of related operations or models change and evolve until the best one emerges.

intelligent agent
Programs and a knowledge base used to perform a specific task for a person, a process, or another program; also called *intelligent robot* or *bot*.

agents are used to help with online training. The software can look ahead in training materials and know what to start next. Staples uses intelligent agents to find job candidates.[26] The software searches a 12-million person database by title, company, gender, and other factors to find the best job candidates for companies. Intelligent agents have been used by the U.S. Army to route security clearance information for soldiers to the correct departments and individuals.[27] What used to take days when done manually now takes hours.

A new hearing aid with artificial intelligence (shown next to a fingernail) contains a tiny microprocessor that works the way the brain does in detecting and distinguishing sounds while filtering out distractions.

(Source: AP/Wide World Photos.)

AN OVERVIEW OF EXPERT SYSTEMS

As mentioned earlier, an expert system behaves similarly to a human expert in a particular field. Charles Bailey, one of the original members of the Library and Information Technology Association, developed one of the first expert systems in the mid-1980s to search the University of Houston's library to retrieve requested resources and citations.[28] Computerized expert systems have been developed to diagnose problems, predict future events, and solve energy problems. They have also been used to design new products and systems, develop innovative insurance products, determine the best use of lumber, and increase the quality of healthcare. Like human experts, computerized expert systems use heuristics, or rules of thumb, to arrive at conclusions or make suggestions. Expert systems have also been used to determine credit limits for credit cards. Soquimich, an agricultural company, uses expert systems to determine the best fertilizer mix to use on certain soils to improve crops while minimizing costs.[29] The research conducted in AI during the past two decades is resulting in expert systems that explore new business possibilities, increase overall profitability, reduce costs, and provide superior service to customers and clients.

Credit card companies often use expert systems to determine credit limits for credit cards.

(Source: © Ron Fehling/Masterfile.)

Characteristics and Limitations of an Expert System

Expert systems have a number of characteristics, including the following:

- *Can explain their reasoning or suggested decisions.* A valuable characteristic of an expert system is the capability to explain how and why a decision or solution was reached. For example, an expert system can explain the reasoning behind the conclusion to approve a particular loan application. The ability to explain its reasoning processes can be the most valuable feature of a computerized expert system. The user of the expert system thus gains access to the reasoning behind the conclusion.

- *Can display "intelligent" behavior.* Considering a collection of data, an expert system can propose new ideas or approaches to problem solving. A few of the applications of expert systems are an imaginative medical diagnosis based on a patient's condition, a suggestion to explore for natural gas at a particular location, and providing job counseling for workers.

- *Can draw conclusions from complex relationships.* Expert systems can evaluate complex relationships to reach conclusions and solve problems. For example, one proposed expert system would work with a flexible manufacturing system to determine the best use of tools. Another expert system can suggest ways to improve quality-control procedures.

- *Can provide portable knowledge.* One unique capability of expert systems is that they can be used to capture human expertise that might otherwise be lost. A classic example of this is the expert system called DELTA (Diesel Electric Locomotive Troubleshooting Aid), which was developed to preserve the expertise of the retiring David Smith, the only engineer competent to handle many highly technical repairs of such machines.

- *Can deal with uncertainty.* One of an expert system's most important features is its ability to deal with knowledge that is incomplete or not completely accurate. The system deals with this problem through the use of probability, statistics, and heuristics.

Even though these characteristics of expert systems are impressive, there are limitations, including the following:

- *Not widely used or tested.* Even though successes occur, expert systems are not used in a large number of organizations. In other words, they have not been widely tested in corporate settings.

- *Difficult to use.* Some expert systems are difficult to control and use. In some cases, the assistance of computer personnel or individuals trained in the use of expert systems is required to help the user get the most from these systems. Today's challenge is to make expert systems easier to use by decision makers who have limited computer programming experience.

- *Limited to relatively narrow problems.* Whereas some expert systems can perform complex data analysis, others are limited to simple problems. Also, many problems solved by expert systems are not that beneficial in business settings. An expert system designed to provide advice on how to repair a machine, for example, is unable to assist in decisions about when or whether to repair it. In general, the narrower the scope of the problem, the easier it is to implement an expert system to solve it.

- *Cannot readily deal with "mixed" knowledge.* Expert systems cannot easily handle knowledge that has a mixed representation. Knowledge can be represented through defined rules, through comparison with similar cases, and in various other ways. An expert system in one application might not be able to deal with knowledge that combined both rules and cases.

- *Possibility of error.* Although some expert systems have limited abilities to learn from experience, the primary source of knowledge is a human expert. If this knowledge is incorrect or incomplete, it will affect the system negatively. Other development errors involve poor programming practices. Because expert systems are more complex than other information systems, the potential for such errors is greater.

- *Cannot refine its own knowledge.* Expert systems are not capable of acquiring knowledge directly. A programmer must provide instructions to the system that determine how the system is to learn from experience. Also, some expert systems cannot refine their own knowledge—such as eliminating redundant or contradictory rules.

- *Difficult to maintain.* Related to the preceding point is the fact that expert systems can be difficult to update. Some are not responsive or adaptive to changing conditions. Adding new knowledge and changing complex relationships may require sophisticated programming skills. In some cases, a spreadsheet used in conjunction with an expert system shell can be used to modify the system. In others, upgrading an expert system can be too difficult for the typical manager or executive. Future expert systems are likely to be easier to maintain and update.
- *May have high development costs.* Expert systems can be expensive to develop when using traditional programming languages and approaches. Development costs can be greatly reduced through the use of software for expert system development. **Expert system shells**, collections of software packages and tools used to develop expert systems, can be implemented on most popular PC platforms to reduce development time and costs.
- *Raise legal and ethical concerns.* People who make decisions and take action are legally and ethically responsible for their behavior. A person, for example, can be taken to court and punished for a crime. When expert systems are used to make decisions or help in the decision-making process, who is legally and ethically responsible—the human experts used to develop the knowledge on which the system relies, the expert system developer, the user, or someone else? For example, if a doctor uses an expert system to make a diagnosis and the diagnosis is wrong, who is responsible? These legal and ethical issues have not been completely resolved.

expert system shell
A collection of software packages and tools used to develop expert systems.

When to Use Expert Systems

Sophisticated expert systems can be difficult, expensive, and time-consuming to develop. This is especially true for large expert systems implemented on mainframes. The following is a list of factors that normally make expert systems worth the expenditure of time and money. Develop an expert system if it can:

- Provide a high potential payoff or significantly reduce downside risk
- Capture and preserve irreplaceable human expertise
- Solve a problem that is not easily solved using traditional programming techniques
- Develop a system more consistent than human experts
- Provide expertise needed at a number of locations at the same time or in a hostile environment that is dangerous to human health
- Provide expertise that is expensive or rare
- Develop a solution faster than human experts can
- Provide expertise needed for training and development to share the wisdom and experience of human experts with a large number of people

Components of Expert Systems

An expert system consists of a collection of integrated and related components, including a knowledge base, an inference engine, an explanation facility, a knowledge base acquisition facility, and a user interface. A diagram of a typical expert system is shown in Figure 7.2. In this figure, the user interacts with the user interface, which interacts with the inference engine. The inference engine interacts with the other expert system components. These components must work together to provide expertise.

The Knowledge Base

The **knowledge base** stores all relevant information, data, rules, cases, and relationships used by the expert system. As shown in Figure 7.3, a knowledge base is a natural extension of a database (presented in Chapter 3) and an information and decision support system (presented in Chapter 6). As discussed in Chapter 3, raw facts can be used to perform basic business transactions but are seldom used without manipulation in decision making. As we move to information and decision support, data is filtered and manipulated to produce a variety of reports to help managers make better decisions. With a knowledge base, we try to understand patterns and relationships in data as a human expert does in making intelligent decisions.

knowledge base
A component of an expert system that stores all relevant information, data, rules, cases, and relationships used by the expert system.

Figure 7.2

Components of an Expert System

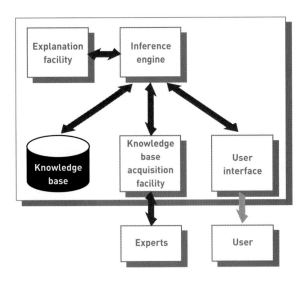

Figure 7.3

The Relationships Among Data, Information, and Knowledge

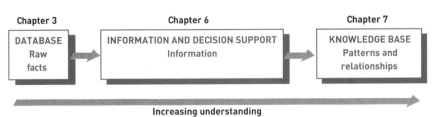

A knowledge base must be developed for each unique application. For example, a medical expert system contains facts about diseases and symptoms. The knowledge base can include generic knowledge from general theories that have been established over time and specific knowledge that comes from more recent experiences and rules of thumb. Knowledge bases, however, go far beyond simple facts, also storing relationships, rules or frames, and cases. For example, certain telecommunications network problems might be related or linked; one problem might cause another. In other cases, rules suggest certain conclusions, based on a set of given facts. In many instances, these rules are stored as **if-then statements**, such as "If a certain set of network conditions exists, then a certain network problem diagnosis is appropriate." Cases can also be used. This technique involves finding instances, or cases, that are similar to the current problem and modifying the solutions to these cases to account for any differences between the previously solved cases stored in the computer and the current situation or problem.

Assembling Human Experts. One challenge in developing a knowledge base is to assemble the knowledge of multiple human experts. Typically, the objective in building a knowledge base is to integrate the knowledge of individuals with similar expertise (e.g., many doctors might contribute to a medical diagnostics knowledge base). A knowledge base that contains information from numerous experts can be extremely powerful and accurate in terms of its predictions and suggestions. Unfortunately, human experts can disagree on important relationships and interpretations of data, presenting a dilemma for designers and developers of knowledge bases and expert systems in general. Some human experts are more expert than others; their knowledge, experience, and information are better developed and more accurately represent reality. When human experts disagree on important points, it can be difficult for expert systems developers to determine which rules and relationships to place in the knowledge base.

The Use of Fuzzy Logic. Another challenge for expert system designers and developers is capturing knowledge and relationships that are not precise or exact. Computers typically work with numerical certainty; certain input values will always result in the same output. In the real world, as you know from experience, this is not always the case. To handle this dilemma, a special research area in computer science, called **fuzzy logic**, has been developed. Fuzzy logic was first applied in Japan in an automatic control system for trains. The control

if-then statements
Rules that suggest certain conclusions.

fuzzy logic
A special research area in computer science that allows shades of gray and does not require everything to be simple black/white, yes/no, or true/false.

system allowed each train to stop within 7 centimeters (about 3 inches) of the right spot on the platform. Fuzzy logic also made train travel smoother and more efficient, saving about 10 percent in energy compared with human-controlled trains.

The Use of Rules. A **rule** is a conditional statement that links given conditions to actions or outcomes. As we saw earlier, a rule can consist of if-then statements. If certain conditions exist, then specific actions are taken or certain conclusions are reached. In an expert system for a weather forecasting operation, for example, the rules could state that if certain temperature patterns exist with a given barometric pressure and certain previous weather patterns over the last 24 hours, then a specific forecast will be made, including temperatures, cloud coverage, and the wind-chill factor. Rules are often combined with probabilities, such as if the weather has a particular pattern of trends, then there is a 65 percent probability that it will rain tomorrow. Likewise, rules relating data to conclusions can be developed for any knowledge base. Most expert systems prevent users from entering contradictory rules. Figure 7.4 shows the use of expert system rules in helping to determine whether a person should receive a mortgage loan from a bank. In general, as the number of rules an expert system knows increases, the precision of the expert system increases.

rule
A conditional statement that links given conditions to actions or outcomes.

Figure 7.4

Rules for a Credit Application

Mortgage Application for Loans from $100,000 to $200,000

If there are no previous credit problems and

If monthly net income is greater than 4 times monthly loan payment and

If down payment is 15% of the total value of the property and

If net assets of borrower are greater than $25,000 and

If employment is greater than three years at the same company

Then accept loan application

Else check other credit rules

The Use of Cases. As mentioned previously, an expert system can use cases in developing a solution to a current problem or situation. This process involves (1) finding cases stored in the knowledge base that are similar to the problem or situation at hand and (2) modifying the solutions to the cases to fit or accommodate the current problem or situation. Cases stored in the knowledge base can be identified and selected by comparing the parameters of the new problem with the cases stored in the computer system. For example, a company might use an expert system to determine the best location of a new service facility in the state of New Mexico. Labor and transportation costs might be the most important factors. The expert system might identify two previous cases involving the location of a service facility where labor and transportation costs were also important—one in the state of Colorado and the other in the state of Nevada. The expert system will modify the solution to these two cases to determine the best location for a new facility in New Mexico. The result might be to locate the new service facility in the city of Santa Fe.

The Inference Engine

The overall purpose of an **inference engine** is to seek information and relationships from the knowledge base and to provide answers, predictions, and suggestions the way a human expert would. In other words, the inference engine is the component that delivers the expert advice.

The process of retrieving relevant information and relationships from the knowledge base is not simple. As you have seen, the knowledge base is a collection of facts, interpretations, and rules. The inference engine must find the right facts, interpretations, and rules and assemble them correctly. In other words, the inference engine must make logical sense out of the information contained in the knowledge base, the way the human mind does when sorting out a complex situation. The inference engine has a number of ways to accomplish its tasks, including backward and forward chaining.

Backward Chaining. **Backward chaining** is the process of starting with conclusions and working backward to the supporting facts. If the facts do not support the conclusion, another conclusion is selected and tested. This process is continued until the correct conclusion is identified. Consider an expert system that forecasts product sales for next month. Backward chaining starts with a conclusion, such as "Sales next month will be 25,000 units." Given this conclusion, the expert system searches for rules in the knowledge base that support the conclusion, such as "IF sales last month were 21,000 units and sales for competing products were 12,000 units, THEN sales next month should be 25,000 units or greater." The expert system verifies the rule by checking sales last month for the company and its competitors. If the facts are not true—in this case, if last month's sales were not 21,000 units or 12,000 units for competitors—the expert system would start with another conclusion and proceed until rules, facts, and conclusions match.

Forward Chaining. **Forward chaining** starts with the facts and works forward to the conclusions. Consider the expert system that forecasts future sales for a product. Forward chaining starts with a fact, such as "The demand for the product last month was 20,000 units." With the forward-chaining approach, the expert system searches for rules that contain a reference to product demand. For example, "IF product demand is over 15,000 units, THEN check the demand for competing products." As a result of this process, the expert system might use information on the demand for competitive products. Next, after searching additional rules, the expert system might use information on personal income or national inflation rates. This process continues until the expert system can reach a conclusion using the data supplied by the user and the rules that apply in the knowledge base.

Comparison of Backward and Forward Chaining. Forward chaining can reach conclusions and yield more information with fewer queries to the user than backward chaining, but this approach requires more processing and a greater degree of sophistication. Forward chaining is often used by more expensive expert systems. Some systems also use mixed chaining, which is a combination of backward and forward chaining.

The Explanation Facility

An important part of an expert system is the **explanation facility**, which allows a user or decision maker to understand how the expert system arrived at certain conclusions or results. A medical expert system, for example, may have reached the conclusion that a patient has a defective heart valve given certain symptoms and the results of tests on the patient. The explanation facility allows a doctor to find out the logic or rationale of the diagnosis made by the expert system. The expert system, using the explanation facility, can indicate all the facts and rules that were used in reaching the conclusion. This facility allows doctors to determine whether the expert system is processing the data and information correctly and logically.

The Knowledge Acquisition Facility

A difficult task in developing an expert system is the process of creating and updating the knowledge base.[30] In the past, when more traditional programming languages were used, developing a knowledge base was tedious and time-consuming. Each fact, relationship, and rule had to be programmed into the knowledge base. In most cases, an experienced programmer had to create and update the knowledge base.

Today, specialized software allows users and decision makers to create and modify their own knowledge bases through the knowledge acquisition facility (see Figure 7.5). The overall

---

inference engine
The part of the expert system that seeks information and relationships from the knowledge base and provides answers, predictions, and suggestions the way a human expert would.

backward chaining
The process of starting with conclusions and working backward to the supporting facts.

forward chaining
The process of starting with the facts and working forward to the conclusions.

explanation facility
The component of an expert system that allows a user or decision maker to understand how the expert system arrived at certain conclusions or results.

purpose of the **knowledge acquisition facility** is to provide a convenient and efficient means for capturing and storing all components of the knowledge base. Knowledge acquisition software can present users and decision makers with easy-to-use menus. After filling in the appropriate attributes, the knowledge acquisition facility correctly stores information and relationships in the knowledge base, making the knowledge base easier and less expensive to set up and maintain. Knowledge acquisition can be a manual process or a mixture of manual and automated procedures. Regardless of how the knowledge is acquired, it is important to validate and update the knowledge base frequently to make sure it is still accurate.

knowledge acquisition facility
The part of the expert system that provides convenient and efficient means of capturing and storing all the components of the knowledge base.

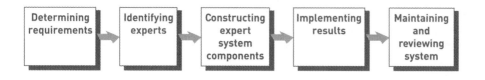

Figure 7.5

The knowledge acquisition facility acts as an interface between experts and the knowledge base.

The User Interface

Specialized user interface software is employed for designing, creating, updating, and using expert systems. The main purpose of the user interface is to make the development and use of an expert system easier for users and decision makers. At one time, skilled computer personnel created and operated most expert systems; today, the user interface permits decision makers to develop and use their own expert systems. Because expert systems place more emphasis on directing user activities than do other types of systems, text-oriented user interfaces (using menus, forms, and scripts) may be more common in expert systems than the graphical interfaces often used with DSSs. A typical user interface between a user and the expert system for the credit application shown in Figure 7.4 follows.

EXPERT SYSTEM: Are there any previous credit problems?
USER: No.
EXPERT SYSTEM: Is the applicant's net income greater than four times the monthly loan payment?
USER: Yes.
EXPERT SYSTEM: Is the down payment greater than 15% of the property value?
USER: Yes.
EXPERT SYSTEM: Are the net assets of the borrower greater than $25,000?
USER: Yes.
EXPERT SYSTEM: Has the borrower worked more than three years at the same company?
USER: Yes.
EXPERT SYSTEM: **Accept the loan application.**

Expert Systems Development

Like other computer systems, expert systems require a systematic development approach for best results (see Figure 7.6). This approach includes determining the requirements for the expert system, identifying one or more experts in the area or discipline under investigation, constructing the components of the expert system, implementing the results, and maintaining and reviewing the complete system.

Determining requirements → Identifying experts → Constructing expert system components → Implementing results → Maintaining and reviewing system

Figure 7.6

Steps in the Expert System Development Process

The Development Process

Specifying the requirements for an expert system begins with identifying the system's objectives and its potential use. Identifying experts can be difficult. In some cases, a company has human experts on hand; in other cases, experts outside the organization will be required. Developing the expert system components requires special skills. Implementing the expert system involves placing it into action and making sure it operates as intended. Like other computer systems, expert systems should be periodically reviewed and maintained to make sure they are delivering the best support to decision makers and users.

Many companies are only now beginning to use and develop expert systems. Expert system development is a team effort, but experienced personnel and users may be in high demand within an organization. Because development can take months or years, the cost of bringing in consultants for development can be high. It is critical, therefore, to find and assemble the right people to assist with development.

Participants in Developing and Using Expert Systems

Typically, several people are involved in developing and using an expert system (see Figure 7.7).

Figure 7.7

Participants in Expert Systems Development and Use

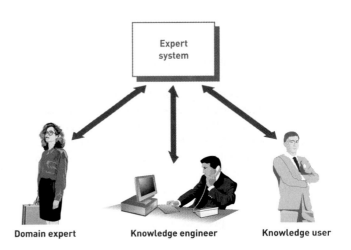

Domain expert   Knowledge engineer   Knowledge user

domain
The area of knowledge addressed by the expert system.

domain expert
The individual or group who has the expertise or knowledge one is trying to capture in the expert system.

The Domain Expert. Because of the time and effort involved in the task, an expert system is developed to address only a specific area of knowledge. This area of knowledge is called the **domain**. The **domain expert** is the individual or group who has the expertise or knowledge one is trying to capture in the expert system. In most cases, the domain expert is a group of human experts. The domain expert (individual or group) usually has the ability to do the following:

- Recognize the real problem
- Develop a general framework for problem solving
- Formulate theories about the situation
- Develop and use general rules to solve a problem
- Know when to break the rules or general principles
- Solve problems quickly and efficiently
- Learn from experience
- Know what is and is not important in solving a problem
- Explain the situation and solutions of problems to others

knowledge engineer
An individual who has training or experience in the design, development, implementation, and maintenance of an expert system.

knowledge user
The individual or group who uses and benefits from the expert system.

The Knowledge Engineer and Knowledge Users. A **knowledge engineer** is an individual who has training or experience in the design, development, implementation, and maintenance of an expert system, including training or experience with expert system shells. The **knowledge user** is the individual or group who uses and benefits from the expert system. Knowledge users do not need any previous training in computers or expert systems.

Expert Systems Development Tools and Techniques

Theoretically, expert systems can be developed from any programming language. Since the introduction of computer systems, programming languages have become easier to use, more powerful, and increasingly able to handle specialized requirements. In the early days of expert systems development, traditional high-level languages, including Pascal, FORTRAN, and COBOL, were used (see Figure 7.8). LISP was one of the first special languages developed and used for artificial intelligence applications. PROLOG, a more recent language, was also developed for AI applications. Since the 1990s, however, other expert system products (such as shells) are available that remove the burden of programming, allowing nonprogrammers to develop and benefit from the use of expert systems.

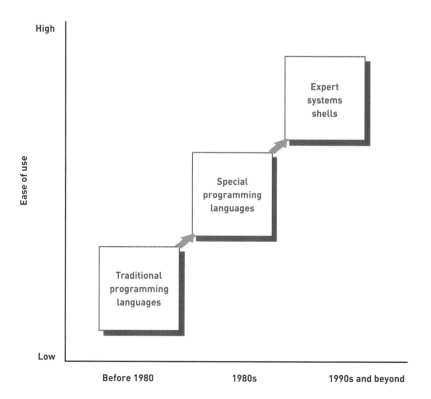

Figure 7.8

Software for expert systems development has evolved greatly since 1980, from traditional programming languages to expert system shells.

Expert System Shells and Products. As discussed, an expert system shell is a collection of software packages and tools used to design, develop, implement, and maintain expert systems. Expert system shells exist for both personal computers and mainframe systems. Some shells are inexpensive, costing less than $500. In addition, off-the-shelf expert system shells are available that are complete and ready to run. The user enters the appropriate data or parameters, and the expert system provides output to the problem or situation.

A number of expert system shells and products are available to analyze LANs, monitor air quality in commercial buildings, and analyze oil and drilling operations. A few expert system shells are summarized in Table 7.2.

Expert Systems Development Alternatives

Expert systems can be developed from scratch by using an expert system shell or by purchasing an existing expert system package. A graph of the general cost and time of development alternatives is shown in Figure 7.9. It is usually faster and less expensive to develop an expert system using an existing package or an expert system shell. Note that there will be an additional cost of developing an existing package or acquiring an expert system shell if the organization does not already have this type of software.

Name of Shell	Application and Capabilities
Financial Advisor	Analyzes financial investments in new equipment, facilities, and the like; requests the appropriate data and performs a complete financial analysis.
G2	Assists in oil and gas operations. Transco, a British company, uses it to help in the transport of gas to more than 20 million commercial and domestic customers.
RAMPART	Analyzes risk. The U.S. General Services Administration uses it to analyze risk to the approximately 8,000 federal buildings it manages.
HazMat Loader	Analyzes hazardous materials in truck shipments.
MindWizard	Enables development of compact expert systems ranging from simple models that incorporate business decision rules to highly sophisticated models; PC based and inexpensive.
LSI Indicator	Helps determine property values; developed by one of the largest residential title and closing companies.

Table 7.2

Popular Expert System Shells

In-House Development: Develop from Scratch. Developing an expert system from scratch is usually more costly than the other alternatives, but an organization has more control over the features and components of the system. Such customization also has a downside; it can result in a more complex system, with higher maintenance and updating costs.

In-House Development: Develop from a Shell. As you have seen, an expert system shell consists of one or more software products that assist in the development of an expert system. In some instances, the same shell can be used to develop many expert systems. Developing an expert system from a shell can be less complex and easier to maintain than developing one from scratch. However, the resulting expert system may need to be modified to tailor it to specific applications. In addition, the capabilities and features of an expert system can be more difficult to control.

Figure 7.9

Some Expert System Development Alternatives and Their Relative Cost and Time Values

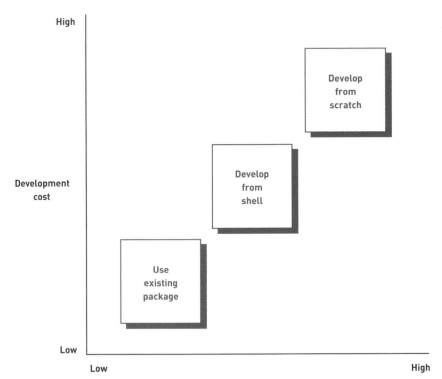

Off-the-Shelf Purchase: Use Existing Packages. Using an existing expert system package is the least expensive and fastest approach in most cases. An existing expert system package is one that has been developed by a software or consulting company for a specific field or area, such as the design of a new computer chip or a weather forecasting and prediction system. The advantages of using an existing package can go beyond development time and cost. These systems can also be easy to maintain and update over time. A disadvantage of using an off-the-shelf package is that it might not be able to satisfy the unique needs or requirements of an organization.

Applications of Expert Systems and Artificial Intelligence

Expert systems and artificial intelligence are being used in a variety of ways. For example, expert systems have been used to help power plants reduce pollutants while maintaining profits.[31] They have also been used to determine the best way to distribute weight in a ferryboat to reduce the risk of capsizing or sinking.[32] Some other applications of these systems are summarized next.

- *Credit granting and loan analysis.* Many banks employ expert systems to review an individual's credit application and credit history data from credit bureaus to make a decision on whether to grant a loan or approve a transaction. KPMG Peat Marwick uses an expert system called Loan Probe to review its reserves to determine whether sufficient funds have been set aside to cover the risk of some uncollectible loans.

- *Catching cheats and terrorists.* Some gambling casinos use expert system software to catch gambling cheats. The CIA is testing the software to see whether it can be used to detect possible terrorists when they make hotel or airline reservations.

- *Information management and retrieval.* The explosive growth of information available to decision makers has created a demand for devices to help manage the information. Expert systems can aid this process through the use of bots. Businesses might use a bot to retrieve information from large distributed databases or a vast network like the Internet. Expert system agents help managers find the right data and information while filtering out irrelevant facts that might impede timely decision making. The Burlington Northern Santa Fe Railroad uses speech-recognition technology to help it retrieve and store shipping data automatically, as the "Information Systems @ Work" feature discusses.

- *AI and expert systems embedded in products.* The antilock braking system on today's automobiles is an example of a rudimentary expert system. A processor senses when the tires are beginning to skid and releases the brakes for a fraction of a second to prevent the skid. AI researchers are also finding ways to use neural networks and robotics in everyday devices, such as toasters, alarm clocks, and televisions.

- *Plant layout and manufacturing.* FLEXPERT is an expert system that uses fuzzy logic to perform plant layout. The software helps companies determine the best placement for equipment and manufacturing facilities. Expert systems can also be used in manufacturing.[33] According to an industry observer, "A few giants will emerge. They'll handle their customers' full logistics needs at the lowest unit cost, using a series of expert systems." Expert systems can also be used to spot defective welds during the manufacturing process.[34] The expert system analyzes radiographic images and suggests which welds could be flawed.

- *Hospitals and medical facilities.* Some hospitals use expert systems to determine a patient's likelihood of contracting cancer or other diseases. Hospitals, pharmacies, and other healthcare providers can use CaseAlert by MEDecision to determine possible high-risk or high-cost patients. MYCIN is an expert system developed at Stanford University to analyze blood infections. UpToDate is another expert system used to diagnose patients. A medical expert system used by the Harvard Community Health Plan allows members of the HMO to get medical diagnoses via home personal computers. For minor problems, the system gives uncomplicated treatments; for more serious conditions, the system schedules appointments. The system is highly accurate, diagnosing 97 percent of the patients correctly (compared with the doctors' 78 percent accuracy rating).

Moving Data—and Freight—Efficiently

The Burlington Northern and Santa Fe Railway Co., more conveniently referred to as BNSF, operates one of the largest railroad networks in North America, with about 32,500 route miles covering 28 states and 2 Canadian provinces. The railway moves more freight than any other rail system in the world, is America's largest grain-hauling railroad, transports the mineral components of many of the products we depend on daily, and hauls enough coal to generate more than 10 percent of the electricity produced in the United States. Keeping track of the freight being hauled in 190,000 freight cars pulled by 5,000 locomotives is no small task and until recently was slow and error prone.

While BNSF maintained a state-of-the-art information system in its network operations center in Fort Worth, Texas, its method of retrieving information from the train crews was antiquated. As freight cars were picked up or dropped off, the crew would radio the information to the dispatcher, who would then type the information into the IBM mainframe system. Crews would start their day with written work orders and turn them in at the end of the day, interspersed with periodic calls to report cars dropped off or picked up. Jeff Campbell, BNSF's CIO, viewed this approach as outdated, cumbersome, and incapable of meeting the demands of customers and railroad management for near-real-time data. BNSF set out to design a system that could provide up-to-the-moment shipment information similar to systems used by UPS and FedEx.

BNSF developed a system, based on speech-recognition software by ScanSoft, Inc., that automatically turned voice radio calls into data that could be integrated into the company's computer systems. Although speech-recognition systems are popular for automated phone systems, using speech recognition for voice radio is a new application for the technology—one that presented some challenges. "Two-way radio systems have lower fidelity than the phone lines traditionally used with IVR [interactive voice recognition]. The fidelity problem was compounded by the noisy environment of a locomotive cab," said Rob Kassel, ScanSoft's senior product manager for network speech.

ScanSoft employed special noise-filtering software to "clean up" the signal and sampled numerous engineer radio calls to teach the software to recognize speech generated in such a noisy environment. In addition to speech-recognition technology, ScanSoft also used voice-recognition software to verify the identity of the person speaking. BNSF did not want to risk an outsider's hacking into its private microwave radio system and feeding the system false information.

After the voice of an incoming caller is recognized and verified, the system prompts the crew for information using an interactive audio menu. The responses from the crew are digitized, interpreted, and translated into text that is automatically entered into the mainframe database. The system supports both radio calls and cell phone calls and is managed on PCs by dispatchers at BNSF's network operations center.

BNSF recently rolled out its new system, which it calls the Radio Telephony Interface (RTI), and plans to take it systemwide by 2005. Although the company declines to disclose the exact cost of the system, BNSF's IT budget increased from $1.5 million in 2003 to $274 million in 2004.

But the resulting system was worth it. BNSF was recognized by *Speech Technology* magazine as winner of its "Most Innovative Solutions Awards." Candidates for the award were judged on their ability to increase customer self-service, improve worker efficiency, and utilize speech in a creative format for an automated solution. Jeff Campbell says that the new system "improves customer satisfaction" by allowing BNSF to update its Transportation Support System in near-real time. The RTI allows BNSF to provide customers with more frequent information on car moves "and closer expected time of arrival."

Dan Miller, an analyst at Zelos Group, Inc., in San Francisco, says radio-to-data interfaces are the next frontier for IVR systems with huge growth potential within many industries, including trucking, utilities, and field service fleet firms.

Discussion Questions

1. What effect has speech recognition had on the daily routines of BNSF crews and dispatchers?
2. What types of return on its investment could BNSF earn with its speech-recognition system?

Critical Thinking Questions

3. Many businesses in the transportation and delivery industries rely on radio communications to keep dispatchers, and even air traffic controllers, in touch with drivers, engineers, and pilots. Which of these businesses might benefit from speech recognition? How?
4. What general communication scenarios are ideal for interactive voice-recognition systems? What is it about these scenarios that make them ideal?

SOURCES: Bob Brewin, "A Railroad Finds Its Voice," *Computerworld*, January 26, 2004, *www.computerworld.com*; "ScanSoft Speech System Deployments Receive Industry Innovation Awards," *Business Wire*, May 19, 2004, *www.lexis-nexis.com*; Scansoft Inc. Web site, accessed June 2, 2004, *www.scansoft.com*; BNSF's Web site, accessed June 2, 2004, *www.bnsf.com*.

- *Help desks and assistance.* Customer service help desks use expert systems to provide timely and accurate assistance. Kaiser Permanente, a large HMO, uses an expert system and voice response to automate its help desk function. The automated help desk frees up staff to handle more complex needs, while still providing more timely assistance for routine calls.
- *Employee performance evaluation.* An expert system developed by Austin-Hayne, called Employee Appraiser, provides managers with expert advice for use in employee performance reviews and career development.
- *Virus detection.* IBM is using neural network technology to help create more advanced software for eradicating computer viruses, a major problem in American businesses. IBM's neural network software deals with "boot sector" viruses, the most prevalent type, using a form of artificial intelligence that mimics the human brain and generalizes by looking at examples. It requires a vast number of training samples, which in the case of antivirus software are three-byte virus fragments.
- *Shipping.* CARGEX cargo expert system is used by Lufthansa, a German airline, to help determine the best shipping routes.
- *Marketing.* CoverStory is an expert system that extracts marketing information from a database and automatically writes marketing reports.

VIRTUAL REALITY

The term *virtual reality* was initially coined by Jason Lanier, founder of VPL Research, in 1989. Originally, the term referred to *immersive virtual reality* in which the user becomes fully immersed in an artificial, three-dimensional world that is completely generated by a computer. Immersive virtual reality can represent any three-dimensional setting, real or abstract, such as a building, an archaeological excavation site, the human anatomy, a sculpture, or a crime scene reconstruction. Through immersion, the user can gain a deeper understanding of the virtual world's behavior and functionality.

A **virtual reality system** enables one or more users to move and react in a computer-simulated environment. Virtual reality simulations require special interface devices that transmit the sights, sounds, and sensations of the simulated world to the user. These devices can also record and send the speech and movements of the participants to the simulation program, enabling users to sense and manipulate virtual objects much as they would real objects. This natural style of interaction gives the participants the feeling that they are immersed in the simulated world. DaimlerChrysler uses virtual reality to help it simulate and design factories.[35] According to the CIO of DaimlerChrysler, "We are piloting a digital plant. As the engineers and designers are developing new products, the digital factory is simulating production." The company believes that this use of virtual reality will reduce the time it takes to move from an idea to production by about 30 percent.

virtual reality system
A system that enables one or more users to move and react in a computer-simulated environment.

Interface Devices

To see in a virtual world, often the user will wear a head-mounted display (HMD) with screens directed at each eye. The HMD also contains a position tracker to monitor the location of the user's head and the direction in which the user is looking. Using this information, a computer generates images of the virtual world—a slightly different view for each eye—to match the direction the user is looking and displays these images on the HMD.

With current technology, virtual-world scenes must be kept relatively simple so that the computer can update the visual imagery quickly enough (at least ten times a second) to prevent the user's view from appearing jerky and from lagging behind the user's movements.

Alternative concepts—BOOM and CAVE—were developed for immersive viewing of virtual environments to overcome the often uncomfortable intrusiveness of a head-mounted display. The BOOM (Binocular Omni-Orientation Monitor) from Fakespace Labs is a head-coupled stereoscopic display device. Screens and optical systems are housed in a box that is attached to a multilink arm. The user looks into the box through two holes, sees the virtual world, and can guide the box to any position within the virtual environment. Head tracking is accomplished via sensors in the links of the arm that holds the box.

The BOOM, a head-coupled display device.

(Source: Courtesy of University of Michigan Virtual Reality Laboratory.)

The Electronic Visualization Laboratory at the University of Illinois at Chicago introduced a room constructed of large screens on which the graphics are projected onto the three walls and the floor. The CAVE, as this room is called, provides the illusion of immersion by projecting stereo images on the walls and floor of a room-sized cube. Several persons wearing lightweight stereo glasses can enter and walk freely inside the CAVE. A head-tracking system continuously adjusts the stereo projection to the current position of the leading viewer.

Viewing the Detroit Midfield Terminal in an immersive CAVE system.

(Source: Courtesy of University of Michigan Virtual Reality Laboratory.)

Users hear sounds in the virtual world through earphones. The information reported by the position tracker is also used to update audio signals. When a sound source in virtual space is not directly in front of or behind the user, the computer transmits sounds to arrive at one ear a little earlier or later than at the other and to be a little louder or softer and slightly different in pitch.

The *haptic* interface, which relays the sense of touch and other physical sensations in the virtual world, is the least developed and perhaps the most challenging to create. Currently, with the use of a glove and position tracker, the computer locates the user's hand and measures finger movements. The user can reach into the virtual world and handle objects; however, it is difficult to generate the sensations that are felt when a person taps a hard surface, picks up an object, or runs a finger across a textured surface. Touch sensations also have to be synchronized with the sights and sounds users experienced.

Forms of Virtual Reality

Aside from immersive virtual reality, virtual reality can also refer to applications that are not fully immersive, such as mouse-controlled navigation through a three-dimensional environment on a graphics monitor, stereo viewing from the monitor via stereo glasses, stereo projection systems, and others.

Some virtual reality applications allow views of real environments with superimposed virtual objects. Motion trackers monitor the movements of dancers or athletes for subsequent studies in immersive virtual reality. Telepresence systems (e.g., telemedicine, telerobotics) immerse a viewer in a real world that is captured by video cameras at a distant location and allow for the remote manipulation of real objects via robot arms and manipulators. Many believe that virtual reality will reshape the interface between people and information technology by offering new ways to communicate information, visualize processes, and express ideas creatively.

Computer-generated image technology and simulation are used by companies to determine plant capacity, manage bottlenecks, and optimize production rates.

(Source: Courtesy of Flexsim Software Products, Inc.)

Virtual Reality Applications

There are literally hundreds of applications of virtual reality, with more being developed as the cost of hardware and software declines and people's imaginations are opened to the potential of virtual reality. Here is a summary of some of the more interesting ones.

Medicine

Surgeons in France performed the first successful closed-chest coronary bypass operation. Instead of cutting open the patient's chest and breaking his breastbone, as is usually done, surgeons used a virtual reality system that enabled them to operate through three tiny half-inch incisions between the patient's ribs. They inserted thin tubes to tunnel to the operating area and protect the other body tissue. Then three arms were inserted into the tubes. One was for a 3-D camera; the other two held tiny artificial wrists to which a variety of tools—scalpels, scissors, needle—were attached. The virtual reality system mimicked the movements of the surgeon's shoulders, elbows, and wrists. The surgeon sat at a computer workstation several feet from the operating table. The instruments inside the patient were so accurate that it was possible to sew up a coronary artery as thin as a thread. The surgeon watched his progress on a screen that enlarged the artery in 3-D to the size of a garden hose.

Virtual reality technology can also link stroke patients to their physical therapists. Patients put on special gloves and other virtual-reality devices at home that are linked to the physical therapist's office. The physical therapist can then see whether the patient is performing the correct exercises without having to travel to the patient's home or hospital room. Use of virtual reality can cut travel time and costs.

Education and Training

Virtual environments are used in education to bring exciting new resources into the classroom. Students can stroll among digital bookshelves, learn anatomy on a simulated cadaver, or participate in historical events—all virtually.

Virtual technology has also been applied by the military. To help with aircraft maintenance, a virtual reality system has been developed to simulate an aircraft and give a user a sense of touch, while computer graphics give the senses of sight and sound. The user sees, touches, and manipulates the various parts of the virtual aircraft during training. Also, the Pentagon is using a virtual reality training lab to prepare for a military crisis. The virtual reality system simulates different war scenarios.

Real Estate Marketing and Tourism

Virtual reality has been used to increase real estate sales in several powerful ways. From Web publishing to laptop display to a potential buyer, virtual reality provides excellent exposure for properties and attracts potential clients. Clients can take a virtual walk through properties on disks or via the Web and save valuable time. Realatrends Real Estate Service, which offers homes for sale in Orange County, California (*www.realatrends.com/virtual_tours.htm*), is just one of many real estate firms offering this service. In another Web application, the U.S. government created a virtual tour of the White House. The virtual tour allowed people to see a 360-degree view of rooms on the Internet.

Entertainment

Computer-generated image technology, or CGI, has been around since the 1970s. A number of movies used this technology to bring realism to the silver screen, including *Finding Nemo*, *Spider-Man II*, *Star Wars Episode II—Attack of the Clones*, and many others. CGI can also be used for games to enhance the viewers' knowledge and enjoyment. Disney's Toontown (*www.toontown.com*) is a virtual reality Internet site for children.[36] Second Life (*www.secondlife.com*) allows people to play games, interact with on-screen characters called avatars, and build structures, such as homes.

OTHER SPECIALIZED SYSTEMS

In addition to artificial intelligence, expert systems, and virtual reality, a number of other interesting and exciting specialized systems have appeared. Segway, for example, is an electric scooter that uses sophisticated software, sensors, and gyro motors to transport people through warehouses, offices, downtown sidewalks, and other spaces.[37] Originally designed to transport people around a factory or around town, more recent versions are being tested by the military for gathering intelligence and transporting wounded soldiers to safety.[38] Cyberkinetics is in a medical trial for a small microchip that could be embedded in the brain of patients with spinal cord injuries and wired to a computer.[39] In the trial, patients will be asked to think about moving the cursor on a computer screen while the system tries to record the physical responses to their thoughts. If successful, the chip might be able to move computer cursors or perform other tasks to help quadriplegics perform tasks they couldn't perform otherwise.[40]

As mentioned previously, *radio-frequency identification (RFID)* tags that contain small chips with information about products or packages can be quickly scanned to perform inventory control or trace a package as it moves from a supplier to a company to its customers. Delta Airlines is testing RFID to track luggage through airports.[41] RFID has given the company an accuracy rate that is greater than 95 percent. RFID is also being used by Metro Group, a German retailer with thousands of stores in 28 countries, to track products from warehouses to stores.[42] The Seattle Seahawks football team uses RFID in the Club Seat section to increase sales and speed food sales and purchases of other products.[43] Fans pass food and other products with RFID tags over a PowerPay reader to ring up sales. Sales are up by about 18 percent as a result of the RFID tags.

Another technology is being used to create "smart containers" for ships, railroads, and trucks.[44] NaviTag and other companies are developing communications systems that would allow containers to broadcast the contents, location, and condition of shipments to shipping and cargo managers.

One special application of computer technology is game theory. **Game theory** involves the use of information systems to develop competitive strategies for people, organizations, or even countries. Two competing businesses in the same market can use game theory to determine the best strategy to achieve their goals. Individual investors could use game theory to determine the best strategies when competing against other investors in a government auction of bonds. Groundbreaking work on game theory was pioneered by John Nash, the mathematician whose life was profiled in the book and film *A Beautiful Mind*.[45] Game theory has also been used to develop approaches to deal with terrorism.

Informatics, another specialized system, combines traditional disciplines, such as science and medicine, with computer systems and technology. *Bioinformatics*, for example, combines biology and computer science. Also called *computational biology*, bioinformatics has been used to help map the human genome and conduct research on biological organisms. Stanford University has a course on bioinformatics and offers bioinformatics certification. Medical informatics combines traditional medical research with computer science. Journals, such as *Healthcare Informatics*, report current research on applying computer systems and technology to reduce medical errors and improve healthcare.

Many other specialized devices are used by companies for a variety of purposes. "Smart dust" was developed at the University of California at Berkeley with Pentagon funding.[46] Smart dust involves small networks powered by batteries and an operating system, called TinyOS. The technology can be used to monitor temperature, light, vibration, or even toxic chemicals. Small radio transceivers can be placed in other products, such as cell phones. The radio transceivers allow cell phones and other devices to connect to the Internet, cellular phone service, and other devices that use the technology. The radio transceivers could save companies hundreds of thousands of dollars annually. Microsoft's Smart Personal Objects Technology (SPOT) allows small devices to transmit data and messages over the air.[47] SPOT is being used in wristwatches to transmit data and messages over FM radio broadcast bands. The new technology requires a subscription to Microsoft's MSN Direct information service. Automotive software allows cars and trucks to be connected to the Internet. The software can track a driver's speed and location, allow gas stations to remotely charge for fuel and related services, and more. Manufacturing experiments are also being done with ink-jet printers to allow them to "print" 3-D parts.[48] The technology is being used in Iowa City to print new circuit boards using a specialized ink-jet printer. The printer sprays layers of polymers onto circuit boards to form transistors and other electronic components.

Segway, a human transport device, uses sophisticated software, sensors, and gyro motors to transport people in an upright position.

(Source: © Reuters NewMedia Inc./ CORBIS.)

game theory
The use of information systems to develop competitive strategies for people, organizations, or even countries.

informatics
A specialized system that combines traditional disciplines, such as science and medicine, with computer systems and technology.

SUMMARY

Principle

Artificial intelligence systems form a broad and diverse set of systems that can replicate human decision making for certain types of well-defined problems.

The term *artificial intelligence* is used to describe computers with the ability to mimic or duplicate the functions of the human brain. Intelligent behavior encompasses several characteristics, including the abilities to learn from experience and apply this knowledge to new experiences; handle complex situations and solve problems for which pieces of information may be missing; determine relevant information in a given situation, think in a logical and rational manner, and give a quick and correct response; and understand visual images and processing symbols. Computers are better than humans at transferring information, making a series of calculations rapidly and accurately, and making complex calculations, but humans are better than computers at all other attributes of intelligence.

Artificial intelligence is a broad field that includes several key components, such as expert systems, robotics, vision systems, natural language processing, learning systems, and neural networks. There are many other artificial intelligence applications also. A genetic algorithm is an approach to solving large, complex problems in which a number of related operations or models change and evolve until the best one emerges. The approach is similar to the theory of evolution that is based on variation and natural selection. Intelligent agents consist of programs and a knowledge base used to perform a specific task for a person, a process, or another program.

Principle

Expert systems can enable a novice to perform at the level of an expert but must be developed and maintained very carefully.

Expert systems can explain their reasoning or suggested decisions, display intelligent behavior, manipulate symbolic information and draw conclusions from complex relationships, provide portable knowledge, and deal with uncertainty. They are not yet widely used; some are difficult to use, are limited to relatively narrow problems, cannot readily deal with mixed knowledge, present the possibility for error, cannot refine their own knowledge base, are difficult to maintain, and may have high development costs. Their use also raises legal and ethical concerns.

An expert system consists of a collection of integrated and related components, including a knowledge base, an inference engine, an explanation facility, a knowledge acquisition facility, and a user interface. The knowledge base contains all the relevant data, rules, and relationships used in the expert system. Fuzzy logic allows expert systems to incorporate facts and relationships into expert system knowledge bases that might be imprecise or unknown. The inference engine processes the rules, data, and relationships stored in the knowledge base to provide answers, predictions, and suggestions the way a human expert would. Two common methods for processing include backward and forward chaining. Mixed chaining is a combination of backward and forward chaining. The explanation facility of an expert system allows the user to understand what rules were used in arriving at a decision. The knowledge acquisition facility helps the user add or update knowledge in the knowledge base. The user interface makes it easier to develop and use the expert system.

The individuals involved in the development of an expert system include the domain expert, the knowledge engineer, and the knowledge users. The domain expert is the individual or group who has the expertise or knowledge being captured for the system. The knowledge engineer is the developer whose job is to extract the expertise from the domain expert. The knowledge user is the individual who benefits from the use of the developed system.

The following is a list of factors that normally make expert systems worth the expenditure of time and money: a high potential payoff or significantly reduced downside risk, the ability to capture and preserve irreplaceable human expertise, the ability to develop a system more consistent than human experts, situations in which expertise is needed at a number of locations at the same time, and situations in which expertise is needed in a hostile environment that is dangerous to human health.

The steps involved in the development of an expert system include determining requirements, identifying experts, constructing expert system components, implementing results, and maintaining and reviewing the system.

Expert systems can be implemented in several ways. LISP and PROLOG are two languages specifically developed for creating expert systems from scratch. A faster and less expensive way to acquire an expert system is to purchase an expert system shell or existing package. The approach selected depends on the benefits compared with cost, control, and complexity considerations.

The benefits of using an expert system are that they can display "intelligent" behavior, manipulate symbolic information and draw conclusions, provide portable knowledge, and can deal with uncertainty. Expert systems can be used to solve problems in many fields or disciplines and can assist in all stages of the problem-solving process.

Principle

Virtual reality systems have the potential to reshape the interface between people and information technology by offering new ways to communicate information, visualize processes, and express ideas creatively.

A virtual reality system enables one or more users to move and react in a computer-simulated environment. Virtual reality simulations require special interface devices that transmit the sights, sounds, and sensations of the simulated world to the user. These devices can also record and send the speech and movements of the participants to the simulation program. This natural style of interaction gives the participants the feeling that they are immersed in the simulated world.

Virtual reality can also refer to applications that are not fully immersive, such as mouse-controlled navigation through a 3-D environment on a graphics monitor, stereo viewing from the monitor via stereo glasses, stereo projection systems, and others. Some virtual reality applications allow views of real environments with superimposed virtual objects.

Virtual reality applications are found in medicine, education and training, real estate and tourism, and entertainment.

Principle

Specialized systems can help organizations and individuals achieve their goals.

A number of specialized systems have recently appeared to assist organizations and individuals in new and exciting ways. Segway is an electric scooter that uses sophisticated software, sensors, and gyro motors to transport people through warehouses, offices, downtown sidewalks, and other spaces. Radio-frequency identification (RFID) tags are used in a variety of settings. Game theory involves the use of information systems to develop competitive strategies for people, organizations, and even countries. Informatics combines traditional disciplines, such as science and medicine, with computer science. Bioinformatics and medical informatics are examples. A number of special-purpose telecommunications systems can be placed in products for varied uses.

CHAPTER 7: SELF-ASSESSMENT TEST

Artificial intelligence systems form a broad and diverse set of systems that can replicate human decision making for certain types of well-defined problems.

1. The field of artificial intelligence (AI) was developed in the 1980s after the development of the personal computer. True or False?

2. _____ are rules of thumb arising from experience or even guesses.

3. What is an important attribute of artificial intelligence?

   a. the ability to use sensors
   b. the ability to learn from experience
   c. the ability to be creative
   d. the ability to acquire a large amount of external information

4. _____ involves mechanical or computer devices that can paint cars, make precision welds, and perform other tasks that require a high degree of precision or are tedious or hazardous for humans.

5. What branch of artificial intelligence involves a computer system that can simulate the functioning of a human brain?

   a. expert systems
   b. neural networks
   c. natural language processing
   d. vision systems

6. A(An) _____ is a combination of software and hardware that allows the computer to change how it functions or reacts to situations based on feedback it receives.

Expert systems can enable a novice to perform at the level of an expert but must be developed and maintained very carefully.

7. What is a disadvantage of an expert system?

   a. the inability to solve complex problems
   b. the inability to deal with uncertainty
   c. limitations to relatively narrow problems
   d. the inability to draw conclusions from complex relationships

8. A(An) _____ is a collection of software packages and tools used to develop expert systems that can be implemented on most popular PC platforms to reduce development time and costs.

9. An expert system heuristic consists of a collection of software and tools used to develop an expert system to reduce development time and costs. True or False?

10. What stores all relevant information, data, rules, cases, and relationships used by the expert system?

    a. the knowledge base
    b. the data interface
    c. the database
    d. the acquisition facility

11. A disadvantage of an expert system is the inability to provide expertise needed at a number of locations at the same time or in a hostile environment that is dangerous to human health. True or False?

12. What is NOT used in the development and use of expert systems?

    a. fuzzy logic
    b. the use of rules
    c. the use of cases
    d. the use of natural language processing

13. An important part of an expert system is the _____, which allows a user or decision maker to understand how the expert system arrived at certain conclusions or results.

14. In an expert system, the domain expert is the individual or group who has the expertise or knowledge one is trying to capture in the expert system. True or False?

Virtual reality systems have the potential to reshape the interface between people and information technology by offering new ways to communicate information, visualize processes, and express ideas creatively.

15. A(An) _____ enables one or more users to move and react in a computer-simulated environment.

16. What type of virtual reality is used to make humans feel as though they are in a three-dimensional setting, such as a building, an archaeological excavation site, the human anatomy, a sculpture, or a crime scene reconstruction?

    a. chaining
    b. relative
    c. immersive
    d. visual

Specialized systems can help organizations and individuals achieve their goals.

17. _____ combines traditional disciplines, such as science and medicine, with computer science.

CHAPTER 7: SELF-ASSESSMENT TEST ANSWERS

(1) False (2) Heuristics (3) d (4) Robotics (5) b (6) learning system (7) c (8) expert system shell (9) False (10) a (11) False (12) d (13) explanation facility (14) True (15) virtual reality system (16) c (17) Informatics

REVIEW QUESTIONS

1. Define the term *artificial intelligence*.
2. What is a vision system? Discuss two applications of such a system.
3. What is natural language processing? What are the three levels of voice recognition?
4. Describe three examples of the use of robotics. How can a microrobot be used?
5. What is a learning system? Give a practical example of such a system.
6. What is a neural network? Describe two applications of neural networks.
7. What is an expert system shell?
8. Under what conditions is the development of an expert system likely to be worth the effort?
9. Identify the basic components of an expert system and describe the role of each.

10. What is fuzzy logic?
11. How are rules used in expert systems?
12. Expert systems can be built based on rules or cases. What is the difference between the two?
13. Describe the roles of the domain expert, the knowledge engineer, and the knowledge user in expert systems.
14. What are the primary benefits derived from the use of expert systems?
15. Identify three approaches for developing an expert system.
16. Describe three applications of expert systems or artificial intelligence.
17. Identify three special interface devices developed for use with virtual reality systems.
18. Identify and briefly describe three specific virtual reality applications.
19. Give three examples of other specialized systems.

DISCUSSION QUESTIONS

1. What are the requirements for a computer to exhibit human-level intelligence? How long will it be before we have the technology to design such computers? Do you think we should push to try to accelerate such a development? Why or why not?

2. What are some of the tasks at which robots excel? Which human tasks are difficult for them to master? What fields of AI are required to develop a truly perceptive robot?

3. Describe how natural language processing could be used in a university setting.

4. You have been hired to capture the knowledge of a brilliant attorney who has an outstanding track record for selecting jury members favorable to her clients during the pretrial jury selection process. This knowledge will be used as the basis for an expert system to enable other attorneys to have similar success. Is this system a good candidate for an expert system? Why or why not?

5. You have been hired to develop an expert system for a university career placement center. Develop five rules a student could use in selecting a career.

6. What is the purpose of a knowledge base? How is one developed?

7. What is the relationship between a database and a knowledge base?

8. Imagine that you are developing the rules for an expert system to select the strongest candidates for a medical school. What rules or heuristics would you include?

9. What skills does it take to be a good knowledge engineer? Would knowledge of the domain help or hinder the knowledge engineer in capturing knowledge from the domain expert?

10. Which interface is the least developed and most challenging to create in a virtual reality system? Why do you think this is so?

11. What application of virtual reality has the most potential to generate increased profits in the future?

PROBLEM-SOLVING EXERCISES

1. You are a senior vice president of a company that manufactures kitchen appliances. You are considering using robots to replace up to ten of your skilled workers on the factory floor. Using a spreadsheet, analyze the costs of acquiring several robots to paint and assemble some of your products versus the cost savings in labor. How many years would it take to pay for the robots from the savings in fewer employees? Assume that the skilled workers make $20 per hour, including benefits.

2. Assume you have just won a lottery worth $100,000. You have decided to invest half the amount in the stock market.

Develop a simple expert system to pick ten stocks to consider. Using your word processing program, create seven or more rules that could be used in such an expert system. Create five cases and use the rules you developed to determine the best stocks to pick.

3. Using a graphics program, diagram the components of a virtual reality system that could be used to market real estate. Carefully draw and label each component. Use the same graphics program to make a one-page outline of a presentation to a real estate company interested in your virtual reality system.

TEAM ACTIVITIES

1. With two or three of your classmates, do research to identify three real examples of natural language processing in use. Discuss the problems solved by each of these systems. Which has the greatest potential for cost savings? What are the other advantages of each system?

2. Form a team to debate other teams from your class on the following topic: "Are expert systems superior to humans

when it comes to making objective decisions?" Develop several points supporting either side of the debate.

3. With members of your team, think of an idea for a virtual reality system for a new, exciting game. What are the main features of the game that make it unique and highly marketable?

WEB EXERCISES

1. Use the Internet to get information about the use of neural networks. Describe three examples of how this technology is used.

2. This chapter discussed several examples of expert systems. Search the Internet for two examples of the use of expert systems. Which one has the greatest potential to increase profits for the firm? Explain your choice.

3. Use the Internet to get more information about one of the specialized systems discussed at the end of chapter. Write a report about what you found. Give an example of a new special-purpose system that has great promise in the future.

CAREER EXERCISES

1. Using the Internet or a library, explore how expert systems can be used in a business career. How can expert systems be used in a nonprofit company?

2. Select and describe a special-purpose system, other than an expert system, that would be the most beneficial in your career of choice. Use the Internet to find how this special-purpose system is actually being used.

VIDEO QUESTIONS

Watch the video clip **Predicting the Future of AI—Marvin Minsky** and answer these questions:

1. What reasons does Dr. Minsky provide for the lengthy amount of time that it is taking to build an intelligent computer?

2. What does Dr. Minsky feel is necessary for a computer to be able to do to achieve human intelligence?

CASE STUDIES

Case One

French Burgundy Wines: The Sweet Smell of Success

Virtual reality (VR) systems strive to simulate real-world experiences by reproducing the details of an environment for our five senses. Of these senses, vision and touch tend to get the most attention. Data helmets and data gloves allow VR users to visually enter virtual environments and manipulate objects within those environments. Some virtual world designers have gone as far as including 3-D sound in their worlds. Few, however, have endeavored to re-create smells. The French trade group, the Bureau Interprofessionnel des Vins de Bourgogne (BIVB), has been working with communications giant France Télécom to produce just such a technology.

The Bureau Interprofessionnel des Vins de Bourgogne, literally translated as the Interprofessional Office of the Wines of Burgundy, wants people to be able to experience the aromas of Burgundy (a region of France), as well as the bouquet of their wines by visiting its Web page. The fruits of its IS labor have produced a computer peripheral called the Olfacom. The Olfacom consists of two foot-tall plastic columns that can be plugged into a PC's USB port to pump out aromas such as hay, flowers, and fruit, using a combination of essential oils stored in its several tanks.

The BIVB developed a Web site that could take advantage of the scent capability in a virtual tour. For example, after clicking on the "winemaking" icon, animation begins on screen, showing workers in a winery unloading pinot noir grapes. As a voice states "the first and vital task is to sort the grapes," the smell of black currants wafts from the Olfacom. An icon at the bottom of the screen notifies users of what they smell.

BIVB has taken its product on the road to demonstrate at a variety of trade shows, including the Comdex and Apple expos. Steven Meyer-Rassow, who works for a British wine retailer and tried the Olfacom at a trade show, likened it to a remote-control air freshener.

Besides providing a more sensual environment for selling wine, the Olfacom has been used in marketing perfume. It was paired with a marketing campaign for Mania, a fragrance from Giorgio Armani. The device was said to have perfectly reproduced the fragrance from the famous Italian designer. Some artistic endeavors have also incorporated the device. Artist Alex Sandover is touring galleries with a show entitled Synaesthia. The show re-creates life in the 1950s using video, photos, and aromas. One life-size scene shows a woman at work in a kitchen. Throughout the scene the audience can smell kitchen aromas from the 1950s, such as natural gas, a freshly lit match, fried chicken, apples, and kitchen cleaning products.

BIVB feels that the Olfacom is destined for success. With the interest of French vineyards and the tourism industry, and with the financial backing of France Télécom, the product has real potential in the French market. The device is being redesigned smaller so that it can be attached to the side of a computer. BIVB anticipates that within a few years, all French tourism sites will have Olfacom capability. "Many [wine] professionals are also showing interest," claims Nelly Blau-Picard, BIVB export marketing manager. "They want to be a part of something that they have never seen available on the market." Just as visual virtual reality has provided a boost to e-commerce by allowing customers to manipulate 3-D products on the display, olfactory virtual reality may soon provide a similar boost to products that depend on their aroma.

Discussion Questions

1. What products do you think could benefit from the use of the Olfacom?

2. How might a device such as the Olfacom be applied to the entertainment industry?

Critical Thinking Questions

3. What are the hazards of dealing with virtual sensory information? Could you imagine an unpleasant Olfacom experience?

4. What types of devices (goggles, headphones, etc.) might be necessary for us to enter a fully simulated world as our senses experience it? What might be the most efficient method for entering a virtual world, requiring the least amount of devices?

SOURCES: Jacob Gaffney, "Smell While You Surf: Burgundy Web Site Hopes to Offer Virtual Tour with Wine Aromas," *Wine Spectator*, May 28, 2004, *www.winespectator.com*; the Olfacom Web site, visited June 2, 2004, *www.olfacom.com*; Bureau Interprofessionnel des Vins de Bourgogne Web site, visited June 2, 2004, *www.bivb.com*; Synaesthia, by Alex Sandover, Web site, accessed June 2, 2004, *www.alexsandover.com/peterborough.html*.

Case Two
Expert System Provides Safety in Nuclear Power Plants

TEPCO, the Tokyo Electric Power Corporation, is Japan's largest electric utility, serving more than 27 million customers in the Tokyo metropolitan region. Even with its $40 billion average annual revenue, TEPCO is being challenged to increase efficiency throughout its business. After streamlining many of its business processes and procedures, TEPCO began evaluating methods to reduce costs and improve efficiency in the reliability and safety of its power plants.

TEPCO uses a technique known as probabilistic safety analysis (PSA), which enables engineers to assess the probability and consequences of potential plant safety-related problems. PSA is a popular assessment technique in the nuclear power, chemical, and aerospace industries. PSA creates and interprets fault-tree diagrams, a graphical model illustrating the pathways within a system with logical constructs such as AND and OR. Following the pathways through the system can identify weaknesses and points of failure. A TEPCO safety engineer might take days or even weeks to work through a fault-tree analysis. This tedious process struck TEPCO management as one that was ideal for automation. TEPCO turned to Gensym Corporation to assist it in developing an expert system to assist with PSA.

Gensym Corporation, out of Burlington, Massachusetts, provides software products and services that enable organizations to automate aspects of their operations that have historically required the direct attention of human experts. Gensym worked together with TEPCO to deploy a system they named FT-Free within the TEPCO nuclear power plant facility. The system automatically creates PSA fault-tree models that represent potential failure modes for critical plant processes. From a model, engineers apply FT-Free's built-in expertise to quickly assess the risks of various types of problems that can adversely affect the plant's reliability and safety. Through this assessment, engineers drive critical decisions about a plant's process configuration and its operational safety procedures.

The system can do in minutes what used to take days or weeks. Not only is it a huge time-saver, but it also provides consistency that was previously lacking. A consistent result is reached independently of which engineer is providing the analysis. This enables TEPCO to best utilize its safety engineers, each of whom typically requires one year of specialized training to be qualified to complete a PSA.

The new system provides "unique abilities to intuitively capture process engineering knowledge using such techniques as connectable object models and rule-based logic," said Koichi Miyata, a project manager with TEPCO Systems. "The result is that we created a solution that greatly reduces the time and effort it takes to make important plant decisions and that is broadly applicable across industries in which safety analysis is critical."

TEPCO Systems has successfully deployed FT-Free within a TEPCO nuclear power plant facility, is expanding its use within TEPCO, and is now marketing it to the electric utility industry.

Discussion Questions

1. What features are essential in an expert system responsible for the safety of something as potentially dangerous as a nuclear power plant?

2. What characteristics of the process of probabilistic safety analysis do you think led TEPCO to believe that it could be automated in an expert system?

Critical Thinking Questions

3. What other industries would benefit from a product such as FT-Free? Why?

4. Why do you think TEPCO decided to package and market its FT-Free system to other power companies? Is this a wise move in terms of competitive advantage?

SOURCES: "TEPCO Systems Expands Use of Gensym's G2 for Assessing Reliability and Safety of Electric Power Plants," *Business Wire*, February 19, 2004, *www.lexis-nexis.com*; the Gensym Web site, accessed June 2, 2004, *www.gensym.com*; TEPCO, Tokyo Electric Power Co., Web site, accessed June 2, 2004, *www.tepco.co.jp/index-e.html*.

NOTES

Sources for the opening vignette: "Amazon.com Calls on SAS for Fraud Detection," *Citigate ICT PR*, May 20, 2003, *www.lexis-nexis.com*; "Growth of Internet Fraud Is Driving New Technologies to Safeguard Online Payments," *PR Newswire*, December 10, 2003; Amazon.com Web site, accessed May 31, 2004, *www.amazon.com*.

1. Sourbut, Elizabeth, "An Agreeable Android," *New Scientist*, February 28, 2004, p. 49.
2. Excell, John, "Interview—Robo Doc," *The Engineer*, March 5, 2004, p. 30.
3. Bentley, Ross, "Man Versus Machine," *Computer Weekly*, March 20, 2003, p. 24.
4. Gray, Madison, "No Winner in Chess Match Pitting Man against Machine," *Rocky Mountain News*, February 8, 2003, p. 31A.
5. Ulanoff, Lance, "Cognitive Machines," *PC Magazine*, July 2003, p. 118.
6. Begley, Sharon, "This Robot Can Design, Perform, and Interpret a Genetic," *The Wall Street Journal (Eastern Edition)*, January 16, 2004, p. A7.
7. Pope, Martin, "Robo Recon," *Rocky Mountain News*, February 3, 2003, p. 11B.
8. Staff, "Firm Cheers Loss of Robot in Iraq," *CNN Online*, April 13, 2004.
9. Lunsford, Lynn et al., "Shuttle Crash Raises Questions about Future Manned Flights," *The Wall Street Journal*, February 3, 2003, p. A1.
10. Staff, "Get You a Beer," *PC World*, March 2003, p. 83.
11. Rosencrance, Linda, "AI Loves Lucy," *Computerworld*, November 10, 2003, p. 36.
12. Walton, Marsha, "Meet Paro, the Therapeutic Robot Seal," *CNN Online*, November 20, 2003.
13. Staff, "Rat's Whiskers Help to Make Better Robots," *Factory Equipment News*, May 2003.
14. Scanlon, Bill, "Robotic da Vinci Sculpts First Success On Gall Bladder," *Rocky Mountain News*, August 12, 2003, p. 8A.
15. Wysocki, Bernard, "Robots in the OR," *The Wall Street Journal*, February 26, 2004, p. B1.
16. Staff, "Robot Football Players Will Give Boot to Beckham," *Electronics Weekly*, May 14, 2003, p. 1.
17. Staff, "Machine Vision and Infrared Lighting Track Production," *Vision Systems Design*, April 2003, p. 29.
18. Lunt, Penny, "Online Retailer Restores Web Sales," *Transform Magazine*, January 1, 2004, p. 32.
19. Ball, Richard, "Fujitsu Walking Robots Gets in Brains from Neural Networks," *Electronics Weekly*, April 9, 2003, p. 19.
20. Shihav, AI et al., "Distributed Intelligence for Multiple-Camera Visual Surveillance," *Pattern Recognition*, April 2004, p. 675.
21. Fielding, Rachel, "NHS Fraud Team Calls in SAS," *Accountancy Age*, February 27, 2003, p.10.
22. Johnson, Colin, "Neural Software Could Become Soldier's Best Friend," *Electronic Engineering Times*, February 2, 2004, p. 51.
23. Ke-zhang et al., "Recognition of Digital Curves Scanned from Paper Drawings Using Genetic Algorithms," *Pattern Recognition*, January 2003, p. 123.
24. Hyung, Rim et al., "An Agent for Selecting Optimal Order Set in EC Marketplace," *Decision Support Systems*, March 2004, p. 371.
25. Sexton, Randall et al., "Improving Decision Effectiveness of Artificial Neural Networks: A Modified Genetic Algorithm Approach," *Decision Sciences*, Summer 2003, p. 421.
26. Tischelle, George, "Searching for That One," *Information Week*, February 17, 2003, p. 58.
27. Overby, Stephanie, "The New, New Intelligence," *CIO Magazine*, January 1, 2003, p. 35.
28. Staff, "The Imagineer," *Library Journal*, March 15, 2003, p. 18.
29. Angel, Ana Maria et al., "Soquimich Uses a System Based on Mixed-Integer Linear Programming and Expert Systems to Improve Customer Service," *Interfaces*, July-August 2003, p. 41.
30. Lenard, Mary Jane, "Knowledge Acquisition and Memory Effects Involving an Expert System Designed as a Learning Tool for Internal Control Assessment," *Decision Sciences*, Spring 2003, p. 23.
31. Huang, Z. et al., "Development of an Intelligent Decision Support System for Air Pollution Control," *Expert Systems with Applications*, April 2004, p. 335.
32. Shaalan, K. et al., "An Expert System for the Best Weight Distribution on Ferryboarts," *Expert Systems with Applications*, April 2004, p. 397.
33. Raymond, Charles, "Horizon Lines LLC," *Journal of Commerce*, January 12, p. 1.
34. Liao, T.W., "Fuzzy Reasoning Based Automatic Inspection of Radiographic Welds," *Journal of Intelligent Manufacturing*, February 2004, p. 69.
35. Saran, Cliff, "Digital Factories Use Virtual Reality to Track Car Production," *Computer Weekly*, March 2, 2004, p. 18.
36. Costa, Dan, "Virtual Worlds," *PC Magazine*, October 28, 2003, p. 158.
37. Armstrong, David, "The Segway," *The Wall Street Journal*, February 12, 2004, p. B1.
38. Staff, "Will Segways Become Battlefield Bots," *CNN Online*, December 2, 2003.
39. Moukheiber, Zina, "Mind Over Matter," *Forbes*, March 15, 2004, p. 186.
40. Pollack, Andrew, "With Tiny Brain Implants, Just Thinking May Make It So," *The New York Times*, April 13, 2004, *www.nytimes.com*.
41. Brewin, Bob, "Delta Says Radio Frequency ID Devices Pass First Bag-Tag Test," *Computerworld*, December 22, 2003, p. 7.
42. Sliwa, Carol, "German Retailer's RFID Effort Rivals Wal-Marts," *Computerworld*, January 13, 2004, p. 10.
43. Alan Cohen, "Fast Food," *PC Magazine*, May 4, 2004, p. 76.
44. Machalaba, Daniel et al., "Thinking Inside the Box," *The Wall Street Journal*, January 15, 2004, p. B1.
45. Begley, Sharon, "A Beautiful Science: Getting the Math Right Can Thwart Terrorism," *The Wall Street Journal*, May 10, 2003, p. B1.
46. Boyle, Matthew, "Smart Dust Kicks Up a Storm," *Fortune*, February 23, 2004, p. 76.
47. Manes, Stephen, "New Twist for the Wrist," *Forbes*, February 2, 2004, p. 94.
48. Weiss, Todd, "Printer Majic," *Computerworld*, January 26, 2004, p. 31.

Kulula.com: The Trials and Tribulations of a South African Online Airline

Anesh Maniraj Singh
University of Durban

Kulula.com was launched in August 2001 as the first online airline in South Africa. Kulula is one of two airlines that are operated by Comair Ltd. British Airways (BA), the other airline that Comair runs, is a full-service franchise operation that serves the South African domestic market. Kulula, unlike BA, is a limited-service operation aimed at providing low fares to a wider domestic market using five aircraft. Since its inception, Kulula has reinvented air travel in South Africa, making it possible for more people to fly than ever before.

Kulula is a true South African e-commerce success. The company boasts as one of its successes the fact that it has been profitable from day one. It is recognised internationally among the top low-cost airlines and participated in a conference attended by other such internationally known low-cost carriers as Virgin Blue, Ryanair, and easyJet. Kulula also received an award from the South African Department of Trade and Industry for being a Technology Top 100 company.

Kulula's success is based on its clearly defined strategy of being the lowest-cost provider in the South African domestic air travel industry. To this end, Kulula has adopted a no-frills approach. Staff and cabin crew wear simple uniforms, and the company has no airport lounges. There are no business class seats and no frequent-flyer programs. Customers pay for their food and drinks. In addition, Kulula does not issue paper tickets, and very few travel agents book its flights—90 percent of tickets are sold directly to customers. Furthermore, customers have to pay for ticket changes, and the company has a policy of "no fly, no refund." Yet, in its drive to keep costs down, Kulula does not compromise on maintenance and safety, and it employs the best pilots and meets the highest safety standards. Like all B2C companies, Kulula aims to create customer value by reducing overhead costs, including salaries, commissions, rent, and consumables such as paper and paper-based documents. Furthermore, by cutting out the middleman such as travel agents, Kulula is able to keep prices low and save customers the time and inconvenience of having to pick up tickets from travel agents. Instead, customers control the entire shopping experience.

Kulula was the sole provider of low-cost flights in South Africa until early 2004, when One Time launched a no-frills service to compete head-on with Kulula. Due to the high price elasticity of demand within the industry, any lowering of price stimulates a higher demand for flights. The increase in competition in the low-price end of the market has seen Kulula decrease fares by up to 20 percent whilst increasing passengers by over 40 percent. There, however, has been no brand switching. Kulula has grown in the market at the expense of others.

Apart from its low-cost strategy, Kulula is successful because of its strong B2C business model. As previously mentioned, 90 percent of its revenue is generated from direct sales. However, Kulula has recently ventured into the B2B market by collaborating with Computicket and a few travel agents, who can log in to the Kulula site from their company intranets. Kulula offers fares at substantial reductions to businesses that use it regularly. Furthermore, Kulula bases its success on three simple principles: Any decision taken must bring in additional revenue, save on costs, and/or enhance customer service. Technology contributes substantially to these three principles.

In its first year, Kulula used a locally developed reservation system, which soon ran out of functionality. The second-generation system was AirKiosk, which was developed in

Boston for Kulula. The system change resulted in an improvement of functionality for passengers. For example, in 2003, Kulula ran a promotion during which tickets were sold at ridiculously low prices, and the system was overwhelmed. Furthermore, Kulula experienced a system crash that lasted a day and a half, which severely hampered sales and customer service. As a result, year two saw a revamp in all technology: All the hardware was replaced, bandwidth was increased, new servers and database servers were installed, and Web hosting was changed. In short, the entire system was replaced. According to IT Director Carl Scholtz, "Our success depends on infrastructural stability; our current system has an output that is four times better than the best our systems could ever produce." Kulula staff members are conscious of the security needs of customers and have invested in 128-bit encryption, giving customers peace of mind that their transactions and information are safe.

The success behind Kulula's systems lies in its branding—its strong identity in the marketplace, which includes its name and visual appeal. The term *kulula* means "easy," and Kulula's Web site has been designed with a simple, no-fuss, user-friendly interface. When one visits the Kulula site, one is immediately aware that an airline ticket can be purchased in three easy steps. The first step allows customers to choose destinations and dates. The second step allows customers to choose the most convenient or cheapest flight based on their need. Kulula also allows customers to book cars and accommodations in step two. Step three is the transaction stage, which allows customers to choose the most suitable payment method. The confirmation and ticket can be printed once payment has been settled. Kulula has not embraced mobile commerce yet, because the technology does not support the ability to allow customers to purchase a ticket in three easy steps. Unlike other e-commerce sites, Kulula is uncluttered and simple to understand, enhancing customer service. Kulula is a fun brand—with offbeat advertising campaigns and bright green and blue corporate and aircraft colours—but behind the fun exterior is a group of people who are serious about business.

Kulula's future is extremely promising. Technology changes continually, and Kulula strives to have the best technology in place at all times. B2B e-commerce will continue to be a major focus of the company in order to develop additional distribution channels with little or no cost. In conjunction with bank partners, Kulula is developing additional methods of payment to replace credit card payments, allowing more people the opportunity to fly. These transactions will be free. Kulula is also involving customers in its marketing efforts by obtaining their permission to promote special offers by e-mail and short message service to customers' cell phones. The Kulula Web site will soon serve as a ticketing portal, where customers can also purchase British Airways tickets, in three easy steps. The company has many other developments in the pipeline that will enhance customer service. According to Scholtz, "We are not an online airline, just an e-tailer that sells airline tickets."

Discussion Questions

1. This case does not mention any backup systems, either electronic or paper based. What would you recommend to ensure that the business runs 24/7/365?
2. It is clear from this case that Kulula is a low-cost provider. What else could Kulula do with its technology to bring in additional revenue, save on cost, and enhance customer service?
3. Does the approach taken by Kulula in terms of its strategy, its business model, and the three principles of success lend itself to other businesses wanting to engage in e-commerce?

4. Kulula flights are almost always full. Do you think that by partnering with a company such as Lastminute.com the airline could fly to capacity at all times? What are the risks related to such a collaboration?

Critical Thinking Questions

5. Kulula initially developed its systems in-house, which it later outsourced to AirKiosk in Boston. Do you think it is wise for an e-business to outsource its systems development? Is it strategically sound to outsource systems development to a company in a different country?
6. With the current trends in mobile commerce, could Kulula offer its services on mobile devices such as cellular phones? Would the company have to alter its strategic thinking to accommodate such a shift? Is it possible to develop a text-based interface that could facilitate a purchase in three easy steps?

Systems Development and Social Issues

STORAGE DATA

BYTE 0

1   2   3   4   5   6

Chapter 8   Systems Development
Chapter 9   Security, Privacy, and Ethical Issues in Information Systems
and the Internet

CHAPTER
· 8 ·

Systems Development

PRINCIPLES	LEARNING OBJECTIVES
■ Effective systems development requires a team effort of stakeholders, users, managers, systems development specialists, and various support personnel, and it starts with careful planning.	■ Identify the key participants in the systems development process and discuss their roles. ■ Define the term *information systems planning* and discuss the importance of planning a project.
■ Systems development often uses different approaches and tools such as traditional development, prototyping, rapid application development, end-user development, computer-aided software engineering, and object-oriented development to select, implement, and monitor projects.	■ Discuss the key features, advantages, and disadvantages of the traditional, prototyping, rapid application development, and end-user systems development life cycles. ■ Discuss the use of computer-aided software engineering (CASE) tools and the object-oriented approach to systems development.
■ Systems development starts with investigation and analysis of existing systems.	■ State the purpose of systems investigation. ■ State the purpose of systems analysis and discuss some of the tools and techniques used in this phase of systems development.
■ Designing new systems or modifying existing ones should always be aimed at helping an organization achieve its goals.	■ State the purpose of systems design and discuss the differences between logical and physical systems design.
■ The primary emphasis of systems implementation is to make sure that the right information is delivered to the right person in the right format at the right time.	■ State the purpose of systems implementation and discuss the various activities associated with this phase of systems development.
■ Maintenance and review add to the useful life of a system but can consume large amounts of resources.	■ State the importance of systems and software maintenance and discuss the activities involved. ■ Describe the systems review process.

INFORMATION SYSTEMS IN THE GLOBAL ECONOMY ▶

BMW, GERMANY

Designing a New System to Ramp Up Production

The BMW Group is reputed to be the only manufacturer of automobiles and motorcycles worldwide that concentrates single-mindedly on premium manufacturing standards and outstanding quality for all its brands. Its brands include BMW, MINI, Rolls-Royce Motor Cars, and BMW Motorcycles. The BMW manufacturing plant in South Carolina is part of BMW Group's global manufacturing network and is the exclusive manufacturing plant for the company's hot-selling Z3 Roadster and X5 Sports Activity Vehicles.

The Z3 and X5 are so popular, in fact, that BMW was selling them faster than it could build them. "Customer demand for our new X5 Sport Activity Vehicle and our popular Z3 models required a solution that would successfully speed up and support streamlined production," says Hans Nowak, program manager for BMW Manufacturing Company in Spartanburg. Manufacturing each of these vehicles requires hundreds of parts from hundreds of suppliers, so coordinating purchasing and acquisition of these parts can be more complicated than constructing the vehicles. Executives at the plant recognized that increasing the efficiency of parts procurement could be key to ramping up production to meet demand.

To lay the groundwork for developing a lean and mean procurement system, BMW organized a team to investigate the problem and gather information. At the time, BMW was using a custom-built information system to manage its supplier and logistics requirements. The old system monitored parts inventory and placed orders with suppliers well before they were needed to avoid halting production. As a result, the plant had large amounts of excess parts on hand and unreliable delivery schedules. The old system was wasteful—and worrisome. The team sat down with the results of their investigation to establish objectives for a new and improved system.

An ideal system, they decided, would need to include a tight supplier network that kept parts coming to the assembly lines in just-in-time (JIT) mode to produce cars to meet customer demand. The challenge was in dealing with hundreds of different suppliers, each of which required varying amounts of time to produce and deliver the parts. It was soon clear to the team that the demands of their dream system far exceeded the capabilities of their current custom-built system. They decided to look to external software vendors for a solution.

After contacting a number of software vendors, issuing requests for proposals (RFPs), and analyzing the results, the team found one company with what appeared to be an ideal solution. Software giant SAP offered a custom-designed automotive procurement solution called mySAP Automotive. mySAP Automotive overcame the challenge of creating schedules for each of BMW's suppliers by automatically generating custom schedules for each part; the system generated schedules to match BMW's assembly line planning and sequencing directives. Once BMW decided on the SAP solution, it was able to complete its implementation within one month.

The new system provides two types of reports to suppliers: long-horizon forecasts and short-horizon JIT delivery schedules. Larger suppliers receive the information via electronic data interchange (EDI). Other suppliers access the mySAP Automotive Supplier Portal, where BMW posts the requirements to provide up-to-date information on its delivery needs. Using only an Internet browser, suppliers can view this information in real time, including release schedules, purchasing documents, invoices, and engineering documents.

As parts are shipped, BMW receives advance shipping notifications (ASNs), which provide exact information on parts counts and delivery dates. Parts arriving at the BMW dock are then received and transferred directly to the line.

The new system goes beyond procurement to monitor production status in real time. mySAP Automotive registers production confirmation and parts consumption information every three minutes. Parts consumed during assembly are removed from the inventory count, and costs are posted to calculate the value of work in process. Production supervisors are able to monitor the ebb and flow of parts inventory in real time, as well as look ahead and know when incoming shipments will be arriving. Hans Nowak sums it up this way: "mySAP Automotive helps us reduce order-to-delivery time, strengthens our supply chain activities in the areas of demand planning and tracking and tracing of material deliveries, and improves inventory accuracy across our Spartanburg plant—enabling us to significantly reduce time-to-customer for our popular X5 and Z3 models."

BMW also values SAP's standardized and scalable platform. The system can be expanded and modified to embrace new functionality or changes in business processes. BMW has already integrated other business processes with mySAP Automotive. Information housed in BMW systems can be easily accessed, including forecast requirements for components, sales orders, and vehicle bills of material. In addition to the industry-specific solution, BMW is using SAP applications for logistics, financials, and human resources.

As you read this chapter, consider the following:

- What is the value of continuously evaluating and improving the information systems that drive a company and connect it with its suppliers and customers?
- What are important considerations when deciding whether to implement a new information system?

Why Learn About Systems Development?

Throughout this book, you have seen many examples of the use of information systems in a variety of careers. A manager at a hotel chain can use an information system to look up client preferences. An accountant at a manufacturing company can use an information system to analyze the costs of a new plant. A sales representative for a music store can use an information system to determine which CDs to order and which to discount because they are not selling. Information systems have been designed and implemented for almost every career and industry. But where do you start to acquire these systems or have them developed? How can you work with IS personnel, such as systems analysts and computer programmers, to get what you need to succeed on the job? This chapter gives you the answer. You will see how you can initiate the systems development process and analyze your needs with the help of IS personnel. This chapter discusses how your project can be planned, aligned with corporate goals, rapidly developed, and much more. We start with an overview of the systems development process.

When an organization needs to accomplish a new task or change a work process, how does it do it? It develops a new system or modifies an existing one. Systems development is the activity of creating new systems or modifying existing ones. It refers to all aspects of the process—from identifying problems to be solved or opportunities to be exploited to the implementation and refinement of the chosen solution. Even governmental agencies use systems development. The Naval Sea Systems Command used systems development to streamline its procurement process.[1] The military operation has an annual budget of $19 billion. As a result of the project, the naval department was able to cut buying time by 85 percent—from roughly 270 days to about 30 days. In another systems development project, the U.S. Congress allocated $1.35 billion to the Internal Revenue Service (IRS) to upgrade its computer systems.[2] Congress and U.S. citizens hope the upgrade will make it easier for people to contact and interact with the IRS.

AN OVERVIEW OF SYSTEMS DEVELOPMENT

In today's businesses, managers and employees in all functional areas work together and use business information systems. As a result, users of all types are helping with development and, in many cases, leading the way. This chapter provides you with a deeper appreciation of the systems development process and helps you avoid costly failures.

Participants in Systems Development

Effective systems development requires a team effort. The team usually consists of stakeholders, users, managers, systems development specialists, and various support personnel. This team, called the *development team*, is responsible for determining the objectives of the information system and delivering a system that meets these objectives. Many development teams use a project manager to head the systems development effort and the project management approach to help coordinate the systems development process. A *project* is a planned collection of activities that achieves a goal, such as constructing a new manufacturing plant or developing a new decision support system. All projects have a defined starting point and ending point, normally expressed as dates such as August 4th and November 11th. Most have a budget, such as $150,000. A *project manager* is the individual responsible for coordinating all people and resources needed to complete a project on time. In systems development, the project manager can be an IS person inside the organization or an external consultant hired to complete the project. Project managers need technical, business, and people skills. "It's a delicate balancing act," says Jack Probst, vice president of IT process and governance at Nationwide Insurance.[3] In addition to completing the project on time and within the specified budget, the project manager is usually responsible for controlling a project's quality, training personnel, facilitating communications, managing risks, and acquiring any necessary equipment, including office supplies and sophisticated computer systems. Research studies have shown that project management success factors include good leadership from executives and project managers, a high level of trust in the project and its potential benefits, and the commitment of the project team and organization to successfully complete the project and implement its results.[4]

In the context of systems development, **stakeholders** are individuals who, either themselves or through the area of the organization they represent, ultimately benefit from the systems development project. One systems development methodology, called *agile modeling*, calls for very active participation of customers and other stakeholders in the systems development process. **Users** are individuals who will interact with the system regularly. They can be employees, managers, or suppliers. For large-scale systems development projects, where the investment in and value of a system can be quite high, it is common to have senior-level managers, including the company president and functional vice presidents (of finance, marketing, and so on), be part of the development team.

Depending on the nature of the systems project, the development team might include systems analysts and programmers, among others. A **systems analyst** is a professional who specializes in analyzing and designing business systems. Systems analysts play various roles while interacting with the stakeholders and users, management, vendors and suppliers, external companies, software programmers, and other IS support personnel (see Figure 8.1). Like an architect developing blueprints for a new building, a systems analyst develops detailed

stakeholders
Individuals who, either themselves or through the organization they represent, ultimately benefit from the systems development project.

users
Individuals who will interact with the system regularly.

systems analyst
A professional who specializes in analyzing and designing business systems.

programmer
The specialist responsible for modifying or developing programs to satisfy user requirements.

plans for the new or modified system. The **programmer** is responsible for modifying or developing programs to satisfy user requirements. Like a contractor constructing a new building or renovating an existing one, the programmer takes the plans from the systems analyst and builds or modifies the necessary software.

Figure 8.1

The systems analyst plays an important role in the development team and is often the only person who sees the system in its totality. The one-way arrows in this figure do not mean that there is no direct communication between other team members. Instead, these arrows just indicate the pivotal role of the systems analyst—an individual who is often called on to be a facilitator, moderator, negotiator, and interpreter for development activities.

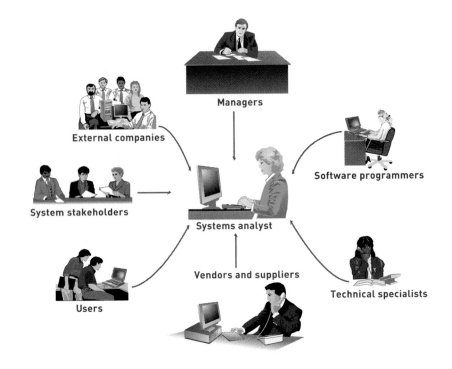

Information Systems Planning and Aligning Corporate and IS Goals

information systems planning
The translation of strategic and organizational goals into systems development initiatives.

The term **information systems planning** refers to the translation of strategic and organizational goals into systems development initiatives (see Figure 8.2). The Marriott hotel chain, for example, invites its chief information officer to board meetings and other top-level management meetings. Proper IS planning ensures that specific systems development objectives support organizational goals.

Figure 8.2

Information systems planning transforms organizational goals outlined in the strategic plan into specific systems development activities.

Aligning organizational goals and IS goals is critical for any successful systems development effort. Because information systems support other business activities, both IS staff and people in other departments need to understand each other's responsibilities and tasks, as the "Information Systems @ Work" feature explains. Determining whether organizational and IS goals are aligned can be difficult, so researchers have increasingly tackled the problem. One measure of alignment uses five levels ranging from ad hoc processes (Level 1 Alignment) to optimized processes (Level 5 Alignment).

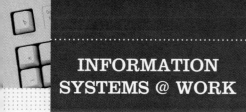

INFORMATION SYSTEMS @ WORK

Fakta Designs Financial Management System to Support Corporate Goals

The leading Danish discount retail chain, Fakta, recently became frustrated with its inability to maintain its budget. The retailer had grown beyond the capabilities of its antiquated budgeting system, and something needed to change. "We live in a world where it is crucial for us to be able to react quickly to changing circumstances because our profit margins are so small," says C. F. Thorhauge, Financial Director. "If we are to earn money, we must be really tough in controlling our costs." But the old system for establishing the corporate budget was not supporting the goals of the organization.

As with all systems development projects, Fakta's IS department began by defining the goals of the system in light of the goals of the organization. They studied the details of the current system to see why it was failing. Fakta has retail outlets in 16 regions in Denmark, each of which has a manager who is financially responsible for the shops in his or her region. The system in use required each regional manager to work out his or her own budget and send it to the company headquarters in Vejle. From there, Fakta's IS department collated the budgets from all the regional managers and passed the combined information on to the finance department. Then, the finance department at the company's headquarters manually determined the overall budget by using spreadsheets or other forms of documentation, sometimes even resorting to pencil and paper.

"It was already clear to us several years ago that the situation was becoming untenable," says Jens Brinkmann, Fakta's financial controller. "It was extremely difficult to control the vast quantity of information that came from the 16 regions into head office. Often, it took days or even weeks to collect all the different budget information, and there was a major risk of errors occurring during the process."

Fakta decided that the new system should provide a centralized server accessible to all regional managers to allow them to store information directly in the central system. The system would automate the calculations, create the budget, and provide useful reports to all managers within the organization. Because budgeting is common to all companies, Fakta guessed that such a solution was probably already on the market. It found its solution in a product from SAS Corporation.

SAS worked with Fakta's IS department to develop a customized system that met the organization's goals. The result? A central, user-controlled system that contains all budget-related functions from inventory turnover to production overhead costs. All regional managers input their budgets into the new system through an easy-to-use Web-based system delivered over the company's intranet. The budget is then finalized, and the numbers are reliable. As a result, everyone can quickly get an overview of the entire budget.

Fakta trained its regional managers and its finance staff on the new system to ensure that they understood the new budgeting process. Fakta's IS department's job has been simplified, too. They no longer have to worry about budgeting. The days of manually collecting and checking data are over. Today, budgeting is automated.

With its new financial management system in place, the national discount chain can control business operations and handle budgeting quickly. "Where it had previously taken days or weeks to determine the budget for the whole discount chain, it can now be accomplished in just a few hours. We've put an end to all the poor excuses for mistakes in the budget," concludes Thorhauge.

Discussion Questions

1. How has Fakta's new system changed the workload of its regional managers, its IS staff, and its finance department?
2. Besides streamlining the process of collecting regional data, how has the new system improved the effectiveness of the organization?

Critical Thinking Questions

3. Now that Fakta's finance department isn't strapped with manually calculating the company's budget, how can finance employees use their additional time more productively?
4. Provide some examples of how financial management software such as that in use at Fakta might influence and change the investment and purchasing style of a company.

SOURCES: "SAS Success Stories: Fakta," accessed June 18, 2004, *www.sas.com/success/fakta.html*; SAS Financial Intelligence Web site, accessed June 18, 2004, *www.sas.com/solutions/financial/index.html*; FDB (owner of Fakta) Web site, accessed June 18, 2004, *www.fdb.dk/default.asp?id=44*.

Importance of IS Planning

One of the primary benefits of IS planning and alignment of goals is a long-range view of information technology's use in the organization. Specific systems development initiatives may spring from the IS plan, but the IS plan must also provide a broad framework for future success. The IS plan should guide development of the IS infrastructure over time. Another benefit of IS planning is that it ensures better use of IS resources—including funds, personnel, and time for scheduling specific projects. The steps of IS planning are shown in Figure 8.3.

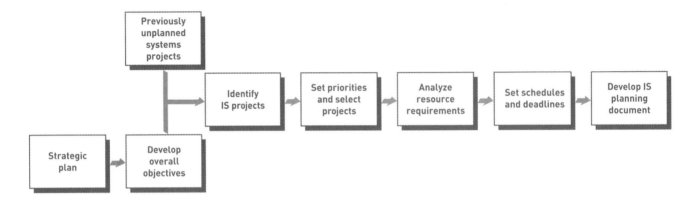

Figure 8.3

The Steps of IS Planning

Some projects are identified through overall IS objectives, whereas additional projects, called *unplanned projects*, are identified from other sources. All identified projects are then evaluated in terms of their organizational priority.

Overall IS objectives are usually distilled from the relevant aspects of the organization's strategic plan. IS projects can be identified either directly from the objectives determined in the first step or identified by others, such as managers within the various functional areas. Setting priorities and selecting projects typically requires the involvement and approval of senior management. When objectives are set, planners consider the resources necessary to complete the projects, including employees (systems analysts, programmers, and others), equipment (computers, network servers, printers, and other devices), expert advice (specialists and other consultants), and software, among others.

SYSTEMS DEVELOPMENT LIFE CYCLES

The systems development process is also called a *systems development life cycle (SDLC)* because the activities associated with it are ongoing. Several common systems development life cycles exist: traditional, prototyping, rapid application development (RAD), end-user development, and others. In addition, companies can outsource the systems development process. With some companies, these approaches are formalized and documented so that system developers have a well-defined process to follow; in other companies, less formalized approaches are used.

The Traditional Systems Development Life Cycle

Traditional systems development efforts can range from a small project, such as purchasing an inexpensive computer program, to a major undertaking. The steps of traditional systems development might vary from one company to the next, but most approaches have five common phases: investigation, analysis, design, implementation, and maintenance and review (see Figure 8.4).

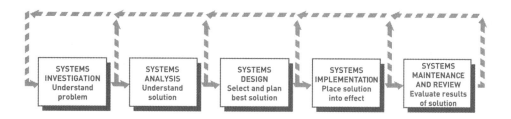

Figure 8.4

The Traditional Systems Development Life Cycle

Sometimes, information learned in a particular phase requires cycling back to a previous phase.

In the **systems investigation** phase, potential problems and opportunities are identified and considered in light of the goals of the business. Systems investigation attempts to answer the question "What is the problem, and is it worth solving?" The primary result of this phase is a defined development project for which business problems or opportunity statements have been created, to which some organizational resources have been committed, and for which systems analysis is recommended. **Systems analysis** attempts to answer the question "What must the information system do to solve the problem?" This phase involves the study of existing systems and work processes to identify strengths, weaknesses, and opportunities for improvement. The major outcome of systems analysis is a list of requirements and priorities. **Systems design** seeks to answer the question "How will the information system do what it must do to obtain the problem solution?" The primary result of this phase is a technical design that either describes the new system or describes how existing systems will be modified. The system design details system outputs, inputs, and user interfaces; specifies hardware, software, database, telecommunications, personnel, and procedure components; and shows how these components are related. **Systems implementation** involves creating or acquiring the various system components detailed in the systems design, assembling them, and placing the new or modified system into operation. An important task during this phase is to train the users. Systems implementation results in an installed, operational information system that meets the business needs for which it was developed. The purpose of **systems maintenance and review** is to ensure that the system operates and to modify the system so that it continues to meet changing business needs. As shown in Figure 8.4, a system under development moves from one phase of the traditional SDLC to the next.

Prototyping

Prototyping takes an iterative approach to the systems development process. During each iteration, requirements and alternative solutions to the problem are identified and analyzed, new solutions are designed, and a portion of the system is implemented. Users are then encouraged to try the prototype and provide feedback (see Figure 8.5). Prototyping begins with the creation of a preliminary model of a major subsystem or a scaled-down version of the entire system. For example, a prototype might be developed to show sample report formats and input screens. After being developed and refined, the prototypical reports and input screens are used as models for the actual system, which can be developed using an end-user programming language such as Visual Basic. The first preliminary model is refined to form the second- and third-generation models, and so on until the complete system is developed (see Figure 8.6).

Rapid Application Development, Agile Development, Joint Application Development, and Other Systems Development Approaches

Rapid application development (RAD) employs tools, techniques, and methodologies designed to speed application development. Vendors such as Computer Associates International, IBM, and Oracle market fourth-generation languages and other products targeting the RAD market. Rational Software, a division of IBM, has a RAD tool, called Rational Rapid Developer, to make developing large Java programs and applications easier and faster.[5]

RAD reduces paper-based documentation, automatically generates program source code, and facilitates user participation in design and development activities. It makes adapting to changing system requirements easier. Other approaches to rapid development, such as *agile*

systems investigation
The systems development phase during which problems and opportunities are identified and considered in light of the goals of the business.

systems analysis
The systems development phase involving the study of existing systems and work processes to identify strengths, weaknesses, and opportunities for improvement.

systems design
The systems development phase that defines how the information system will do what it must do to obtain the problem's solution.

systems implementation
The systems development phase involving the creation or acquisition of various system components detailed in the systems design, assembling them, and placing the new or modified system into operation.

systems maintenance and review
The systems development phase that ensures the system operates and also modifies the system so that it continues to meet changing business needs.

prototyping
An iterative approach to the systems development process.

rapid application development (RAD)
A systems development approach that employs tools, techniques, and methodologies designed to speed application development.

Prototyping Is an Iterative Approach to Systems Development

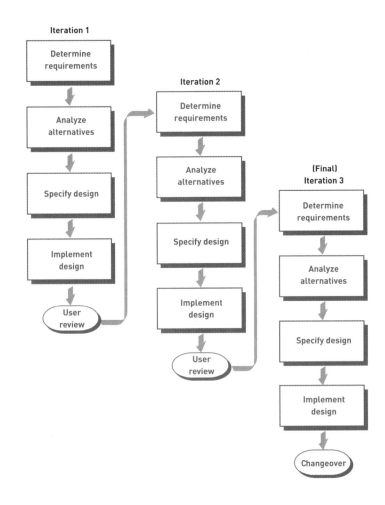

Prototyping is a popular technique in systems development. Each generation of prototype is a refinement of the previous generation based on user feedback.

development or *extreme programming (XP)*, allow the systems to change as they are being developed. Agile development requires frequent face-to-face meetings with the systems developers and users as they modify, refine, and test system needs and capabilities. Some predict that agile programming will eventually be used by most IT departments because of the volatility of the business environment—the length of the traditional development life cycle can reduce a system's usefulness after it is finally completed. The agile development process is more fluid and flexible, but it can be complex and time-consuming. XP uses pairs of programmers who work together to design, test, and code parts of the systems they develop. The iterative nature of XP helps companies develop robust systems, with fewer errors. Sabre Airline Solutions, a $2 billion airline travel company, used XP to eliminate programming errors and shorten program development times.[6] The company manages 13 million lines of programming code in 62 software products. Sabre uses the Java programming language and XP to rapidly develop its applications.

RAD makes extensive use of the **joint application development (JAD)** process for data collection and requirements analysis. Originally developed by IBM Canada in the 1970s, JAD involves group meetings in which users, stakeholders, and IS professionals work together to analyze existing systems, propose possible solutions, and define the requirements of a new or modified system. A group normally requires one or more top-level executives who initiate the JAD process, a group leader for the meetings, potential users, and one or more individuals who act as secretaries and clerks to record what is accomplished and to provide general support for the sessions. Many companies have found that groups can develop better requirements than individuals working independently. Today, JAD often uses *group support systems (GSS)* software (discussed in Chapter 6) to foster positive group interactions, while suppressing negative group behavior.

The End-User Systems Development Life Cycle

The term **end-user systems development** describes any systems development project in which the primary effort is undertaken by a combination of business managers and users. Rather than ignoring these initiatives, astute IS professionals encourage them by offering guidance and support. Technical assistance, communication of standards, and the sharing of "best practices" throughout the organization are just some of the ways IS professionals work with motivated managers and employees undertaking their own systems development. In this way, end-user-developed systems can be structured as complementary to, rather than in conflict with, existing and emerging information systems. In addition, this open communication among IS professionals, managers of the affected business area, and users allows the IS professionals to identify specific initiatives so that additional organizational resources, beyond those available to business managers or users, are provided for its development.

End-user-developed systems range from the very small (e.g., a software routine to merge form letters) to those of significant organizational value (such as customer contact databases for the Web). Adnan Osmani from Dublin, Ireland, for example, developed his own Internet browser.[7] "I just wanted it faster for myself," Osmani said. Like all projects, some end-user-developed systems fail, and others are successful. Initially, IS professionals discounted the value of these projects. As the number and magnitude of these projects increased, however, IS professionals began to realize that for the good of the entire organization, their involvement with these projects needed to increase.

Outsourcing and On Demand Computing

Many companies hire an outside consulting firm or computer company that specializes in systems development to take over some or all of their development and operations activities. As mentioned in Chapter 1, *outsourcing* and *on demand computing* are often used.[8] Sears, Roebuck, and Co., for example, looked for an outside company to perform many of its systems development and operations activities.[9] Sears is hoping that the outsourcing company it selected will hire some of its 200 people who will be laid off. Outsourcing has also become an important economic and political issue in today's economy for companies that outsource overseas.[10] A group of U.S. companies and organizations have formed the Coalition for Economic Growth and American Jobs to try to combat the outsourcing backlash. Companies can spend millions or even billions of dollars to hire other companies to manage Web sites, network servers, data storage devices, and help desk operations. In one outsourcing deal, a major phone company agreed to pay an outsourcing company about $3 billion to provide guidance on cutting costs and improving efficiency. One U.S. phone company was given an outsourcing deal worth $20 million to develop a cellular phone system in Baghdad, Iraq, by the U.S. Department of Defense.[11] Companies such as Goodyear and Saks have used on demand computing to increase services and reduce costs.[12]

A report by the U.S. Commerce Department, however, reports that the United States imports more private-sector jobs from other countries than it exports.[13] Private-sector jobs

joint application development (JAD)
A process for data collection and requirements analysis in which users, stakeholders, and IS professionals work together to analyze existing systems, propose possible solutions, and define the requirements of a new or modified system.

end-user systems development
Any systems development project in which the primary effort is undertaken by a combination of business managers and users.

Many end users today are already demonstrating their systems development capability by designing and implementing their own PC-based systems.

(Source: Photodisc/Getty Images.)

include computer programming, telecommunications, management consulting, legal work, and related jobs. The surplus of jobs coming into the United States from other countries is greater than $53 billion, according to the report. Outsourcing can also involve a large number of countries and companies in bringing new products and services to market.[14] The idea for a new computer server can originate in Singapore, be approved in Houston, be designed in India, be engineered in Taiwan, and be assembled in Australia. The chain of events can be complex.

A number of companies offer outsourcing and on demand computing services—from general systems development to specialized services. IBM's Global Services, for example, is one of the largest full-service outsourcing and consulting services.[15] IBM has consultants located in offices around the world. Electronic Data Systems (EDS) is another large company that specializes in consulting and outsourcing.[16] EDS has approximately 140,000 employees in almost 60 countries and more than 9,000 clients worldwide. In one year, the company signed $31.4 billion in new contracts for consulting and outsourcing. Accenture is another company that specializes in consulting and outsourcing.[17] The company has more than 75,000 employees in 47 countries.

Use of Computer-Aided Software Engineering (CASE) Tools

computer-aided software engineering (CASE)

Tools that automate many of the tasks required in a systems development effort and enforce adherence to the SDLC.

Computer-aided software engineering (CASE) tools automate many of the tasks required in a systems development effort and enforce adherence to the SDLC, thus instilling a high degree of rigor and standardization to the entire systems development process. VRCASE, for example, is a CASE tool that can be used by a team of developers to assist in developing applications in C++ and other languages.[18] Prover Technology has developed a CASE tool that searches for programming bugs. The CASE tool searches for all possible design scenarios to make sure that the program is error free. Other CASE tools include Visible Systems (*www.visible.com*) and Popkin Software (*www.popkin.com*). Popkin Software, for example, can generate code in fourth-generation programming languages, such as C++, Java, and Visual Basic. Other CASE-related tools include Rational Rose (part of IBM) and Visio, a charting and graphics program from Microsoft. Companies that produce CASE tools are Accenture, Microsoft, Oracle, and others. Oracle Designer and Developer CASE tools, for example, can help systems analysts automate and simplify the development process for database systems. See Table 8.1 for a list of some CASE tools and their providers. The advantages and disadvantages of CASE tools are listed in Table 8.2.

Table 8.1

Typical CASE Tools

CASE Tool	Vendor
Oracle Designer	Oracle Corporation, www.oracle.com
Visible Analyst	Visible Systems Corporation, www.visible.com
Rational Rose	Rational Software, www-306.ibm.com/software/rational
Embarcadero Describe	Embarcadero Describe, www.embarcadero.com

Object-Oriented Systems Development

In addition to the CASE tools a development team uses, the success of a systems development effort can also depend on the specific programming tools and approaches they select. Object-oriented (OO) programming languages allow the interaction of programming objects—that is, an object consists of both data and the actions that can be performed on the data. So, an object could be data about an employee and all the operations (such as payroll, benefits, and tax calculations) that might be performed on the data.

Advantages	Disadvantages
Produce systems with a longer effective operational life	Produce initial systems that are more expensive to build and maintain
Produce systems that more closely meet user needs and requirements	Require more extensive and accurate definition of user needs and requirements
Produce systems with excellent documentation	May be difficult to customize
Produce systems that need less systems support	Require more training of maintenance staff
Produce more flexible systems	May be difficult to use with existing systems

Table 8.2

Advantages and Disadvantages of CASE Tools

Developing programs and applications using OO programming languages involves constructing modules and parts that can be reused in other programming projects.[19] Chapter 2 discussed a number of programming languages that use the object-oriented approach, including Visual Basic, C++, and Java. These languages allow systems developers to take the OO approach, making program development faster and more efficient, resulting in lower costs. Modules can be developed internally or obtained from an external source. After a company has the programming modules, programmers and systems analysts can modify them and integrate them with other modules to form new programs. GE Power developed a new system to bridge the gap between new Internet-based applications and older, legacy systems. The new system allows its managers and employees to seamlessly share data and information. GE Power used object-oriented languages and approaches, such as Java classes and JavaBeans, to reuse computer code and reduce systems development time and costs.[20]

Object-oriented systems development (OOSD) combines the logic of the systems development life cycle with the power of object-oriented modeling and programming. OOSD follows a defined systems development life cycle, much like the SDLC. The life cycle phases can be, and usually are, completed with many iterations. Object-oriented systems development typically involves the following:

- *Identifying potential problems and opportunities within the organization that would be appropriate for the OO approach.* This process is similar to traditional systems investigation. Ideally, these problems or opportunities should lend themselves to the development of programs that can be built by modifying existing programming modules.
- *Defining what kind of system users require.* This analysis means defining all the objects that are part of the user's work environment (object-oriented analysis). The OO team must study the business and build a model of the objects that are part of the business (such as a customer, an order, or a payment). Many of the CASE tools discussed in the previous section can be used, starting with this step of OOSD.
- *Designing the system.* This process defines all the objects in the system and the ways they interact (object-oriented design). Design involves developing logical and physical models of the new system by adding details to the object model started in analysis.
- *Programming or modifying modules.* This implementation step takes the object model begun during analysis and completed during design and turns it into a set of interacting objects in a system. Object-oriented programming languages are designed to allow the programmer to create classes of objects in the computer system that correspond to the objects in the actual business process. Objects such as customer, order, and payment are redefined as computer system objects—a customer screen, an order-entry menu, or a dollar sign icon. Programmers then write new modules or modify existing ones to produce the desired programs.
- *Evaluation by users.* The initial implementation is evaluated by users and improved. Additional scenarios and objects are added, and the cycle repeats. Finally, a complete, tested, and approved system is available for use.

object-oriented systems development (OOSD)
The approach that combines the logic of the systems development life cycle with the power of object-oriented modeling and programming.

- *Periodic review and modification.* The completed and operational system is reviewed at regular intervals and modified as necessary.

SYSTEMS INVESTIGATION

As discussed earlier in the chapter, systems investigation is the first phase in the traditional SDLC of a new or modified business information system. The purpose is to identify potential problems and opportunities and consider them in light of the goals of the company. In general, systems investigation attempts to uncover answers to the following questions:

- What primary problems might a new or enhanced system solve?
- What opportunities might a new or enhanced system provide?
- What new hardware, software, databases, telecommunications, personnel, or procedures will improve an existing system or are required in a new system?
- What are the potential costs (variable and fixed)?
- What are the associated risks?

Initiating Systems Investigation

systems request form
A formal document to initiate systems investigation.

Because systems development requests can require considerable time and effort to implement, many organizations have adopted a formal procedure for initiating systems development, beginning with systems investigation. The **systems request form** is a document that is filled out by someone who wants the IS department to initiate systems investigation. This form typically includes the following information:

- Problems in or opportunities for the system
- Objectives of systems investigation
- Overview of the proposed system
- Expected costs and benefits of the proposed system

The information in the systems request form helps to rationalize and prioritize the activities of the IS department. Based on the overall IS plan, the organization's needs and goals, and the estimated value and priority of the proposed projects, managers make decisions regarding the initiation of each systems investigation for such projects.

Feasibility Analysis

feasibility analysis
Assessment of the technical, economic, legal, operational, and schedule feasibility of a project.

technical feasibility
Assessment of whether the hardware, software, and other system components can be acquired or developed to solve the problem.

economic feasibility
Determination of whether the project makes financial sense and whether predicted benefits offset the cost and time needed to obtain them.

legal feasibility
Determination of whether laws or regulations might prevent or limit a systems development project.

operational feasibility
Measure of whether the project can be put into action or operation.

A key step of the systems investigation phase is the **feasibility analysis**, which assesses technical, economic, legal, operational, and schedule feasibility. **Technical feasibility** is concerned with whether the hardware, software, and other system components can be acquired or developed to solve the problem. Technical problems, for example, were encountered in developing an automatic tool system for long-haul trucks on the German autobahn.[21] The satellite-based systems development project may never be fully implemented.

Economic feasibility determines whether the project makes financial sense and whether predicted benefits offset the cost and time needed to obtain them. A securities company, for example, investigated the economic feasibility of sending research reports electronically instead of through the mail. Economic analysis revealed that the new approach could save the company up to $500,000 per year. Economic feasibility can involve cash flow analysis such as that done in net present value or internal rate of return (IRR) calculations.

Legal feasibility determines whether laws or regulations might prevent or limit a systems development project. For example, an Internet site that allowed users to share music without paying musicians or music producers was sued. Legal feasibility involves an analysis of existing and future laws to determine the likelihood of legal action against the systems development project and the possible consequences.

Operational feasibility is a measure of whether the project can be put into action or operation. It can include logistical and motivational (acceptance of change) considerations. Motivational considerations are very important because new systems affect people and data flows and might have unintended consequences. As a result, power and politics might come into play, and some people might resist the new system. Because of deadly hospital errors, a healthcare consortium looked into the operational feasibility of developing a new computerized physician order entry system to require that all prescriptions and every order a doctor gives to staff be entered into the computer. The computer then checks for drug allergies and interactions between drugs. If operationally feasible, the new system could save lives and lawsuits.

Schedule feasibility determines whether the project can be completed in a reasonable amount of time—a process that involves balancing the time and resource requirements of the project with other projects.

schedule feasibility
Determination of whether the project can be completed in a reasonable amount of time.

Object-Oriented Systems Investigation

The object-oriented approach can be used during all phases of systems development, from investigation to maintenance and review. In addition to identifying key participants and performing basic feasibility analysis, key objects can be identified during systems investigation. Consider a kayak rental business in Maui, Hawaii, in which the owner wants to computerize its operations. There are many system objects for this business, including the kayak rental clerk, renting kayaks to customers, and adding new kayaks into the rental program. These objects can be diagrammed in a use case diagram (see Figure 8.7). As you can see, the kayak rental clerk rents kayaks to customers and adds new kayaks to the current inventory of kayaks available for rent. The stick figure is an example of an *actor*, and the ovals each represent an event, called a *use case*. In our example, the actor (the kayak rental clerk) interacts with two use cases (rent kayaks to customers and add new kayaks to inventory). The use case diagram is part of the Unified Modeling Language that is used in object-oriented systems development.

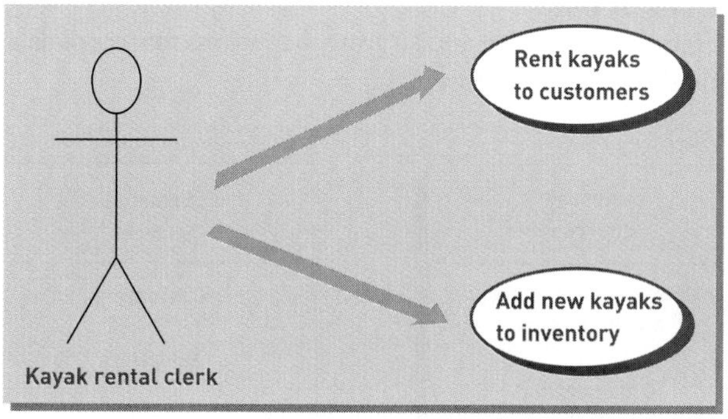

Figure 8.7

Use Case Diagram for a Kayak Rental Application

The Systems Investigation Report

The primary outcome of systems investigation is a **systems investigation report**. This report summarizes the results of systems investigation and the process of feasibility analysis and recommends a course of action: continue on into systems analysis, modify the project in some manner, or drop it. A typical table of contents for the systems investigation report is shown in Figure 8.8.

systems investigation report
A summary of the results of the systems investigation and the process of feasibility analysis and recommendations for a course of action.

SYSTEMS ANALYSIS

After a project has been approved for further study, the next step is to answer the question, "What must the information system do to solve the problem?" The process needs to go beyond mere computerization of existing systems. The entire system, and the business process with which it is associated, should be evaluated. Often, a firm can make great gains if it restructures both business activities and the related information system simultaneously. The

Figure 8.8

A Typical Table of Contents for a Systems Investigation Report

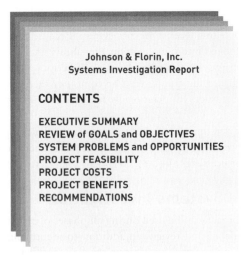

Johnson & Florin, Inc.
Systems Investigation Report

CONTENTS

EXECUTIVE SUMMARY
REVIEW of GOALS and OBJECTIVES
SYSTEM PROBLEMS and OPPORTUNITIES
PROJECT FEASIBILITY
PROJECT COSTS
PROJECT BENEFITS
RECOMMENDATIONS

overall emphasis of analysis is gathering data on the existing system, determining the requirements for the new system, considering alternatives within these constraints, and investigating the feasibility of the solutions. The primary outcome of systems analysis is a prioritized list of systems requirements.

Data Collection

The purpose of data collection is to seek additional information about the problems or needs identified in the systems investigation report. During this process, the strengths and weaknesses of the existing system are emphasized.

Identifying Sources of Data

Data collection begins by identifying and locating the various sources of data, including both internal and external sources (see Figure 8.9).

Figure 8.9

Internal and External Sources of Data for Systems Analysis

Internal sources
Users, stakeholders, and managers
Organization charts
Forms and documents
Procedure manuals and policies
Financial reports
IS manuals
Other measures of business process

External sources
Customers
Suppliers
Stockholders
Government agencies
Competitors
Outside groups
Journals, etc.
Consultants

Collecting Data

After data sources have been identified, data collection begins. Figure 8.10 shows the steps involved. Data collection often requires a number of tools and techniques, such as interviews, direct observation, and questionnaires.

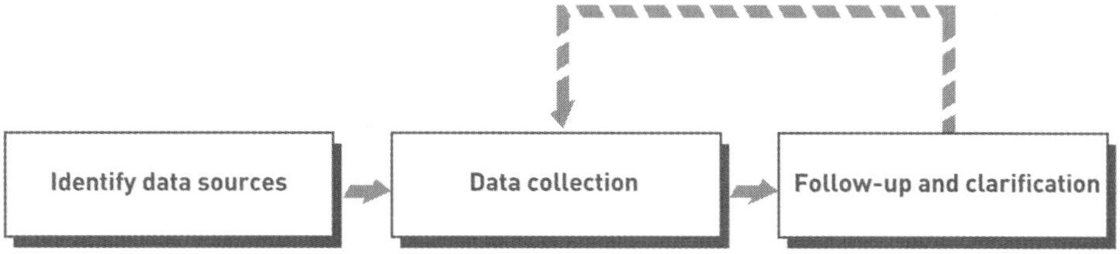

Identify data sources → Data collection → Follow-up and clarification

Figure 8.10

The Steps in Data Collection

Interviews are either structured or unstructured. In a **structured interview**, the questions are written in advance. In an **unstructured interview**, the questions are not written in advance; the interviewer relies on experience in asking the best questions to uncover the inherent problems of the existing system. With **direct observation**, one or more members of the analysis team directly observe the existing system in action. When many data sources are spread over a wide geographic area, **questionnaires** might be the best approach. Like interviews, questionnaires can be either structured or unstructured.

Data Analysis

The data collected in its raw form is usually not adequate to determine the effectiveness and efficiency of the existing system or the requirements for the new system. The next step is to manipulate the collected data so that it is usable for the development team who is participating in systems analysis. This manipulation is called **data analysis**. Data and activity modeling, using data-flow diagrams and entity-relationship diagrams, are useful during data analysis to show data flows and the relationships among various objects, associations, and activities.

Data Modeling

Data modeling, first introduced in Chapter 3, is a commonly accepted approach to modeling organizational objects and associations that employ both text and graphics. The exact way data modeling is employed, however, is governed by the specific systems development methodology.

Data modeling is most often accomplished through the use of *entity-relationship (ER) diagrams*. Recall from Chapter 3 that an entity is a generalized representation of an object type—such as a class of people (employee), events (sales), things (desks), or places (Philadelphia)—and that entities possess certain attributes. Objects can be related to other objects in numerous ways. An entity-relationship diagram, such as the one shown in Figure 8.11a, describes a number of objects and the ways they are associated. An ER diagram is not capable by itself of fully describing a business problem or solution because it lacks descriptions of the related activities. It is, however, a good place to start, because it describes object types and attributes about which data might need to be collected for processing.

Activity Modeling

To fully describe a business problem or solution, it is necessary to describe the related objects, associations, and activities. Activities in this sense are events or items that are necessary to fulfill the business relationship or that can be associated with the business relationship in a meaningful way.

Activity modeling is often accomplished through the use of data-flow diagrams. A **data-flow diagram (DFD)** models objects, associations, and activities by describing how data can flow between and around various objects. DFDs work on the premise that for every activity there is some communication, transference, or flow that can be described as a data element. DFDs describe what activities are occurring to fulfill a business relationship or accomplish a business task, not how these activities are to be performed. That is, DFDs show the logical sequence of associations and activities, not the physical processes. A system modeled with a DFD could operate manually or could be computer based; if computer based, the system could operate with a variety of technologies. DFDs are easy to develop and easily understood by nontechnical people (see Figure 8.11b). Figure 8.11c provides a brief description of the business relationships for clarification.

structured interview
An interview in which the questions are written in advance.

unstructured interview
An interview in which the questions are not written in advance.

direct observation
The process of watching the existing system in action by one or more members of the analysis team.

questionnaires
A method of gathering data when the data sources are spread over a wide geographic area.

data analysis
Manipulation of the collected data so that it is usable for the development team members who are participating in systems analysis.

data-flow diagram (DFD)
A model of objects, associations, and activities by describing how data can flow between and around various objects.

Figure 8.11

Data and Activity Modeling

(a) An entity-relationship diagram.
(b) A data-flow diagram. (c) A
semantic description of the
business process.

(Source: G. Lawrence Sanders, *Data
Modeling*. Danvers, MA: Boyd &
Fraser Publishing, 1995.)

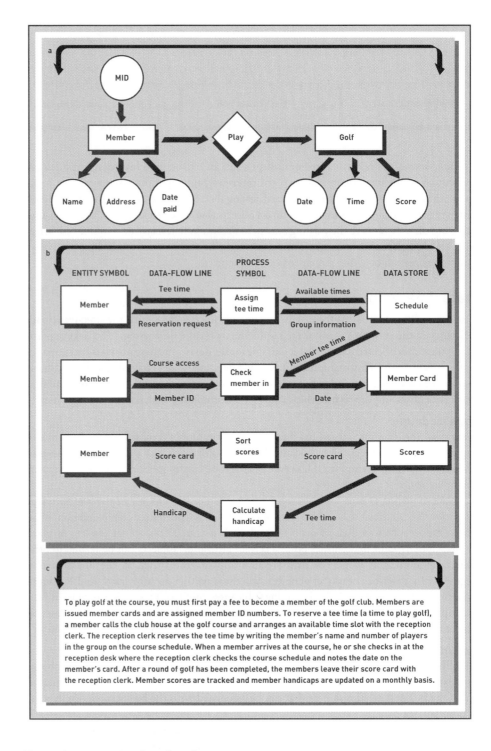

Requirements Analysis

The overall purpose of **requirements analysis** is to determine user, stakeholder, and organizational needs. For an accounts payable application, the stakeholders could include suppliers and members of the purchasing department. Questions that should be asked during requirements analysis include the following:

- Are these stakeholders satisfied with the current accounts payable application?
- What improvements could be made to satisfy suppliers and help the purchasing department?

requirements analysis
Determination of user, stakeholder, and organizational needs.

Asking Directly

One the most basic techniques used in requirements analysis is asking directly. **Asking directly** is an approach that asks users, stakeholders, and other managers about what they want and expect from the new or modified system. This approach works best for stable systems in which stakeholders and users clearly understand the system's functions. The role of the systems analyst during the analysis phase is to critically and creatively evaluate needs and define them clearly so that the systems can best meet them.

Critical Success Factors

Another approach uses critical success factors (CSFs). Managers and decision makers are asked to list only the factors that are critical to the success of their area of the organization. A CSF for a production manager might be adequate raw materials from suppliers; a CSF for a sales representative could be a list of customers currently buying a certain type of product. Starting from these CSFs, the system inputs, outputs, performance, and other specific requirements can be determined.

The IS Plan

As you have seen, the IS plan translates strategic and organizational goals into systems development initiatives. The IS planning process often generates strategic planning documents that can be used to define system requirements. Working from these documents ensures that requirements analysis will address the goals set by top-level managers and decision makers (see Figure 8.12). There are unique benefits to applying the IS plan to define systems requirements. Because the IS plan takes a long-range approach to using information technology within the organization, the requirements for a system analyzed in terms of the IS plan are more likely to be compatible with future systems development initiatives.

Figure 8.12

Converting Organizational Goals into Systems Requirements

Requirements Analysis Tools

A number of tools can be used to document requirements analysis. Again, CASE tools are often employed. As requirements are developed and agreed on, entity-relationship diagrams, data-flow diagrams, and other types of documentation will be stored in the CASE repository. These requirements might also be used later as a reference during the rest of systems development or for a different systems development project.

Object-Oriented Systems Analysis

The object-oriented approach can also be used during systems analysis. Like traditional analysis, problems or potential opportunities are identified during object-oriented analysis. Identifying key participants and collecting data is still performed. But instead of analyzing the existing system using data-flow diagrams and flowcharts, an object-oriented approach is used.

The section titled "Object-Oriented Systems Investigation" introduced a kayak rental example. A more detailed analysis of that business reveals that there are two classes of kayaks: single kayaks for one person and tandem kayaks that can accommodate two people. With the OO approach, a class is used to describe different types of objects, such as single and tandem kayaks. The classes of kayaks can be shown in a generalization/specialization hierarchy diagram (see Figure 8.13). KayakItem is an object that will store the kayak identification number (ID) and the date the kayak was purchased (datePurchased).

Of course, there could be subclasses of customers, life vests, paddles, and other items in the system. For example, price discounts for kayak rentals could be given to seniors (people over 65 years) and students. Thus, the Customer class could be divided into regular, senior, and student customer subclasses.

Figure 8.13

Generalization/Specialization Hierarchy Diagram for Single and Tandem Kayak Classes

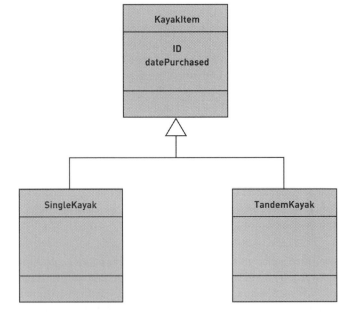

The Systems Analysis Report

Systems analysis concludes with a formal systems analysis report. It should cover the following elements:

- The strengths and weaknesses of the existing system from a stakeholder's perspective
- The user/stakeholder requirements for the new system (also called the *functional requirements*)
- The organizational requirements for the new system
- A description of what the new information system should do to solve the problem

Suppose analysis reveals that a marketing manager thinks a weakness of the existing system is its inability to provide accurate reports on product availability. These requirements and a preliminary list of the corporate objectives for the new system will be in the systems analysis report. Particular attention is placed on areas of the existing system that could be improved to meet user requirements. The table of contents for a typical report is shown in Figure 8.14. After the formal report is prepared, the team can move from systems analysis to perform the next step of the systems development life cycle—systems design.

Figure 8.14

A Typical Table of Contents for a Report on an Existing System

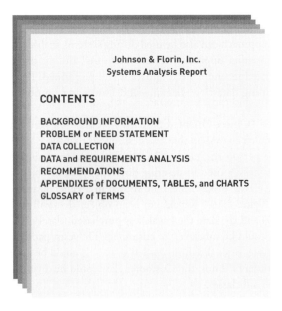

Johnson & Florin, Inc.
Systems Analysis Report

CONTENTS

BACKGROUND INFORMATION
PROBLEM or NEED STATEMENT
DATA COLLECTION
DATA and REQUIREMENTS ANALYSIS
RECOMMENDATIONS
APPENDIXES of DOCUMENTS, TABLES, and CHARTS
GLOSSARY of TERMS

SYSTEMS DESIGN

▼

The purpose of *systems design* is to answer the question, "How will the information system solve a problem?" The primary result of the systems design phase is a technical design that details system outputs, inputs, and user interfaces; specifies hardware, software, databases, telecommunications, personnel, and procedures; and shows how these components are related. The new system should overcome shortcomings of the existing system and help the organization achieve its goals. The system must also meet certain guidelines, including user and stakeholder requirements and the objectives defined during previous development phases. There, Inc. (*www.there.com*), a California company that lets people meet and interact on a 3-D Web site, spent more than $17 million to design an online simulation that lets people chat, play cards, and flirt with avatars, lifelike characters that appear on Internet sites.[22] The company hopes that the four-year systems development project will generate revenues from individual subscribers and companies that want to advertise their products on its Web site.

As discussed earlier in the database chapter (Chapter 3), design has two dimensions: logical and physical. The **logical design** refers to what the system will do. Logical design describes the functional requirements of a system. That is, it conceptualizes what the system will do to solve the problems identified through earlier analysis. Without this step, the technical details of the system (such as which hardware devices should be acquired) often obscure the best solution. Logical design involves planning the purpose of each system element, independent of hardware and software considerations. The logical design specifications that are determined and documented include output, input, process, telecommunications, procedures, controls and security, and personnel and job requirements. The **physical design** refers to how the tasks are accomplished, including how the components work together and what each component does. Physical design specifies the characteristics of the system components necessary to put the logical design into action. In this phase, the characteristics of hardware, software, database, telecommunications, personnel, procedures, and control specifications must be detailed.

logical design
Description of the functional requirements of a system.

physical design
Specification of the characteristics of the system components necessary to put the logical design into action.

Object-Oriented Design

Logical and physical design can be accomplished using either the traditional structured approach or the objected-oriented approach to systems development. Both approaches use a variety of design models to document the new system's features and the development team's understandings and agreements. Many organizations today are turning to OO development because of its increased flexibility. So, this section outlines a few OO design considerations and diagrams.

Using the OO approach, system developers design key objects and classes of objects in the new or updated system. This process includes consideration of the problem domain, the operating environment, and the user interface. The problem domain involves the classes of objects related to solving a problem or realizing an opportunity. In the Maui, Hawaii, kayak rental shop example introduced earlier in this chapter and referring back to the generalization/specialization hierarchy showing classes presented there, KayakItem in Figure 8.13 is an example of a problem domain object that we will use to store information on kayaks in the rental program. The operating environment for the rental shop's system includes objects that interact with printers, system software, and other software and hardware devices. The user interface for the system includes objects with which users interact, such as buttons and scroll bars in a Windows program.

During the design phase, we also need to consider the sequence of events that must happen for the system to function correctly. For example, we might want to design the sequence of events that are needed to add a new kayak to the rental program. A sequence of events is often called a *scenario*, which can be diagrammed in a sequence diagram (see Figure 8.15).

A Sequence Diagram to Add a New KayakItem Scenario

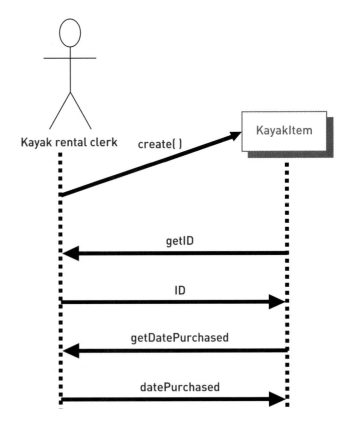

A sequence diagram is read starting from the top and moving down.

1. The Create arrow at the top is a message from the kayak rental clerk to the KayakItem object to create information on a new kayak to be placed into the rental program.
2. The KayakItem object knows that it needs the ID for the kayak and sends a message to the clerk requesting the information. See the getID arrow.
3. The clerk then types the ID into the computer. This is shown with the ID arrow. The data is stored in the KayakItem object.
4. Next, KayakItem requests the purchase date. This is shown with the getDatePurchased arrow.
5. Finally, the clerk types the purchase date into the computer. The data is also transferred to KayakItem object. This is shown with the datePurchased arrow at the bottom of Figure 8.15.

This scenario is only one example of a sequence of events. Other scenarios might include entering information about life jackets, paddles, suntan lotion, and other accessories. The same types of use case and generalization/specialization hierarchy diagrams discussed earlier in this chapter can be created for each, and additional sequence diagrams will also be needed.

Generating Systems Design Alternatives

When individuals or organizations require a system to perform additional functions that an existing system cannot support, they often turn to outside vendors to design and supply their new systems. Such purchases involve both hardware and software expertise. Whether an individual is purchasing a personal computer or a company is acquiring an expensive main-frame computer, the system is often obtained from one or more vendors.[23] Companies can use multiple vendors or a single vendor in acquiring information systems and technology. Florida-based Fidelity National Financial, for example, used the single-vendor approach to acquire a centralized information system.[24] The company selected IBM for its multimillion-dollar project to speed processing for the $8 trillion of mortgages and loans it processes for large banks every day. In some cases, the vendor simply provides hardware or software. In other cases, the vendor provides additional services.

When additional hardware and software are not required, alternative designs are often generated without input from vendors. If the new system is complex, the original development team might want to involve other personnel in generating alternative designs. If new hardware and software are to be acquired from an outside vendor, a formal request for proposal (RFP) should be made. The **request for proposal (RFP)** is one of the most important documents generated during systems development. It often results in a formal bid that is used to determine who gets a contract for new or modified systems. The RFP specifies in detail the required resources, such as hardware and software. Although it can take time and money to develop a high-quality RFP, it can save a company in the long run. Companies that frequently generate RFPs can automate the process. The Table of Contents for a typical RFP is shown in Figure 8.16.

request for proposal (RFP)
A document that specifies in detail required resources such as hardware and software.

Johnson & Florin, Inc.
Systems Investigation Report

Contents

COVER PAGE (with company name and contact person)
BRIEF DESCRIPTION of the COMPANY
OVERVIEW of the EXISTING COMPUTER SYSTEM
SUMMARY of COMPUTER-RELATED NEEDS and/or PROBLEMS
OBJECTIVES of the PROJECT
DESCRIPTION of WHAT IS NEEDED
HARDWARE REQUIREMENTS
PERSONNEL REQUIREMENTS
COMMUNICATIONS REQUIREMENTS
PROCEDURES to BE DEVELOPED
TRAINING REQUIREMENTS
MAINTENANCE REQUIREMENTS
EVALUATION PROCEDURES (how vendors will be judged)
PROPOSAL FORMAT (how vendors should respond)
IMPORTANT DATES (when tasks are to be completed)
SUMMARY

Figure 8.16

A Typical Table of Contents for a Request for Proposal

When it comes to acquiring computer systems, several choices are available, including purchasing, leasing, or renting. Cost objectives and constraints set for the system play a significant role in the alternative chosen, as do the advantages and disadvantages of each. Table 8.3 summarizes the advantages and disadvantages of these financial options.

Evaluating and Selecting a System Design

The final step in systems design is to evaluate the various alternatives and select the one that will offer the best solution for organizational goals. Evaluating and selecting the best design involves a balance of system objectives that will best support organizational goals. Normally, evaluation and selection involves both a preliminary and a final evaluation before a design is selected. A *preliminary evaluation* begins after all proposals have been submitted. The purpose of this evaluation is to dismiss unwanted proposals. The *final evaluation* begins with a detailed investigation of the proposals offered by the remaining vendors. The vendors should be asked to make a final presentation and to fully demonstrate the system. The demonstration should be as close to actual operating conditions as possible. Such applications as payroll, inventory control, and billing should be conducted using a large amount of test data. After the final presentations and demonstrations have been given, the organization makes the final evaluation and selection. Cost comparisons, hardware performance, delivery dates, price, flexibility, backup facilities, availability of software training, and maintenance factors are considered.

Table 8.3

Advantages and Disadvantages of Acquisition Options

Renting (Short-Term Option)

Advantages	Disadvantages
No risk of obsolescence	No ownership of equipment
No long-term financial investment	High monthly costs
No initial investment of funds	Restrictive rental agreements
Maintenance usually included	

Leasing (Longer-Term Option)

Advantages	Disadvantages
No risk of obsolescence	High cost of canceling lease
No long-term financial investment	Longer time commitment than renting
No initial investment of funds	No ownership of equipment
Less expensive than renting	

Purchasing

Advantages	Disadvantages
Total control over equipment	High initial investment
Can sell equipment at any time	Additional cost of maintenance
Can depreciate equipment	Possibility of obsolescence
Low cost if owned for a number of years	Other expenses, including taxes and insurance

The Design Report

System specifications are the final results of systems design. They include a technical description that details system outputs, inputs, and user interfaces, as well as all hardware, software, databases, telecommunications, personnel, and procedure components and the way these components are related. The specifications are contained in a **design report**, which is the result of systems design. The design report reflects the decisions made for systems design and prepares the way for systems implementation. The contents of the design report are summarized in Figure 8.17.

design report
The result of systems design, reflecting the decisions made for system design and preparing the way for systems implementation.

SYSTEMS IMPLEMENTATION

After the information system has been designed, a number of tasks must be completed before the system is installed and ready to operate. This process, called *systems implementation*, includes hardware acquisition, software acquisition or development, user preparation, hiring and training of personnel, site and data preparation, installation, testing, start-up, and user acceptance. The typical sequence of systems implementation activities is shown in Figure 8.18. Companies can reap great rewards after implementing new systems. United

Figure 8.17

A Typical Table of Contents for a Systems Design Report

Johnson & Florin, Inc.
Systems Design Report

Contents

PREFACE
EXECUTIVE SUMMARY of SYSTEMS
DESIGN
REVIEW of SYSTEMS ANALYSIS
MAJOR DESIGN RECOMMENDATIONS
      Hardware design
      Software design
      Personnel design
      Communications design
      Database design
      Procedures design
      Training design
      Maintenance design
SUMMARY of DESIGN DECISIONS
APPENDIXES
GLOSSARY of TERMS
INDEX

Parcel Service (UPS), for example, implemented a $30 million project to improve how packages flow and are tracked from pickup through delivery.[25] UPS's new efficiencies should cut more than 100 million delivery miles and save the company about 14 million gallons of fuel. In the banking industry, FleetBoston Financial Corporation implemented a $10 million IS project to centralize all of its network operations into one room. The project will allow FleetBoston to cut its IS staff and save about 3,500 square feet of office space that was used to store networking equipment.

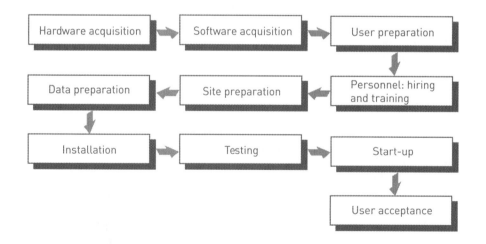

Figure 8.18

Typical Steps in Systems Implementation

Acquiring Hardware from an IS Vendor

To obtain the components for an information system, organizations can purchase, lease, or rent computer hardware and other resources from an IS vendor. eBay, the popular Internet auction site, for example, acquired hundreds of hardware servers to increase the speed and reliability of its computer system.[26] The old hardware couldn't prevent a crash that lasted about a day and cost eBay about $4 million in lost sales and $5 billion in the value of its stock. The newly acquired hardware is available 99.94 percent of the time.

An IS vendor is a company that offers hardware, software, telecommunications systems, databases, IS personnel, and/or other computer-related resources. Types of IS vendors include general computer manufacturers (e.g., IBM and Hewlett-Packard), small computer

Computer dealers, such as CompUSA, manufacture build-to-order computer systems and sell computers and supplies from other vendors.

(Source: Courtesy of CompUSA, Inc.)

make-or-buy decision
The decision of whether to purchase software from external developers or develop it in-house.

manufacturers (e.g., Dell and Gateway), peripheral equipment manufacturers (e.g., Epson and Canon), computer dealers and distributors (e.g., Radio Shack and CompUSA), and leasing companies (e.g., National Computer Leasing and Paramount Computer Rentals, plc).

In addition to buying, leasing, or renting computer hardware, it is possible to pay only for the computing services that a company uses.[27] Called "pay as you go," "on demand," or "utility" computing, this approach requires an organization to pay only for the computer power it uses, similar to paying for a utility such as electricity. JPMorgan Chase, for example, is buying only the computer resources it needs from IBM. Hewlett-Packard offers its clients a "capacity-on-demand" approach, in which organizations pay according to the computer resources actually used, including processors, storage devices, and network facilities.[28] It is also possible to purchase used computer equipment.

Acquiring Software: Make or Buy?

As with hardware, application software can be acquired several ways. It can be purchased from external developers or developed in-house. This decision is often called the **make-or-buy decision**. Alaska Airlines, for example, decided to purchase software for its Internet search engine for fares and prices.[29] The purchased software allows its customers more choices for flight schedules and prices. To share the cost of in-house software development, Coca-Cola Enterprises and SAP, a large ERP software company, have decided to jointly develop software for beverage companies.[30] According to the CIO of Coca-Cola, "This should improve market execution, and the consumer will experience better service." In some cases, companies use a blend of external and internal software development. That is, off-the-shelf or proprietary software programs are modified or customized by in-house personnel. Software can also be rented.[31] Salesforce.com, for example, rents software online that helps organizations manage their sales force and internal staff. Increasingly, software is being viewed as a utility or service, not a product you purchase. It is also possible to reuse software from other development efforts to reduce the time it takes to deliver in-house software. Bank of America, for example, reuses previously developed software to deliver new software in 90 days or fewer.

Acquiring Database and Telecommunications Systems

Acquiring or upgrading database systems can be one of the most important steps of a systems development effort. Although most companies use a relational database, some are starting to use object-oriented database systems, such as Object Store from Excelon Corporation and Objectivity from Objectivity, Inc.

Because databases are a blend of hardware and software, many of the approaches discussed earlier for acquiring hardware and software also apply to database systems. For example, an upgraded inventory control system might require database capabilities, including more hard disk storage or a new DBMS. If so, additional storage hardware will have to be acquired from an IS vendor. New or upgraded software might also be purchased or developed in-house.

With the increased use of e-commerce, the Internet, intranets, and extranets, telecommunications is one of the fastest-growing applications for today's businesses and individuals. The Nasdaq Stock Market, for example, is investing $50 million in a new network system to streamline operations and cut costs.[32] According to the chief information officer for Nasdaq, "We're going to be able to take a lot of circuits out of our network and save costs." Like database systems, telecommunications systems require a blend of hardware and software. For personal computer systems, the primary piece of hardware is a modem. For client/server and mainframe systems, the hardware can include multiplexers, concentrators, communications processors, and a variety of network equipment. Communications software will also have to be acquired from a software company or developed in-house. The Internal Revenue

Service (IRS) hired a new chief information officer to oversee its large systems development effort to improve network security and other telecommunications aspects of the IRS computer system.[33] Again, the earlier discussion on acquiring hardware and software also applies to the acquisition of telecommunications hardware and software.

User Preparation

User preparation is the process of readying managers, decision makers, employees, other users, and stakeholders for the new systems. This activity is an important but often ignored area of systems implementation. A small airline might not do adequate employee training with a new software package. The result could be a grounding of most of its flights and the need to find hotel rooms to accommodate unhappy travelers who were stranded.

Providing users with proper training can help ensure that the information system is used correctly, efficiently, and effectively.

(Source: Photodisc/Getty Images.)

IS Personnel: Hiring and Training

Depending on the size of the new system, an organization might have to hire and, in some cases, train new IS personnel. An IS manager, systems analysts, computer programmers, data-entry operators, and similar personnel might be needed for the new system.

As with users, the eventual success of any system depends on how it is used by the personnel within the organization. Training programs should be conducted for the IS personnel who will be using the computer system. These programs are similar to those for the users, although they might be more detailed in the technical aspects of the systems. Effective training will help IS personnel use the new system to perform their jobs and support other users in the organization.

Site Preparation

The location of the new system needs to be prepared in a process called **site preparation**. For a small system, site preparation can be as simple as rearranging the furniture in an office to make room for a computer. With a larger system, this process is not so easy because it might require special wiring and air-conditioning. One or more rooms might have to be completely renovated, and additional furniture might have to be purchased. A special floor might have to be built, under which the cables connecting the various computer components are placed, and a new security system might be needed to protect the equipment. For larger systems, additional power circuits might also be required.

site preparation
Preparation of the location of the new system.

data preparation, or data conversion
The process of ensuring all files and databases are ready to be used with new computer software and systems.

Data Preparation

Data preparation, or **data conversion**, involves making sure that all files and databases are ready to be used with new computer software and systems. If an organization is installing a new payroll program, the old employee-payroll data might have to be converted into a format that can be used by the new computer software or system. After the data has been prepared or converted, the computerized database system or other software will then be used to maintain and update the computer files.

installation
The process of physically placing the computer equipment on the site and making it operational.

Installation

Installation is the process of physically placing the computer equipment on the site and making it operational. Although normally the manufacturer is responsible for installing computer equipment, someone from the organization (usually the IS manager) should oversee the process, making sure that all equipment specified in the contract is installed at the proper location. After the system is installed, the manufacturer performs several tests to ensure that the equipment is operating as it should.

Testing

Good testing procedures are essential to ensure that the new or modified information system operates as intended. Inadequate testing can result in mistakes and problems. A popular tax preparation company, for example, implemented a Web-based tax preparation system, but people could see one another's tax returns. The president of the tax preparation company called it "our worst-case scenario." Better testing can prevent these types of problems.

Several forms of testing should be used, including testing each of the individual programs (unit testing), testing the entire system of programs (system testing), testing the application with a large amount of data (volume testing), and testing all related systems together (integration testing), as well as conducting any tests required by the user (acceptance testing). In addition to these tests, there are additional types. Alpha testing involves testing an incomplete or early version of the system, and beta testing involves testing a complete and stable system by end users. Alpha-unit testing, for example, is testing an individual program before it is completely finished. Beta-unit testing, on the other hand, is performed after alpha testing, when the individual program is complete and ready to use by end users.

Start-Up

start-up
The process of making the final tested information system fully operational.

direct conversion (also called *plunge* or *direct cutover*)
The process of stopping the old system and starting the new system on a given date.

phase-in approach, or piecemeal approach
The process of slowly replacing components of the old system with those of the new one; this process is repeated for each application until the new system is running every application and performing as expected.

pilot start-up
The process of running the new system for one group of users rather than for all users.

Start-up begins with the final tested information system. When start-up is finished, the system is fully operational. Various start-up approaches are available (see Figure 8.19). **Direct conversion** (also called *plunge* or *direct cutover*) involves stopping the old system and starting the new system on a given date. Direct conversion is usually the least desirable approach because of the potential for problems and errors when the old system is shut off and the new system is turned on at the same instant. The Tennessee Valley Authority (TVA) implemented a large IS project to improve efficiency and reduce costs. The implementation was a direct cutover that went live after a week-long implementation process. "The big-bang approach was very scary," said the senior vice president of information systems for TVA. "This meant getting everything going at the same time."[34] The **phase-in approach**, or **piecemeal approach**, is a popular technique preferred by many organizations. In this approach, components of the new system are slowly phased in while components of the old one are slowly phased out. When everyone is confident that the new system is performing as expected, the old system is completely phased out. This gradual replacement is repeated for each application until the new system is running every application.

Pilot start-up involves running the new system for one group of users rather than all users. For example, a manufacturing company with a number of retail outlets throughout the country could use the pilot start-up approach and install a new inventory control system at one of the retail outlets. When this pilot retail outlet runs without problems, the new inventory control system can be implemented at other retail outlets. Carnival Cruise Lines, for example, is using a pilot start-up for a systems development project to remotely manage

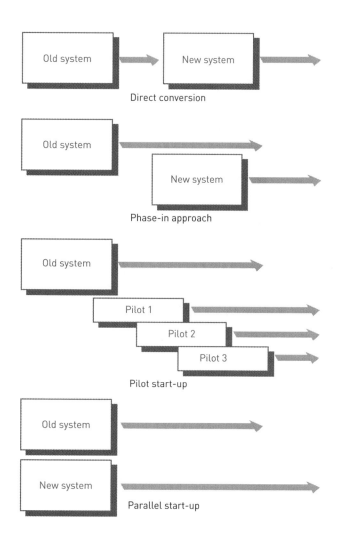

Figure 8.19

Start-Up Approaches

PCs.[35] When fully implemented, the new system will remotely manage about 4,000 PCs, including about 1,700 PCs on the company's 19 ships. The new system will allow Carnival to remotely upgrade software and perform some fixes without the need to fly a technician to a remote location or a ship.

Parallel start-up involves running both the old and new systems for a period of time. The output of the new system is compared closely with the output of the old system, and any differences are reconciled. When users are comfortable that the new system is working correctly, the old system is eliminated.

User Acceptance

Most mainframe computer manufacturers employ a **user acceptance document**—a formal agreement signed by the user that states that a phase of the installation or the complete system is approved. This is a legal document that usually removes or reduces the IS vendor from liability for problems that occur after the user acceptance document has been signed. Because this document is so important, many companies get legal assistance before they sign the acceptance document. Stakeholders might also be involved in acceptance to make sure that the benefits to them are indeed realized.

parallel start-up
The process of running both the old and new systems for a period of time, comparing the new system's output closely with the old system's, reconciling any differences, and finally eliminating the old system.

user acceptance document
A formal agreement signed by the user that states that a phase of the installation or the complete system is approved.

SYSTEMS OPERATION AND MAINTENANCE

systems operation
Use of a new or modified system.

Systems operation involves all aspects of using the new or modified system. Throughout this book, you have seen many examples of information systems operating in a variety of settings and industries. Thus, there is no need to cover the operation of an information system in detail in this section. The operation of any information system, however, does require adequate user training before the system is used and continual support while the system is being operated. This training and support is required for all stakeholders, including employees, customers, and others. Companies typically provide training through seminars, manuals, and online documentation. To provide adequate support, many companies use a formal help desk. A *help desk* consists of people with technical expertise, computer systems, manuals, and other resources needed to solve problems and give accurate answers to questions. With today's advances in telecommunications, help desks can be located around the world. If you are having trouble with your PC and call a toll-free number for assistance, you might reach a help desk in India, China, or another country.

Systems maintenance involves checking, changing, and enhancing the system to make it more useful in achieving user and organizational goals. This process can be especially difficult for older software. A *legacy system* is an old system that may have been patched or modified repeatedly over time. An old payroll program in COBOL developed decades ago and frequently changed is an example of a legacy system. Legacy systems can be very expensive to maintain. At some point, it becomes less expensive to switch to new programs and applications than to repair and maintain the legacy system.

Software maintenance is a major concern for organizations.[36] Hardware maintenance is also important. Companies such as IBM are investigating *autonomic computing,* in which computers will be able to manage and maintain themselves.[37] The "Ethical and Societal Issues" special feature explores how software companies are trying to enhance the trustworthiness of information systems.

SYSTEMS REVIEW

Systems review, the final step of systems development, is the process of analyzing systems to make sure they are operating as intended.[38] This process often compares the performance and benefits of the system as it was designed with the actual performance and benefits of the system in operation. Hymans Robertson, an actuarial firm, reviewed an existing financial system and discovered that it no longer met its needs.[39] According to the finance director for the firm, "We had been using a finance system that had been implemented three years ago. We soon realized we had outgrown this system as it was not flexible enough to fit our changing needs." Problems and opportunities uncovered during systems review will trigger systems development and begin the process anew. For example, as the number of users of an interactive system increases, system response time often increases. If the increased response time is too great, it might be necessary to redesign some of the system, modify databases, or increase the power of the computer hardware.

event-driven review
A review triggered by a problem or opportunity, such as an error, a corporate merger, or a new market for products.

There are two types of review procedures: event driven and time driven (see Table 8.4). An **event-driven review** is triggered by a problem or opportunity, such as an error, a corporate merger, or a new market for products. In some cases, companies wait until a large problem or opportunity occurs before a change is made, ignoring minor problems. In contrast, some companies use a continuous improvement approach to systems development. With this approach, an organization makes changes to a system even when small problems or opportunities occur. Although continuous improvement can keep the system current and responsive, doing the repeated design and implementation can be both time-consuming and expensive.

Event Driven	Time Driven
A problem with an existing system	Monthly review
A merger	Yearly review
A new accounting system	Review every few years
An executive decision that an upgraded Internet site is needed to stay competitive	Five-year review

Table 8.4

Examples of Review Types

A **time-driven review** is performed after a specified amount of time. Many application programs are reviewed every six months to a year. With this approach, an existing system is monitored on a schedule. If problems or opportunities are uncovered, a new systems development cycle might be initiated. A payroll application, for example, might be reviewed once a year to make sure it is still operating as expected. If it is not, changes are made.

Many companies use both approaches. A billing application, for example, might be reviewed once a year for errors, inefficiencies, and opportunities to reduce operating costs. This is a time-driven approach. In addition, the billing application might be redone if there is a corporate merger, if one or more new managers require different information or reports, or if federal laws on bill collecting and privacy change. This is an event-driven approach.

time-driven review
A review performed after a specified amount of time.

Finding Trust in Computer Systems

In June 2004, in what was heralded as "the worst mess since banks put their faith in computers," the Royal Bank of Canada was unable to tell its 10 million Canadian customers exactly how much money was in their accounts. Canada's largest bank had a problem that kept tens of millions of transactions, including every direct payroll deposit it handles, from showing up in accounts.

You can imagine the furor as millions of bank customers discovered that the paycheck they counted on hadn't yet been deposited. Although many people were just inconvenienced, others were in a panic. Vacations were postponed, bill payments became delinquent, and customers were lined up at banks to try to get money that they were due for basic living expenses.

The nightmare began during what was intended to be a routine programming update. Soon afterward, the bank's entire nationwide system failed to register withdrawals and deposits against customer balances for several days. The more days that passed, the larger the backlog became, and the worse the chance of the system's being able to recover.

Over the past few years, computer system failures have drawn increased attention as cybercriminals have created serious problems. Worms and viruses have ravaged the Internet's operating systems and other software. Corporate networks have been shut down, and valuable and private information has been stolen and sold on the black market. All of these catastrophes have brought increasing pressure on systems developers and software manufacturers to invest more in the quality of their products.

Software developers are confronting the issue and working to regain customer trust. Microsoft recently instituted its "Trustworthy Computing Initiative." Microsoft Enterprise Technologies Director Greg Stone says, "I see Trustworthy Computing having three key elements: capability to do what you say you will, consistency of quality, and the commitment to act in the customer's best interests." Other Microsoft executives characterize the principles of Trustworthy Computing as being reliability, business integrity, privacy, and security.

Some doubt Microsoft's ability to deliver trustworthy systems in the near future. They fear that the complexity of Microsoft's programming code is beyond human ability to secure. "Microsoft has done vastly more than what they've done historically, but is the state of security any better?" asks Lloyd Hession, CSO of Radianz, which operates extranets for 5,000 financial institutions. "What Microsoft has done hasn't had much of an impact. Because of the complexity of the code, you find more vulnerabilities once you start to tinker with the code."

Microsoft is investing an increasing proportion of the company's $6.8 billion research and development fund on proactive security projects, such as creating self-healing software, often called *autonomic computing*. Autonomic computing appears to be a solution that most of the big software companies are pursuing. The major systems players, HP, IBM, and Sun Microsystems, are offering products that are self-managing and self-healing. Autonomic technology is being applied to networks, databases, software, and nearly all components of information systems. The hope is that systems will be able to react on the fly to changes in demand or unexpected problems, providing a constant level of service, much like a public utility.

Without doubt, we will see increasing efforts from all systems and software developers to deliver trustworthy systems. As our lives become increasingly dependent on information systems, the future of our society could be jeopardized. It is clear that today's systems, in all of their complexity, will require a high degree of automation and artificial intelligence capabilities to successfully manage them. Computer system stability, dependability, and security will remain a major goal for governments, industries, and technology specialists to work together to achieve.

Critical Thinking Questions

1. What social factors have arisen to make computer system stability and security an increasing concern?
2. Whose responsibility is it to ensure that computer systems are dependable and secure? Why?

What Would You Do?

As the owner of TicketBlaster, an online service that sells tickets for concerts, shows, and sporting events, you depend on technology. Over the past month, your company has experienced a total of six hours of sporadic downtime—10 minutes here, 20 minutes there. Sales figures have dropped, and you fear that customers might be turning to your competitor out of frustration.

3. What steps could you take to uncover the cause of your downtime?
4. What actions could you take to renew your customer's faith in your service?

SOURCES: John Saunders and Richard Bloom, "Bank's Clients in Limbo," *The Globe and Mail*, June 4, 2004, *www.theglobeandmail.com*; Lawrence Walsh, "Microsoft's Paradox," *Information Security*, January 2004, *www.lexis-nexis.com*; Mark Hollands, "Microsoft Struggles to Build Trust," *The Australian*, March 30, 2004, *www.lexis-nexis.com*; Ann Bednarz, "Autonomic Authority," *Network World*, March 22, 2004, *www.lexis-nexis.com*; "Evident Software Partners with IBM to Further Autonomic Computing Initiative," *Business Wire*, March 8, 2004, *www.lexis-nexis.com*.

SUMMARY

Principle

Effective systems development requires a team effort of stakeholders, users, managers, systems development specialists, and various support personnel, and it starts with careful planning.

The systems development team consists of stakeholders, users, managers, systems development specialists, and various support personnel. The development team is responsible for determining the objectives of the information system and delivering to the organization a system that meets its objectives. A systems analyst is a professional who specializes in analyzing and designing business systems. The programmer is responsible for modifying or developing programs to satisfy user requirements. Other support personnel on the development team include technical specialists. Depending on the magnitude of the systems development project and the number of IS systems development specialists on the team, the team might also include one or more IS managers.

Information systems planning refers to the translation of strategic and organizational goals into systems development initiatives. Benefits of IS planning include a long-range view of information technology use and better use of IS resources. Planning requires developing overall IS objectives; identifying IS projects; setting priorities and selecting projects; analyzing resource requirements; setting schedules, milestones, and deadlines; and developing the IS planning document.

Principle

Systems development often uses different approaches and tools such as traditional development, prototyping, rapid application development, end-user development, computer-aided software engineering, and object-oriented development to select, implement, and monitor projects.

The five phases of the traditional SDLC are investigation, analysis, design, implementation, and maintenance and review. Systems investigation involves identifying potential problems and opportunities and considering them in light of organizational goals. Systems analysis seeks a general understanding of the solution required to solve the problem; the existing system is studied in detail and weaknesses are identified. Systems design involves creating new or modified system requirements. Systems implementation encompasses programming, testing, training, conversion, and operation of the system. Systems operation involves running the system after it is implemented. Systems maintenance and review entails monitoring the system and performing enhancements or repairs.

Prototyping is an iterative development approach that involves defining the problem, building the initial version, having users utilize and evaluate the initial version, providing feedback, and incorporating suggestions into the second version. Rapid application development (RAD) uses tools and techniques designed to speed application development. Its use reduces paper-based documentation, automates program source code generation, and facilitates user participation in development activities. An agile, or extreme programming, approach allows systems to change as they are being developed. RAD makes extensive use of the joint application development (JAD) process to gather data and perform requirements analysis. JAD involves group meetings in which users, stakeholders, and IS professionals work together to analyze existing systems, propose possible solutions, and define the requirements for a new or modified system. The end-user SDLC is used to support projects in which the primary effort is undertaken by a combination of business managers and users. End-user SDLC is becoming increasingly important as more users develop systems for their personal computers.

The use of automated tools enables detailed development, tracking, and control of the project schedule. Effective use of these tools enables a project manager to deliver a high-quality system and to make intelligent trade-offs among cost, schedule, and quality. CASE tools can automate many of the systems development tasks, thus reducing the time and effort required to complete them while ensuring good documentation. With the object-oriented systems development (OOSD) approach, a project can be broken down into a group of objects that interact. Instead of requiring thousands or millions of lines of detailed computer instructions or code, the systems development project might require a few dozen or maybe a hundred objects.

Principle

Systems development starts with investigation and analysis of existing systems.

In most organizations, a systems request form initiates the investigation process. This form typically includes the problems in or opportunities for the system, objectives of systems investigation, overview of the proposed system, and expected costs and benefits of the proposed system. The systems investigation is designed to assess the feasibility of implementing solutions for business problems. An investigation team follows up on the request and performs a feasibility analysis that addresses technical, economic, legal, operational, and schedule feasibility. Object-oriented systems investigation is being used to a greater extent today. As a final step in the investigation process, a systems investigation report should be prepared to document relevant findings.

Systems analysis is the examination of existing systems, which begins after approval for further study is received from management. Additional study of a selected system allows those involved to further understand the system's weaknesses and potential improvement areas. An analysis team is assembled to collect and analyze data on the existing system.

Data collection methods include observation, interviews, and questionnaires. Data analysis manipulates the collected data to provide information. Data modeling is used to model organizational objects and associations using text and graphical diagrams. It is most often accomplished through the use of entity-relationship (ER) diagrams. Activity modeling is often accomplished through the use of data-flow diagrams (DFDs), which model objects, associations, and activities by describing how data can flow between and around various objects. DFDs use symbols for data flows, processing, entities, and data stores. The overall purpose of requirements analysis is to determine user and organizational needs. Object-oriented systems analysis also involves diagramming techniques, such as a generalization/specialization hierarchy diagram.

Principle

Designing new systems or modifying existing ones should always be aimed at helping an organization achieve its goals.

The purpose of systems design is to prepare the detailed design needs for a new system or modifications to an existing system. Logical systems design refers to the way the various components of an information system will work together. Physical systems design refers to the specification of the actual physical components.

If new hardware or software will be purchased from a vendor, a formal request for proposal (RFP) is needed. The RFP outlines the company's needs; in response, the vendor provides a written reply. Organizations have three alternatives for acquiring computer systems: purchasing, leasing, or renting. RFPs from various vendors are reviewed and narrowed down to the few most likely candidates. After the vendor is chosen, contract negotiations can begin. One of the most important steps in systems design is to develop a good contract if new computer facilities are being acquired. The final step is to develop a design report that details the outputs, inputs, and user interfaces. It also specifies hardware, software, databases, telecommunications, personnel, and procedure components and the way these components are related.

Principle

The primary emphasis of systems implementation is to make sure that the right information is delivered to the right person in the right format at the right time.

The purpose of systems implementation is to install a system and make everything, including users, ready for its operation. Systems implementation includes hardware acquisition, software acquisition or development, user preparation, hiring and training of IS personnel, site and data preparation, installation, testing, start-up, and user acceptance. Hardware acquisition requires purchasing, leasing, or renting computer resources from a vendor. Increasingly, companies are using service providers to acquire software, Internet access, and other IS resources.

Software can be purchased from external vendors or developed in-house—a decision termed the *make-or-buy decision*. Implementation must also address database and telecommunications systems, user preparation, and IS personnel requirements. User preparation involves readying managers, employees, and other users for the new system. New IS personnel might need to be hired, and users must be well trained in the system's functions. The physical site of the system must be prepared, and any existing data to be used in the new system must be converted to the new format. Hardware is installed during the implementation step. Testing includes program (unit) testing, systems testing, volume testing, integration testing, and acceptance testing.

There are a number of different start-up approaches. Direct conversion (also called *plunge* or *direct cutover*) involves stopping the old system and starting the new system on a given date. With the phase-in approach, components of the new system are slowly phased in while components of the old one are slowly phased out. Pilot start-up involves running the new system for one group of users rather than all users. Parallel start-up involves running both the old and new systems for a period of time. The final step of implementation is user acceptance.

Principle

Maintenance and review add to the useful life of a system but can consume large amounts of resources.

Systems operation is the use of a new or modified system. Systems maintenance involves checking, changing, and enhancing the system to make it more useful in obtaining user and organizational goals. Some major reasons for maintenance are changes in business processes; new requests from stakeholders, users, and managers; bugs or errors in the program; technical and hardware problems; corporate mergers and acquisitions; government regulations; change in the operating system or hardware; and unexpected events, such as terrorist attacks.

Systems review is the process of analyzing systems to ensure that they are operating as intended. The two types of review procedures are event-driven review and time-driven review. An event-driven review is triggered by a problem or opportunity. A time-driven review is started after a specified amount of time.

CHAPTER 8: SELF-ASSESSMENT TEST

Effective systems development requires a team effort of stakeholders, users, managers, systems development specialists, and various support personnel, and it starts with careful planning.

1. _____ is the activity of creating or modifying existing business systems. It refers to all aspects of the process—from identifying problems to be solved or opportunities to be exploited to the implementation and refinement of the chosen solution.

2. Which of the following individuals ultimately benefit from a systems development project?

    a. computer programmers
    b. systems analysts
    c. stakeholders
    d. senior-level managers

3. Like a contractor constructing a new building or renovating an existing one, the programmer takes the plans from the systems analyst and builds or modifies the necessary software. True or False?

Systems development often uses different approaches and tools such as traditional development, prototyping, rapid application development, end-user development, computer-aided software engineering, and object-oriented development to select, implement, and monitor projects.

4. Joint application development employs tools, techniques, and methods designed to speed application development. True or False?

5. _____ takes an iterative approach to the systems development process. During each iteration, requirements and alternative solutions to the problem are identified and analyzed, new solutions are designed, and a portion of the system is implemented.

Systems development starts with investigation and analysis of existing systems.

6. Feasibility analysis is typically done during which systems development stage?

    a. investigation
    b. analysis
    c. design
    d. implementation

7. Data modeling is most often accomplished through the use of _____, and activity modeling is often accomplished through the use of _____.

Designing new systems or modifying existing ones should always be aimed at helping an organization achieve its goals.

8. Determining the needed hardware and software for a new system is an example of _____.

    a. logical design
    b. physical design
    c. interactive design
    d. object-oriented design

9. The _____ often results in a formal bid that is used to determine who gets a contract for designing new or modifying existing systems. It specifies in detail the required resources such as hardware and software.

The primary emphasis of systems implementation is to make sure that the right information is delivered to the right person in the right format at the right time.

10. Software can be purchased from external developers or developed in-house. This decision is often called the _____ decision.

11. The phase-in approach to conversion involves running both the old system and the new system for three months or longer. True or False?

Maintenance and review add to the useful life of a system but can consume large amounts of resources.

12. A systems review that is caused by a problem with an existing system is called _____.

    a. object review
    b. structured review
    c. event-driven review
    d. critical factors review

13. Monitoring a system after it has been implemented to make it more useful in achieving user and organizational goals is called _____.

CHAPTER 8: SELF-ASSESSMENT TEST ANSWERS

(1) Systems development (2) c (3) True (4) False (5) Prototyping (6) a (7) entity-relationship (ER) diagrams, data-flow diagrams (8) b (9) request for proposal (RFP) (10) make-or-buy (11) False (12) c (13) systems maintenance

REVIEW QUESTIONS

1. What is an information system stakeholder?
2. What is the goal of information systems planning? What steps are involved in IS planning?
3. What are the steps of the traditional systems development life cycle?
4. What is the difference between systems investigation and systems analysis? Why is it important to identify and remove errors early in the systems development life cycle?
5. List the different types of feasibility.
6. What is the purpose of systems analysis?
7. How does the JAD technique support the RAD systems development life cycle?
8. What is the purpose of systems design?
9. What are the steps of object-oriented systems development?
10. What is an RFP? What is typically included in one? How is it used?
11. What is systems operation?
12. What activities go on during the user preparation phase of systems implementation?
13. Give three examples of a computer system vendor.
14. What are the financial options of acquiring hardware?
15. What are some of the reasons for program maintenance?

DISCUSSION QUESTIONS

1. Why is it important for business managers to have a basic understanding of the systems development process?
2. Briefly describe the role of a system user in the systems investigation and systems analysis stages of a project.
3. For what types of systems development projects might prototyping be especially useful? What are the characteristics of a system developed with a prototyping technique?
4. Imagine that your firm has never developed an information systems plan. What sort of issues between the business functions and IS organization might exist?
5. Assume that you are responsible for a new payroll program. What steps would you take to ensure a high-quality payroll system?
6. Briefly describe when you would use the object-oriented approach to systems development instead of the traditional systems development life cycle.
7. How important are communications skills to IS personnel? Consider this statement: "IS personnel need a combination of skills—one-third technical skills, one-third business skills, and one-third communications skills." Do you think this is true? How would this affect the training of IS personnel?
8. Imagine that you are a highly paid consultant who has been retained to evaluate an organization's systems development processes. With whom would you meet? How would you make your assessment?
9. You are a senior manager of a functional area in which a critical system is being developed. How can you safeguard this project from mushrooming out of control?
10. Assume that you are the owner of a company that is about to start marketing and selling bicycles over the Internet. Describe your top three objectives in developing a new Web site for this systems development project.
11. Assume that you want to start a new video-rental business for students at your college or university. Go through logical design for a new information system to help you keep track of the videos in your inventory.
12. Identify some of the advantages and disadvantages of purchasing versus leasing hardware.
13. Identify the various forms of testing used. Why are there so many different types of tests?
14. What is the goal of conducting a systems review? What factors need to be considered during systems review?
15. How would you go about evaluating a software vendor?
16. Assume that you have a personal computer that is several years old. Describe the steps you would use to perform a systems review to determine whether you should acquire a new PC.

PROBLEM-SOLVING EXERCISES

1. You are developing a new information system for The Fitness Center, a company that has five fitness centers in your metropolitan area, with about 650 members and 30 employees in each location. This system will be used by

both members and fitness consultants to track participation in various fitness activities, such as free weights, volleyball, swimming, stair climbers, and yoga and aerobic classes. One of the performance objectives of the system is that it helps members plan a fitness program to meet their particular needs. The primary purpose of this system, as envisioned by the director of marketing, is to assist The Fitness Center in obtaining a competitive advantage over other fitness clubs.

Use word processing software to prepare a brief memo to the required participants in the development team for this systems development project. Be sure to specify what roles these individuals will play and what types of information you hope to obtain from them. Assume that the relational database model will be the basis for building this system. Use a database management system to define the various tables that will make up the database.

2. You have been hired to develop a new computer system for a video-rental business using the object-oriented approach.

Using a graphics program, develop a use case diagram for the business.

3. You are considering purchasing a new PC. Using a database program, create a table titled "PC" that includes columns on all the important costs, including all hardware, software, Internet, printer and ink, and other costs. The primary key should be an order number. There should be a separate row for each vendor. Your database should have at least four vendors. Create a second table titled "VENDOR." This table should include the order number, sales representative, vendor name, vendor phone number, vendor address, and related information for each vendor. Using the database program, create a report that selects the PC system that minimizes total costs. The report should include all the information about the new computer in the PC table and the sales representative, phone number, and vendor name from the VENDOR table.

TEAM ACTIVITIES

1. Systems development is more of an art and less of a science, with a wide variety of approaches in how companies perform this activity. You and the members of your team are to interview members of an IS organization's development group. List the steps that the IS group uses in developing a new system or modifying an existing one. How does the organization's approach compare with the techniques discussed in the chapter, including the traditional systems development life cycle and the object-oriented approach? Prepare a short report on your findings.

2. Your team has been hired to analyze the potential of developing a database of job openings and descriptions for the

companies visiting your campus this year. Describe the tasks your team would perform to complete systems analysis.

3. Assume you work for a medium-sized company that trades treasury bonds in New York. Your firm has 500 employees in a downtown location. The firm is considering the purchase of a local area network (hardware and software) that is tied into a global trading network with other firms. Develop a brief request for proposal (RFP) using your word processing program for this new LAN system.

WEB EXERCISES

1. Use the Internet to find two different companies that have recently implemented a new information system. Describe the specific steps the companies used. You might be asked to develop a report or send an e-mail message to your instructor about what you found.

2. Locate a company on the Internet that sells products, such as books or clothes. Write a report describing the strengths and weaknesses of the Web pages you encountered. In your opinion, what are the most important steps of the systems development process that could be used to improve the Internet site?

CAREER EXERCISES

1. Pick a career that you are considering. What type of information system would help you on the job? Perform technical, economic, legal, operational, and schedule feasibility for an information system you would like developed for you.

2. Describe the type of information system you would like to have on the job. Your description should include logical and physical design.

VIDEO QUESTIONS

Watch the video clip **Design Your Own Video Game** and answer these questions:

1. Because of his enthusiasm for computer video games, 16-year-old Maneesh Sethi began programming in grade school. What might this story suggest as to trends in computer programming? Might computer programming some day become part of our elementary school's curriculum? Why or why not?

2. According to the video clip, what skills are required for developing computer video games besides computer programming?

CASE STUDIES

Case One

PepsiCo Implements New Procurement System to Minimize Costs

PepsiCo is a world leader in manufacturing convenience foods and beverages, with annual revenues of about $25 billion and more than 142,000 employees. The company consists of the snack businesses of Frito-Lay North America and Frito-Lay International; the beverage businesses of Pepsi-Cola North America, Gatorade/Tropicana North America, and PepsiCo Beverages International; and Quaker Foods North America, manufacturer and marketer of ready-to-eat cereals and other food products. PepsiCo brands are available in nearly 200 countries and territories.

The volume of supplies and ingredients purchased by PepsiCo's Frito-Lay division alone is huge. The company purchases raw materials from hundreds of vendors, which deal in everything from ingredients for potato chips to office products. To manage procurement processes across its divisions, PepsiCo implemented a system it named Purchase to Pay. Purchase to Pay tracks a variety of processes from product purchases to procurement management to vendor selection and payment.

In the company's ongoing mission to reduce costs, PepsiCo turned to the Purchase to Pay system to see whether it could assist staff in negotiating the best deal with suppliers. The investigation uncovered a considerable amount of waste in the procurement process. PepsiCo was not getting good deals on supply purchases and sometimes was being overcharged. PepsiCo needed to improve the system to control its spending. The company's IS staff set out to develop a standardized system to allow them to better track and analyze purchases. PepsiCo wanted to negotiate volume discounts with vendors and control individual, or "maverick," purchases that were above negotiated prices.

In exploring existing procurement solutions from vendors, PepsiCo decided on a system from BusinessObjects Corporation. That vendor's system would store procurement data in a database and provide business intelligence information through a Web-based user interface over the corporate intranet. "Ease of use, scalability, and support are some of the reasons why we chose the BusinessObjects solutions to be an integral part of PepsiCo's Business Intelligence strategy," said Tien Nguyen, vice president of application services at PepsiCo Business Solutions Group. "The real value in Purchase to Pay comes in the ability to analyze our spending patterns and identify cost saving opportunities," explains Yelak Biru, Business Intelligence and Integration team member. "BusinessObjects is the ideal solution for this, and we can replicate the benefits as we extend it across our corporation."

One example of how BusinessObjects has improved the Purchase to Pay system can be found in PepsiCo's raw materials payment system. PepsiCo typically pays vendors upon receipt of goods. During any month, the company may receive multiple deliveries from a vendor and will cut a check for each of those deliveries. Using the BusinessObject solution, PepsiCo can make one monthly payment that provides vendors with an itemized statement detailing each bill of lading, invoice number, the amount of each check, and the grand total, all via an extranet.

PepsiCo has minimized its expenses by streamlining the Purchase to Pay process. The company anticipates a savings of more than $10 million in the system's first year of operation and more than $43 million over the next three years. Within the next few years, the majority of business intelligence reporting will be done using solutions from BusinessObjects. As users learn the system and as PepsiCo develops more reports, users are expected to experiment with ad hoc queries and dig deeper into the data.

PepsiCo's new and improved Purchase to Pay system is a perfect example of the benefits of continuous improvement through the systems development process. By implementing a new system that is flexible and scalable, PepsiCo has simplified and improved its procurement processes both now and for the future as it changes to meet new challenges.

Discussion Questions

1. What stage of the systems development life cycle led PepsiCo to discover the need for improvement in its previous Purchase to Pay system?
2. How was the previous Purchase to Pay system not aligned with the goals of PepsiCo's organization?

Critical Thinking Questions

3. What are the benefits of implementing a system that allows you to custom design reports and experiment with ad hoc queries?
4. Relate the characteristics of valuable data (provided in Chapter 1, Table 1.2) to the information provided by, and the method of delivery used by, PepsiCo's new system.

SOURCES: "Customers in the Spotlight: PepsiCo," accessed June 19, 2004, *www.businessobjects.com/customers/spotlight/pepsico.asp*, WebIntelligence Web site, accessed June 19, 2004, *www.businessobjects.com/products/query analysis/webi.asp*; PepsiCo Web site, accessed June 19, 2004, *www. pepsico.com.*

Case Two

Segway Stays Light and Nimble with Outsourced Systems

Dean Kamen holds more than 150 U.S. and foreign patents related to medical devices, climate control systems, and helicopter design. In 2001, he developed a business to manufacture and market a "human transport" device that he believed would revolutionize travel. The Segway, Kamen declared, "would be to the car what the car was to the horse and buggy."

Although the device has been a bit slower in taking off than Kamen had hoped, it has captured the attention of the media and transportation industry analysts. The two-wheeled, Segway Human Transporter (HT) employs a unique patented "dynamic stabilization" technology.

With the assistance of optimistic investors, Segway hired a seasoned CIO to design the information systems for the new company. Patrick Zilvitis, who was CIO at The Gillette Co. before leaving to work part-time for Segway, decided that this unique company required a unique approach to information management. Zilvitis said in an interview that when he joined the company in fall 2000, he quickly decided that outsourcing would be the wisest path for a start-up that had a minimal IT infrastructure. He thought that would be the best way to hold down its technology costs.

Zilvitis thought that it would be a costly mistake for Segway to build a traditional data center and hire a big dedicated IT staff—a mistake that would slow the company's growth. Instead, he decided Segway should use software under a hosted environment that could grow with the company's needs. Outsourcing "allows us to upsize or downsize our IT infrastructure as needed," noted Scott Frock, Segway's director of finance. "For a small-to-medium-size company, there are a lot of advantages." Outsourcing also lets companies like Segway avoid up-front investments in servers, software, and technical support staffers, said Terry Jost, a Dallas-based consultant at Cap Gemini Ernst & Young LLP.

Before Zilvitis came on board, Segway had been using Intuit Inc.'s QuickBooks accounting software. To support the company's expansion, Zilvitis believed that it should upgrade to either an accounting package designed for small and mid-sized businesses or a larger system that it could grow into. The company decided to look for a larger system to avoid a "painful and expensive" conversion down the road.

After evaluating software from Oracle and SAP AG, the company chose an uncustomized version of an Oracle suite geared to small manufacturers. The system includes manufacturing and order-management modules in addition to the finance applications. Workers at Segway's headquarters and its manufacturing plant access the software via Windows 2000 PCs.

In addition, Zilvitis chose a finance system hosted by Fremont, California–based Appshop Inc., which manages the Oracle applications from a Sprint Corporation data center in Denver. So, Segway's information system includes components supplied by three different vendors in three different parts of the country.

Three years after the system's implementation, Segway is reaping the benefits of the larger Oracle system. The company recently upgraded to a newer version of its Oracle system and added software that will let workers at its customer service partner, Frazer, Pennsylvania–based

Decision-One, access product warranty data and other information via Web browsers. The Oracle system has made the integration and expansion a breeze to implement.

Maintaining data and systems off-site has provided some unanticipated telecommunications costs and savings. To speed the data from the application service provider to Segway's headquarters and manufacturing plant, Segway replaced its private branch exchange switches with Voice-over-Internet-Protocol (VoIP) equipment costing "considerably less than $20,000," Zilvitis said. He added that the transition to VoIP has provided several benefits, including the ability to set up low-cost switchboard extensions for remote employees, extra voice and data bandwidth for future growth, and cheap yet reliable connections to Sprint's data center. For instance, Segway's Internet connection to Denver costs the company $1,000 to $2,000 per month, compared with the monthly tab for a T-1 connection that would have been $10,000 to $12,000, he said.

By outsourcing its systems, Segway has been able to focus on its product rather than data centers, servers, and an extensive IT staff. Instead, the company relies on one part-time veteran CIO to manage its outsourced resources. Because the future of the company is uncertain, it makes sense for Segway to select a solution that is flexible, nimble, and can turn on a dime—just like its product.

Discussion Questions

1. What benefits is Segway enjoying by outsourcing its information infrastructure and services?
2. What benefits can a company gain by managing its own information infrastructure and services?

Critical Thinking Questions

3. What factors might influence a company to outsource its IT infrastructure and services?
4. What frustrations do you think Segway endures in dealing with outsourced vendors?

SOURCES: Thomas Hoffman, "Segway's Tech Plans Look Down the Road to Growth," *Computerworld*, January 26, 2004, *www.computerworld.com*; Segway's Web site, *www.segway.com*, accessed June 19, 2004; "A Long Road Ahead of It," *The Economist*, June 12, 2004, *www.lexis-nexis.com*.

NOTES

Sources for the opening vignette: "SAP Customer Success Story: BMW Group," accessed June 18, 2004, *www.sap.com/solutions/industry/automotive/customersuccess*; BMW Group Web site, accessed June 18, 2004, *www.bmwgroup.com*; mySAP Automotive Web site, accessed June 18, 2004, *www.sap.com/solutions/industry/automotive*.

1. Hamblen, Matt, "NavSea's ROI Ship Comes in," *Computerworld*, March 24, 2003, p. 46.
2. Weiss, Todd, "New CIO Takes Reins of IRS Tech Upgrade," *Computerworld*, June 2, 2003, p. 14.
3. Hoffman, Thomas, "The Resourceful Project Manager," *Computerworld*, February 6, 2004, p. 35.
4. Carter, Pamela, "The Management of Meaning: Toward a New Model of Project Management," *Research Colloquium, Information and Management Science Department, College of Business, Florida State University*, September 21, 2003.
5. Sliwa, Carol, "Rational Software Set to Roll Out Rapid Development Tool," *Computerworld*, May 19, 2003. p. 7.
6. Anthes, Gary, "Sabre Takes Extreme Measures," *Computerworld*, March 29, 2004, p. 28.
7. Marks, Debra, "Teenage Web Wiz Aspires to Create a Better Browser," *The Wall Street Journal*, May 14, 2003, p. B5.
8. Rappa, M. A. "The Utility Business Model and the Future of Computing Services," *IBM Systems Journal*, Volume 43, No. 1, 2004, p. 32.
9. Sliwa, Carol, "Sears Plans to Outsource Part of IT Infrastructure," *Computerworld*, January 19, 2004, p. 1.
10. Schroeder, Michael, "Business Coalition Battles Outsourcing Backlash," *The Wall Street Journal*, March 1, 2004, p. A1.
11. Brewin, Bob, "WorldCom Wins $20M Bid to Build Baghdad Cell Network," *Computerworld*, May 26, 2003, p. 14.
12. Staff, "On Demand Vision," *www.IBM.com*, accessed on January 25, 2004.
13. Phillips, Michael, "More Work Is Outsourced to U.S. Than Away from It, Data Shows," *The Wall Street Journal*, March 15, 2004, p. A2.
14. Buckman, Rebecca, "HP Outsourcing: Beyond China," *The Wall Street Journal*, February 23, 2004, p. A14.
15. IBM Web page at *www-IBM.com.services/strategies*, accessed on July 29, 2003.
16. EDS Web page at *www.eds.com*, accessed on July 29, 2003.
17. Accenture Web page at *www.accenture.com*, accessed on July 29, 2003.
18. Liu, Xiaohua, "Multiuser Collaborative Work in a Virtual Environment Based CASE Tool," *Information and Software Technology*, April 1, 2003, p. 253.
19. Lehmann, Hans, "An Object-Oriented Architecture Model for International Information Systems," *Journal of Global Information Management*, July 2003, p. 1.
20. Vijayan, Jaikumar, "Best in Class: Application Framework Allows Easy Portal Access," *Computerworld*, February 24, 2003, p. 51.
21. Karnitsching, Matthew, "Autobahn Plan Is Stuck in Slow Lane," *The Wall Street Journal*, October 21, 2003, p. A11.
22. Clark, Don, "The Affluent Avatar," *The Wall Street Journal*, January 8, 2003, p. B1.
23. Melymuka, Kathleen, "How Will You Manage Your Vendors," *Computerworld*, January 6, 2003, p. 32.
24. Mearian, Lucas, "Fidelity National Revamps IT with Single-Vendor Track," *Computerworld*, March 8, 2004, p. 18.
25. Brewin, Bob, "UPS Invests $30 Million in IT to Speed Package Delivery," *Computerworld*, September 29, 2003, p. 14.
26. Murphy, Victoria, "Control Freak," *Forbes*, March 29, 2004, p. 79.
27. McWilliams, Gary, "Pay as You Go," *The Wall Street Journal*, March 31, 2003, p. R8.
28. Hoffman, Thomas, "HP Takes New Pricing Path," *Computerworld*, May 26, 2003, p. 1.
29. Staff, "Alaska Airlines Selects ITA Software," *Telecomworldwire*, January 22, 2004.

30. Songini, Marc, "Coke, SAP Codevelop Bottling App," *Computerworld*, February 23, 2004, p. 8.

31. Clark, Don, "Renting Software Online," *The Wall Street Journal*, June 3, 2003, p. B1.

32. Mearian, Lucas, "Nasdaq's CIO Looks to Streamline Systems," *Computerworld*, June 2, 2003, p. 19.

33. Weiss, Todd, "New CIO Takes Reins of IRS Tech Upgrade," *Computerworld*, June 2, 2003, p. 14.

34. Songini, Marc, "Best in Class: Re-engineering Drives Down Cost of Power," *Computerworld*, February 24, 2003, p. 46.

35. Brewin, Bob, "Carnival Cruise Lines Piloting Remote Management of PCs," *Computerworld*, August 18, 2003, p. 7.

36. Scheier, Robert, "Surviving Software Upgrades," *Computerworld*, May 26, 2003, p. 44.

37. Ganek, A.G. et al., "The Dawning of the Autonomic Computing Era," *IBM Systems Journal*, Volume 42, 2003, p. 5.

38. Purushothaman, D. et al., "Branch and Price Methods for Prescribing Profitable Upgrades of High Technology Products with Stochastic Demand," *Decision Sciences*, Winter 2004, p. 55.

39. Staff, "Hymans Selects CMS, Net Software," *Pensions Week*, January 26, 2004.

CHAPTER

· 9 ·

Security, Privacy, and Ethical Issues in Information Systems and the Internet

<table>
<tr><th>PRINCIPLES</th><th>LEARNING OBJECTIVES</th></tr>
<tr>
<td>

- Policies and procedures must be established to avoid computer waste and mistakes.

</td>
<td>

- Describe some examples of waste and mistakes in an IS environment, their causes, and possible solutions.
- Identify policies and procedures useful in eliminating waste and mistakes.

</td>
</tr>
<tr>
<td>

- Computer crime is a serious and rapidly growing area of concern requiring management attention.

</td>
<td>

- Explain the types and effects of computer crime.
- Identify specific measures to prevent computer crime.
- Discuss the principles and limits of an individual's right to privacy.

</td>
</tr>
<tr>
<td>

- Working conditions must be designed to avoid negative ethical consequences.

</td>
<td>

- Outline criteria for the ethical use of information systems.

</td>
</tr>
</table>

INFORMATION SYSTEMS IN THE GLOBAL ECONOMY
COMPUTER ASSISTED PASSENGER PRESCREENING SYSTEM, UNITED STATES

Data Privacy Concerns Ground Security System

Since the terrorist attacks of 9/11, the U.S. government and the airline industry have been urgently searching for ways to identify and thwart potential attacks before they occur. Unfortunately, these initiatives give the airlines conflicting goals—their duty and desire to help prevent terrorism and their need to maintain the privacy of customer data. The Computer Assisted Passenger Prescreening System (CAPPS) was one such initiative that highlights the dilemma.

The original CAPPS system relied on the airlines' own reservation systems to check passenger information against a government-supplied watch list. Under the proposed CAPPS II system, the passenger-screening process would be turned over to the federal government. Passengers' identities would be authenticated by matching airline passenger data such as name, address, phone number, and birth date against a Transportation Security Administration (TSA) database. Passengers would then be checked against both a federal terrorism database and lists of individuals who have outstanding warrants for violent crimes. CAPPS II would ultimately assign each airline passenger a threat level. Threat-level data would be deleted for most passengers once they reached their destinations. Data for travelers deemed high risk would be retained for an unspecified length of time. Details of exactly how the TSA would decide whether a passenger should be allowed to board a plane or questioned were never revealed.

The Air Transport Association, the trade organization for the major U.S. airlines, estimated that the new system could cost the airline industry $1 billion to change their reservation systems to provide the data required by CAPPS II. In addition to costs, the air carriers faced many other thorny issues, not the least of which was maintaining the privacy of their passengers. Precautions to protect passenger privacy would have included installing private networks between the TSA and the airlines that would pass only encrypted data, requiring the data to pass through a multitier firewall before entering the TSA system, and implementing a 24-hour audit trail that documents all access to data.

The airlines would not voluntarily turn over the data needed to implement CAPPS II, however. Although the airlines' Air Transport Association supported the concept of CAPPS II, its members wanted more privacy guarantees before they supplied data for any purpose. They wanted assurances that the TSA-collected information would pertain only to aviation security, that the information would be securely stored, that it would be destroyed as soon as travel is completed, and that passengers could access their own data and correct any errors. Airlines were especially concerned with "mission creep," in which information intended for one purpose is used for another.

The Senate Governmental Affairs Committee learned in June 2004 that at least eight airlines and airline-reservation services provided passenger data to contractors building CAPPS II for the TSA. The contractors acquired information about passengers who had reserved tickets on various airlines and who booked flights through online travel Web sites. In providing this information to the government agency, the contractors violated the Privacy Act of 1974 by not notifying the public of what type of information their screening systems would collect and how individual passengers could find out whether their data was included in the test systems. The airline industry is now facing expensive customer class action lawsuits as a result.

In July 2004, then Homeland Security Secretary Tom Ridge cited data privacy issues and system-interoperability issues as the basis for stopping further work on the CAPPS II system. Just two months later, though, the Secure Flight passenger-screening system was announced. This system would compare passenger data with watch lists held in the Terrorist Screening Database in an attempt to keep suspicious passengers from boarding domestic flights.

As you read this chapter, consider the following:

- Which issue should take precedence—maintaining the physical safety of travelers or protecting their rights?
- What actions can organizations take when the government requires them to do activities that will upset or inconvenience their customers?

Why Learn About Security, Privacy, and Ethical Issues in Information Systems and the Internet?

A wide range of "nontechnical" issues associated with the use of information systems and the Internet provide both opportunities and threats to modern organizations. The issues span the full spectrum—from preventing computer waste and mistakes, to avoiding violations of privacy, to complying with laws on collecting data about customers, to monitoring employees. If you become a member of a human resource, information systems, or legal department within an organization, you will likely be charged with leading the rest of the organization in dealing with these and other issues covered in this chapter. Also, as a user of information systems and the Internet, it is in your own self-interest to become well versed on these issues. You need to know about the topics in this chapter to help avoid or recover from crime, fraud, privacy invasion, and other potential problems. We begin with a discussion of preventing computer waste and mistakes.

Earlier chapters detailed the amazing benefits of computer-based information systems in business, including increased profits, superior goods and services, and higher quality of work life. Computers have become such valuable tools that today's businesspeople would have difficulty imagining work without them. Yet the information age has also brought some potential problems for workers, companies, and society in general (see Table 9.1).

Table 9.1

Social Issues in Information Systems

• Computer waste and mistakes	• Health concerns
• Computer crime	• Ethical issues
• Privacy	• Patent and copyright violations

To a large extent, this book has focused on the solutions—not the thorny issues—presented by information systems. In this chapter we discuss some of the issues as a reminder of the social and ethical considerations underlying the use of computer-based information systems. No business organization, and hence no information system, operates in a vacuum. All IS professionals, managers, and users have a responsibility to see that the potential consequences of IS use are fully considered.

Managers and users at all levels play a major role in helping organizations achieve the positive benefits of IS. These individuals must also take the lead in helping to minimize or eliminate the negative consequences of poorly designed and improperly utilized information systems. For managers and users to have such an influence, they must be properly educated.

Many of the issues presented in this chapter, for example, should cause you to think back to some of the systems design and systems control issues we have already discussed. They should also help you look forward to how these issues and your choices might affect your future IS management considerations.

COMPUTER WASTE AND MISTAKES

Computer-related waste and mistakes are major causes of computer problems, contributing as they do to unnecessarily high costs and lost profits. Computer waste involves the inappropriate use of computer technology and resources. Computer-related mistakes refer to errors, failures, and other computer problems that make computer output incorrect or not useful, caused mostly by human error. In this section we explore the damage that can be done as a result of computer waste and mistakes.

Computer Waste

The U.S. government is the largest single user of information systems in the world. It should come as no surprise then that it is also perhaps the largest misuser. The government is not unique in this regard—the same type of waste and misuse found in the public sector also exists in the private sector. Some companies discard old software and even complete computer systems when they still have value. Others waste corporate resources to build and maintain complex systems never used to their fullest extent. A less-dramatic, yet still relevant, example of waste is the amount of company time and money employees may waste playing computer games, sending unimportant e-mail, or accessing the Internet. Junk e-mail, also called *spam*, and junk faxes also cause waste. Read the "Ethical and Societal Issues" special feature to find out about more about efforts to slow the spread of spam.

Computer-Related Mistakes

Despite many people's distrust, computers themselves rarely make mistakes. Even the most sophisticated hardware cannot produce meaningful output if users do not follow proper procedures. Mistakes can be caused by unclear expectations and a lack of feedback. Or a programmer might develop a program that contains errors. In other cases, a data-entry clerk might enter the wrong data. Unless errors are caught early and prevented, the speed of computers can intensify mistakes. As information technology becomes faster, more complex, and more powerful, organizations and individuals face increased risks of experiencing the results of computer-related mistakes. Take, for example, these cases from recent news.

Since the disintegration of the space shuttle Columbia in February 2003 put NASA manned space flights on hold, the Russian Soyuz capsules have been the linchpin of the international space station's supply program. A May 2003 space mission ended in a wild ride, with the American and Russian crew going some 250 miles off course due to a computer error. A computer malfunction sent the capsule's occupants on such a steep reentry trajectory that their tongues rolled back in their mouths. Indeed, the landing was so far off target that more than two gut-wrenching hours passed before the recovery team knew the men were safe.[1]

A Japanese company, Catena Corporation, was deluged with thousands of orders for more than 100 million Apple eMac PCs after a glitch caused the computers to be listed on an online shopping site for a price of $25.45. The company said a code number assigned to a set of five 8X DVD-R discs, which were the products it was intending to sell, was sent to Yahoo! Japan; however, a product information database matched that code with details for the eMac computer and with the price for the DVDs. The result was a listing for the computer, part number M9461J/A, at a price of $25 rather than the usual price of more than $916.[2]

CAN-SPAM: Deterrent or Accelerant?

Every day, millions of people worldwide receive dozens of unsolicited commercial e-mails, known popularly as spam. An individual spam e-mail may be sent to a distribution list containing millions of addresses, with the sender expecting that only a tiny number of readers will respond to the offer. For some, spam represents a minor annoyance, but many of us become so overwhelmed with spam that we are forced to switch e-mail addresses. The Yankee Group estimates that some two-thirds of e-mail qualifies as spam. The Radicati Group estimates spam costs an employer of 10,000 people nearly $500,000 just for the additional e-mail servers required to handle the load.

Spammers often take advantage of Internet technologies to conceal their identities and the location of their operations. They even resort to including the e-mail addresses of innocent third parties in the reply-to addresses of their unwanted messages or simply forging e-mail headers. Known as *spoofing*, this tactic is employed by spammers to inflict the burden of bounce-back messages (generated when spam is sent to a nonworking address) on someone other than their mail provider. The result for the recipients is effectively a denial-of-service attack, as their e-mail in-boxes become overwhelmed with tens of thousands of messages.

The Controlling the Assault of Non-Solicited Pornography and Marketing Act (CAN-SPAM), which went into effect in January 2004, does not outlaw spam; it simply prohibits certain practices. Outlawed practices include using false or misleading transmission information, deceptive subject headings, and automated methods of registering multiple e-mail accounts for spamming. Senders of unsolicited commercial e-mail also must include their physical address and a way for recipients to opt out of future mailings. In addition, companies must update their marketing databases to note whether a person has opted out of solicitations. Companies that rent lists from third parties can run afoul of the law if the list lacks time-date stamps to document an individual's decision to receive messages. Although many spammers use computers outside the United States to send the spam, they can still be prosecuted under the CAN-SPAM law because their spam causes damage to U.S. Internet service providers and consumers.

America Online, EarthLink, Microsoft, and Yahoo! filed civil actions under the law in February 2004 against four individuals for sending e-mail pitches for weight-loss products that don't work and that violated the CAN-SPAM Act. The accused face up to five years in prison for illegal spamming and up to an additional 20 years for mail fraud. The alleged spammers should at least be able to afford excellent legal counsel, because it is estimated that they grossed an average of $100,000 per month from August 2003 to January 2004. Scott Richter has been called the Spam King, labeled one of the most prolific spammers in the world, and been sued for spamming by both Microsoft and the New York State attorney general. His company, OptInRealBig.com, sends more than 100 million e-mail messages every day. While it is still too soon to tell, some believe that the CAN-SPAM Act may have actually increased the volume of spam. Antispam vendor Commtouch Software Ltd. estimates that only 1 percent of spam messages in January 2004 complied with CAN-SPAM. That percentage increased to 9.5 percent in May 2004. But, coupled with a continuing rise in the volume of spam in recent months, that statistic suggests that the law has increased the amount of spam by effectively defining what "legal spam" is.

Critical Thinking Questions

1. What negative impact does spam have on corporate America? Is it simply an annoyance or a serious problem?
2. Monitor each of your e-mail accounts for a week. What percentage of your e-mail messages is spam? What percentage of your spam messages appear to conform to the CAN-SPAM law?

What Would You Do?

Your legitimate small business has relied heavily on e-mail to sell fresh organic fruit and vegetables. Recently, your company's computers were taken over by a hacker and used to send spam and viruses. As a result, your company was "blacklisted" by the antispammers and now appears on a directory of alleged offenders circulated to Internet service providers, security companies, and other businesses. As a result, most of your outgoing direct marketing messages are blocked.

3. What actions can you take to reestablish the good name of your firm and get it off the blacklist?
4. What sort of action might you take to recover your firm's losses?

SOURCES: Thomas Claburn, "Spam Law Changes Game," *InformationWeek*, December 15, 2003, *www.informationweek.com*; The Associated Press, "Anti-Spam Law Goes into Force in Europe," *InformationWeek*, October 31, 2003, *www.informationweek.com*; Thomas Claburn, "War against Spam Rages On," *InformationWeek*, May 24, 2004, *www.informationweek.com*; Thomas Claburn, "U.S. Charges Four under Can-Spam Law," *InformationWeek*, April 29, 2004, *www.informationweek.com*; Thomas Claburn, "Does Can-Spam Act Lead to More Spam?" *InformationWeek*, June 3, 2004, *www.informationweek.com*; and Liane Cassavoy, "Three Minutes: The So-Called Spam King Sounds Off," *PC World*, August 2004, *www.pcworldcom*.

California Macy's stores agreed to a $1.2 million settlement for overcharging shoppers in the city of San Diego, Los Angeles County, and three other counties. Investigators found that although all the company's scanners were 100 percent accurate, consumers were being charged from a few cents to more than $10 over the advertised and shelf prices on some clothing and household items. The problem was tracked to difficulty in synchronizing the advertised prices, the prices on the shelf, and the prices in the (checkout register) computer. No one alleged that Macy's intentionally overcharged consumers, but the store "had a real pattern of inaccuracy all across the state," according to the Los Angeles County prosecuting attorney.[3]

Various errors in the code of Internet Explorer (IE) have given hackers a means to compromise personal computers using the Microsoft browser. Attackers can exploit the vulnerabilities to bypass a security check in IE or to download and execute a malicious file on a user's computer. These problems with Explorer are serious, and the vulnerabilities can enable hackers to place code in a machine and run it.[4]

PREVENTING COMPUTER-RELATED WASTE AND MISTAKES

To remain profitable in a competitive environment, organizations must use all resources wisely. Preventing computer-related waste and mistakes like those just described should therefore be a goal. Today, nearly all organizations use some type of CBIS. To employ IS resources efficiently and effectively, employees and managers alike should strive to minimize waste and mistakes. Preventing waste and mistakes involves (1) establishing, (2) implementing, (3) monitoring, and (4) reviewing effective policies and procedures. "What does CBIS stand for?"

Establishing Policies and Procedures

The first step to prevent computer-related waste is to establish policies and procedures regarding efficient acquisition, use, and disposal of systems and devices. Computers permeate organizations today, and it is critical for organizations to ensure that systems are used to their full potential. As a result, most companies have implemented stringent policies on the acquisition of computer systems and equipment, including requiring a formal justification statement before computer equipment is purchased, definition of standard computing platforms (operating system, type of computer chip, minimum amount of RAM, etc.), and the use of preferred vendors for all acquisitions.

Prevention of computer-related mistakes begins by identifying the most common types of errors, of which there are surprisingly few (see Table 9.2). To control and prevent potential problems caused by computer-related mistakes, companies have developed policies and procedures that cover the following:

- Acquisition and use of computers, with a goal of avoiding waste and mistakes
- Training programs for individuals and workgroups
- Manuals and documents on how computer systems are to be maintained and used
- Approval of certain systems and applications before they are implemented and used to ensure compatibility and cost-effectiveness
- Requirement that documentation and descriptions of certain applications be filed or submitted to a central office, including all cell formulas for spreadsheets and a description of all data elements and relationships in a database system; such standardization can ease access and use for all personnel

Once companies have planned and developed policies and procedures, they must consider how best to implement them.

Sometimes computer error combines with human procedural errors to lead to the loss of human life. In March 2003, a Patriot missile battery on the Kuwait border accidentally shot

Table 9.2

Types of Computer-Related Mistakes

- Data entry or capture errors
- Errors in computer programs
- Errors in handling files, including formatting a disk by mistake, copying an old file over a newer one, and deleting a file by mistake
- Mishandling of computer output
- Inadequate planning for and control of equipment malfunctions
- Inadequate planning for and control of environmental difficulties (electrical problems, humidity problems, etc.)
- Installing computing capacity inadequate for the level of activity on corporate Web sites
- Failure to provide access to the most current information by not adding new and deleting old URL links

down a British Royal Air Force Tornado GR-4 aircraft that was returning from a mission over Iraq. Two British pilots were killed in the incident. Defense industry experts disagreed about the possibility of a software problem being solely responsible for downing a friendly aircraft. A likely scenario combined problems with the Patriot's radar with human error to result in friendly fire.[5]

Implementing Policies and Procedures

Implementing policies and procedures to minimize waste and mistakes varies according to the business conducted. Most companies develop such policies and procedures with advice from the firm's internal auditing group or its external auditing firm. The policies often focus on the implementation of source data automation and the use of data editing to ensure data accuracy and completeness, and the assignment of clear responsibility for data accuracy within each information system. Table 9.3 lists some useful policies to minimize waste and mistakes.

Table 9.3

Useful Policies to Eliminate Waste and Mistakes

- Changes to critical tables, HTML, and URLs should be tightly controlled, with all changes authorized by responsible owners and documented.
- A user manual should be available that covers operating procedures and that documents the management and control of the application.
- Each system report should indicate its general content in its title and specify the time period it covers.
- The system should have controls to prevent invalid and unreasonable data entry.
- Controls should exist to ensure that data input, HTML, and URLs are valid, applicable, and posted in the right time frame.
- Users should implement proper procedures to ensure correct input data.

Training is another key aspect of implementation. Many users are not properly trained in developing and implementing applications, and their mistakes can be very costly. Because more and more people use computers in their daily work, it is important that they understand how to use them. Training is often the key to acceptance and implementation of policies and procedures. Because of the importance of maintaining accurate data and of people understanding their responsibilities, companies converting to ERP and e-commerce systems invest weeks of training for key users of the system's various modules.

Monitoring Policies and Procedures

To ensure that users throughout an organization are following established procedures, the next step is to monitor routine practices and take corrective action if necessary. By understanding what is happening in day-to-day activities, organizations can make adjustments or develop new procedures. Many organizations implement internal audits to measure actual results against established goals, such as percentage of end-user reports produced on time, percentage of data input errors rejected, number of input transactions entered per eight-hour shift, and so on.

The passage of the Sarbanes-Oxley Act has caused many companies to monitor their policies and procedures and to plan changes in financial information systems. These changes could profoundly affect many business activities. As mentioned in Chapter 5, the act requires public companies to implement procedures to ensure that their audit committees can document underlying financial data to validate earnings reports. Companies that fail to comply could find their top execs behind bars. In October 2004, SunTrust Banks disclosed that it was restating its earnings upward for the first two quarters of 2004 and delaying its third-quarter earnings statement because of improper accounting procedures in its auto financing division. As a result of the error, SunTrust underreported earnings for the first two quarters by $17 million and $5 million, respectively. Two executives, the chief credit officer and the controller, were put on paid leave.[6]

Reviewing Policies and Procedures

The final step is to review existing policies and procedures and determine whether they are adequate. During review, people should ask the following questions:

- Do current policies cover existing practices adequately? Were any problems or opportunities uncovered during monitoring?
- Does the organization plan any new activities in the future? If so, does it need new policies or procedures on who will handle them and what must be done?
- Are contingencies and disasters covered?

Information systems professionals and users still need to be aware of the misuse of resources throughout an organization. Preventing errors and mistakes is one way to do so. Another is implementing in-house security measures and legal protections to detect and prevent a dangerous type of misuse: computer crime.

COMPUTER CRIME

Even good IS policies may not be able to predict or prevent computer crime. A computer's ability to process millions of pieces of data in less than a second can help a thief steal data worth thousands or millions of dollars. Compared with the physical dangers of robbing a bank or retail store with a gun, a computer criminal with the right equipment and know-how can steal large amounts of money from the privacy of a home. Computer crime often defies detection, the amount stolen or diverted can be substantial, and the crime is "clean" and nonviolent.

Here is a sample of recent computer crimes.

- Cyberattacks against corporations aren't under control, nor will they go away soon. The Computer Emergency Response Team (CERT) reported 137,529 security incidents in 2003, and damages attributed to Mydoom, to date the fastest and most pervasive worm attack on record, have been pegged at $40 billion worldwide.[7]
- Each year, nearly 700,000 people are victims of identity theft and other forms of computer fraud. Such a wide range of methods are used by the perpetrators of these crimes that it makes investigating them difficult. In one case, corrupt security guards at New York's Macy's department store were stopping customers as they exited the store and stealing their credit card numbers off their receipts. In another case, clerks at Bloomingdale's

connected credit card scanners to handheld devices, allowing them to surreptitiously swipe the cards and collect thousands of credit card numbers.[8] A Michigan man pleaded guilty to four counts of wire fraud and unauthorized access to a computer after he and two accomplices used a vulnerable wireless network at a Lowe's store in Michigan to attempt to steal credit card numbers from the company's main computer systems in North Carolina and other U.S. Lowe's stores.[9]

- Fourteen members of an Italian hacker group known as the Reservoir Dogs were arrested by the Italian police in what became known as Operation Rootkit. The group was responsible for a series of hacking incidents that compromised more than 1,000 systems spanning the globe, affecting the likes of NASA, the U.S. Army and Navy, and various financial companies in the United States and abroad. Some members of the group were working as information security managers in big consulting firms and Internet service providers and were even employed in Italian branches of U.S. companies.[10]

Although no one really knows how pervasive cybercrime is, the number of IT-related security incidents is increasing dramatically. The Computer Emergency Response Team Coordination Center (CERT/CC) is located at the Software Engineering Institute (SEI), a federally funded research and development center at Carnegie Mellon University in Pittsburgh, Pennsylvania. It is charged with coordinating communication among experts during computer security emergencies and helping to prevent future incidents. CERT employees study Internet security vulnerabilities, handle computer security incidents, publish security alerts, research long-term changes in networked systems, develop information and training to help organizations improve security at their sites, and conduct an ongoing public awareness campaign. The number of security problems reported to CERT increased sixfold between 2000 and 2003, as shown in Figure 9.1. As many as 60 percent of all attacks go undetected, according to security experts. What's more, of the attacks that are exposed, only an estimated 15 percent are reported to law enforcement agencies. Why? Companies don't want the bad press. Such publicity makes the job even tougher for law enforcement. Most companies that have been electronically attacked won't talk to the press. A big concern is loss of public trust and image—not to mention the fear of encouraging copycat hackers.

Figure 9.1

Number of Incidents Reported to CERT

(Source: Data from CERT Web site at *www.CERT.org/stats/#incients*, accessed on May 29, 2004.)

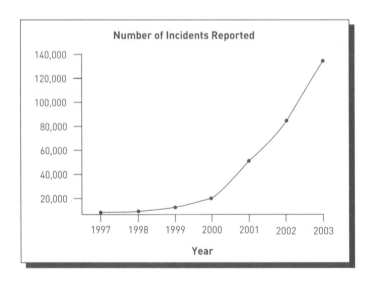

Today, computer criminals are a new breed—bolder and more creative than ever. With the increased use of the Internet, computer crime is now global. It's not just on U.S. shores that law enforcement has to battle cybercriminals. Regardless of its nonviolent image, computer crime is different only because a computer is used. It is still a crime. Part of what makes computer crime so unique and hard to combat is its dual nature—the computer can be both the tool used to commit a crime and the object of that crime.

THE COMPUTER AS A TOOL TO COMMIT CRIME

A computer can be used as a tool to gain access to valuable information and as the means to steal thousands or millions of dollars. It is, perhaps, a question of motivation—many individuals who commit computer-related crime claim they do it for the challenge, not for the money. Credit card fraud—whereby a criminal illegally gains access to another's line of credit with stolen credit card numbers—is a major concern for today's banks and financial institutions. In general, criminals need two capabilities to commit most computer crimes. First, the criminal needs to know how to gain access to the computer system. Sometimes obtaining access requires knowledge of an identification number and a password. Second, the criminal must know how to manipulate the system to produce the desired result. Frequently, a critical computer password has been talked out of an individual, a practice called **social engineering**. Or, the attackers simply go through the garbage—**dumpster diving**—for important pieces of information that can help crack the computers or convince someone at the company to give them more access. In addition, there are more than 2,000 Web sites that offer the digital tools—for free—that will let people snoop, crash computers, hijack control of a machine, or retrieve a copy of every keystroke.

Also, with today's sophisticated desktop publishing programs and high-quality printers, crimes involving counterfeit money, bank checks, traveler's checks, and stock and bond certificates are on the rise. As a result, the U.S. Treasury Department redesigned and printed new currency that is much more difficult to counterfeit.

Cyberterrorism

Government officials and IS security specialists have documented a significant increase in Internet probes and server scans since early 2001. There is a growing concern among federal officials that such intrusions are part of an organized effort by cyberterrorists, foreign intelligence services, or other groups to map potential security holes in critical systems. A **cyberterrorist** is someone who intimidates or coerces a government or organization to advance his or her political or social objectives by launching computer-based attacks against computers, networks, and the information stored on them.

Even before the September 11, 2001, terrorist attacks, the U.S. government considered the potential threat of cyberterrorism serious enough that it established the National Infrastructure Protection Center in February 1998. This function was transferred to the Homeland Security Department's Information Analysis and Infrastructure Protection Directorate to serve as a focal point for threat assessment, warning, investigation, and response for threats or attacks against our country's critical infrastructure, which provides telecommunications, energy, banking and finance, water systems, government operations, and emergency services. Successful cyberattacks against the facilities that provide these services could cause widespread and massive disruptions to the normal function of our society.

Identity Theft

Identity theft is a crime in which an imposter obtains key pieces of personal identification information, such as Social Security or driver's license numbers, in order to impersonate someone else. The information is then used to obtain credit, merchandise, and services in the name of the victim or to provide the thief with false credentials.

In some cases, the identity thief uses personal information to open new credit accounts, establish cellular phone service, or open a new checking account to obtain blank checks. In other cases, the identity thief uses personal information to gain access to the person's existing accounts. Typically, the thief will change the mailing address on an account and run up a huge bill before the person whose identity has been stolen realizes there is a problem. The Internet has made it easier for an identity thief to use the stolen information because transactions can be made without any personal interaction.

social engineering
The practice of talking a critical computer password out of an individual.

dumpster diving
Searching through the garbage for important pieces of information that can help crack an organization's computers or be used to convince someone at the company to give someone access to the computers.

cyberterrorist
Someone who intimidates or coerces a government or organization to advance his or her political or social objectives by launching computer-based attacks against computers, networks, and the information stored on them.

identity theft
A crime in which an imposter obtains key pieces of personal identification information, such as Social Security or driver's license numbers, in order to impersonate someone else. The information is then used to obtain credit, merchandise, and services in the name of the victim or to provide the thief with false credentials.

The Federal Trade Commission received 516,740 identity-theft complaints in 2003, up from 404,000 in 2002. An FTC report issued that year estimates that more than 27 million Americans have been victims of identity theft during the past five years.[11]

According to studies at Michigan State University's identity-theft research center, at least half of identity theft now results from the theft of personal information stored on business databases. Potentially as much as 70 percent of identity thefts originate in the workplace by employees or people impersonating employees, so the majority of identity thefts can be considered inside jobs. The research also showed that identities were stolen most often from healthcare-related institutions and then from financial institutions.[12]

A high-profile case of insider identity theft broke in late 2002, when the Department of Justice charged a help desk worker at financial data company Teledata Communications with fraud and conspiracy in connection with an identity-theft scheme involving more than 30,000 victims. The worker allegedly used his insider status to access thousands of credit reports, which he sold for $60 apiece through a coconspirator.[13]

The U.S. Congress passed the Identity Theft and Assumption Deterrence Act of 1998 to fight identity theft. Under this act, the Federal Trade Commission (FTC) is assigned responsibility to help victims restore their credit and erase the impact of the imposter. It also makes identity theft a federal felony punishable by a prison term ranging from 3 to 25 years.

THE COMPUTER AS THE OBJECT OF CRIME

A computer can also be the object of the crime, rather than the tool for committing it. Tens of millions of dollars of computer time and resources are stolen every year. Each time system access is illegally obtained, data or computer equipment is stolen or destroyed, or software is illegally copied, the computer becomes the object of crime. These crimes fall into several categories: illegal access and use, data alteration and destruction, information and equipment theft, software and Internet piracy, computer-related scams, and international computer crime.

Illegal Access and Use

hacker
A person who enjoys computer technology and spends time learning and using computer systems.

criminal hacker (cracker)
A computer-savvy person who attempts to gain unauthorized or illegal access to computer systems.

script bunnies
Wannabe crackers with little technical savvy who download programs—scripts—that automate the job of breaking into computers.

insider
An employee, disgruntled or otherwise, working solo or in concert with outsiders to compromise corporate systems.

Crimes involving illegal system access and use of computer services are a concern to both government and business. Since the outset of information technology, computers have been plagued by criminal hackers. A **hacker** is a person who enjoys computer technology and spends time learning and using computer systems. A **criminal hacker**, also called a **cracker**, is a computer-savvy person who attempts to gain unauthorized or illegal access to computer systems to steal passwords, corrupt files and programs, or even transfer money. In many cases, criminal hackers are people who are looking for fun and excitement—the challenge of beating the system. **Script bunnies** are wannabe crackers with little technical savvy—crackers who download programs called *scripts*—that automate the job of breaking into computers. **Insiders** are employees, disgruntled or otherwise, working solo or in concert with outsiders to compromise corporate systems.

Catching and convicting criminal hackers remains a difficult task. The method behind these crimes is often hard to determine. Even if the method behind the crime is known, tracking down the criminals can take a lot of time. It took years for the FBI to arrest one criminal hacker for the alleged "theft" of almost 20,000 credit card numbers that had been sent over the Internet. Table 9.4 provides some guidelines to follow in the event of a computer security incident.

Data Alteration and Destruction

Data and information are valuable corporate assets. The intentional use of illegal and destructive programs to alter or destroy data is as much a crime as destroying tangible goods. The most common of these programs are viruses and worms, which are software programs that, when loaded into a computer system, will destroy, interrupt, or cause errors

Table 9.4

- Follow your site's policies and procedures for a computer security incident. (They are documented, aren't they?)

- Contact the incident response group responsible for your site as soon as possible.

- Inform others, following the appropriate chain of command.

- Further communications about the incident should be guarded to ensure intruders do not intercept information.

- Document all follow-up actions (phone calls made, files modified, system jobs that were stopped, etc.).

- Make backups of damaged or altered files.

- Designate one person to secure potential evidence.

- Make copies of possible intruder files (malicious code, log files, etc.) and store them offline.

- Evidence, such as tape backups and printouts, should be secured in a locked cabinet, with access limited to one person.

- Get the National Computer Emergency Response Team involved if necessary.

- If you are unsure of what actions to take, seek additional help and guidance before removing files or halting system processes.

Table 9.4

How to Respond to a Security Incident

in processing. There are more than 60,000 known computer viruses today, with more than 5,000 new viruses and worms being discovered each year.

A **virus** is a computer program file capable of attaching to disks or other files and replicating itself repeatedly, typically without the user's knowledge or permission. Some viruses attach to files, so when the infected file executes, the virus also executes. Other viruses sit in a computer's memory and infect files as the computer opens, modifies, or creates the files. They are often disguised as games or images with clever or attention-grabbing titles such as "Boss, nude." Some viruses display symptoms, and some viruses damage files and computer systems. Computer viruses are written for several operating systems, including Windows, Macintosh, UNIX, and others.

Worms are parasitic computer programs that replicate but, unlike viruses, do not infect other computer program files. Worms can create copies on the same computer or can send the copies to other computers via a network. Worms often spread via IRC (Internet relay chat). The Mydoom worm, also known as Shimgapi and Novarg, started spreading in January 2004 and quickly became the most virulent e-mail worm ever. The worm arrives as an e-mail with an attachment that has various names and extensions, including .exe, .scr, .zip, and .pif. When the attachment is executed, the worm starts sending copies of itself to other e-mail addresses stored in the infected computer.

A **Trojan horse** program is a malicious program that disguises itself as a useful application and purposefully does something the user does not expect. Trojans are not viruses, because they do not replicate, but they can be just as destructive. Many people use the term to refer only to nonreplicating malicious programs, thus making a distinction between Trojans and viruses.

A **logic bomb** is a type of Trojan horse that executes when specific conditions occur. Triggers for logic bombs can include a change in a file by a particular series of keystrokes or at a specific time or date. In May 2004, authorities in Taiwan arrested 30-year-old computer engineer Wang Ping-an, who was accused of creating the Peep Trojan. Taiwan's Internet crime investigation task force, called "CIB," made the arrest. The existence of the Peep Trojan was uncovered when Taiwanese authorities discovered the theft of confidential information from government agencies, schools, and companies. Ping-an did not steal the data himself. Apparently, when he was unable to sell his data-stealing virus, he posted it on hackers' Web sites for free. Then, hackers from mainland China used the program to steal the data. Ping-an faces up to five years in prison, if convicted. More and more hackers are getting

virus
A computer program capable of attaching to disks or other files and replicating itself repeatedly, typically without the user's knowledge or permission.

worm
An independent program that replicates its own program files until it interrupts the operation of networks and computer systems.

Trojan horse
A program that appears to be useful but actually masks a destructive program.

logic bomb
An application or system virus designed to "explode" or execute at a specified time and date.

caught as the global law-enforcement community is taking these crimes more seriously. "The laws are getting better, the penalties steeper, and there is greater cooperation now that people have seen the damage these things can do," says Panda Software CTO Patrick Hinojosa.[14]

In some cases, a virus or a worm can completely halt the operation of a computer system or network for days or longer until the problem is found and repaired. In other cases, a virus or a worm can destroy important data and programs. If backups are inadequate, the data and programs may never be fully functional again. The costs include the effort required to identify and neutralize the virus or worm and to restore computer files and data, as well as the value of business lost because of unscheduled computer downtime.

United Parcel Service can't allow computer viruses to disrupt the information systems that are key to the operation of its Worldport distribution hub in Louisville, Kentucky, so the company works to hunt viruses and worms down *before* an attack. The UPS technical support group, TSG, actively monitors security and hacker Internet newsgroups to check for new viruses and acts quickly when they discover a new one. When the Blaster worm hit earlier this year, a TSG technician came in to work at 4 A.M. after spotting news of the worm on newsgroups while working at home. This preemptive approach prevented Worldport systems from getting hit by the Slammer virus.[15] Table 9.5 lists the computer incidents that have had the greatest economic impact.

Table 9.5

Computer Incidents with the Greatest Worldwide Economic Impact

Year	Code Name	Worldwide Economic Impact
2000	I Love You	$8.75 billion
2001	Code Red	$2.62 billion
2001	SirCam	$1.15 billion
1999	Melissa	$1.10 billion

Macintosh computers are immune to infection from high-profile Windows worms like SoBig and Mydoom, which exploit security flaws and architectural shortcomings in Windows operating systems and software applications to cause problems. That software only runs on Windows machines. As of early 2004, experts were unaware of any Mac OS X–specific virus or worm, but it's likely only a matter of time before hackers begin developing Mac OS X viruses and worms.[16] Table 9.6 lists the most active viruses in May 2004.

Table 9.6

Most Active Viruses in the World—May 30, 2004

(Source: "F-Secure Virus Statistics," *www.f-secure.com/ virus-info/statistics*, accessed May 30, 2004. Used with permission.)

Place	Virus Name
1	W32/Sober.G@mm
2	W32/Netsky.P@mm
3	W32/Netsky.D@mm
4	W32/Bagle.Z@mm
5	W32/Lovgate.W@mm
6	W32/Netsky.B@mm
7	W32/Netsky.Z@mm
8	W32/Netsky.Q@mm
9	Html_Netsky.P
10	W32/Netsky.C@mm

Using Antivirus Programs

As a result of the increasing threat of viruses and worms, most computer users and organizations have installed **antivirus programs** on their computers. Such software runs in the background to protect your computer from dangers lurking on the Internet and other possible sources of infected files. Some antivirus software is even capable of repairing common virus infections automatically, without interrupting your work. The latest virus definitions are downloaded automatically when you connect to the Internet, ensuring that your PC's protection is current. To safeguard your PC and prevent it from spreading viruses to your friends and coworkers, some antivirus software scans and cleans both incoming and outgoing e-mail messages. Table 9.7 lists some of the most popular antivirus software.

antivirus program
Program or utility that prevents viruses and recovers from them if they infect a computer.

Table 9.7

Antivirus Software

Antivirus Software	Software Manufacturer	Web Site
Symantec's Norton AntiVirus 2005	Symantec	www.symantec.com
McAfee Virus Scan	McAfee	www.mcafee.com
Panda Antivirus Platinum	Panda Software	www.pandasoftware.com
Vexira Antivirus	Central Command	www.centralcommand.com
Sophos Antivirus	Sophos	www.sophos.com
PC-cillin	Trend Micro	www.trendmicro.com

Antivirus software should be used and updated often.

Norton AntiVirus

Scan Progress: Scan for Threats

1 **Scan Progress**

2 **Repair Wizard**

Fix

Quarantine

Delete

Exclude

3 **Summary**

Scanning for threats:

Current Item
C:\WINDOWS\$xpsp1hfm$\Q819696\update

Action	Files
Scanned	949
Detected	0
Fixed	0
Deleted	0

[Pause] [Stop Scan]

Proper use of antivirus software requires the following steps:

1. *Install antivirus software and run it often.* Many of these programs automatically check for viruses each time you boot up your computer or insert a diskette or CD, and some even monitor all e-mail and file transmissions and copying operations.

2. *Update antivirus software often.* New viruses are created all the time, and antivirus software suppliers are constantly updating their software to detect and take action against these new viruses.

3. *Scan all diskettes and CDs before copying or running programs from them.* Hiding on diskettes or CDs, viruses often move between systems. If you carry document or program files on diskettes or CDs between computers at school or work and your home system, always scan them.

4. *Install software only from a sealed package or secure Web site of a known software company.* Even software publishers can unknowingly distribute viruses on their program disks or software downloads. Most scan their own systems, but viruses may still remain.

5. *Follow careful downloading practices.* If you download software from the Internet or a bulletin board, check your computer for viruses immediately after completing the transmission.

6. *If you detect a virus, take immediate action.* Early detection often allows you to remove a virus before it does any serious damage.

Despite careful precautions, viruses can still cause problems. They can elude virus-scanning software by lurking almost anywhere in a system. Future antivirus programs may incorporate "nature-based models" that check for unusual or unfamiliar computer code. The advantage of this type of virus program is the ability to detect new viruses that are not part of an antivirus database.

Information and Equipment Theft

Data and information are assets or goods that can also be stolen. Individuals who illegally access systems often do so to steal data and information. To obtain illegal access, criminal hackers require identification numbers and passwords. Some criminals try different identification numbers and passwords until they find ones that work. Using password sniffers is another approach. A **password sniffer** is a small program hidden in a network or a computer system that records identification numbers and passwords. In a few days, a password sniffer can record hundreds or thousands of identification numbers and passwords. Using a password sniffer, a criminal hacker can gain access to computers and networks to steal data and information, invade privacy, plant viruses, and disrupt computer operations. A hacker secretly installed sniffer software in 14 Kinko's stores in New York City that allowed him to capture 450 usernames and passwords to access and even open bank accounts online. The hacker was caught when he used one of the stolen passwords to access a computer with GoToMyPC software, which lets individuals remotely access their own computers from elsewhere. The GoToMyPC subscriber happened to be home at the time and suddenly saw the cursor on his computer move around the screen and files open as if by themselves. To his amazement, he then saw an account being opened in his name at an online payment transfer service.[17]

In addition to theft of data and software, all types of computer systems and equipment have been stolen from offices. Computer theft is now second only to automobile theft, according to recent U.S. crime statistics. In the United Kingdom more than 30 percent of all reported thefts are computer related. Printers, desktop computers, and scanners are often targets. Portable computers such as laptops (and the data and information stored in them) are especially easy for thieves to take. In some cases, the data and information stored in these systems are more valuable than the equipment. Without adequate protection and security measures, equipment can easily be stolen. According to the Brigadoon software 2003 Computer Theft Survey, more than 1.6 million computers were stolen in the United States in the past three years. Based on the responses of the 676 participants in the survey, 48 percent of the devices stolen were laptop computers, 27 percent were desktop computers, and 13 percent were some form of PDA. Unfortunately, the overwhelming majority of theft victims (88 percent) did not encrypt proprietary data on these devices.[18]

password sniffer
A small program hidden in a network or a computer system that records identification numbers and passwords.

To fight computer crime, many companies use devices that disable the disk drive and lock the computer to the desk.

(Source: Courtesy of Kensington Technology Group.)

Software and Internet Software Piracy

Each time you use a word processing program or access software on a network, you are taking advantage of someone else's intellectual property. Like books and movies—other intellectual properties—software is protected by copyright laws. Often, people who would never think of plagiarizing another author's written work have no qualms about using and copying software programs they have not paid for. Such illegal duplicators are called *pirates*; the act of illegally duplicating software is called **software piracy**.

Technically, software purchasers are granted the right only to use the software under certain conditions; they don't really own the software. Licenses vary from program to program and may authorize as few as one computer or one individual to use the software or as many as several hundred network users to share the application across the system. Making additional copies, or loading the software onto more than one machine, may violate copyright law and be considered piracy.

It is estimated that the software industry loses between $11 and $12 billion in revenue to software piracy annually. Half the loss comes from Asia, where China and Indonesia are the biggest offenders. In Western Europe, annual piracy losses range between $2.5 and $3 billion. Although the rate of software piracy is quite high in Latin America and Central Europe, those software markets are so small that the dollar losses are considerably lower. About $2 billion in annual piracy losses come from North America. Continuing education and enforcement efforts are making a difference, however. In the United States, for example, the level of piracy was reduced from 48 percent in 1989 to 25 percent in 2002.[19]

Internet software piracy occurs when software is illegally downloaded from the Internet. It is the most rapidly expanding type of software piracy and the most difficult form to combat. The same purchasing rules apply to online software purchases as for traditional purchases. Internet piracy can take several forms including the following:

- Pirate Web sites that make software available for free or in exchange for uploaded programs
- Internet auction sites that offer counterfeit software, which infringes copyrights
- Peer-to-peer networks, which enable unauthorized transfer of copyrighted programs

Penalties for software piracy can be severe. If the copyright owner brings a civil action against someone, the owner can seek to stop the person from using its software immediately and can also request monetary damages. The copyright owner may then choose between compensation for actual damages—which includes the amount it has lost because of the person's infringement, as well as any profits attributable to the infringement—and statutory damages, which can be as much as $150,000 for each program copied. In addition, the government can prosecute software pirates in criminal court for copyright infringement. If convicted, they could be fined up to $250,000 or sentenced to jail for up to five years, or both.[20]

Computer-Related Scams

People have lost hundreds of thousands of dollars on real estate, travel, stock, and other business scams. Today, many of these scams are being perpetrated with computers. Using the Internet, scam artists offer get-rich-quick schemes involving bogus real estate deals, tout "free" vacations with huge hidden costs, commit bank fraud, offer fake telephone lotteries, sell worthless penny stocks, and promote illegal tax-avoidance schemes.

Over the past few years, credit card customers of various banks have been targeted by scam artists trying to get personal information needed to use their credit cards. The scam works by sending customers an e-mail including a link that seems to direct users to their bank's Web site. Once at the site, they are greeted with a pop-up box asking them for their full debit card numbers, their personal identification numbers, and their credit card expiration dates. The problem is that the Web site customers are directed to a fake site operated by someone trying to gain access to that information.[21] As discussed previously, this form of scam is called *phishing*. The Anti-Phishing Working Group received reports of more than 1,100 unique phishing campaigns in April 2004, a 178 percent increase from the previous month.[22] A 2004 study by the Gartner Consulting Group surveyed 5,000 adult Internet

software piracy
The act of illegally duplicating software.

Internet software piracy
Illegally downloading software from the Internet.

users and found that roughly 3 percent reported giving up financial data or other personal information after being drawn into phishing scams. The results suggest that as many as 30 million adults have experienced a phishing attack and that 1.78 million adults could have fallen victim to the scams. The 6 percent success rate is more than enough to ensure a continuation of such scams.[23]

The U.S. Federal Trade Commission (FTC) filed suit against a company that promotes creation of a firm's Web site over the phone and then charges the victims' phone bills without their authorization. The company calls small businesses and gets someone to agree to a 15-day trial offer to look at a Web site designed for that business. Unbeknownst to the small business, the look at the site results in a $29.95 addition to the firm's monthly telephone bill. Often the charges on the phone bill appear as "MIS Int Serv," which people often mistake for Internet access charges. The scam has been going on for years and targets businesses all across the United States.[24]

Here is a list of tips to help you avoid becoming a scam victim.

- Don't agree to anything in a high-pressure meeting or seminar. Insist on having time to think it over and to discuss things with your spouse, your partner, or even your lawyer. If a company won't give you the time you need to check it out and think things over, you don't want to do business with it. A good deal now will be a good deal tomorrow; the only reason for rushing you is if the company has something to hide.
- Don't judge a company based on appearances. Flashy Web sites can be created and put up on the Net in a matter of days. After a few weeks of taking money, a site can vanish without a trace in just a few minutes. You may find that the perfect money-making opportunity offered on a Web site was a money maker for the crook and a money loser for you.
- Avoid any plan that pays commissions simply for recruiting additional distributors. Your primary source of income should be your own product sales. If the earnings are not made primarily by sales of goods or services to consumers or sales by distributors under you, you may be dealing with an illegal pyramid.
- Beware of shills, people paid by a company to lie about how much they've earned and how easy the plan was to operate. Check with an independent source to make sure that you aren't having the wool pulled over your eyes.
- Beware of a company's claim that it can set you up in a profitable home-based business but that you must first pay up front to attend a seminar and buy expensive materials. Frequently, seminars are high-pressure sales pitches, and the material is so general that it is worthless.
- If you are interested in starting a home-based business, get a complete description of the work involved before you send any money. You may find that what you are asked to do after you pay is far different from what was stated in the ad. You should never have to pay for a job description or for needed materials.
- Get in writing the refund, buy-back, and cancellation policies of any company you deal with. Do not depend on oral promises.
- Do your homework. Check with your state attorney general and the National Fraud Information Center before getting involved, especially when the claims about a product or potential earnings seem too good to be true.

If you need advice about an Internet or online solicitation, or if you want to report a possible scam, use the Online Reporting Form or Online Question & Suggestion Form features on the Web site for the National Fraud Information Center at *http://fraud.org*, or call the NFIC hotline at 1-800-876-7060.

International Computer Crime

Computer crime is also an international issue, and it becomes more complex when it crosses borders. As already mentioned, the software industry loses about $11 to $12 billion in revenue to software piracy annually, with about $9 billion of that occurring outside the United States.[25]

With the increase in electronic cash and funds transfer, some are concerned that terrorists, international drug dealers, and other criminals are using information systems to launder illegally obtained funds. Computer Associates International developed software called CleverPath for Global Compliance for customers in the finance, banking, and insurance industries to eliminate money laundering and fraud. Companies that are required to comply with legislation such as the USA Patriot Act and Sarbanes-Oxley Act may lack the resources and processes to do so. The software automates manual tracking and auditing processes that are required by regulatory agencies and helps companies handle frequently changing reporting regulations. The application can drill into a company's transactions and automatically detect fraud or other illegal activities based on built-in business rules and predictive analysis. Suspected fraud cases are identified and passed on to the appropriate personnel for action to thwart criminals and help companies avoid paying fines.[26]

PREVENTING COMPUTER-RELATED CRIME

Because of increased computer use today, greater emphasis is placed on the prevention and detection of computer crime. Although all states have passed computer crime legislation, some believe that these laws are not effective because companies do not always actively detect and pursue computer crime, security is inadequate, and convicted criminals are not severely punished. However, all over the United States, private users, companies, employees, and public officials are making individual and group efforts to curb computer crime, and recent efforts have met with some success.

Crime Prevention by State and Federal Agencies

State and federal agencies have begun aggressive attacks on computer criminals, including criminal hackers of all ages. In 1986, Congress enacted the Computer Fraud and Abuse Act, which mandates punishment based on the victim's dollar loss. The Department of Defense also supports the Computer Emergency Response Team (CERT), which responds to network security breaches and monitors systems for emerging threats. Law enforcement agencies are also increasing their efforts to stop criminal hackers, and many states are now passing new, comprehensive bills to help eliminate computer crimes. Recent court cases and police reports involving computer crime show that lawmakers are ready to introduce newer and tougher computer crime legislation. Several states have passed laws in an attempt to outlaw spamming. For example, an Arizona law enacted in May 2003 requires that unsolicited commercial e-mail messages include a label ("ADV:") at the beginning of the subject line, and contain an opt-out mechanism. In September 2003, legislation was approved in California that made it the second state (after Delaware) to adopt an opt-in for e-mail advertising. Under this legislation, it is illegal to send unsolicited commercial e-mail from California or to a California e-mail address.

Crime Prevention by Corporations

Companies are also taking crime-fighting efforts seriously. Many businesses have designed procedures and specialized hardware and software to protect their corporate data and systems. Specialized hardware and software, such as encryption devices, can be used to encode data and information to help prevent unauthorized use. As discussed in Chapter 4, encryption is the process of converting an original electronic message into a form that can be understood only by the intended recipients. A key is a variable value that is applied using an algorithm to a string or block of unencrypted text to produce encrypted text or to decrypt encrypted text. Encryption methods rely on the limitations of computing power for their effectiveness—if breaking a code requires too much computing power, even the most determined code crackers will not be successful. The length of the key used to encode and decode messages determines the strength of the encryption algorithm.

public key infrastructure (PKI)
A means to enable users of an unsecured public network such as the Internet to securely and privately exchange data through the use of a public and a private cryptographic key pair that is obtained and shared through a trusted authority.

biometrics
The measurement of a person's trait, whether physical or behavioral.

Public key infrastructure (PKI) enables users of an unsecured public network such as the Internet to securely and privately exchange data through the use of a public and a private cryptographic key pair that is obtained and shared through a trusted authority. PKI is the most common method on the Internet for authenticating a message sender or encrypting a message. PKI uses two keys to encode and decode messages. One key of the pair, the message receiver's public key, is readily available to the public and is used by anyone to send that individual encrypted messages. The second key, the message receiver's private key, is kept secret and is known only by the message receiver. Its owner uses the private key to *decrypt* messages—convert encoded messages back into the original message. Knowing an individual's public key does not enable you to decrypt an encoded message to that individual.

Using biometrics is another way to protect important data and information systems. **Biometrics** involves the measurement of one of a person's traits, whether physical or behavioral. Biometric techniques compare a person's unique characteristics against a stored set to detect differences between them. Biometric systems can scan fingerprints, faces, handprints, irises, and retinal images to prevent unauthorized access to important data and computer resources. Most of the interest among corporate users is in fingerprint technology, followed by face recognition. Fingerprint scans hit the middle ground between price and effectiveness. Iris and retina scans are more accurate, but they are more expensive and involve more equipment.

In June 2004, the U.S. Department of Homeland Security awarded a $10 billion, five-year contract to global management consulting and technology services company Accenture to oversee the creation of a comprehensive border-control system. The system, known as the United States Visitor and Immigration Status Indicator Technology, or US-Visit, employs a combination of biometric technologies to identify visitors. Radio-frequency identification, voice- and facial-recognition, retinal- or iris-scanning, and digital-fingerprinting systems are all being tested and evaluated for inclusion in the program. A visitor to the United States might be required to undergo digital fingerprinting at a consulate in his or her home country to apply for a visa. Once here, the visitor would present a smart card that is encoded with the digital fingerprint to an immigration official and undergo a second digital fingerprint scan to ensure a match. The data can be instantly cross-referenced against a database containing digital descriptors of known terrorists, criminals, and other undesirables.[27]

As employees move from one position to another at a company, they can build up access to multiple systems because security procedures often fail to revoke access privileges. It is clearly not appropriate for people who have changed positions and responsibilities still to have access to systems they no longer use. To avoid this problem, many organizations are creating role-based system access lists so that only people filling a particular job function can access a specific system.

Crime-fighting procedures usually require additional controls on the information system. Before designing and implementing controls, organizations must consider the types of computer-related crime that might occur, the consequences of these crimes, and the cost and complexity of needed controls. In most cases, organizations conclude that the trade-off between crime and the additional cost and complexity weighs in favor of better system controls. Having knowledge of some of the methods used to commit crime is also helpful in preventing, detecting, and developing systems resistant to computer crime (see Table 9.8). Some companies actually hire former criminals to thwart other criminals. Table 9.9 provides a set of useful guidelines to protect your computer from hackers.

Companies are also joining together to fight crime. The Software Publisher's Association (SPA) was the original antipiracy organization, formed and financed by many of the large software publishers. Microsoft financed the formation of a second antipiracy organization, the Business Software Alliance (BSA). The BSA, through intense publicity, has become the more prominent organization. Other software companies, including Apple, Adobe, Hewlett-Packard, and IBM, now contribute to the BSA.

Fingerprint authentication devices provide security in the PC environment by using fingerprint information instead of passwords.

(Source: Courtesy of DigitalPersona.)

Methods	Examples
Add, delete, or change inputs to the computer system.	Delete records of absences from class in a student's school records.
Modify or develop computer programs that commit the crime.	Change a bank's program for calculating interest to make it deposit rounded amounts in the criminal's account.
Alter or modify the data files used by the computer system.	Change a student's grade from C to A.
Operate the computer system in such a way as to commit computer crime.	Access a restricted government computer system.
Divert or misuse valid output from the computer system.	Steal discarded printouts of customer records from a company trash bin.
Steal computer resources, including hardware, software, and time on computer equipment.	Make illegal copies of a software program without paying for its use.
Offer worthless products for sale over the Internet.	Send e-mail requesting money for worthless hair growth product.
Blackmail executives to prevent release of harmful information.	Eavesdrop on organization's wireless network to capture competitive data or scandalous information.
Blackmail company to prevent loss of computer-based information.	Plant logic bomb and send letter threatening to set it off unless paid considerable sum.

Table 9.8

Common Methods Used to Commit Computer Crimes

Even though the number of potential computer crimes appears to be limitless, the actual methods used to commit crime are limited.

- Install strong user authentication and encryption capabilities on your firewall.
- Install the latest security patches, which are often available at the vendor's Internet site.
- Disable guest accounts and null user accounts that let intruders access the network without a password.
- Do not provide overfriendly login procedures for remote users (e.g., an organization that used the word *welcome* on their initial logon screen found they had difficulty prosecuting a hacker).
- Give an application (e-mail, file transfer protocol, and domain name server) its own dedicated server.
- Restrict physical access to the server and configure it so that breaking into one server won't compromise the whole network.
- Turn audit trails on.
- Consider installing caller ID.
- Install a corporate firewall between your corporate network and the Internet.
- Install antivirus software on all computers and regularly download vendor updates.
- Conduct regular IS security audits.
- Verify and exercise frequent data backups for critical data.

Table 9.9

How to Protect Your Corporate Data from Hackers

The Business Software Alliance required Red Bull North America, Inc., to pay $105,000 to settle claims that it had more copies of Adobe, Microsoft, and Symantec software programs on its computers than it had licenses to support. Red Bull is based in Germany and markets a nonalcoholic energy drink in more than 70 countries. The software infringement was reported through a call to the BSA hotline, and then BSA attorneys contacted Red Bull. The company cooperated with the BSA and voluntarily conducted a self-audit. To avoid similar problems, all companies should draft policies and implement procedures to maintain an effective software management program.[28]

Using Intrusion Detection Software

intrusion detection system (IDS)

Software that monitors system and network resources and notifies network security personnel when it senses a possible intrusion.

An **intrusion detection system (IDS)** monitors system and network resources and notifies network security personnel when it senses a possible intrusion. Examples of suspicious activities include repeated failed login attempts, attempts to download a program to a server, and access to a system at unusual hours. Such activities generate alarms that are captured on log files. Intrusion detection systems send an alarm, often by e-mail or pager, to network security personnel when they detect an apparent attack. Unfortunately, many IDSs frequently provide false alarms that result in wasted effort. If the attack is real, then network security personnel must make a decision about what to do to resist the attack. Any delay in response increases the probability of damage from a hacker attack. Use of an IDS provides another layer of protection in the event that an intruder gets past the outer security layers—passwords, security procedures, and corporate firewall.

The following story is true, but the company's name has been changed to protect its identity. The ABCXYZ company employs more than 25 IDS sensors across its worldwide network, enabling it to monitor 90 percent of the company's internal network traffic. The remaining 10 percent comes from its engineering labs and remote sales offices, which are not monitored because of a lack of resources. The company's IDS worked very well in providing an early warning of an impending SQL Slammer attack. The Slammer worm had entered the network via a server in one of the engineering labs. The person monitoring the IDS noticed outbound traffic consistent with SQL Slammer at about 7:30 A.M. He contacted the network operations group by e-mail and followed up with a phone call and a voice mail message. Unfortunately, the operations group gets so many e-mails that if a message is not highlighted as URGENT, the message may be missed. That is exactly what happened—the e-mail alert wasn't read, and the voice message wasn't retrieved in time to block the attack. A few hours later, the ABCXYZ company found itself dealing with a massive number of reports of network and server problems.[29]

A firm called Internet Security Systems (ISS) manages security for other organizations through its Security Incident Prevention Service. The company's IDSs are designed to recognize 30 of the most-critical threats, including worms that go after Microsoft software and those that exploit Apache Web servers and other programs. When an attack is detected, the service automatically blocks it without requiring human invention. Taking the manual intervention step out of the process enables a faster response and minimizes damage from a hacker. To encourage customers to adopt its service, ISS guaranteed up to $50,000 in cash if the prevention service failed.[30]

Using Managed Security Service Providers (MSSPs)

managed security service provider (MSSP)

An organization that monitors, manages, and maintains network security hardware and software for its client companies.

Keeping up with computer criminals—and with new regulations—can be daunting for organizations. Hackers are constantly poking and prodding, trying to breach the security defenses of companies. Also, such recent legislation as HIPAA, the Sarbanes-Oxley Act, and the USA Patriot Act requires businesses to prove that they are securing their data. For most small and midsized organizations, the level of in-house network security expertise needed to protect their business operations can be quite costly to acquire and maintain. As a result, many are outsourcing their network security operations to **managed security service providers (MSSPs)** such as Counterpane, Guardent, Internet Security Services, Riptech, and Symantec. MSSPs monitor, manage, and maintain network security for both hardware and software. These companies provide a valuable service for IS departments drowning in reams of alerts and false alarms coming from virtual private networks (VPNs); antivirus, firewall, intrusion detection systems; and other security monitoring systems. In addition, some provide vulnerability scanning and Web blocking/filtering capabilities.

Internet Laws for Libel and Protection of Decency

To help parents control what their children see on the Internet, some companies provide software called *filtering software* to help screen Internet content. Many of these screening programs also prevent children from sending personal information over e-mail or through chat groups. This stops children from broadcasting their name, address, phone number, or other personal information over the Internet. The two approaches used are filtering, which blocks certain Web sites, and rating, which places a rating on Web sites. According to the 2004 Internet Filter Review, the five top-rated filtering software packages are, in order: ContentProtect, Cybersitter, Net Nanny, CyberPatrol, and FilterPack.[31]

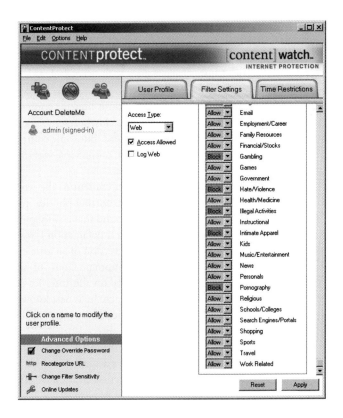

ContentProtect is a filtering software program that helps block unwanted Internet content from children and young adults.

(Source: Courtesy of ContentWatch Inc.)

The Internet Content Rating Association (ICRA) is a nonprofit organization whose members include Internet industry leaders such as America Online, Bell South, British Telecom, IBM, Microsoft, UUNet, and Verizon. Its specific goals are to protect children from potentially harmful material while also safeguarding free speech on the Internet. Using the **ICRA rating system**, Web authors fill out an online questionnaire describing the content of their site—what is and isn't present. The broad topics covered are the following: chat capabilities, the language used on the site, the nudity and sexual content of a site, the violence depicted on the site, and other areas such as alcohol, drugs, gambling, and suicide. Based on the authors' responses, ICRA then generates a content label (a short piece of computer code) that the authors add to their site. Internet users (and parents) can then set their browser to allow or disallow access to Web sites based on the objective rating information declared in the content label and their own subjective preferences. Reliance on Web site authors to do their own rating has its weaknesses, though. Web site authors can lie when completing the ICRA questionnaire so that their site receives a content label that doesn't accurately reflect the site's content. In addition, many hate groups and sexually explicit sites don't have an ICRA rating, so they will not be blocked unless a browser is set to block all unrated sites.

ICRA rating system
A system to protect individuals from harmful or objectionable Internet content, while safeguarding the free speech rights of others.

Also, this option would block out so many acceptable sites that it could make Web surfing useless. For these reasons, at this time, site labeling is at best a complement to other filtering techniques.

The Children's Internet Protection Act (CIPA) is a federal law passed in December 2000 that required federally funded libraries to use some form of prevention measure (such as Internet filters) to block access to obscene material and other material considered harmful to minors. Opponents of the law feared that it transferred power over the education process to private software companies that develop the Internet filters and define which sites are to be blocked. In June 2003, in a ruling on the consolidated cases *U.S. v. Multnomah County Library et al.* and *U.S. v. American Library Association*, the Supreme Court upheld the law, but ruled that librarians can disable the software entirely on request and that patrons do not have to provide a reason why they want a site unblocked. The ruling also implies that patrons would not have to identify themselves to request unblocking. Some observers believe that the court's ruling left unclear the unblocking rules with respect to children.[32]

With the increased popularity of networks and the Internet, libel becomes an important legal issue. A publisher, such as a newspaper, can be sued for libel, which involves publishing a written statement that is damaging to a person's reputation. Generally, a bookstore cannot be held liable for statements made in newspapers or other publications it sells. Online services, such as CompuServe and America Online, may exercise some control over who puts information on their service but may not have direct control over the content of what is published by others on their service. So, can online services be sued for libel for content that someone else publishes on their service? Do online services more closely resemble a newspaper or a bookstore? This legal issue has not been completely resolved, but some court cases have been decided. The *Cubby, Inc. v. CompuServe* case ruled that CompuServe was similar to a bookstore and not liable for content put on its service by others.

Companies should be aware that publishing Internet content to the world may subject them to different countries' laws in the same way that exporting physical products does. A December 2003 ruling by the High Court of Australia found that a story published by Dow Jones & Co. on a U.S.-hosted Web site was grounds for a defamation lawsuit in Australia. The suit was brought by an Australian over the Internet version of an article in Dow Jones's *Barron's* magazine. The individual filed the suit in the Supreme Court of his home state of Victoria in Australia, saying that the article's appearance on the Internet enabled it to be accessed by people in Victoria, thereby defaming him where he is best known. "The torts of libel and slander are committed when and where comprehension of the defamatory matter occurs," agreed the Australian High Court, citing several precedents.[33]

Geolocation tools match the user's IP address with outside information to determine the actual geographic location of the online user where the customer's computer signal enters the Internet. This enables someone to identify the user's actual location within approximately 50 miles. As such tools become broadly available, Internet publishers will be able to limit the reach of their published speech to avoid potential legal risks. Use of such technology may also result in a division of the global Internet into separate content regions, with readers in Brazil, Japan, and the United States all receiving variations of the same information from the same publisher.

Individuals, too, must be careful what they post on the Internet to avoid libel charges. There have been many cases of disgruntled former employees being sued by their former employees for material posted on the Internet.

Preventing Crime on the Internet

As mentioned in Chapter 4, Internet security can include firewalls and a number of methods to secure financial transactions. A firewall can include both hardware and software that act as a barrier between an organization's information systems and the outside world. A number of systems have been developed to safeguard against crime on the Internet.

To help prevent crime on the Internet, the following steps can be taken:

1. Develop effective Internet usage and security policies for all employees.
2. Use a stand-alone firewall (hardware and software) with network monitoring capabilities.
3. Deploy intrusion detection systems, monitor them, and follow up on their alarms.

4. Monitor managers and employees to make sure that they are using the Internet for business purposes.

5. Use Internet security specialists to perform audits of all Internet and network activities.

Even with these precautions, computers and networks can never be completely protected against crime. One of the biggest threats is from employees. Although firewalls provide good perimeter control to prevent crime from the outside, procedures and protection measures are needed to protect against computer crime by employees. Passwords, identification numbers, and tighter control of employees and managers also help prevent Internet-related crime.

Mohegan Sun, a casino in Connecticut owned by the Mohegan Tribe, operates a large network with a number of critical systems that track its more than 2.5 million customers, their winnings, and their creditworthiness, as well as loyalty program points that can be used to purchase items at the casino and its shops. The casino is trying out new software from Intrusic, Inc., in an attempt to detect illegal behavior of rogue employees. Intrusic's Zephon software analyzes communications between users and the computer network, looking for violations of what it calls the "physics of networks" or fundamental laws that govern the way legitimate network traffic looks. The casino already employs perimeter-defense products such as firewalls and intrusion detection systems, but it is increasingly concerned about insider threats or external compromise as a result of fast-moving worms and viruses.[34]

PRIVACY ISSUES

Another important social issue in information systems involves privacy. In 1890, U.S. Supreme Court Justice Louis Brandeis stated that the "right to be left alone" is one of the most "comprehensive of rights and the most valued by civilized man." Basically, the issue of privacy deals with this right to be left alone or to be withdrawn from public view. With information systems, privacy deals with the collection and use or misuse of data. Data is constantly being collected and stored on each of us. This data is often distributed over easily accessed networks and without our knowledge or consent. Concerns of privacy regarding this data must be addressed.

With today's computers, the right to privacy is an especially challenging problem. More data and information are produced and used today than ever before. When someone is born, takes certain high school exams, starts working, enrolls in a college course, applies for a driver's license, purchases a car, serves in the military, gets married, buys insurance, gets a library card, applies for a charge card or loan, buys a house, or merely purchases certain products, data is collected and stored somewhere in computer databases. Read the "Information Systems @ Work" special feature to learn more about privacy issues associated with fundamental medical and DNA data of individuals. A difficult question to answer is, "Who owns this information and knowledge?" If a public or private organization spends time and resources to obtain data on you, does the organization own the data, and can it use the data in any way it desires? Government legislation answers these questions to some extent for federal agencies, but the questions remain unanswered for private organizations.

Privacy and the Federal Government

The U.S. federal government is perhaps the largest collector of data. Close to 4 billion records exist on individuals, collected by about 100 federal agencies, ranging from the Bureau of Alcohol, Tobacco, and Firearms to the Veterans Administration. Other data collectors include state and local governments and profit and nonprofit organizations of all types and sizes. In recent years, a number of actions by the federal government have caused concern for the privacy of individuals' data.

UK BioBank Raises Privacy Issues

The Biobank is a £45 million ($80 million U.S.) project dedicated to collecting genetic, medical, and lifestyle data from 500,000 middle-aged people in the United Kingdom. The goal is to enable scientists to pinpoint specific factors involved in certain diseases, ultimately leading to new and better treatments. Volunteers chosen at random will be asked to donate DNA and answer a detailed questionnaire on their health and lifestyle. Their health will be followed for decades to help discover the roles of genes and the environment in developing cancer, heart disease, and other illnesses. Ultimately, doctors could use the databases to discover a new treatment that will work on anyone with a certain set of genetic traits. The information's true value may not be realized for some 30 years.

However, before a single piece of data has been collected, the UK Biobank has become embroiled in controversy over its goals, its costs, its underlying science, and its possibility for commercial exploitation.

Critics of the project claim that there is no evidence to suggest that investing millions of pounds in the gene bank would be as beneficial as spending it on well-targeted public health initiatives to prevent illness. They fear that the database could effectively become a permanent medical rap sheet, analogous to biological credit reports that could haunt a person's children, grandchildren, and other relatives. Ian Gibson, a member of Parliament within Britain's governing Labour party and chairman of the House of Commons Science and Technology Committee, said he has concerns that industry sectors such as insurance, which could have a strong interest in the bank's database, might eventually obtain the medical information. Currently, in the UK there are no laws that prevent discrimination by insurers and employers.

Many practical questions remain, but the Biobank project's leaders say that the public has already expressed wide support. They have identified many people who want to participate and be part of developments in a better life.

Discussion Questions

1. What are the most controversial aspects of the Biobank project?
2. What are the potential negative impacts of these concerns on the Biobank project?

Critical Thinking Questions

3. Imagine that you are the member of a public relations firm hired to improve the public's impression of the Biobank project and overcome the arguments of its strongest critics. What specific actions would you recommend be taken by the administrators and scientists associated with the Biobank project to win over public opinion?
4. What additional actions might people outside the Biobank project take to alleviate some of the privacy concerns?

SOURCES: Kevin Davies, "First Base: Behold—BioBank," *Bio-IT World*, August 7, 2003, *www.bio-ITworld.com*; Scarlet Pruitt, "U.K. Prepares to Open World's Largest Biobank," *Bio-IT World*, November 14, 2003, *www.bio-ITWorld.com*; Pat Hagan, "Biobank Debate Heats Up," *The Scientist*, April 8, 2003, *www. biomedcentral.com*; and Pat Hagan, "UK Biobank Reveals Ethics Framework," *The Scientist*, September 23, 2003, *www.biomedcentral.com*.

In response to the September 11 terrorist attacks, the U.S. Department of Defense and the Defense Advanced Research Projects Agency (DARPA) launched the Total Information Awareness (TIA) program to conduct research on technology's use for largescale information gathering and analysis. The program's intent was to access a mix of government, intelligence, and commercial databases to mine electronic transactions, such as airline-ticket purchases and car rentals to look for potential terrorists and terrorist threats. In July 2003, the Senate, concerned about the lack of limits on the sources of such data gathering, voted to cancel funding for TIA. Since its inception, TIA had drawn criticism from privacy rights advocates who feared it would allow authorities to rummage though the electronic transactions of millions of law-abiding U.S. citizens in an effort to uncover the activities of suspected terrorists.[35]

Also launched in response to the September 11 terrorist attacks, the Matrix system lets states share criminal, prison, and vehicle information and cross-reference it with databases held by Seisint. Seisint is a provider of information products that enable organizations to extract useful knowledge from huge amounts of data. Its products are created by integrating supercomputer technology, tens of billions of data records on individuals and businesses, and patent-pending data-linking methods.[36] The information contained within Matrix includes civil court records, voter registration information, and address histories going back as far as 30 years. Current participants in the project include Connecticut, Florida, Michigan, New York, Ohio, Pennsylvania, and Utah. Several other states considered the program before dropping out, citing concerns about privacy or the long-term costs.[37]

Although privacy concerns prompted Congress to kill both the CAPPS II and TIA programs, government computers are still scanning a vast array of databases for clues about criminal or terrorist activity, according to the General Accounting Office (GAO), the investigative arm of Congress. The U.S. government is managing 199 data-mining efforts, 36 of which collect personal information from the private sector. While data mining can be a useful tool for the government, "safeguards should be put in place to ensure that information isn't abused," said Nuala O'Connor Kelly, chief privacy officer at the Department of Homeland Security. Senator Daniel Akaka of Hawaii said he has asked the GAO to examine some projects more closely. "The federal government collects and uses Americans' personal information and shares it with other agencies to an astonishing degree, raising serious privacy concerns," Akaka said in a statement.[38] A growing number of privacy advocates say the Matrix database seems to be a substitute for the TIA data-mining program scrapped by the Pentagon.[39]

Most companies and computer vendors are wary of having the federal government dictate Internet privacy standards. A group called the Online Privacy Alliance is developing a voluntary code of conduct. It is backed by companies such as Apple, AT&T, Boeing, Dell, DoubleClick, eBay, IBM, Microsoft, Time Warner, Verizon Communications, and Yahoo! The alliance's guidelines call on companies to notify users when they are collecting data at Web sites to gain consent for all uses of that data, to provide for the enforcement of privacy policies, and to have a clear process in place for receiving and addressing user complaints. The alliance's policy can be found at *www.privacyalliance.org/resources/ppguidelines.shtml.*

The European Union has already passed a data-protection directive that requires firms transporting data across national boundaries to have certain privacy procedures in place. This directive affects virtually any company doing business in Europe, and it is driving much of the attention being given to privacy in the United States.

Privacy at Work

The right to privacy at work is also an important issue. Currently, the rights of workers who want their privacy and the interests of companies that demand to know more about their employees are in conflict. Recently, companies that have been monitoring have raised their employee's concerns. For example, workers may find that they are being closely monitored via computer technology. These computer-monitoring systems tie directly into workstations; specialized computer programs can track every keystroke made by a user. This type of system can determine what workers are doing while at the keyboard. The system also knows when the worker is not using the keyboard or computer system. These systems can estimate what

E-mail has changed how workers and managers communicate in the same building or around the world. E-mail, however, can be monitored and intercepted. As with other services—such as cellular phones—the convenience of e-mail must be balanced with the potential of privacy invasion.

(Source: © Gary Conner/Photo Edit.)

Platform for Privacy Preferences (P3P)

A screening technology that shields users from Web sites that don't provide the level of privacy protection they desire.

a person is doing and how many breaks he or she is taking. Needless to say, many workers consider this close supervision very dehumanizing.

E-Mail Privacy

E-mail also raises some interesting issues about work privacy. Federal law permits employers to monitor e-mail sent and received by employees. Furthermore, e-mail messages that have been erased from hard disks may be retrieved and used in lawsuits because the laws of discovery demand that companies produce all relevant business documents. On the other hand, the use of e-mail among public officials may violate "open meeting" laws. These laws, which apply to many local, state, and federal agencies, prevent public officials from meeting in private about matters that affect the state or local area.

Privacy and the Internet

Some people assume that there is no privacy on the Internet and that you use it at your own risk. Others believe that companies with Web sites should have strict privacy procedures and be accountable for privacy invasion. Regardless of your view, the potential for privacy invasion on the Internet is huge. People wanting to invade your privacy could be anyone from criminal hackers to marketing companies to corporate bosses. Your personal and professional information can be seized on the Internet without your knowledge or consent. E-mail is a prime target, as discussed previously. Sending an e-mail message is like having an open conversation in a large room—people can listen to your messages. When you visit a Web site on the Internet, information about you and your computer can be captured. When this information is combined with other information, companies can know what you read, what products you buy, and what your interests are. According to an executive of an Internet software monitoring company, "It's a marketing person's dream."

Most people who buy products on the Web say it's very important for a site to have a policy explaining how personal information is used, and the policy statement must make people feel comfortable and be extremely clear about what information is collected and what will and will not be done with it. However, many Web sites still do not prominently display their privacy policy or implement practices completely consistent with that policy. The real issue that Internet users need to be concerned with is: What do content providers want with their personal information? If a site requests that you provide your name and address, you have every right to know why and what will be done with it. If you buy something and provide a shipping address, will it be sold to other retailers? Will your e-mail address be sold on a list of active Internet shoppers? And if so, you should realize that it's no different than the lists compiled from the orders you place with catalog retailers. You have the right to be taken off any mailing list.

A potential solution to some consumer privacy concerns is the screening technology called the **Platform for Privacy Preferences (P3P)** being proposed to shield users from sites that don't provide the level of privacy protection they desire. Instead of forcing users to find and read through the privacy policy for each site they visit, P3P software in a computer's browser will download the privacy policy from each site, scan it, and notify the user if the policy does not match his or her preferences. (Of course, unethical marketers can post a privacy policy that does not accurately reflect the manner in which the data is treated.) The World Wide Web Consortium, an international industry group whose members include Apple, Commerce One, Ericsson, and Microsoft, is supporting the development of P3P. Version 1.0 of the P3P was released in April 2002 and can be found at *www.w3.org/TR/P3P*.

The Children's Online Privacy Protection Act (COPPA) was passed by Congress in October 1998. This act was directed at Web sites catering to children, requiring them to post comprehensive privacy policies on their sites and to obtain parental consent before they collect any personal information from children under 13 years of age. Web site operators who violate the rule could be liable for civil penalties of up to $11,000 per violation.[40]

Fairness in Information Use

Selling information to other companies can be so lucrative that many companies will continue to store and sell the data they collect on customers, employees, and others. When is this information storage and use fair and reasonable to the individuals whose data is stored and sold? Do individuals have a right to know about data stored about them and to decide what data is stored and used? As shown in Table 9.10, these questions can be broken down into four issues that should be addressed: knowledge, control, notice, and consent.

Fairness Issues	Database Storage	Database Usage
The right to know	Knowledge	Notice
The ability to decide	Control	Consent
Knowledge. Should individuals have knowledge of what data is stored on them? In some cases, individuals are informed that information on them is stored in a corporate database. In others, individuals do not know that their personal information is stored in corporate databases.		
Control. Should individuals have the ability to correct errors in corporate database systems? This is possible with most organizations, although it can be difficult in some cases.		
Notice. Should an organization that uses personal data for a purpose other than the original purpose notify individuals in advance? Most companies don't do this.		
Consent. If information on individuals is to be used for other purposes, should these individuals be asked to give their consent before data on them is used? Many companies do not give individuals the ability to decide if information on them will be sold or used for other purposes.		

Table 9.10

The Right to Know and the Ability to Decide

Federal Privacy Laws and Regulations

In the past few decades, significant laws have been passed regarding an individual's right to privacy. Other laws relate to business privacy rights and the fair use of data and information. These laws are summarized in Table 9.11.

State Privacy Laws and Regulations

State legislatures have been considering and passing privacy legislation that is far-reaching and potentially more burdensome to business than existing federal legislation. The use of Social Security numbers, access to medical records, the disclosure of unlisted telephone numbers, the sharing of credit reports by credit bureaus, the disclosure of bank and personal financial information, and the use of criminal files are some of the issues being considered by state legislators. These state proposals could have an enormous effect on companies that do business within their borders.

- In Georgia and California, businesses are prohibited from discarding records containing personal information without shredding, erasing, or modifying the documents to make the sensitive data unreadable.
- Federal "opt out" regulations force consumers to take action if they don't want the company to sell or share information about them. In Vermont, California, New Mexico, and North Dakota, however, "opt-in" is the default, and businesses must get consumers' permission to share their data.
- In a countermeasure against identify thieves, California allows its residents to "freeze" their credit reports to prevent lenders from accessing the files and granting credit. Consumers who take this step are given a special password to unfreeze the accounts when they want to apply for new loans or credit cards.

Such state-by-state and even county-by-county exceptions to the federal law greatly complicate financial record keeping and data sharing.

Law	Provisions
Fair Credit Reporting Act of 1970 (FCRA)	Regulates operations of credit-reporting bureaus, including how they collect, store, and use credit information
Tax Reform Act of 1976	Restricts collection and use of certain information by the Internal Revenue Service
Electronic Funds Transfer Act of 1979	Outlines the responsibilities of companies that use electronic funds transfer systems, including consumer rights and liability for bank debit cards
Right to Financial Privacy Act of 1978	Restricts government access to certain records held by financial institutions
Freedom of Information Act of 1970	Guarantees access for individuals to personal data collected about them and about government activities in federal agency files
Education Privacy Act	Restricts collection and use of data by federally funded educational institutions, including specifications for the type of data collected, access by parents and students to the data, and limitations on disclosure
Computer Matching and Privacy Act of 1988	Regulates cross-references between federal agencies' computer files (e.g., to verify eligibility for federal programs)
Video Privacy Act of 1988	Prevents retail stores from disclosing video rental records without a court order
Telephone Consumer Protection Act of 1991	Limits telemarketers' practices
Cable Act of 1992	Regulates companies and organizations that provide wireless communications services, including cellular phones
Computer Abuse Amendments Act of 1994	Prohibits transmissions of harmful computer programs and code, including viruses
Gramm-Leach-Bliley Act of 1999	Requires all financial institutions to protect and secure customers' nonpublic data from unauthorized access or use
USA Patriot Act of 2001	Requires Internet service providers and telephone companies to turn over customer information, including numbers called, without a court order, if the FBI claims that the records are relevant to a terrorism investigation

Table 9.11

Federal Privacy Laws and Their Provisions

Corporate Privacy Policies

Even though privacy laws for private organizations are not very restrictive, most organizations are very sensitive to privacy issues and fairness. They realize that invasions of privacy can hurt their business, turn away customers, and dramatically reduce revenues and profits. Consider a major international credit card company. If the company sold confidential financial information on millions of customers to other companies, the results could be disastrous. In a matter of days, the firm's business and revenues could be reduced dramatically. Thus, most organizations maintain privacy policies, even though they are not required by law. Some companies even have a privacy bill of rights that specifies how the privacy of employees, clients, and customers will be protected. Corporate privacy policies should address a customer's knowledge, control, notice, and consent over the storage and use of information. They may also cover who has access to private data and when it may be used.

Multinational companies face an extremely difficult challenge in implementing data-collection and dissemination processes and policies because of the multitude of differing country or regional statutes. For example, Australia requires companies to destroy customer data (including backup files) or make it anonymous once it's no longer needed. Firms that transfer customer and personnel data out of Europe must comply with European privacy laws that allow customers and employees to access data about them and let them determine how that information can be used.

A good database design practice is to assign a single unique identifier to each customer —so that each has a single record describing all relationships with the company across all its business units. That way, the organization can apply customer privacy preferences consistently throughout all databases. Failure to do so can expose the organization to legal risks—aside from upsetting customers who opted out of some collection practices. The 1999 Gramm-Leach-Bliley Financial Services Modernization Act required all financial service institutions to communicate their data privacy rules and honor customer preferences.

Individual Efforts to Protect Privacy

Although numerous state and federal laws deal with privacy, privacy laws do not completely protect individual privacy. In addition, not all companies have privacy policies. As a result, many people are taking steps to increase their own privacy protection. Some of the steps that individuals can take to protect personal privacy include the following:

- *Find out what is stored about you in existing databases.* Call the major credit bureaus to get a copy of your credit report. You can obtain one free if you have been denied credit in the last 60 days. The major companies are Equifax (800-685-1111, *www.equifax.com*), TransUnion (800-916-8800, *www.transunion.com*), and Experian (888-397-3742, *www.experian.com*). You can also submit a Freedom of Information Act request to a federal agency that you suspect may have information stored on you.

- *Be careful when you share information about yourself.* Don't share information unless it is absolutely necessary. Every time you give information about yourself through an 800, 888, or 900 call, your privacy is at risk. Be vigilant in insisting that your doctor, bank, or financial institution not share information about you with others without your written consent.

- *Be proactive to protect your privacy.* You can get an unlisted phone number and ask the phone company to block caller ID systems from reading your phone number. If you change your address, don't fill out a change-of-address form with the U.S. Postal Service; you can notify the people and companies that you want to have your new address. Destroy copies of your charge card bills and shred monthly statements before disposing of them in the garbage. Be careful about sending personal e-mail messages over a corporate e-mail system. You can also get help in avoiding junk mail and telemarketing calls by visiting the Direct Marketing Association Web site at *www.the-dma.org.* Go to the Web site and look under Consumer Help—Remove Name from Lists.

- *When purchasing anything from a Web site, make sure that you safeguard your credit card numbers, passwords, and personal information.* Do not do business with a site unless you know that it handles credit card information securely (with Netscape Navigator, look for a solid blue key in a small blue rectangle; with Microsoft Explorer, look for the words "Secure Web Site"). Do not provide personal information without reviewing the site's data privacy policy.

In addition to privacy concerns, another issue has been the focus of recent headlines. That issue is the ethical use of information systems, which we discuss next.

ETHICAL ISSUES IN INFORMATION SYSTEMS

As you've seen throughout the book in our "Ethical and Societal Issues" boxes, ethical issues deal with what is generally considered right or wrong. Some IS professionals believe that their field offers many opportunities for unethical behavior but that unethical behavior can be reduced by top-level managers developing, discussing, and enforcing codes of ethics. Information systems professionals are usually more satisfied with their jobs when top management stresses ethical behavior.

According to one view of business ethics, the "old contract" of business, the only responsibility of business is to its stockholders and owners. According to another view, the "social contract" of business, businesses are responsible to society. At one point or another in their operations, businesses may have employed one or both philosophies.

Various organizations and associations promote ethically responsible use of information systems and have developed codes of ethics. These organizations include the following:

- The Association of Information Technology Professionals (AITP), formerly the Data Processing Management Association (DPMA)
- The Association for Computing Machinery (ACM)
- The Institute of Electrical and Electronics Engineers (IEEE)
- Computer Professionals for Social Responsibility (CPSR)

The AITP Code of Ethics

The AITP has developed a code of ethics, standards of conduct, and enforcement procedures that give broad responsibilities to AITP members (see Figure 9.2). In general, the code of ethics is an obligation of every AITP member in the following areas:

- Obligation to management
- Obligation to fellow AITP members
- Obligation to society
- Obligation to college or university
- Obligation to the employer
- Obligation to country

For each area of obligation, standards of conduct describe the specific duties and responsibilities of AITP members. In addition, enforcement procedures stipulate that any complaint against an AITP member must be in writing, signed by the individual making the complaint, properly notarized, and submitted by certified or registered mail. Charges and complaints may be initiated by any AITP member in good standing.

Figure 9.2

AITP Code of Ethics

(Source: Courtesy of AITP— *www.aitp.org.*)

Code of Ethics

I acknowledge:

That I have an obligation to management, therefore, I shall promote the understanding of information processing methods and procedures to management using every resource at my command.

That I have an obligation to my fellow members, therefore, I shall uphold the high ideals of AITP as outlined in the Association Bylaws. Further, I shall cooperate with my fellow members and shall treat them with honesty and respect at all times.

That I have an obligation to society and will participate to the best of my ability in the dissemination of knowledge pertaining to the general development and understanding of information processing. Further, I shall not use knowledge of a confidential nature to further my personal interest, nor shall I violate the privacy and confidentiality of information entrusted to me or to which I may gain access.

That I have an obligation to my College or University, therefore, I shall uphold its ethical and moral principles.

That I have an obligation to my employer whose trust I hold, therefore, I shall endeavor to discharge this obligation to the best of my ability, to guard my employer's interests, and to advise him or her wisely and honestly.

That I have an obligation to my country, therefore, in my personal, business, and social contacts, I shall uphold my nation and shall honor the chosen way of life of my fellow citizens.

I accept these obligations as a personal responsibility and as a member of this Association. I shall actively discharge these obligations and I dedicate myself to that end.

The ACM Code of Professional Conduct

The ACM has developed a number of specific professional responsibilities. These responsibilities include the following:

- Strive to achieve the highest quality, effectiveness, and dignity in both the process and products of professional work
- Acquire and maintain professional competence
- Know and respect existing laws pertaining to professional work
- Accept and provide appropriate professional review
- Give comprehensive and thorough evaluations of computer systems and their impacts, including analysis of possible risks
- Honor contracts, agreements, and assigned responsibilities
- Improve public understanding of computing and its consequences
- Access computing and communication resources only when authorized to do so

The mishandling of the social issues discussed in this chapter—including waste and mistakes, crime, privacy, health, and ethics—can devastate an organization. The prevention of these problems and recovery from them are important aspects of managing information and information systems as critical corporate assets. Increasingly, organizations are recognizing that people are the most important component of a computer-based information system and that long-term competitive advantage can be found in a well-trained, motivated, and knowledgeable workforce.

SUMMARY

Principle

Policies and procedures must be established to avoid computer waste and mistakes.

Computer waste is the inappropriate use of computer technology and resources in both the public and private sectors. Computer mistakes relate to errors, failures, and other problems that result in output that is incorrect and without value. Waste and mistakes occur in government agencies as well as corporations. At the corporate level, computer waste and mistakes impose unnecessarily high costs for an information system and drag down profits. Waste often results from poor integration of IS components, leading to duplication of efforts and overcapacity. Inefficient procedures also waste IS resources, as do thoughtless disposal of useful resources and misuse of computer time for games and personal processing jobs.

Preventing waste and mistakes involves establishing, implementing, monitoring, and reviewing effective policies and procedures. Careful programming practices, thorough testing, flexible network interconnections, and rigorous backup procedures can help an information system prevent and recover from many kinds of mistakes. Companies should develop manuals and training programs to avoid waste and mistakes. Company policies should specify criteria for new resource purchases and user-developed processing tools to help guard against waste and mistakes.

Principle

Computer crime is a serious and rapidly growing area of concern requiring management attention.

Although no one really knows how pervasive cybercrime is, the number of IT-related security incidents is increasing dramatically—sixfold between 2000 and 2003. Some crimes use computers as tools (e.g., to manipulate records, counterfeit money and documents, commit fraud via telecommunications links, and make unauthorized electronic transfers of money). Identity theft is a crime in which an imposter obtains key pieces of personal information in order to impersonate someone else and obtain credit, merchandise, and services in the name of the victim, or to provide the thief with false credentials.

A cyberterrorist is someone who intimidates or coerces a government or organization to advance his or her political or social objectives by launching computer-based attacks against computers, networks, and the information stored on them. A criminal hacker, also called a *cracker*, is a computer-savvy person who attempts to gain unauthorized or illegal access to computer systems to steal passwords, corrupt files and programs, and even transfer money. Script bunnies are wannabe crackers with little technical savvy. Insiders are employees, disgruntled or otherwise, working solo or in concert with outsiders to compromise corporate systems.

Computer crimes target computer systems and include illegal access to computer systems by criminal hackers, alteration and destruction of data and programs by viruses, and simple theft of computer resources. A virus is a program that attaches itself to other programs. A worm functions as an independent program, replicating its own program files until it destroys other systems and programs or interrupts the operation of computer systems and networks. A Trojan horse program is a malicious program that disguises itself as a useful application and purposefully does something the user does not expect. A logic bomb is designed to "explode" or execute at a specified time and date.

Because of increased computer use, greater emphasis is placed on the prevention and detection of computer crime. Antivirus software is used to detect the presence of viruses, worms, and logic bombs. Use of an intrusion detection system (IDS) provides another layer of protection in the event that an intruder gets past the outer security layers—passwords, security procedures, and corporate firewall. It monitors system and network resources and notifies network security personnel when it senses a possible intrusion. Many small and midsized organizations are outsourcing their network security operations to managed security service providers (MSSPs), which monitor, manage, and maintain network security hardware and software.

Software and Internet piracy may represent the most common computer crime. It is estimated that the software industry loses nearly $12 billion in revenue each year to software piracy. Computer scams have cost individuals and companies thousands of dollars. Computer crime is also an international issue.

Many organizations and people help prevent computer crime, among them state and federal agencies, corporations, and individuals. Security measures, such as using passwords, identification numbers, and data encryption, help to guard against illegal computer access, especially when supported by effective control procedures. Public key infrastructure (PKI) enables users of an unsecured public network such as the Internet to securely and privately exchange data through the use of a public and a private cryptographic key pair that is obtained and shared through a trusted authority. The use of biometrics, involving the measurement of a person's unique characteristics such as iris, retina, or voice pattern, is another way to protect important data and information systems. Virus scanning software identifies and removes damaging computer programs. Law enforcement agencies armed with new legal tools enacted by Congress now actively pursue computer criminals.

Although most companies use data files for legitimate, justifiable purposes, opportunities for invasion of privacy abound. Privacy issues are a concern with government agencies, e-mail use, corporations, and the Internet. The Children's Internet Protection Act was enacted to protect minors using the Internet. Other laws have been enacted also, such as the Privacy Act of 1974, the USA Patriot Act, and the Gramm-Leach-Bliley Act.

Some states supplement federal protections and limit private organizations' activities within their jurisdictions. A business should develop a clear and thorough policy about privacy rights for customers, including database access. That policy should also address the rights of employees, including electronic monitoring systems and e-mail. Fairness in information use for privacy rights emphasizes knowledge, control, notice, and consent for people profiled in databases. Individuals should have knowledge of the data that is stored about them and have the ability to correct errors in corporate database systems. If information on individuals is to be used for other purposes, these individuals should be asked to give their consent beforehand. Each individual has the right to know and the ability to decide. Platform for Privacy Preferences (P3P) is a screening technology that shields users from Web sites that don't provide the level of privacy protection they desire.

Principle

Working conditions must be designed to avoid negative ethical consequences.

Ethics determine generally accepted and discouraged activities within a company and society at large. Ethical computer users define acceptable practices more strictly than just refraining from committing crimes; they also consider the effects of their IS activities, including Internet usage, on other people and organizations. The Association for Computing Machinery and the Association of Information Technology Professionals have developed guidelines and a code of ethics. Many IS professionals join computer-related associations and agree to abide by detailed ethical codes.

CHAPTER 9: SELF-ASSESSMENT TEST

Policies and procedures must be established to avoid computer waste and mistakes.

1. All IS professionals, managers, and users have a responsibility to see that the potential consequences of IS use are fully considered and addressed. True or False?

2. Computer-related waste and mistakes are major causes of computer problems, contributing to unnecessarily high _____ and lost _____.

3. The first step to prevent computer-related waste is to:

   a. establish policies and procedures regarding efficient acquisition, use, and disposal of systems and devices

   b. implement policies and procedures to minimize waste and mistakes according to the business conducted

   c. monitor routine practices and take corrective action if necessary

   d. review existing policies and procedures and determine whether they are adequate

Computer crime is a serious and rapidly growing area of concern requiring management attention.

4. The number of security problems reported to CERT between 2000 and 2003:

    a. increased sixfold
    b. decreased 50 percent
    c. doubled
    d. stayed about the same

5. _____ is the fastest and most pervasive worm attack on record, with damages estimated at $40 billion.

6. _____ is a crime in which an imposter obtains key pieces of personal identification information, such as Social Security or driver's license numbers, in order to impersonate someone else.

7. A person who enjoys computer technology and spends time learning and using computer systems is called a:

    a. script bunny
    b. hacker
    c. criminal hacker or cracker
    d. social engineer

8. A logic bomb is a type of Trojan horse that executes when specific conditions occur. True or False?

9. A program capable of attaching to disks or other files and replicating itself repeatedly, typically without user knowledge or permission is called a:

    a. logic bomb
    b. Trojan horse
    c. virus
    d. worm

10. Half the loss from software piracy comes from Asia. True or False?

11. Phishing is a computer scam that seems to direct users to their bank's Web site but actually captures key personal information about its victims. True or False?

Working conditions must be designed to avoid negative ethical consequences.

12. Ethical issues deal with what is generally considered to be legal and illegal. True or False?

13. Which of the following is an information systems-related organization that also promotes the ethically responsible use of information systems and has developed a code of ethics?

    a. Association of Information Technology Professionals
    b. Association for Computing Machinery
    c. Institute of Electrical and Electronic Engineers
    d. All of the above

CHAPTER 9: SELF-ASSESSMENT TEST ANSWERS

(1) True (2) costs, profits (3) a (4) a (5) Mydoom (6) Identity theft (7) b (8) True (9) d (10) True (11) True (12) False (13) d

---

REVIEW QUESTIONS

1. What is the CAPPS II program and what was its purpose?
2. Give two recent examples of computer mistakes causing serious repercussions.
3. Identify four broad actions that can be taken to prevent computer waste and mistakes.
4. Give two examples of recent, major computer crimes.
5. What is a cyberterrorist? What evidence is there of increased cyberterrorism activity?
6. What is identity theft? What actions can you take to reduce the likelihood that you will be a victim of this crime?
7. What is a virus? What is a worm? How are they different?
8. Outline measures you should take to protect yourself against viruses and worms.
9. What are the penalties for copyright infringement? How can a software user be guilty of this crime?
10. Identify at least five tips to follow to avoid becoming a victim of a computer scam.
11. What is biometrics, and how can it be used to protect sensitive data?
12. What is the difference between antivirus software and an intrusion detection system?
13. What is the Children's Internet Protection Act?
14. What is a code of ethics? Give an example.
15. What is the role of the Computer Emergency Response Team/Coordination Center?
16. What is social engineering?

DISCUSSION QUESTIONS

1. How does the CAN-SPAM Act affect those who send out spam? How might it affect those who receive spam?
2. Outline an approach, including specific techniques (e.g., dumpster diving, phishing, social engineering) that you could employ to gain personal data about the members of your class.
3. Identify and describe the different classes of computer criminals. What do you think motivates each?
4. Imagine that you are a hacker and have developed a Trojan horse program. What tactics might you use to get unsuspecting victims to load the program onto their computer?
5. Your marketing department has just opened a Web site and is requesting visitors to register at the site to enter a promotional contest where the chances of winning a prize are better than one in three. Visitors must provide the information necessary to contact them and fill out a brief survey about the use of your company's products. What data privacy issues may arise?
6. Briefly discuss the potential for cyberterrorism to cause a major disruption in our daily life. What are some likely targets of a cyberterrorist? What sort of action could a cyberterrorist take against these targets?
7. Imagine that you are a manager in a small business organization. Identify three good reasons why your firm should hire an MSSP. Can you think of any reasons why this may not be a good idea?
8. Compare and contrast the use of the ICRA rating system and Internet filters for protecting minors from viewing inappropriate material on the Internet.
9. During 2002, a number of corporations were forced to restate earnings because they had used unethical accounting practices. Briefly discuss the extent to which you think that these problems were caused by a failure in the firm's accounting information systems.
10. Using information presented in this chapter on federal privacy legislation, identify which federal law regulates the following areas and situations: cross-checking IRS and Social Security files to verify the accuracy of information, customer liability for debit cards, individuals' right to access data contained in federal agency files, the IRS obtaining personal information, the government obtaining financial records, and employers' access to university transcripts.
11. Briefly discuss the difference between acting morally and acting legally. Give an example of acting legally and yet immorally.

PROBLEM-SOLVING EXERCISES

1. Access the Web site of one of the antivirus software providers. Get statistics on the number of viruses and their cost impact for the past four years. Use the graphics routine in your spreadsheet software to graph the increase over time.
2. Using your word processing software, write a few brief paragraphs summarizing the trends you see from reviewing the virus data for the past few years. Then, cut and paste the graph from exercise 1 into your report.
3. Draft a letter to the Webmaster of the antivirus software Web site you chose in exercise 1. Share your report and request an estimate of the number of viruses for the next three years.

TEAM ACTIVITIES

1. Visit your local library and interview the librarians about their experience with the use of Internet filters. What software is used? Who is responsible for updating the list of sites off limits to minors? What level and kinds of complaints are made about the use of this technology? Do some patrons insist that the filter should be removed? Overall, what is the librarian's opinion of the use of Internet filters? Write a brief paper summarizing the interview.
2. Have each member of your team access ten different Web sites and summarize their findings in terms of the existence of data privacy policy statements: Did the site have such a policy? Was it easy to find? Was it complete and easy to understand? Did you find any sites using the P3P standard or ICRA rating method?

WEB EXERCISES

1. Search the Web for a site that provides software to detect and remove spyware. Write a short report for your instructor summarizing your findings.
2. Do research on the Web to find evidence of an increase or decrease in the amount of spam. To what is the change attributed? Write a brief memo to your instructor identifying your sources and summarizing your findings.
3. Echelon is a top-secret electronic eavesdropping system managed by the U.S. National Security Agency that is capable of intercepting and decrypting almost any electronic message sent anywhere in the world. It may have been in operation as early as the 1970s, but it wasn't until the 1990s that journalists were able to confirm its existence and gain insight into its capabilities. Do Web research to find out more about this system and its capabilities. Write a paragraph or two summarizing your findings.

CAREER EXERCISES

1. Computer forensics is a relatively new but growing field, which involves the discovery of computer-related evidence and data. It relies on formal computer evidence-processing protocols. Its findings may be presented in a court of law. Cases involving trade secrets, commercial disputes, employment discrimination, misdemeanor and felony crimes, and personal injury can be won or lost solely with the introduction of recovered e-mail messages and other electronic files and records. Computer forensics tools and methods are used extensively by law enforcement, military, intelligence agencies, and businesses. Do research to identify the experience and training necessary to become a certified computer forensics specialist. Would you consider this as a possible career field? Why or why not?

2. Your marketing organization is establishing a Web site to promote, market, and sell your firm's products. Make a list of the laws and regulations that might apply to the design and operation of this Web site. How would you confirm that your Web site is in conformance with all of them?

VIDEO QUESTIONS

Watch the video clip **Controversial Digital Bouncers** and answer these questions:

1. Would you be comfortable with a business establishment collecting and storing identifying information about you from your driver's license, along with a snapshot of your face for security use? Why or why not?
2. Are there any laws that restrict a club from distributing the information it collects about you to other businesses?

CASE STUDIES

Case One

Working to Reduce the Number of Software Vulnerabilities

Exploiting programming flaws (called *vulnerabilities*) is the primary means by which hackers gain access to networks, computers, and applications today. Examples of commonly exploited vulnerabilities include buffer overflows, invalidated parameters, format string errors, and broken access control. The number of reported vulnerabilities increased from 171 in 1995 to 3,784 in 2003, according to Carnegie Mellon's CERT Coordination Center. Not surprisingly, there has been a dramatic increase in the number of security breaches.

Software developers and their customers seem to be stuck in an endless cycle of costly development and deployment of faulty software, followed by the release of patches to fix newly discovered vulnerabilities, and then scrambling by end users to install the patches, ideally fast enough that hackers do not take advantage of them.

"Zero-day" attacks are hacker attacks that take advantage of programming vulnerabilities before software makers can identify them and develop patches to correct them. Fortunately, there haven't yet been any major zero-day attacks, but malicious hackers are getting better and faster at exploiting flaws. Summer 2003's Blaster worm, one of the most virulent and widespread ever, hit the Internet barely a month after Microsoft released a patch for the software flaw it exploited. In contrast, January 2003's SQL Slammer worm took eight months to appear after the vulnerability it targeted was first disclosed. The lead time for users to apply fixes is shrinking.

Fortunately, there are a small, but growing group of software vendors offering tools designed to help companies identify and fix flaws in the application development stage and to eliminate software vulnerabilities before the applications are deployed. Cambridge, Massachusetts–based @stake unveiled its SmartRisk Analyzer in May 2004. This modeling and analysis tool scans computer code written in the C, C++, and Java languages for flaws that could pose security risks for customers using finished software products. The software uses a process called *static analysis*, which allows developers to identify and eliminate problems as they are writing the code. This proactive approach is opposed to so-called dynamic analysis tools, which use automated input tests to measure the response of finished applications. The SmartRisk Analyzer product compares binary code (the zeros and ones that are the foundation of all computer languages) to an @stake database of about 400 security and code reliability rules. It can generate reports that list flaws by type or rank them by severity. A remediation module marks erroneous code and suggests ways to fix coding mistakes.

In April 2004, start-up company Fortify Software introduced Fortify Source Code Analysis, a suite of software products that lets companies compare C++ and Java code to a list of more than 500 vulnerabilities published by software quality management company Cigital. The suite also includes a capability that allows project testers and quality assurance teams to test the software just before deployment, including simulating a hacker using every trick in the book to compromise the software. The software is not inexpensive. A 25-person project team using the security platform would pay around $150,000 plus an annual $1,000 subscription to get the latest rules when they come out.

Unlike manual audits and code reviews that are time-consuming and limited in scope, an automated approach accelerates the industry's ability to deliver secure software. Software developers now can run security checks whenever they want—and against as much of the code base as they desire.

Discussion Questions

1. Why is there concern over the shrinking time between identification of a software vulnerability and the launch of an attack exploiting that vulnerability?
2. What is static analysis? What are some advantages of static analysis over dynamic analysis? Why do software developers employ both methods?

Critical Thinking Questions

3. What factors might cause a software vendor to sacrifice the quality of the underlying code of its products?
4. What measures can a software buyer take to ensure the quality of the underlying code of a software package—before purchasing and installing the software?

SOURCES: Paul Roberts, "Secure Coding Attracts Interest, Investment," *Computerworld*, May 24, 2004, *www.computerworld.com*; Jaikumar Vijavan, "Fortify Launches Tools for Security Testing during App Development," *Computerworld*, April 5, 2004, *www.computerworld.com*; Jim Wagner, "A New Approach to Fortify Your Software," *E-Security Planet*, April 5, 2004, *www.esecurityplanet.com*; Jaikumar Vijavan, "InfoSec 2003: 'Zero-Day' Attacks Seen as Growing Threat," *Computerworld*, December 11, 2003, *www.computerworld.com*.

Case Two
Beware Spyware!

Adware is any software application in which advertising banners are displayed while the program is running. The authors of these applications include additional code that delivers the ads, which can be viewed through pop-up windows or through a bar that appears on a computer screen. The developer of the application is paid an advertising fee based on how many users view the ads. This additional payment helps recover the author's programming development cost and hold down the cost for the user.

But adware can include additional code called *spyware* that captures the user's personal information and passes it on to third parties, without the user's explicit consent or knowledge. Although spyware is not illegal, computer security and privacy advocates, including the Electronic Privacy Information Center, denounce its use. Their concern is with the uses of the information that adware and spyware providers collect and the lack of control users have over what those providers "feed" them.

Spyware and adware both enter a computer system when a user opens an attachment or clicks on a Web page that allows a program to be executed on the system without the user's knowledge. Both make similar modifications to a system, such as changing Windows registry settings, adding services, and installing and executing applications. This process usually involves the tracking and sending of data and statistics via a server installed on the user's PC and the use of the user's Internet connection in the background. Even though the name may indicate it, spyware is not illegal. However, it raises certain privacy issues. The spyware may include surveillance programs, key loggers, remote-control tools, and Trojan horses that run in the background without the users' knowledge, downloading information on Web-surfing activities and uploading advertising in the background for use in pop-up ads.

Initially, organizations viewed spyware as a nuisance that was best handled by desktop support groups. But the key issue isn't system performance or productivity-sapping pop-ups—it's the uneasy feeling that these programs have opened an unauthorized communication channel that could put sensitive information at risk. CIOs worry that, in addition to downloading data on Web-surfing activity, a spyware program may capture user login and password information or that a benign adware program may provide a communications pathway that could be hijacked for uploading more malicious software or sending sensitive data to competitors.

Unfortunately, the current generation of antivirus and firewall software does not control or eliminate spyware effectively. Indeed, most users aren't even aware that they've been infected until they start noticing slow system performance or odd computer behavior. Once a system is infected, often the only way to remove the problem is to rebuild the system from scratch, a tedious and time-consuming process.

Discussion Questions

1. A number of software applications are available as freeware to help computer users search for and remove suspected spyware programs. Is it possible that a purveyor of spyware could attract unsuspecting users by promoting free spyware-removal software?
2. Do a Web search on spyware and identify specific actions recommended to identify and remove spyware.

Critical Thinking Questions

3. You are short on funds, and a friend approaches you with an opportunity to work for a company creating Web links for spyware downloads. He has been receiving checks for this work every couple of weeks. Make a list of the ethical and privacy factors you would consider in deciding whether or not to take on this opportunity.
4. Would you take a job helping companies to download spyware? Why or why not?

SOURCES: Robert L. Mitchell, "Spyware Sneaks into the Desktop," *Computerworld*, May 3, 2004, *www.computerworld.com*; John Soat, "IT Confidential: Anti-Spyware's First Step; Accenture OK," *InformationWeek*, June 21, 2004, *www.informationweek.com*; The Advisory Council, "SmartAdvice: Clean and Manage Company Data to Learn What Information You've Got," *InformationWeek*, June 21, 2004, *www.informationweek.com*; Mathias Thurman, "Spyware Gets Top Billing," *Computerworld*, July 5, 2004, *www.computerworld.com*; Emily Kumler, "Who's Seeding the Net with Spyware," *PC World*, June 15, 2004, *www.pcworld.com*.

NOTES

Sources for the opening vignette: Gross,Grant, "U.S. Agencies Defend Data Mining Plans," *Computerworld*, May 12, 2003, *www.computerworld.com*; Greenemeier, Larry, "CAPPS II Progress Raises Privacy Concerns," *InformationWeek*, February 12, 2004, *www.informationweek.com*; Verton, Dan, "Airline Passenger Screening System Faces Deployment Delays," *Computerworld*, February 16, 2004, *www.computerworld.com*; Kontzer, Tony, "Privacy Pressure," *InformationWeek*, March 22, 2004, *www.informationweek.com/story*; Greenemeier, Larry, "Security Agency Used More Passenger Data Than First Thought," *InformationWeek*, June 23, 2004, *www.informationweek.com*; Greenemeier, Larry, "CAPPS II Crash-Lands," *InformationWeek*, July 19, 2004, *www.informationweek.com*.

1. Associated Press, "Soyuz Capsule Lands Safely in Kazakhstan," *FoxNew.com*, October 27, 2003, accessed at *www.foxnews.com*.
2. Williams, Martyn, "Japan Retailer Deluged with Orders for $25 eMacs," *Computerworld*, April 23, 2004, accessed at *www.computerworld.com*.
3. Krasnowski, Matt, "Macy's OKs Settlement in Price-Accuracy Case," *San Diego Union Tribune*, May 22, 2004, accessed at *www.signonsandiego.com*.
4. Wrolstad, Jay, "New IE Flaws Labeled Extremely Critical," *News Factor Network*, November 26, 2003, accessed at *www.newfactor.com*.
5. Roberts, Paul, IDG News Service, "Software Bug May Cause Missile Errors," *PC World*, March 2003, *www.pcworld.com*.

6. Martin, Steve, "Lack of IT Controls Seen as Reason for Earnings Restatement," *InformationWeek*, October 13, 2004, *www.informationweek.com*.
7. "2004 Global Information Security Survey," *InformationWeek*, June 30, 2004, accessed at *www.informationweek.com*.
8. Verton, Dan, "Criminals Using High Tech Methods for Old Style Crimes," *Computerworld*, February 13, 2003, accessed at *www.computerworld.com*.
9. Roberts, Paul, "Michigan Man Pleads Guilty to Wireless Hack into Stores," *Computerworld*, June 7, 2004, accessed at *www.computerworld.com*.
10. Verton, Dan, "Europe Battles Insiders-Turned-Hackers, EU Cyber-cop Says," *Computerworld*, July 31, 2003, accessed at *www.computerworld.com*.
11. Claburn, Thomas, "Feds Want Tougher Penalties for Insider Identity Theft," *InformationWeek*, May 24, 2004, accessed at *www.informationweek.com*.
12. Claburn, Thomas, "Feds Want Tougher Penalties for Insider Identity Theft," *InformationWeek*, May 24, 2004, accessed at *www.informationweek.com*.
13. Claburn, Thomas, "Feds Want Tougher Penalties for Insider Identity Theft," *InformationWeek*, May 24, 2004, accessed at *www.informationweek.com*.
14. Morphy, Erika, "Trojan Virus Author Busted for Making a Peep," *NewsFactor Network*, May 28, 2004, accessed at *www.newsfactor.com*.

15. Brewin, Bob, "Worldport Help Desk Helps Prevent Virus Attacks," *Computerworld*, April 19, 2004, accessed at *www.computerworld.com*.

16. Cohen, Peter, "Macs and Viruses—Are Users as Safe as They Think?" *Computerworld*, February 18, 2004, accessed at *www.computerworld.com*.

17. Jesdandun, Anick, "Kinko's Case Highlights Internet Risks," *InformationWeek*, July 22, 2004, accessed at *www./information.com*.

18. " 2003 BSI Computer Theft Survey," accessed at *www.brigadoonsoftware.com/survey.php* on May 30, 2004.

19. "What Is Piracy?" SIIAA Anti-Piracy Web site, *www.spa.org/piracy/whatis.aspon*, accessed on May 30, 2004.

20. The Business Software Alliance Web site, "Piracy and the Law," *www.bsa.org/usa/antipiracy/Piracy-and-the-Law.cfm*, accessed on May 30, 2004.

21. Rosencrance, Linda, "Citibank Customers Hit with E-Mail Scam," *Computerworld*, October 24, 2003, accessed at *www.computerworld.com*.

22. Roberts, Paul, "Phishing Scams Skyrocket in April," *Computerworld*, May 18, 2004, accessed at *www.computerworld.com*.

23. Roberts, Paul, "Gartner: Phishing Attacks Up against U.S. Consumers," *Computerworld*, May 6, 2004, accessed at *www.computerworld.com*.

24. Evers, Joris, "FTC Cracks Down on Web Page Selling Scam," *Computerworld*, August 13, 2003, accessed at *www.computerworld.com*.

25. Software and Information Industry Association, "Anti-Piracy," accessed at *www.siia.net/piracy/whatis.asp*, on June 29, 2004.

26. Songini, Marc L., "CA to Push Automated Fraud Detection," *Computerworld*, March 10, 2003, accessed at *www.computerworld.com*.

27. McDougall, Paul, "Accenture's 'Virtual Border' Project," *InformationWeek*, June 7, 2004, accessed at *www.informationweek.com*.

28. Rosencrance, Linda, "Red Bull Pays $105,000 to Settle Software Piracy Claim," *Computerworld*, March 17, 2004, accessed at *www.computerworld.com*.

29. Thurman, Mathias, "Failure to Communicate," *Computerworld*, April 26, 2004, *www.computerworld.com*.

30. "ISS Backs New IPS Offering with Cash-or-Credit Guarantee," *Network World*, January 1, 2004, accessed at *www.nwfusion.com*.

31. 2004 Internet Filter Report, found at *www.internet-filter-review.com*, accessed on July 1, 2004.

32. "ACLU Disappointed in Ruling on Internet Censorship in Libraries, But Sees Limited Impact for Adults," the American Civil Liberties Union Web site, *www.aclu.org*, on June 23, 2003.

33. Roberts, Paul, "I'll See Your Web Site in Court," *Computerworld*, March 7, 2003, accessed at *www.computerworld.com*.

34. Roberts, Paul, "Mohegan Sun Won't Gamble on Insider Threats," *IDG News Service*, April 30, 2004, accessed at *www.computerworld.com*.

35. Verton, Dan, "Senate Votes to Kill Funds for Antiterror Data Mining," *Computerworld*, July 21, 2003, accessed at *www.computerworld.com*.

36. From the Seisint Web site, *www2.seisint.com/aboutus/index.html*, accessed on July 3, 2004.

37. Bergstein, Brian, "Seven-State Info Store Is a Potent Repository of Personal Data," *InformationWeek*, January 23, 2004, accessed at *www.informationweek.com*.

38. Sullivan, Andy, "Government Computer Surveillance Rings Alarm Bells," *Computerworld*, May 27, 2004, accessed at *www.computerworld.com*.

39. Bergstein, Brian, "Seven-State Info Store Is a Potent Repository of Personal Data," *InformationWeek*, January 23, 2004, accessed at *www.informationweek.com*.

40. "The Children's Online Privacy Protection Act," Electronic Privacy Information Center, *www.epic.org/privacy/kids*, accessed on July 3, 2004.

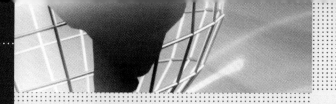

Brandon Trust Develops MIS Capability for Improved Operations and Services

Andy Igonor
University of the West of England

Organizations typically exist to make money and to save money. But what about those organizations registered as charities, whose goals are something other than profit making? Brandon Trust, a major player in the healthcare sector, with an annual turnover exceeding £20 million, is a registered charity in England. It does not exist to make money; rather, in the words of its chief executive officer, Steve Bennett, Brandon Trust is a "people supporting people" organization. As a visionary, Brandon Trust's CEO believes that making a difference is essential to sustaining the organization's aim and purpose.

In the quest to improve service operations, Steve Bennett met with Charles Harvey, a business strategist and expert in leadership and management practices, who also currently serves as dean of the UK's Bristol Business School. The decision to invest in information systems followed from this preliminary study of the Brandon Trust's strategy and business systems. The study revealed a lack of consistency and some redundancy in existing processes stemming from a lack of integration. Brandon Trust's situation is similar to other enterprises and was of both theoretical and practical interest to researchers at the Bristol Business School who were studying business systems and innovation. The outcome of this consultation led to a partnership agreement to help in the strategic streamlining of service operations and delivery. Sponsored by the Knowledge Transfer Partnership (KTP)—a government-funded program—the project involves the automation, rationalization, and reengineering of part of Brandon Trust's activities and procedures, with the goal of eliminating bottlenecks in information production, distribution, and use.

Brandon Trust operates community teams, employment and training units, and day-care centers spread across various areas of Bristol and the South West region of England (of which Bristol is a part). Brandon Trust also provides intensive support and supported living so that people with disabilities can live in a place of their preference. In meeting its obligations, the Trust runs a coordinated operational network, where people, both office and field staff, have to effectively communicate in meeting clients' needs. Apart from its full-time employees, the Trust also recruits part-time, hourly workers to facilitate the efficient flow of services. Brandon Trust exists in a large market for learning disability care in which purchasers demand greater efficiency and effectiveness. Brandon Trust is a market leader in providing learning disability services, and it has grown by taking over contracts. Its service area is mainly confined to Bristol, South Gloucestershire, Bath, Northeast Somerset, and North Somerset. Its competitive advantage lies in service innovation, user-centered provision, and quality management. Its main competitors are local authorities and other charitable trusts, which are typically quite small and less developed in terms of management and systems. The Trust is now trying to improve the quality of its own management systems by using innovative approaches to maximize its effectiveness.

Brandon Trust's Finance and Business Systems Director, Hilary Pearce, believes that information systems can, if efficiently applied and deployed, be used to improve facets of the organization's business. In her words, "I believe that by working with our academic partner and our KTP Associate that the Brandon Trust will derive real benefit from the project. The pooling of ideas and expertise will be vital as we analyse our information needs and data sources and identify key result areas that we need to focus on to improve the management of the Trust. I envisage an innovative and integrated business system solution that will continue to evolve and grow along with the Trust's operational needs." The role of information systems in strengthening the Trust's service level cannot be overemphasized. Its chief executive officer, Steve Bennett, had this to say: "As an organization that has grown 500% in the ten years of its existence, it is important [for Brandon Trust] to take stock and look for continual improvements in our performance. Information and the systems that provide it are fundamental to performance management, and the importance of having systems that utilize available technology in a pragmatic and useable way cannot be overstated. I therefore am excited to address this with our academic partners and KTP associates in a two-year project, the outcome of which will be an integrated business system solution that will continue to evolve and grow with the continuing development of Brandon Trust."

The Trust's services are distributed via 50 sites, and it is essential to improve information flows for efficiency and the management of risk. This development project was therefore focused on implementing new management systems and integrating them with other core business systems, including finance, following a detailed analysis of information requirements. Lack of systems integration currently makes provision of sound information for management controls, budgeting, and costing of contract proposals difficult to achieve. Improving information and business systems across the Trust will enhance its capacity for further growth by ensuring the best possible quality and design in its services. It is anticipated that the program will significantly reduce costs through business process reengineering, resulting in savings in the region of £250,000 per annum.

One KTP associate, Abid Mohammed, was recruited to serve as Industrial Manager and to work with Andy Igonor (Academic Advisor and Project Manager) and Hilary Pearce. Abid's initial task included a review and evaluation of all existing business systems, particularly computer systems at the Trust. While a number of systems existed in the Trust, the systems did not to talk to one another; notably, the crucial financial system needed to manage all finance-related issues from payroll to payment for contract services. The hoped-for result of the development project is a system with the following capabilities:

- An effective system capable of delivering relevant information when needed
- An integrated system with simplified information presentation and elimination of redundancy
- The development of an IS/IT audit report indicating current operating resources, which also serves as guide for future work, including identification of appropriate software and hardware for improved operations, improved access to information, improved management information systems reporting, and improved performance

Discussion Questions

1. What are the key issues to be looked into during systems investigation? How might they affect the overall success of the project?
2. Brandon Trust does not exist to make money. How relevant will this systems development project be to the organization?
3. Why is it necessary for Brandon Trust's systems to be integrated? Of what benefit would this be to the organization?

Critical Thinking Questions

4. In what ways can a management information system benefit an organization?
5. How often should reports be generated within such systems in order to fulfill managerial needs?

Note: All information provided in this case is courtesy of the Brandon Trust. Used with permission.

Strategic Enterprise Management at International Manufacturing Corporation (IMC)

Bernd Heesen
University of Applied Sciences Furtwangen, Germany

The International Manufacturing Corporation (IMC), with 3,000 employees worldwide, is a supplier of automotive parts and is headquartered in the United States. Given the increasing market pressure and the need to operate internationally, IMC has recently made three acquisitions. The acquisition of one company in Great Britain and two companies in Germany is expected to leverage the companies' complementary manufacturing and sales capabilities to gain a stronger market presence in North America and Europe. The subsidiaries are still managed by their former management teams, who are now reporting to the U.S. headquarters. Each of the subsidiaries still operates its own management information system because the system fulfilled the requirements in the past and at the time of the acquisition there was no visible benefit to change what was working. A couple of interfaces were created to facilitate the monthly financial reporting to the headquarters.

Recently, the CIO of IMC realized some disadvantages of the diversity of hardware and software platforms. When one IT expert from one German subsidiary terminated her employment, no other IT expert within IMC had detailed knowledge of the local system, which threatened regular operations in that plant. The company needed to seek external consulting support to maintain the system until a new IT expert could be hired. This dependency on individual experts in each location was expensive and caused problems during periods when the experts were on leave. In addition, the interfaces between the subsidiaries' systems and the headquarters' system that were created right after the acquisition had to be modified whenever a change or upgrade was made to one of the subsidiary systems. Even minor changes, such as the reorganization of product codes or sales organizations, required reprogramming, maintenance of conversion tables, and subsequent testing of the related business processes in the system. The IT departments worked overtime and still did not find the time to invest in new initiatives such as enabling mobile computing for the sales force and developing a strategic enterprise management system that would allow consolidated planning and monitoring of all key performance indicators for the organization, a project long requested by the CEO. The two German subsidiaries had recently upgraded their systems, and problems identified during the testing of the interfaces could not be corrected in time for the submission of the monthly reports to headquarters. The data was finally corrected, but the monthly reporting was several days late. This was not the first time that the monthly financial reporting had been delayed or that the data needed to be corrected because of system-related problems, which caused other problems at headquarters.

The CEO finally requested consolidation of the systems in the coming year to allow for an integrated strategic enterprise management system that would provide current information from all legal entities and support a consolidated budgeting and planning process. He compared the company's current system to the cockpit of an airplane and stated that "no one would expect a pilot to fly an airplane with malfunctioning instruments; hence nobody can expect management to run a company based on incomplete or incorrect information." The CIO was asked to develop a business case for the implementation of such a system, considering the savings from personnel, hardware, and software, as well as process improvements.

After IMC's board approved the business case to implement enterprise resource planning (ERP) software from SAP, the CIO established a project in January 1999. The project schedule called for completion of the financial, human resource, supply chain management, and customer relationship management functions for the complete organization—called the Big Bang—in January 2000. A prototype of the application was planned to be ready by July 1999 so that the departments could test all functionality with a set of converted data extracted from the current systems. The plan also called for complete conversion of live data by the end of November 1999 so that in December both the current system and the new system could be used in parallel before making the decision to switch systems. Consultants from SAP were hired to support the implementation—for project management and configuration of the system—and an independent training organization was charged with planning the end-user training and developing computer-based training (CBT) programs for ongoing use. The project plan allowed for hiring temporary staff to help with the cutover and redundant data entry in December 1999.

The project started with a kickoff meeting in February, after which nearly all team members began their project work. The complete prototype of the application could only be made available in September, because some of the team members had to fulfill their regular jobs while providing their expertise to the project. Some of the early project decisions had to be revised to accommodate the best-practice business model from IMC, once the IMC team members better understood how the SAP system worked and the consultants had gained a better understanding of how IMC wanted to operate. The CIO still wanted to maintain the timeline and called weekly meetings of the project's team leaders for finance, human resources, supply chain management, and customer relationship management. Those meetings led to the decision to delay rollout of some of the functionality and to reduce the functionality to what was really needed (eliminating nice-to-have functionality). The prototype testing produced additional feedback from the departments that needed to be incorporated in the design of the final solution prior to the conversion. Many of the department heads, who had not been part of the core project team, only then became aware of the changes and started to talk about the new software, saying that it was not working properly yet and some of the team members were scared they would not make the deadline. In November, the data conversion revealed some additional problems (e.g., with data entry errors made by the temporary staff who were keying data into the new system in instances in which an automated conversion was not cost-effective). As a result, the parallel use of the current and new system, which depended on all data being available, was delayed. To add to the pressure, the CIO believed that delaying the project's completion could be perceived as his failure. The CIO and steering committee of the project needed to make a decision on how to move forward: (1) continue with the project's completion in January 2000 as planned or (2) delay conversion to the new system until a proper parallel test could be completed.

Discussion Questions

1. How would you develop a business case to justify an investment in an integrated strategic enterprise management system?
2. What are the essential requirements for a strategic management system?

Critical Thinking Questions

3. Which tasks would you consider when developing the original project plan for the implementation?
4. Given IMC's current situation, what would you decide if you were the CIO?

Note: The name International Manufacturing Corporation was selected to protect the identity of the real company featured in the case.

Efforts to Build E-Government and an Information Society in Hungary

Janos Fustos
Metropolitan State College of Denver

Communication between citizens and their government has always been an important goal. State agencies and institutions must ensure open access to documents and forms and provide information about programs, services, and policies. This information must be timely, accurate, complete, and updated regularly and has to be freely and immediately available. Modern communication technology and the application of Internet services can support such governmental efforts and help establish e-government and an information society. These efforts are even more important in the former Communist countries in Eastern Europe, where the citizen-to-government communication had to be rebuilt and the governments have higher obligations to close the information gap. These countries also have to promote the use of high-tech solutions to help bridge the "information divide," keep pace with the rest of the world, and start to build the foundations of the information society.

After living under Soviet influence for 40 years, Hungary had the first free multiparty parliamentary elections in April 1990. The new National Assemblies and the coalition governments formed after the elections committed themselves to the establishment and stabilization of the political, economic, and legal foundations brought about by the systemic change. There was—and still is—a lot to do, but the gap is narrowing, as shown by the spread of telecommunications technology in the European Union (EU):

Phone lines (per 1,000 people):

	1990	1995	2002
USA	550	630	646
EU	422	492	546

Before the changes, communication was under political control—in an effort to avoid free dissemination of information and make sure it closely followed the official view. As Hungary started to move toward the Western world and a free-market economy, communication—phone lines and Internet connections—became focal points for development. By 1997, a vast majority of the previously state-owned companies and assets had been privatized and were sold to international investors that competed for the businesses.

The Hungarian government was one of the early adopters of Internet communication and solutions. In 1991, it established the Inter-Departmental Committee of Informatics and the Coordination Office at the Prime Minister's Office. These two offices coordinated official efforts for and between departments and agencies. In 1995, the government published a Strategy for Informatics, which laid out countrywide communication efforts. In 1999, the government published the *Hungarian Reply to the Challenges of the Information Society*, which discussed the technological background and requirements of an information-based society. Hungary also joined NATO in 1999, which was a major acknowledgment of the country's efforts to rebuild its political system and establish a free society. Then in 2001, the European Commission published the *eEurope + Action Plan* in conjunction with Central and East European countries. The goal of this plan was to foster the development of an information society in the countries that were joining the EU, including Hungary. That same year, the prime minister's office launched Hungary's first e-government portal, eKormanyzat.hu, providing citizens and businesses with a user-friendly entry point to government information and services. In 2002, the Ministry of Informatics and Communications (IHM) was created and assumed responsibility for development of the information society from the Office of the Government. To establish wider involvement, the Interministerial Committee on Information Society was formed to coordinate programs with professional organizations, businesses, and economic partners. The Interministerial Committee also oversees the formation and annual renewal of Hungarian information society strategy (*http://english.itktb.hu/Engine.aspx*). As of May 1, 2004, Hungary became a full member of the EU. Recent statistics on Internet usage show the growth of Hungary's information society, in comparison with the United States and the EU overall:

Number of Internet users (per 1,000 people)

	1992	1997	2002
USA	18	150	538
EU	3	56	204
Hungary	1	19	158

Number of Internet hosts (per 1,000 people)

	1997	2002
USA	56	382
EU	12	40
Hungary	3	20

The Hungarian government also established several programs to promote the use of computers and created opportunities for citizens to use them in education and in everyday life. Higher education institutions were among the first to obtain high-speed Internet access, and it soon became fully integrated into the European academic community. At this time, all Hungarian schools have Internet connections. Students and teachers (in primary education, high school, college, and continuing education) get tax relief when buying computers, digital devices, and software. The Sulinet (school-net) program also supports research, conferences and seminars, developers, content providers, and smart classrooms. The following statistics show the growth of PC usage over a decade:

Number of PCs (per 1,000 people):

	1992	1997	2002
USA	218	409	659
EU	97	194	310
Hungary	19	58	108

The Internet is becoming more and more available as a community service all over the country. Post offices, schools, bus stations, churches, hospitals, libraries, and a network of community centers offer access for free or for a flat rate. Hungarians even invented a traffic sign containing an "@" symbol to identify Internet access points and recommended it for worldwide use.

"E-government is not 'old government' plus the Internet. E-government is the use of new technologies to transform Europe's public administrations and to improve radically the way they work with their customers, be they citizens, enterprises, or other administrations. Furthermore, e-government is now a key vehicle for the implementation and achievement of higher policy objectives," states the Commission of the European Communities.

Discussion Questions

1. Why is it important for governments to introduce and apply high-tech solutions to promote programs, services, and policies?
2. What features are implemented in your state or local government's Web portal?

Critical Thinking Questions

3. How can access to information transform a society?
4. Besides e-government efforts, what are other elements of an information society?

accounting MIS An information system that provides aggregate information on accounts payable, accounts receivable, payroll, and many other applications.

ad hoc DSS A DSS concerned with situations or decisions that come up only a few times during the life of the organization.

antivirus program Program or utility that prevents viruses and recovers from them if they infect a computer.

application program interface (API) The interface that allows applications to make use of the operating system.

application service provider (ASP) A company that provides software, end-user support, and the computer hardware on which to run the software from the user's facilities.

application software The programs that help users solve particular computing problems.

arithmetic/logic unit (ALU) The portion of the CPU that performs mathematical calculations and makes logical comparisons.

ARPANET A project started by the U.S. Department of Defense (DoD) in 1969 as both an experiment in reliable networking and a means to link DoD and military research contractors, including a large number of universities doing military-funded research.

artificial intelligence (AI) The ability of computers to mimic or duplicate the functions of the human brain.

artificial intelligence systems The people, procedures, hardware, software, data, and knowledge needed to develop computer systems and machines that demonstrate the characteristics of intelligence.

asking directly An approach to gather data that asks users, stakeholders, and other managers about what they want and expect from the new or modified system.

attribute A characteristic of an entity.

audit trail Documentation that allows the auditor to trace any output from the computer system back to the source documents.

backbone One of the Internet's high-speed, long-distance communications links.

backward chaining The process of starting with conclusions and working backward to the supporting facts.

batch processing system Method of computerized processing in which business transactions are accumulated over a period of time and prepared for processing as a single unit or batch.

best practices The most efficient and effective ways to complete a business process.

biometrics The measurement of a person's trait, whether physical or behavioral.

bot A software tool that searches the Web for information, products, prices, and so forth.

brainstorming A decision-making approach that often consists of members offering ideas "off the top of their heads."

business continuity planning Identification of the business processes that must be restored first in the event of a disaster and specification of what actions should be taken and who should take them to restore operations.

business intelligence (BI) The process of gathering enough of the right information in a timely manner and usable form and analyzing it to have a positive impact on business strategy, tactics, or operations.

business-to-business (B2B) e-commerce A form of e-commerce in which the participants are organizations.

business-to-consumer (B2C) e-commerce A form of e-commerce in which customers deal directly with the organization, avoiding any intermediaries.

byte (B) Eight bits that together represent a single character of data.

catalog management software Software that automates the process of creating a real-time interactive catalog and delivering customized content to a user's screen.

central processing unit (CPU) The part of the computer that consists of two associated elements: the arithmetic/logic unit and the control unit.

centralized processing The processing alternative in which all processing occurs in a single location or facility.

certification A process for testing skills and knowledge that results in a statement by the certifying authority that says an individual is capable of performing a particular kind of job.

character A basic building block of information, consisting of uppercase letters, lowercase letters, numeric digits, or special symbols.

chat room A facility that enables two or more people to engage in interactive "conversations" over the Internet.

choice stage The third stage of decision making, which requires selecting a course of action.

client/server An architecture in which multiple computer platforms are dedicated to special functions, such as database management, printing, communications, and program execution.

clock speed A series of electronic pulses produced at a predetermined rate that affect machine cycle time.

command-based user interface A user interface that requires that text commands be given to the computer to perform basic activities.

communications protocol A standard set of rules that control a telecommunications connection.

communications software The software that provides a number of important functions in a network, such as error checking and data security.

compact disc read-only memory (CD-ROM) A common form of optical disc on which data, after it has been recorded, cannot be modified.

competitive advantage A significant and (ideally) long-term benefit to a company over its competition.

competitive intelligence A continuous process involving the legal and ethical collection of information about competitors, its analysis, and controlled dissemination of information to decision makers.

competitive local exchange carrier (CLEC) A company that is allowed to compete with the LECs, such as a wireless, satellite, or cable service provider.

computer network The communications media, devices, and software needed to connect two or more computer systems or devices.

computer programs The sequences of instructions for the computer.

computer system platform The combination of a particular hardware configuration and systems software package.

computer-aided software engineering (CASE) Tools that automate many of the tasks required in a systems development effort and enforce adherence to the SDLC.

computer-based information system (CBIS) A single set of hardware, software, databases, telecommunications, people, and procedures that are configured to collect, manipulate, store, and process data into information.

concurrency control A method of dealing with a situation in which two or more people need to access the same record in a database at the same time.

consumer-to-consumer (C2C) e-commerce A form of e-commerce in which the participants are individuals, with one serving as the buyer and the other as the seller.

content streaming A method for transferring multimedia files over the Internet so that the data stream of voice and pictures plays more or less continuously without a break, or very few of them; enables users to browse large files in real time.

contract software The software developed for a particular company.

control unit The part of the CPU that sequentially accesses program instructions, decodes them, and coordinates the flow of data in and out of the ALU, primary storage, and even secondary storage and various output devices.

cookie A text file that an Internet company can place on the hard disk of a computer system to track user movements.

coprocessor The part of the computer that speeds processing by executing specific types of instructions while the CPU works on another processing activity.

counterintelligence The steps an organization takes to protect information sought by "hostile" intelligence gatherers.

criminal hacker (cracker) A computer-savvy person who attempts to gain unauthorized or illegal access to computer systems.

cryptography The process of converting a message into a secret code and changing the encoded message back to regular text.

culture A set of major understandings and assumptions shared by a group.

cybermall A single Web site that offers many products and services at one Internet location.

cyberterrorist Someone who intimidates or coerces a government or organization to advance his or her political or social objectives by launching computer-based attacks against computers, networks, and the information stored on them.

data The raw facts, such as an employee's name and number of hours worked in a week, inventory part numbers, or sales orders.

data analysis Manipulation of the collected data so that it is usable for the development team members who are participating in systems analysis.

data collection The process of capturing and gathering all data necessary to complete transactions.

data correction The process of reentering miskeyed or misscanned data that was found during data editing.

data definition language (DDL) A collection of instructions and commands used to define and describe data and data relationships in a specific database.

data dictionary A detailed description of all the data used in the database.

data editing The process of checking data for validity and completeness.

data integrity The degree to which the data in any one file is accurate.

data item The specific value of an attribute.

data manipulation The process of performing calculations and other data transformations related to business transactions.

data manipulation language (DML) The commands that are used to manipulate the data in a database.

data mart A subset of a data warehouse.

data mining An information-analysis tool that involves the automated discovery of patterns and relationships in a data warehouse or a data mart.

data model A diagram of data entities and their relationships.

data preparation, or data conversion The process of ensuring all files and databases are ready to be used with new computer software and systems.

data redundancy A duplication of data in separate files.

data storage The process of updating one or more databases with new transactions.

data warehouse A database that collects business information from many sources in the enterprise, covering all aspects of the company's processes, products, and customers.

database An organized collection of facts and information.

database administrator (DBA) A skilled IS professional who directs all activities related to an organization's database.

database approach to data management An approach whereby a pool of related data is shared by multiple application programs.

database management system (DBMS) A group of programs that manipulate the database and provide an interface between the database and its users and other application programs.

data-flow diagram (DFD) A model of objects, associations, and activities by describing how data can flow between and around various objects.

decentralized processing The processing alternative in which processing devices are placed at various remote locations.

decision room A room that supports decision making, with the decision makers in the same building, combining face-to-face verbal interaction with technology to make the meeting more effective and efficient.

decision support system (DSS) An organized collection of people, procedures, software, databases, and devices used to support problem-specific decision making.

decision-making phase The first part of problem solving, including three stages: intelligence, design, and choice.

delphi approach A decision-making approach in which group decision makers are geographically dispersed; this approach encourages diversity among group members and fosters creativity and original thinking in decision making.

demand reports The reports that are developed to give certain information at a person's request.

design report The result of systems design, reflecting the decisions made for system design and preparing the way for systems implementation.

design stage The second stage of decision making, in which alternative solutions to the problem are developed.

desktop computer A relatively small, inexpensive single-user computer that is highly versatile.

dialogue manager The user interface that allows decision makers to easily access and manipulate the DSS and to use common business terms and phrases.

digital certificate An attachment to an e-mail message or data embedded in a Web page that verifies the identity of a sender or a Web site.

digital signature The encryption technique used to verify the identity of a message sender for processing online financial transactions.

digital versatile disc (DVD) A storage medium used to store digital video or computer data over a network.

direct access The retrieval method in which data can be retrieved without the need to read and discard other data.

direct access storage device (DASD) The device used for direct access of secondary storage data.

direct conversion (also called *plunge* or *direct cutover*) The process of stopping the old system and starting the new system on a given date.

direct observation The process of watching the existing systems in action by one or more members of the analysis team.

distributed database A database in which the data can be spread across several smaller databases connected via telecommunications devices.

distributed processing The processing alternative in which computers are placed at remote locations but connected to each other via a network.

document production The process of generating output records and reports.

documentation The text that describes the program functions to help the user operate the computer system.

domain (1) The allowable values for data attributes. (2) The area of knowledge addressed by the expert system.

domain expert The individual or group who has the expertise or knowledge one is trying to capture in the expert system.

drill-down reports The reports that provide increasingly detailed data about a situation.

dumpster diving Searching through the garbage for important pieces of information that can help crack an organization's computers or be used to convince someone at the company to give someone access to the computers.

e-commerce Any business transaction executed electronically between parties such as companies (business-to-business, B2B), companies and consumers (business-to-consumer, B2C), consumers and other consumers (consumer-to-consumer, C2C), businesses and the public sector, and consumers and the public sector.

e-commerce software Software that supports catalog management, product configuration, and shopping cart facilities.

economic feasibility Determination of whether the project makes financial sense and whether predicted benefits offset the cost and time needed to obtain them.

electronic bill presentment A method of billing whereby the biller posts an image of your statement on the Internet and alerts you by e-mail that your bill has arrived.

electronic cash An amount of money that is computerized, stored, and used as cash for e-commerce transactions.

electronic data interchange (EDI) An intercompany, application-to-application communication of data in standard format for business transactions.

electronic exchange An electronic forum where manufacturers, suppliers, and competitors buy and sell goods, trade market information, and run back-office operations.

electronic retailing (e-tailing) The direct sale from business to consumer through electronic storefronts, typically designed around an electronic catalog and shopping cart model.

electronic shopping cart A model commonly used by many e-commerce sites to track the items selected for purchase, allowing shoppers to view what is in their cart, add new items to it, and remove items from it.

electronic wallet A computerized stored value that holds credit card information, electronic cash, owner identification, and address information.

encryption The conversion of a message into a secret code.

end-user systems development Any systems development project in which the primary effort is undertaken by a combination of business managers and users.

enterprise data modeling The data modeling done at the level of the entire enterprise.

enterprise resource planning (ERP) system A set of integrated programs capable of managing a company's vital business operations for an entire multisite, global organization.

entity A generalized class of people, places, or things for which data is collected, stored, and maintained.

entity-relationship (ER) diagram The data models that use basic graphical symbols to show the organization of and relationships between data.

event-driven review A review triggered by a problem or opportunity, such as an error, a corporate merger, or a new market for products.

exception reports The reports that are automatically produced when a situation is unusual or requires management action.

executive support system (ESS) A specialized DSS that includes all hardware, software, data, procedures, and people used to assist senior-level executives within the organization.

expandable storage devices Expandable storage devices use removable disk cartridges to provide additional storage capacity.

expert system A system that gives a computer the ability to make suggestions and act like an expert in a particular field.

expert system shell A collection of software packages and tools used to develop expert systems.

explanation facility The component of an expert system that allows a user or decision maker to understand how the expert system arrived at certain conclusions or results.

extranet A network based on Web technologies that allows selected outsiders, such as business partners and customers, to access authorized resources of the intranet of a company.

feasibility analysis Assessment of the technical, economic, legal, operational, and schedule feasibility of a project.

feedback The output that is used to make changes to input or processing activities.

field Typically a name, number, or combination of characters that describes an aspect of a business object or activity.

file A collection of related records.

file server An architecture in which the application and database reside on the one host computer, called the *file server*.

financial MIS An information system that provides financial information to all financial managers within an organization.

firewall A device that sits between an internal network and the Internet, limiting access into and out of a network based on access policies.

five-forces model A widely accepted model that identifies five key factors that can lead to attainment of competitive advantage, including (1) the rivalry among existing competitors, (2) the threat of new entrants, (3) the threat of substitute products and services, (4) the bargaining power of buyers, and (5) the bargaining power of suppliers.

forecasting The process of predicting future events to avoid problems.

forward chaining The process of starting with the facts and working forward to the conclusions.

fuzzy logic A special research area in computer science that allows shades of gray and does not require everything to be simple black/white, yes/no, or true/false.

game theory The use of information systems to develop competitive strategies for people, organizations, or even countries.

genetic algorithm An approach to solving large, complex problems in which a number of related operations or models change and evolve until the best one emerges.

geographic information system (GIS) A computer system capable of assembling, storing, manipulating, and displaying geographic information, that is, data identified according to their locations.

graphical user interface (GUI) An interface that uses icons and menus displayed on screen to send commands to the computer system.

grid computing The use of a collection of computers, often owned by multiple individuals or organizations, to work in a coordinated manner to solve a common problem.

group consensus approach A decision-making approach that forces members in the group to reach a unanimous decision.

group support system (GSS) The software application that consists of most of the elements in a DSS, plus software to provide effective support in group decision-making settings; also called *group decision support system* or *computerized collaborative work system*.

hacker A person who enjoys computer technology and spends time learning and using computer systems.

handheld computer A single-user computer that provides ease of portability because of its small size.

hardware The computer equipment used to perform input, processing, and output activities.

heuristics The commonly accepted guidelines or procedures that usually find a good solution.

hierarchy of data Bits, characters, fields, records, files, and databases.

highly structured problems Problems that are straightforward and require known facts and relationships.

home page A cover page for a Web site that has graphics, titles, and text.

HTML tags The codes that let the Web browser know how to format text—as a heading, as a list, or as body text—and whether images, sound, and other elements should be inserted.

human resource MIS An information system that is concerned with activities related to employees and potential employees of an organization; also called a *personnel MIS*.

hypermedia The tools that connect the data on Web pages, allowing users to access topics in whatever order they want.

Hypertext Markup Language (HTML) The standard page description language for Web pages.

ICRA rating system A system to protect individuals from harmful or objectionable Internet content, while safeguarding the free speech rights of others.

identity theft A crime in which an imposter obtains key pieces of personal identification information, such as Social Security or driver's license numbers, in order to impersonate someone else. The information is then used to obtain credit, merchandise, and services in the name of the victim or to provide the thief with false credentials.

if-then statements Rules that suggest certain conclusions.

implementation stage A stage of problem solving in which a solution is put into effect.

inference engine The part of the expert system that seeks information and relationships from the knowledge base and provides answers, predictions, and suggestions the way a human expert would.

informatics A specialized system that combines traditional disciplines, such as science and medicine, with computer systems and technology.

information A collection of facts organized in such a way that they have additional value beyond the value of the facts themselves.

information center A support function that provides users with assistance, training, application development, documentation, equipment selection and setup, standards, technical assistance, and troubleshooting.

information service unit A miniature IS department.

information system (IS) A set of interrelated components that collect, manipulate, and disseminate data and information and provide a feedback mechanism to meet an objective.

information systems planning The translation of strategic and organizational goals into systems development initiatives.

input The activity of gathering and capturing raw data.

insider An employee, disgruntled or otherwise, working solo or in concert with outsiders to compromise corporate systems.

installation The process of physically placing the computer equipment on the site and making it operational.

instant messaging A method that allows two or more individuals to communicate online using the Internet.

institutional DSS A DSS that handles situations or decisions that occur more than once, usually several times a year or more. An institutional DSS is used repeatedly and refined over the years.

intelligence stage The first stage of decision making, in which potential problems or opportunities are identified and defined.

intelligent agent Programs and a knowledge base used to perform a specific task for a person, a process, or another program; also called *intelligent robot* or *bot*.

intelligent behavior The ability to learn from experiences and apply knowledge acquired from experience, handle complex situations, solve problems when important information is missing, determine what is important, react quickly and correctly to a new situation, understand visual images, process and manipulate symbols, be creative and imaginative, and use heuristics.

international network A network that links systems between countries.

Internet The world's largest computer network, actually consisting of thousands of interconnected networks, all freely exchanging information.

Internet Protocol (IP) The communications standard that enables traffic to be routed from one network to another as needed.

Internet service provider (ISP) Any company that provides individuals or organizations with access to the Internet.

Internet software piracy Illegally downloading software from the Internet.

intranet An internal corporate network built using Internet and World Wide Web standards and products; used by employees to gain access to corporate information.

intrusion detection system (IDS) Software that monitors system and network resources and notifies network security personnel when it senses a possible intrusion.

Java An object-oriented programming language from Sun Microsystems based on C++ that allows small programs (applets) to be embedded within an HTML document.

joining The data manipulation that combines two or more tables.

joint application development (JAD) A process for data collection and requirements analysis in which users, stakeholders, and IS professionals work together to analyze existing systems, propose possible solutions, and define the requirements of a new or modified system.

key A field or set of fields in a record that is used to identify the record.

key-indicator report A summary of the previous day's critical activities; typically available at the beginning of each workday.

knowledge The awareness and understanding of a set of information and the ways it can be used.

knowledge acquisition facility The part of the expert system that provides convenient and efficient means of capturing and storing all the components of the knowledge base.

knowledge base (1) The collection of data, rules, procedures, and relationships that must be followed to achieve value or the proper outcome. (2) A component of an expert system that stores all relevant information, data, rules, cases, and relationships used by the expert system.

knowledge engineer An individual who has training or experience in the design, development, implementation, and maintenance of an expert system.

knowledge management The process of capturing a company's collective expertise wherever it resides—in computers, on paper, in people's heads—and distributing it wherever it can help produce the biggest payoff.

knowledge user The individual or group who uses and benefits from the expert system.

learning systems A combination of software and hardware that allows the computer to change how it functions or reacts to situations based on feedback it receives.

legal feasibility Determination of whether laws or regulations might prevent or limit a systems development project.

linking The data manipulation that relates or links two or more tables using common data attributes.

local area network (LAN) A network that connects computer systems and devices within the same geographic area.

local exchange carrier (LEC) A public telephone company in the United States that provides service to homes and businesses within its defined geographical area called its *local access and transport area (LATA)*.

logic bomb An application or system virus designed to "explode" or execute at a specified time and date.

logical design Description of the functional requirements of a system.

long-distance carrier A traditional long-distance phone provider, such as AT&T, Sprint, or MCI.

magnetic disk A common secondary storage medium, with bits represented by magnetized areas.

magnetic tape A common secondary storage medium; Mylar film coated with iron oxide with portions of the tape magnetized to represent bits.

mainframe computer A large, powerful computer often shared by hundreds of concurrent users connected to the machine via terminals.

make-or-buy decision The decision of whether to purchase software from external developers or develop it in-house.

managed security service provider (MSSP) An organization that monitors, manages, and maintains network security hardware and software for its client companies.

management information system (MIS) An organized collection of people, procedures, software, databases, and devices used to provide routine information to managers and decision makers.

market segmentation The identification of specific markets to target them with advertising messages.

marketing MIS An information system that supports managerial activities in product development, distribution, pricing decisions, promotional effectiveness, and sales forecasting.

metropolitan area network (MAN) A telecommunications network that connects users and their computers within a geographical area larger than that covered by a LAN, but smaller than the area covered by a WAN, such as a city or college campus.

mobile commerce (m-commerce) The use of wireless devices such as PDAs and cell phones to place orders and conduct business anywhere.

model base A part of a DSS that provides decision makers access to a variety of models and assists them in decision making.

model management software (MMS) The software that coordinates the use of models in a DSS.

monitoring stage The final stage of the problem-solving process, in which decision makers evaluate the implementation.

multiprocessing The simultaneous execution of two or more instructions at the same time.

multitasking The capability that allows a user to run more than one application at the same time.

natural language processing Processing that allows the computer to understand and react to statements and commands made in a "natural" language, such as English.

network management software The software that enables a manager on a networked desktop to monitor the use of individual computers and shared hardware (like printers), scan for viruses, and ensure compliance with software licenses.

network operating system (NOS) The systems software that controls the computer systems and devices on a network and allows them to communicate with each other.

networks The connected computers and computer equipment in a building, around the country, or around the world to enable electronic communications.

neural network A computer system that can simulate the functioning of a human brain.

nominal group technique A decision-making approach that encourages feedback from individual group members, and the final decision is made by voting, similar to the way public officials are elected.

nonprogrammed decisions The decisions that deal with unusual or exceptional situations.

object-oriented database The database that stores both data and its processing instructions.

object-oriented database management system (OODBMS) A group of programs that manipulate an object-oriented database and provide a user interface and connections to other application programs.

object-oriented systems development (OOSD) The approach that combines the logic of the systems development life cycle with the power of object-oriented modeling and programming.

object-relational database management system (ORDBMS) A DBMS capable of manipulating audio, video, and graphical data.

off-the-shelf software An existing software program that is purchased.

online analytical processing (OLAP) The software that allows users to explore data from a number of different perspectives.

online transaction processing (OLTP) Computerized processing in which each transaction is processed immediately, without the delay of accumulating transactions into a batch.

operating system (OS) A set of computer programs that controls the computer hardware and acts as an interface with application programs.

operational feasibility Measure of whether the project can be put into action or operation.

optimization model A process to find the best solution, usually the one that will best help the organization meet its goals.

order processing systems Systems that process order entry, sales configuration, shipment planning, shipment execution, inventory control, invoicing, customer relationship management, and routing and scheduling.

organization A formal collection of people and other resources established to accomplish a set of goals.

organizational change The responses that are necessary for for-profit and nonprofit organizations to plan for, implement, and handle change.

organizational culture The major understandings and assumptions for a business, a corporation, or an organization.

output The production of useful information, usually in the form of documents and reports.

parallel processing A form of multiprocessing that speeds processing by linking several processors to operate at the same time, or in parallel.

parallel start-up The process of running both the old and new systems for a period of time, comparing the new system's output closely with the old system's, reconciling any differences, and finally eliminating the old system.

password sniffer A small program hidden in a network or a computer system that records identification numbers and passwords.

perceptive system A system that approximates the way a human sees, hears, and feels objects.

personal area network (PAN) A network that supports the interconnection of information technology within a range of 33 feet or so.

personal productivity software The software that enables users to improve their personal effectiveness, increasing the amount of work they can do and its quality.

phase-in approach The process of slowly replacing components of the old system with those of the new one; this process is repeated for each application until the new system is running every application and performing as expected; also called *piecemeal approach.*

physical design Specification of the characteristics of the system components necessary to put the logical design into action.

pilot start-up The process of running the new system for one group of users rather than for all users.

planned data redundancy A way of organizing data in which the logical database design is altered so that certain data entities are combined, summary totals are carried in the data records rather than calculated from elemental data, and some data attributes are repeated in more than one data entity to improve database performance.

Platform for Privacy Preferences (P3P) A screening technology that shields users from Web sites that don't provide the level of privacy protection they desire.

point-of-sale (POS) device A terminal used in retail operations to enter sales information into the computer system.

Point-to-Point Protocol (PPP) A communications protocol that transmits packets over telephone lines.

portable computer A computer small enough to be carried easily.

predictive analysis A form of data mining that combines historical data with assumptions about future conditions to predict outcomes of events such as future product sales or the probability that a customer will default on a loan.

primary key A field or set of fields that uniquely identifies the record.

problem solving A process that goes beyond decision making to include the implementation stage.

procedures The strategies, policies, methods, and rules for using a CBIS.

process A set of logically related tasks performed to achieve a desired outcome.

processing The activity of converting or transforming data into useful outputs.

product configuration software Software used by buyers to build the product they need online.

productivity A measure of the output achieved divided by the input required.

programmed decisions The decisions made using a rule, procedure, or quantitative method.

programmer The specialist responsible for modifying or developing programs to satisfy user requirements.

programming languages The sets of keywords, symbols, and a system of rules for constructing statements by which humans can communicate instructions to be executed by a computer.

projecting The data manipulation that eliminates columns in a table.

proprietary software A one-of-a-kind program for a specific application, usually developed and owned by a single company.

prototyping An iterative approach to the systems development process.

public key infrastructure (PKI) A means to enable users of an unsecured public network such as the Internet to securely and privately exchange data through the use of a public and a private cryptographic key pair that is obtained and shared through a trusted authority.

questionnaires A method of gathering data when the data sources are spread over a wide geographic area.

radio-frequency identification (RFID) A technology that employs a microchip with an antenna that broadcasts its unique identifier and location to receivers.

random access memory (RAM) A form of memory in which instructions or data can be temporarily stored.

rapid application development (RAD) A systems development approach that employs tools, techniques, and methodologies designed to speed application development.

read-only memory (ROM) A nonvolatile form of memory.

record A collection of related data fields.

redundant array of independent/ inexpensive disks (RAID) A method of storing data that generates extra bits of data from existing data, allowing the system to create a "reconstruction map" so that if a hard drive fails, the system can rebuild lost data.

relational model A database model that describes data in which all data elements are placed in two-dimensional tables, called *relations*, that are the logical equivalent of files.

replicated database A database that holds a duplicate set of frequently used data.

request for proposal (RFP) A document that specifies in detail required resources such as hardware and software.

requirements analysis Determination of user, stakeholder, and organizational needs.

return on investment (ROI) One measure of IS value that investigates the additional profits or benefits that are generated as a percentage of the investment in information systems technology.

robotics Mechanical or computer devices that perform tasks requiring a high degree of precision or that are tedious or hazardous for humans.

rule A conditional statement that links given conditions to actions or outcomes.

satisficing model A model that will find a good—but not necessarily the best—problem solution.

scalability The ability to increase the capability of a computer system to process more transactions in a given period by adding more, or more powerful, processors.

schedule feasibility Determination of whether the project can be completed in a reasonable amount of time.

scheduled reports The reports that are produced periodically, or on a schedule, such as daily, weekly, or monthly.

schema A description of the entire database.

script bunnies Wannabe crackers with little technical savvy who download programs—scripts—that automate the job of breaking into computers.

search engine A Web search tool.

secondary storage (permanent storage) The devices that store larger amounts of data, instructions, and information more permanently than allowed with main memory.

selecting The data manipulation that eliminates rows according to certain criteria.

semistructured or unstructured problems More complex problems in which the relationships among the data are not always clear, the data may be in a variety of formats, and the data is often difficult to manipulate or obtain.

sequential access The retrieval method in which data must be accessed in the order in which it is stored.

sequential access storage device (SASD) The device used to sequentially access secondary storage data.

Serial Line Internet Protocol (SLIP) A communications protocol that transmits packets over telephone lines.

server A computer designed for a specific task, such as network or Internet applications.

site preparation Preparation of the location of the new system.

smart card A credit card–sized device with an embedded microchip to provide electronic memory and processing capability.

social engineering The practice of talking a critical computer password out of an individual.

software The computer programs that govern the operation of the computer.

software bug A defect in a computer program that keeps it from performing in the manner intended.

software piracy The act of illegally duplicating software.

software suite A collection of single application programs packaged in a bundle.

sphere of influence The scope of problems and opportunities addressed by a particular organization.

stakeholders Individuals who, either themselves or through the organization they represent, ultimately benefit from the systems development project.

start-up The process of making the final tested information system fully operational.

storage area network (SAN) The technology that provides high-speed connections between data-storage devices and computers.

strategic alliance (strategic partnership) An agreement between two or more companies that involves the joint production and distribution of goods and services.

strategic planning The process of determining long-term objectives by analyzing the strengths and weaknesses of the organization, predicting future trends, and projecting the development of new product lines.

structured interview An interview in which the questions are written in advance.

subschema A file that contains a description of a subset of the database and identifies which users can view and modify the data items in the subset.

supercomputers The most powerful computer systems, with the fastest processing speeds.

supply chain management A key value chain composed of demand planning, supply planning, and demand fulfillment.

syntax A set of rules associated with a programming language.

systems analysis The systems development phase involving the study of existing systems and work processes to identify strengths, weaknesses, and opportunities for improvement.

systems analyst The professional who specializes in analyzing and designing business systems.

systems design The systems development phase that defines how the information system will do what it must do to obtain the problem's solution.

systems development The activity of creating new systems or modifying existing ones.

systems implementation The systems development phase involving the creation or acquisition of various system components detailed in the systems design, assembling them, and placing the new or modified system into operation.

systems investigation The systems development phase during which problems and opportunities are identified and considered in light of the goals of the business.

systems investigation report A summary of the results of the systems investigation and the process of feasibility analysis and recommendations for a course of action.

systems maintenance and review The systems development phase that ensures the system operates and modifies the system so that it continues to meet changing business needs.

systems operation Use of a new or modified system.

systems request form A formal document to initiate systems investigation.

systems software The set of programs designed to coordinate the activities and functions of the hardware and various programs throughout the computer system.

technical feasibility Assessment of whether the hardware, software, and other system components can be acquired or developed to solve the problem.

technology acceptance model (TAM) A model that describes the factors that can lead to higher acceptance and usage of technology.

technology diffusion A measure of how widely technology is spread throughout the organization.

technology infrastructure All the hardware, software, databases, telecommunications, people, and procedures that are configured to collect, manipulate, store, and process data into information.

technology infusion The extent to which technology is deeply integrated into an area or department.

technology-enabled relationship management The use of detailed information about a customer's behavior, preferences, needs, and buying patterns to set prices, negotiate terms, tailor promotions, add product features, and otherwise customize the entire relationship with that customer.

telecommunications The electronic transmission of signals for communications; enables organizations to carry out their processes and tasks through effective computer networks.

telecommunications medium Anything that carries an electronic signal and interfaces between a sending device and a receiving device.

terminal-to-host An architecture in which the application and database reside on one host computer, and the user interacts with the application and data using a "dumb" terminal.

thin client A low-cost, centrally managed computer with essential but limited capabilities that is devoid of a DVD player, internal disk drive, and expansion slots.

time-driven review A review performed after a specified amount of time.

time-sharing The capability that allows more than one person to use a computer system at the same time.

total cost of ownership (TCO) The measurement of the total cost of owning computer equipment, including desktop computers, networks, and large computers.

traditional approach to data management An approach whereby separate data files are created and stored for each application program.

transaction Any business-related exchange such as payments to employees, sales to customers, and payments to suppliers.

transaction processing cycle The process of data collection, data editing, data correction, data manipulation, data storage, and document production.

transaction processing system (TPS) An organized collection of people, procedures, software, databases, and devices used to record completed business transactions.

transaction processing system audit An examination of the TPS to answer whether the system meets the business need for which it was implemented, what procedures and controls have been established, whether these procedures and controls are being used properly, and whether the information systems and procedures are producing accurate and honest reports.

Transmission Control Protocol (TCP) A widely used Transport-layer protocol that is used in combination with IP by most Internet applications.

Trojan horse A program that appears to be useful but actually masks a destructive program.

tunneling The process by which VPNs transfer information by encapsulating traffic in IP packets over the Internet.

Uniform Resource Locator (URL) An assigned address on the Internet for each computer.

unstructured interview An interview in which the questions are not written in advance.

user acceptance document A formal agreement signed by the user that states that a phase of the installation or the complete system is approved.

user interface The element of the operating system that allows individuals to access and command the computer system.

users Individuals who will interact with the system regularly.

value chain A series (chain) of activities that includes inbound logistics, warehouse and storage, production, finished product storage, outbound logistics, marketing and sales, and customer service.

virtual private network (VPN) A secure connection between two points across the Internet.

virtual reality The simulation of a real or imagined environment that can be experienced visually in three dimensions.

virtual reality system A system that enables one or more users to move and react in a computer-simulated environment.

virtual workgroups Teams of people who are located around the world working on common problems.

virus A computer program capable of attaching to disks or other files and replicating itself repeatedly, typically without the user's knowledge or permission.

vision systems The hardware and software that permit computers to capture, store, and manipulate visual images and pictures.

voice-recognition device An input device that recognizes human speech.

Web auction An Internet site that matches people who want to sell products and services with people who want to purchase those products and services.

Web browser The software that creates a unique, hypermedia-based menu on a computer screen, providing a graphical interface to the Web.

Web log (blog) A Web site that people can create and use to write about their observations, experiences, and feelings on a wide range of topics.

Web page construction software Software that uses Web editors and extensions to produce both static and dynamic Web pages.

Web services The standards and tools that streamline and simplify communication among Web sites for business and personal purposes.

Web site development tools Tools used to develop a Web site, including HTML or visual Web page editor, software development kits, and Web page upload support.

wide area network (WAN) A network that ties together large geographic regions.

wireless application protocol (WAP) A standard set of specifications for Internet applications that run on handheld, wireless devices.

wordlength The number of bits the CPU can process at any one time.

workgroup application software The software that supports teamwork, whether in one location or around the world.

workstation A more powerful personal computer that is used for technical computing, such as engineering, but still fits on a desktop.

World Wide Web (WWW, or W3) A collection of tens of thousands of independently owned computers that work together as one in an Internet service.

worm An independent program that replicates its own program files until it interrupts the operation of networks and computer systems.

INDEX

A

A4Vision, 273
Academic Management Services (AMS), 215
access
    to the Internet, 150, 175
    logical vs. physical, 68
    to system resources, 69
accounting MISs, 242
accounting systems, 208–209
Ace Hardware Corporation, 30, 117
ACM code of professional conduct, 374–375
Act! marketing automation tool, 239
added value, information as, 6
administrators, database (DBA), 97
Advanced Research Projects Agency (ARPA), 148
advertising
    adware, spyware, 380–381
    and e-commerce, 13
agile development, 311–312
AI chess, 267–268
Air Transport Association, 345
airlines safety, passenger screening, 345–346
AITP code of ethics, 374
Alaska Airlines, 328
Albertson's, 190
Allgemeines Rechenzentrum GmbH (ARZ), 48
alliances, strategic, 26
Amazon.com's fraud detection techniques, 265
America Online, 158
American Express, use of smart cards, 203
American Home Products, 86
Ameritrade.com, 197
analysis
    of data in systems development, 319–320
    predictive, 119
    and systems development, 20
Anderson Economic Group, 210
Anti-Phishing Working Group, 14
antivirus programs, using, 357–358
Apple Computer
    AppleTalk feature, 69–70
    Mac OS X Server, 73
    operating systems, 70
application program interfaces (APIs), 67–68
application service providers (ASPs), 76, 87
application software
    described, 65, 86
    types and functions of, 75–81
    workgroup and enterprise, 82–83
applications
    e-commerce, 191–199
    selected DSS (table), 243
    software, 10
    summary of Internet and Web (table), 158
Appshop Inc., 341
architecture
    client/server, 144
    terminal-to-host (fig.), 143
archiving pictures, 91
arithmetic/logic unit (ALU), 49
Armour, Tom, 273
ARPANET project, 148, 170
artificial intelligence (AI)
    applications of, 285, 287
    described, 18–19, 292
    overview of, 266–275
Artificial Intelligence (movie), 267
ASPs (application service providers), 76, 87
Association for Computing Machinery (ACM) code of
    professional conduct, 374–375
Association of Information Technology Professionals (AITP)
    code of ethics, 374
AT&T, 23
ATM cards, 203
ATM devices as input devices, 58
attributes, data, 97–98
auctions, Web, 163, 198–199
audio database systems, 125
audits of transaction processing systems, 211
authentication, technologies for, 202
automated teller machine (ATM) cards, 203
automatic number identification (ANI), 141
automatic teller machines (ATM) devices as input devices, 58
autonomic computing, 332, 334
autonomics, 31
Autonomy, Inc., 273
AutoTradeCenter, Inc., 117
Aviall, 157, 187
AXS0One Email archival system, 120

B

back-end applications, 115
backbone, Internet, 147, 149
background processing, 68
backward chaining, 280
Bailey, Charles, 275
Ballmer, Steve, 10, 72
Bank of New Zealand, 164
BankFinancial Corporation of Chicago, 121
banking
    and magnetic ink coding, 59
    online, 197
    transaction processing systems. See transaction processing
        systems
bar codes
    scanners as input devices, 57
    and smart tags, 58
Barilla Group, The, 135–136
barnesandnoble.com, 157
batch processing systems, 204–205
Bayley, Vida, 179
Ben and Jerry's Ice Cream, 121
Bennett, Steve, 383

Berners-Lee, Tim, 117, 153
best practices described, 214
Best Western Hotels, online reservations, 175
Bill and Melinda Gates Foundation, 151
bill of materials (BOM), 213
bills, electronic payment of, 198
Binocular Omni-Orientation Monitor (BOOM), 287–288
Biobank, 368
bioinformatics, 293
biometrics, 362
Bisker, Janie, 29
Black & Vetch, 34
Blagojevich, Gov. Rod, 192
blogs, 159–160
Blue Cross of Pennsylvania, 17
Bluetooth specification, 146–147
BMW Group, 305
Boehringer Ingelheim, 3–4, 233
Boeing Scan Eagle drone, 10
BOOM (Binocular Omni-Orientation Monitor), 287–288
bots described, 163
brainstorming, 248
Brand Capital, 95
Brandon Trust, 383–384
bridges, 147
Brief History of Time, A (Hawking), 266
Bristol University, 180
British Airways (BA), 299
Broad Institute, 268
browsers, Web, 155
buddy lists, 158
bugs, software, 85, 87
bundled software, 81
Bureau Interprofessionnel de Vins de Bourgogne, 296
Burlington Northern and Santa Fe Railway Co. (BNSF), 57, 286
Burnboch, Bill, 95
business
    continuity planning, 209–210
    electronic (e-business), 15
    functions, traditional vs. database approach to, 97–101
    uses of the Web, 157
business information systems
    electronic and mobile commerce, 12–13, 15
    information and decision support, 16–18
    specialized systems, 18–20
    transaction processing systems and ERP, 15–16
business intelligence (BI), 121, 228
Business Software Alliance, 364
business-to-business (B2B) e-commerce, 12, 186, 188–189, 217
business-to-consumer (B2C) e-commerce, 12, 186, 189, 217
BusinessObjects Corporation, 340
bytes (B), 50, 97

C

cabling, transmission media, 137–139
CAD (computer-aided design), 63
Campbell, Jeff, 286

CAN-SPAM (Controlling the Assault of Non-Solicited Pornography and Marketing Act), 348
Canadian prescription drug Web sites, 192
CAPPS (Computer Assisted Passenger Prescreening System), 273, 345–346
careers
    in information systems, 31–36
    involving the Internet, 35–36
    job searching on the Internet, 159
    job searching online, 161
Carnegie Mellon's www.cert.org site, 168
carriers
    local exchange carriers (LECs), 139–140
    telecommunications, types of, 169
CartaSi S.p.A, 63–64
CASE diagrams, 317
CASE tools, 314–315, 321
CaseMap, 113
cash, electronic, 202
catalog management software, 200–201
Catena Corporation, 347
cathode ray tubes (CRTs), 58
CAVE virtual reality environment, 287–288
CD-ROMs (compact disk read-only memory), 55
CDW marketing, 239
cell phones
    handheld entertainment, 185
    Internet, 159
Cendant travel, real estate services, 222
Centers for Disease Control (CDC), disease databases, 125
central processing unit. *See* CPU
centralized processing, 142
CERN, and development of the Internet, 153
CERT (Computer Emergency Response Team), 351, 352, 361
certification and IS careers, 36
Certified Information Systems Security Professional (CISSP), 36
Certive Corporation, 121
Cessna Aircraft, 21
Chamberlain, D. D., 110
change
    organizational, 22, 37
    sustaining and disruptive, 23
Charles Schwab & Co., 187
chat rooms, 161, 355
Check Clearing for the 21st Century Act, 59, 132
Check Point Certified Security Administrator, 30
chess, AI, 267–268
chief financial officer (CFO), 34, 38
chief information officer (CIO), 34–36
chief technology officer (CTO), 34
Children's Internet Protection Act (CIPA), 366
Cingular, 23, 141
CIPA (Children's Internet Protection Act), 366
ciphertext, 167
Cisco Systems, wireless network implementation, 135
Clark County, Nevada, 144
ClearForest database package, 119
client/server systems, 143, 169
clients, thin, 62

Cline, Davis & Mann, 70
Clinger-Cohen Act of 1996, 34
clip art, 80
clock speed, 50
COBOL programming language, 85, 282
Coca-Cola, 328
code
    and open source software, 72
    programming language, 84–85
cognitive amplifiers, 273
Cohen, Jay, 120
collaborative software, 82
Collard, Tom, 240
Collins, Jim, 24
Comair, Ltd., 299
Comerica, 235
command-based user interface, 67
communications channels, 169
communications protocols, 146–147
communications, software, 146
compact disk read-only memory. *See* CD-ROMs
Compagnie Générale de Géophysique SA, 64
companies
    See also specific company
    competitive use of information systems (table), 24
    crime prevention by, 361–362
    factors and strategies (table), 27
    that have advantageously used the Internet (table), 11
    top places to work in IT (table), 31
competitive advantage, 24–27, 38
competitive intelligence, 122
competitive local exchange carriers (CLECs), 140
CompUSA, 328
computational biology, 291
computed fields, 97
computer-aided design (CAD), 63
computer-aided software engineering (CASE), 314
Computer Assisted Passenger Prescreening System
    CAPPS), 273, 345–346
computer crime
    cyberterrorism, identify theft, 352–353
    ethical issues in IS, 373–375
    introduction to, 351–352, 375–376
    preventing, 361–367
    viruses, theft, scams, 354–361
computer downsizing, 63
Computer Emergency Response Team (CERT), 351, 352, 361
Computer Fraud and Abuse Act, 361
computer incidents
    with greatest impact (table), 356
    responding to (table), 355
computer networks. *See* networks
computer programs, 65
computer systems
    components (fig.), 49
    types of (table), 61
Computer Theft Survey 2003, 358
computer waste, and mistakes, 347–349, 375
computerized information systems (CBISs)
    components of (fig.), 10
    described, 9–12

procedures, 12
comScore Networks, 195–196
concurrency control, 110
connections
    client/server, 144
    file server (fig.), 143
    to the Internet, 150, 170
consumer-to-consumer (C2C) e-commerce, 12, 186–187, 189–190, 217
containers in flat files, 109
content streaming, 162
ContentProtect, 365
contract software, 75
control units, CPU, 49
Controlling the Assault of Non-Solicited Pornography and
    Marketing Act (CAN-SPAM), 157–158, 348
CookieCop, 166
cookies and privacy, 166
coprocessors, 51
copyrights to software, 85
Corporate Express, 13
Corporate Governance Task Force, 253
corporate privacy policies, 372–373
costs
    data-related regulations, 120
    data storage medium comparison (table), 53
    database management systems (DBMSs), 115
    expert systems development, 284
    software, 65
    total cost of ownership (TCO), 30–31
Counter Fraud Service, 274
counterintelligence, 122
Coventry Business School, 179
Coviello, Arthur, 253
Cox Insurance Holdings, 20
CPU (central processing unit), 49–52, 69
crackers, 353
credit cards
    applications, 279
    determining credit limits, 275
    e-commerce security, 13
    fraud, 274, 353
    and online shopping, 203–204
crime, computer
    cyberterrorism, identify theft, 352–353
    introduction to, 351–352, 375–376
    preventing, 361–367
    viruses, theft, scams, 354–361
criminal hackers (crackers), 353
crisis management support, 255
critical success factors (CSFs), 321
CRM (customer relationship management), 22
cryptography and information systems, 167–168
Crystal Decisions, Inc., 132
culture, 22, 37
Cunard Line Ltd., 41
customer
    awareness and satisfaction, 30
    relationship management (CRM), 22, 42, 241
cyberkinetics, 290
cybermalls, 193
cyberterrorism, 351, 353

cycles
systems development life, 310–316
transaction processing, 206

D

DaimlerChrysler, 287
DARPA (Defense Advanced Research Projects Agency), 273, 368
DART advertising targeting technology, 196
data
collection. *See* data collection
correction in transaction processing, 207
described, 36
dictionaries, 107
editing in transaction processing, 207
entities, attributes, keys, 97–98
gloves, 19, 20
hierarchy of, 97
integrity, 100
management approaches, 99–101
manipulation in transaction processing, 207–208
marts, 118
mining, 121, 124, 369
modeling. *See* data modeling
preparation or conversion, 330
processing, transaction processing activities (fig.), 206
redundancy, 99
relationship with information and knowledge (fig.), 278
sequential and direct access, 52–53
storage and transactions, 208
synchronization, 122
turning into information, 6
types of (table), 5
vs. information, 4–5
warehouses, 117–118
data collection
by the federal government, 367, 369
in systems development, 318–319, 336
in transaction processing, 206
data definition language (DDL), 107
Data Encryption Standard (DES), 167
data manipulation language (DML), 110
data modeling
described, 126
and entity-relationship (ER) diagrams, 319
and the relational database model, 101–106
database administrators (DBA), 97, 112
database applications, 79–80
database management systems (DBMSs)
concurrency control, 110
creating, modifying database, 107–109
decision support system component, 245–246
described, 97, 127
managing data with, 100
selecting, 112–114
types, 106–107
upgrading, 132
database systems
special-purpose, 113
use of in organizations, 96–97

visual, audio, and other, 125
databases
described, 10
distributed databases, 122–123
joining and linking tables in, 104–105
relational database model, 101–106
using with other software, 115
dataflow diagrams (DFD), 319–320
DDB advertising agency, 95–96
debit cards, 203
decency protection, 365
decentralized processing, 142
decision making and problem solving, 229–232
decision room, decision support, 250–251
decision support systems (DSSs)
described, 17, 256
essential elements of (fig.), 18
overview of, 243–245
decisions, programmed vs. nonprogrammed, 230, 255
Defense Advanced Research Projects Agency (DARPA), 273, 368
delivery, just-in-time (JIT), 305–306
Dell, Inc., 9, 191, 201
Dell, Michael, 9
delphi approach to decision making, 248
Delta Airlines, 25, 291
demand reports, 234–235
Department of Homeland Security, 253
DES (Data Encryption Standard), 167
design stage, problem solving, 229
design reports, 326
design, systems, 37, 20, 311
desktop computers, 63
developing Web content, 156
development team, systems development, 307
diagrams
dataflow (DFD), 319–320
entity-relationship (ER), 102, 319
sequence, 323–324
dialog managers, 245–247
diffusion, technology, 23
digital certificates, 202, 218
digital signatures, encryption technique, 167
digital subscriber lines (DSLs), 153–157
digital versatile disc players (DVDs), 55
Diller, Barry, 222
direct access, 52
direct access storage devices (DASDs), 53
direct conversion, systems implementation, 330
display monitors as output devices, 58
disruptive changes, 23
distance learning, 163
distributed databases, 122–123
distributed processing, 143
document production in transaction processing, 208
documentation, software, 65
documents, user acceptance, 331
domains
in relational databases, 103
U.S. top-level affiliations (table), 149
dot-com era, 35
DoubleClick advertising, 196

Dow Chemical Company, 195
downloading music, 163, 185
downsizing, computer, 63
Dragon Systems voice recognition software, 272
drill-down reports, 235
drugs
    bar codes on, 120
    prescription drug Web sites, 192
DSS. *See* decision support systems
dumb terminals, 143
DVDs (digital versatile disc players), 55

E

e-commerce
    applications, 191–199
    described, 12, 37
    and enterprise resource planning, 212–217
    introduction to, 186–191
    mobile, 185
    phishing fraud, 14
    purchase order process (fig.), 15
    selecting DBMS for, 114
    and supply chain management, 187–188
    technology, infrastructure and development, 199–204
    transaction processing systems. *See* transaction
        processing systems
e-mail
    and instant messaging, 157–158
    privacy and, 370
    spam, 232, 347, 348
e-tailing, 193, 217
earnings and return on investment (ROI), 30
Eastman Chemical Company, 174
Eastman Kodak Co., 60
eBay, 198
Echelon Corporation, 26
EDI (electronic data interchange), 188
education
    Academic Management Services (AMS), 215
    distance learning, 163
    virtual learning environments, 179
    virtual reality applications, 290
Edwards, J.D., 124
electronic and mobile commerce, 12–13
electronic bill presentment, 198
electronic business (e-business), 15
electronic cash, 202
electronic data interchange (EDI), 188
Electronic Data Systems (EDS), 314
electronic exchanges, 193, 194, 217
electronic management (e-management), 15
electronic payment systems
    bill-paying, 198
    types of, 201–204
electronic procurement (e-procurement), 15
electronic retailing, 193, 217
electronic shopping carts, 201
electronic wallets, 202
embedded operating systems, 74
EMC Corp., 54

encryption issues, 167–168
end-user systems development, 313
enterprise data modeling, 102
enterprise operating systems, 73–74
enterprise resource planning (ERP) systems
    described, 16, 82–83, 218
    and e-commerce, 212–217
    at Porsche AG, 47
enterprise sphere of influence, 66
entertainment, virtual reality applications, 290
entities, data, 97–98
entity-relationship (ER) diagrams, 102, 105, 319
Environmental Protection Agency, 34
Environmental Systems Research Institute, 243
EQL tutorials, 180
ER (entity-relationship) diagrams, 102, 105
Ernst & Young's "Three Cs" rule for groupware, 82
ERP systems. *See* enterprise resource planning systems
eService Center, 42
ESSs. *See* executive support systems
ethical issues
    in information systems, 373–375, 376
    open source software, 72
    phishing fraud, 14
evaluating systems designs, 325–326
Evans, Grant, 273
event-driven review, 332–333
Excel (Microsoft), 231
exception reports, 235
exchange carriers (ECs), 140
executive information systems (EIS), 17, 252
executive support systems (ESSs), 252–255
expandable storage devices, 55, 56
expert systems
    applications of, 287
    characteristics and limitations of, 276–277
    described, 29, 270, 292
    overview of, 275–25
Express Pay, 223
extranets and intranets, 11, 164–165, 170
extreme programming (XP), 312

F

factors leading to competitive advantage (table), 27
Fakta retail chain, 309
Falcon Fraud Manager, 274
family history Web site, 42
fax modems, 139
feasibility analysis in systems investigation, 316–317
Federal Express
    delivering e-commerce, 189–190
    demand reports, 235
    FedEx Technology Institute, 31
    package tracking, 12
    tracking screen (fig.), 204
feedback in information systems, 8
Fidelity National Financial, 324
field, data, 97
Fijitsu Laboratories, 272
file management feature of OSs, 69

file server connections (fig.), 143
filtering software, 365
Final Cut Pro, 177
financial forecasting in ERP systems, 213
financial management information systems (financial MISs), 236–238
Financial Messaging (IrFM), 223
fingerprint authentication, 362
firewalls, 168
five-forces model for competitive advantage, 24
flat file databases, 109
FleetBoston Financial Corporation, 327
FLEXERT expert system, 285
Food and Drug Administration, 120
Ford Motor Company, 51
forecasting
    concept of, 8
    financial, in ERP, 213
forward chaining, 280
Forrester Research, 187
FORTRAN programming language, 85, 282
fraud
    Amazon.com's fraud detection techniques, 265
    Falcon Fraud Manager, 274
    and the Internet, 166–167
    National Fraud Information Center, 360
    phishing, 14
Free Software Foundation, the, 81
front-end applications, 115
Fukuda, Keiji, 227
functions in IS departments, 34–36
Fustos, Janos, 387–388
fuzzy logic, 278–279

G

game theory, 291
Gartner Group, 30
Gartner Inc., 116
Gary, Chares, 112
Gates, Bill, 151
gateways, 147
genetic algorithms, 274
Genex Services, 141
Gensym Corporation, 297
geographic information system (GIS), 242
GetThere.com, 26
Gibson, Ian, 368
Gilbert, Yves, 261
Global Digital Divide Initiative (GDDI), 151
global divide, movement to lessen, 151
global economy
    e-commerce and m-commerce, 191
    organizations in the, 23–24
global positioning systems (GPSs), 232
Global System for Mobile communication (GSM) technology, 141
Goldberg, Larry, 29
Good To Great (Collins), 24
Goode, Sigi, 177
Grand, Steve, 267

graphical user interface (GUI), 67
graphics programs, 80
Green, Gretna, 113
grid computing, 51–52
Grigsby, Paul, 116
group decision support systems, 17
group support systems (GSSs), 247–250, 256
groupware, 17, 82
Grupo Pão de Acucar, 131
GSM networks, 141
GUI (graphical user interface), 67

H

hackers
    protecting against (table), 363
    Reservoir Dogs, 352
    zero-day attacks, 380
Halfords, 119
Handa, Yoshihiko, 227
handheld computers, 61–62, 159, 185
haptic interface, 288
hard disk drives, 23
hard disks, 54
hardware
    acquiring from IS vendor, 327–328
    common operating system functions, 67
    described, 9
    described, components of, 49
    Web servers, 200
Harvard Community Health Plan, 285
Harvey, Charles, 383
Hawking, Steven, 266
head-mounted display (HMD), 20, 287
Health Insurance Portability and Accountability Act (HIPAA), 120, 132
HealthSouth, 62
hearing aids, 275
heating, venting, and air-conditioning (HVAC) equipment, 63
Heesen, Bernd, 385
Heizer, Raymond, 132
help desks, 332
Hession, Lloyd, 334
heuristics, 232, 269
Hewlett-Packard (HP)
    delivering e-commerce, 189–190
    e3000 family of computers, 73
    V-class servers, 47
hierarchy of data, 97
Hilton Hotels' database system of, 96–97
Hinojosa, Patrick, 356
HIPAA (Health Insurance Portability and Accountability Act), 120
Holland American Cruise Lines, 116
home pages, 154
homes, online commerce in, 222
Honda's Asimo robot, 267, 271
Hostmann, Bill, 116
hosts on the Internet, 149
Hotel Commonwealth, 10

Hourihan, Meg, 160
HTML (Hypertext Markup Language), 154
http (Hypertext Transport Protocol), 147
Huber, George, 229
hubs, 147
human resource management information systems, 241
Hungary, building information society in, 387–388
Huntington Bancshares, 132
HVAC equipment, 63
hypermedia, 154
Hypertext Markup Language (HTML), 154
Hypertext Preprocessor (PHP) programming language, 156
Hypertext Transport Protocol (http), 147

I

IBM
    creation of SQL, 110
    Global Services, 314
    job searching online, 159
    mainframe computers, 64
    and MS-DOS, 69
    ParnetWorld, B2B business, 13
    Rational Software, 311
    Red Hat Linux for mainframes, 74
ICRA (Internet Content Rating Association), 365
identity theft, 351, 353–354
IDSs (intrusion detection systems), 364
if-then statements, 278, 280
iMac computers, 63
image replacement document (IRD), 59
images, creating with graphics programs, 80
imaging, laptop, 177
inference engines in knowledge bases, 280
InfoPath (Microsoft), 109–110
informatics, 291
information
    characteristics of valuable (table), 7
    centers, 34
    process of transforming data into (fig.), 6
    relationship with data and knowledge (fig.), 278
    service units, 34
    systems. See information systems
    technology (IT), 9
    theft, 358
    value of, 6, 36
    vs. data, 4–5
Information Security Governance: A Call to Action (report), 253
information systems (ISs)
    business, 12–20
    careers in, 31–36
    components of (fig.), 7
    computerized. See computerized information systems
    described, 4, 36
    introduction to, 2
    IS department. See IS department
    IS plan for systems development, 321
    manual, 9
    OneBeacon Insurance Group's use of, 29
    organizations and, 21–24

performance-based, 27–31
    planning, 308, 310
    primary responsibilities of, 33
    ROI and the value of, 30–31
    social issues in (table), 346
    three stages in business use of (fig.), 28
infrastructure, e-commerce, 9, 199–204
infusion, technology, 23
input
    devices, 56–57, 86
    in information systems, 7–8
    and productivity, 28
insiders, and computer crime, 353
instant messaging, 158
integrity, data, 100
intelligence, natural vs. artificial (table), 269
intelligent agents, 274–275
intelligent behavior, 267
InterActiveCorp, 222
Internal Revenue Service and payroll systems, 16
international
    computer crime, 360–361
    cybercrime measures, 351–352
    privacy laws, 372
    transaction processing issues, 211
International Manufacturing Corporation (IMIC), 385–386
international networks, 145–146
Internet Activities Board (IAB), 166
Internet Content Rating Association (ICRA), 365
Internet Explorer code errors, 349
Internet phones and videoconferencing services, 161–162
Internet Protocol (IP)
    communications standard, 148
    phones, 161–162
Internet Relay Chat (IRC), 161–162
Internet service providers (ISPs), 152–153
Internet Society, 166
Internet software piracy, 359
Internet, the
    accessing the, 150, 152
    careers involving, 35–36
    companies that have effectively used (table), 11
    connecting to, 170
    current issues about, 166–168
    described, 11, 147–148
    growth in Hungary, 387–388
    job searching on, 159
    linking company databases to, 115, 117
    preventing crime on, 366–367
    privacy on, 370
    routing messages on (fig.), 148
intranets and extranets, 11, 164–165, 170
intrusion detection systems (IDSs), 364
inventory control
    data dictionary and, 108
    in ERP systems, 213
    programmed vs. nonprogrammed decisions, 230
    traditional vs. database approach to data management, 99–100
investors, Web sites for (table), 196–197
IRD (image replacement document), 59

IrFM (Financial Messaging) technology, 223
Iron Mountain, 210
IS. *See* information systems
IS departments
    personnel in, 38
    roles, functions, careers in, 32–34
    typical titles, functions in, 34–36
ISPs (Internet service providers), 152–153
items, data, 98

J

JAD (joint application development), 313, 335
Japanese financial messaging technology, 223
Java Web programming languages, 156
Jevans, Dave, 14
job searching online, 161
joining tables in databases, 104–105
joint application development (JAD), 313, 335
Jordan Education Initiative, 151
J.P. Morgan Chase & Co., 113
just-in-time (JIT) supplying, 305–306

K

Kaiser Permante, 287
Kamen, Dean, 341
KDDI Corp., 223
Kelkoo comparison-shopping site, 189
Kennametal, 113
Kennedy Space Center, 248
key-indicator reports, 234
keys in databases, 97–98
Kiku-Masamune sake, 227
knowledge
    acquisition, for expert systems, 280, 281
    described, 6, 36
    management described, 122
    relationship with information and data (fig.), 278
knowledge bases
    described, 19
    expert system component, 277–278
Kodak, 25, 233
Kolb, Ron, 273
Kolhatkar, Jaya, 265
Korea's broadband technology, 191
Kraft, 260–261
Kulula.com, 299–300

L

LAN administrators in IS departments, 35, 38
Lander, Eric, 268
Lanier, Jason, 287
LANs (local area networks), 144, 145, 150
laptop computers, 62
laptop imaging, 177
laser printers, 61
Laurel Pub Company, 235
LCDs (liquid crystal displays) as output devices, 60
learning systems, 19, 271–272

legacy systems, 332
legal issues
    blocking Internet content, 366
    blocking unwanted e-mail, 157
    CAN-SPAM Act, 348
    Check Clearing for the 21st Century Act, 132
    Clinger-Cohen Act of 1996, 34
    Computer Fraud and Abuse Act, 361
    fairness in information use, 371
    feasibility analysis of systems development projects, 316
    Health Insurance Portability and Accountability Act (HIPAA), 120
    Internet issues, 166–168
    privacy, 367–373
    Privacy Act of 1974, 346
    right to know issues, 371–372
    sales tax and e-commerce, 13
    Sarbanes-Oxley Act of 2002, 120, 211, 351
    USA Patriot Act, 364
libel laws, 365
Library and Information Technology Association, 275
licenses, software, 85
life cycles, systems development, 310–316
linking
    company databases to Internet, 115, 117
    tables in databases, 104–105
links, sponsored, 155
Linux operating systems, 70–71
liquid crystal displays (LCDs) as output devices, 60
Liquid Films, 177
LISP programming language, 282
local area decision network, 250–251
local area networks. *See* LANs
local exchange carriers (LECs), 140
locating records in databases, 98
logic bombs, 355
logic, fuzzy, 278–279
logical access path (LAP), 108
logical design, 323
logical records and physical records, 126
long-distance carriers, 140
Lord of the Rings (movie), 9, 71
Lotus Notes, 249

M

m-commerce, 12–13, 190, 191
Mac OS, 70
Mac OS X Server, 73
Macintosh computers, 177
Macy's, 349
magnetic disks, 53
magnetic ink character recognition (MICR)
    banking and, 58
    as input devices, 58
magnetic tape storage, 53
mainframe computers, 63–64
make-or-buy decision, 328, 336
Makimoto, Dr., 271
managed security service providers (MSSPs), 364
management, electronic (e-management), 15

management information systems (MISs)
  compared with DSSs (table), 245
  described, 17, 37
  financial, illustrated (fig.), 237
  outputs of, 232–236
  overview of, 232–236
  report guidelines (table), 236
  reports generated by (figs.), 234–235
MANs (metropolitan area networks), 144
manual information systems, 9
manufacturing management information systems, 238–239
manufacturing, repair, and operations (MRO) good and
  services, 193
MapExtreme, 125
MapInfo, 125
Mariano, John, 215
market share, 30
market forecasts, 6
marketing
  on the Internet, 195–196
  management information systems (MISs), 239–240
markets
  gaining competitive advantage in, 23–27
  segmentation of, 195
MasterCard, 203
Matchmaker System, 235
McCandish, Ian, 113
McCarthy, John, 266
MEDecision, 285
medicine, virtual reality applications, 288–289
memory
  cards described, 55
  characteristics, functions, 50–51, 86
  data storage media compared (table), 53
  devices, processing power, speed, capability, 49–52
  management feature of OSs, 67–68
  storage devices compared (table), 56
Mendes, Andre, 156
Mental Images, 9
Merritt, Denny, 96
MetaGroup, the, 112
metropolitan area networks (MANs), 144
Meyer-Rassow, Steven, 296
Michael Angelo's Gourmet Foods, 213
MICR (magnetic ink character recognition) as input devices,
  58
Microsoft, 10
  certifications, 36
  Internet Explorer code errors, 349
  .NET platform, 156
  NetMeeting, 162
  and open source software, 72
  PC operating systems, 69–70
  ROI values, 30
  Smart Personal Objects Technology (SPOT), 291
  Trustworthy Computing Initiative, 334
Microsoft Access relational database, 105
Microsoft Excel, 231
Microsoft Exchange Server 2003, 82
Microsoft Internet Explorer 6
  Secure Web site signal, 202
  site-viewing, privacy policy, 166

Microsoft Office 2003 suite, 81
Miller, Dan, 286
MISs (management information systems), 17, 232–236
mistakes, computer-related, 347–349
mobile commerce (m-commerce), 12–13, 190
model base, DSS component, 246
model management software (MMS), 247
models
  data, 102
  DSS (fig.), 246
  five-forces, for competitive advantage, 24
  optimization, 231
  of organizations (fig.), 21
  relational database, 101–106
  satisfying, 231
  technology acceptance model (TAM), 23
  telecommunications (fig.), 137
modems, 139
Mohegan Sun, 367
monitoring stage, problem solving, 230
Monsanto, 82
Monster.com, 35, 161
Morreale, Dan, 91
Motion Computing, Inc., 62
mouse as input devices, 56–57
movies, and virtual reality, 290
MP3 music, 163
MPE/iX operating system, 73
MS-DOS, 69
MSN Internet service, 152
MSSPs (managed security service providers), 364
Mueller, Robert, 119
multimedia messaging service (MMC), 159
multiple users, database management systems (DBMSs), 108
multiplexers, 139
multiprocessing, 51, 86
Multiprocessing Executive with integrated POSIX
  (MPE/iX), 73
multitasking, 68
music on the Internet, 163, 185
MyFamily.com, 42
mySAP Automotive, 305
MySQL, 112

N

nanotechnology, 10
Napster, 163
NASA
  Ames Research Center supercomputer, 64
  computer-related mistakes of, 347
  spatial data technology, 126
Nasdaq Stock Market, 328
Natan, Mike, 29
natural language processing, 19, 271
Naval Sea Systems Command, 307
.NET platform, 156
NetMeeting (Microsoft), 162
Netscape Navigator, secure Web site signal, 202
NetSuite's NetERP software, 212
NetWare operating system, 71

network management software, 146
network operating systems (NOSs), 146
networks
        communications software, protocols, 146–147
        described, 10–11, 142
        neural, 266
        telecommunications and, 137–147
        types of, 144–146
        VPNs. *See* virtual private networks
neural networks, 266, 272, 274
New World Pasta, 211
Newton, Isaac, 266
Nguyen, Tien, 340
Nokia cell phones, 74
nominal group technique, 249
North Bronx Healthcare Network, 91
Norton Ghost, 177
notebook computers, 62
Novak, Hans, 305
Novell certifications, 36

O

Object Data Standard, 124
object-oriented database management system (OODBMS), 124, 127
object-oriented databases, 124
object-oriented design, 323–324
object-oriented systems analysis, 321–322
object-oriented systems development (OOSD), 314–316
object-relational database management system (ORDBMS), 125
objects in databases, 126
off-the-shelf application software vs. proprietary (table), 76
Office Depot data warehouses, 117
OLAP (online analytical processing), 123–124, 127
on-demand computing, 313–314
OneBeacon Insurance Group, 29
online analytical processing (OLAP), 123–124, 127
online services, accessing the Internet using, 152
online shopping, electronic payment systems, 201–204
online transaction processing (OLTP) systems, 117, 204–205
OOSD (object-oriented systems development), 315
Open Datalink Interface (ODI), 150
open source software
        ethical, societal issues of, 72
        Hypertext Preprocessor (PHP), 156
        MySQL, 112
operating systems (OSs)
        current, 69–71
        described, 86–87
        embedded, 74
        enterprise, 73–74
        generally, 66–69
        small or special systems, 74–75
        workgroup, 71, 73
operations, IS departments, 32
optical character recognition (OCR), 57
optical data readers as input devices, 57
optimization
        in problem solving, 231

spreadsheet program feature, 79
OptlnRealBig.com, 348
ORDBMS (object-relational database management system), 125
order processing systems, 208–209
organic light-emitting diodes (OLEDs) as output devices, 60
organizational change, culture, 22
organizations
        described, 37
        described, and information systems, 21–24
        information systems in, 2–4
        IS department. *See* IS department
        performance tuning, 114
output
        database (fig.), 111
        devices described, 58–61, 86
        in information systems, 8
        of management information systems (MISs), 232–236
        and productivity, 27–28
outsourcing, 313–314
Owens & Minor, 121
Oxford Bookstore, 17

P

P3P (Platform for Privacy Preferences), 166, 370
Palm OS, 74
PANs (personal area networks), 144
Pantellos Group electronic exchange, 194–195
parallel start-up, systems implementation, 331
partnerships, strategic, 26
password sniffers, 358
passwords on Internet, intranets, extranets, 165
paths, logical and physical access (fig.), 108
PBX systems, 139
PC-DOS, 69
PC memory cards, 55
PCS Vision network (Sony), 185
PDAs. *See* personal digital assistants
Peep Trojan, 355
people
        human experts and expert systems, 278
        in information systems, 12
        IP opportunities in other countries, 32
        in IS departments, 38
PepsiCo, 340
perceptive systems, 268
performance-based information systems, 27–31
performance tuning of organizations, 114
permanent storage, 52–53
Perry Manufacturing, 117
personal application software, types compared, 76–81
personal area networks (PANs), 144
Personal Computer Memory Card International Association (PCMCIA), 55
personal computers as input devices, 56–57
personal digital assistants (PDAs) and wireless networks, 136
personal information managers (PIMs), 80–81
personal productivity software, 66, 81
phase-in approach to systems implementation, 330
phishing fraud, 14, 167

phones
    Internet. *See* Internet phones
    IP. *See* IP phones
physical access path (PAP), 108
physical design, 323
physical and logical records, 126
picture archiving, 91
piecemeal approach to systems implementation, 330
pilot start-up approach to systems implementation, 330
PIMs (personal information managers), 80–81
Ping-an, Wang, 355
piracy, software, 359
Pitney Bowes, 198
Pixar, 9
PKI (public key infrastructure), 362
plaintext, 167
planned data redundancy, 102
planning
    business continuity, 209–210
    strategic, for competitive advantage, 25–27
    and systems development, 308, 310
Platform for Privacy Preferences (P3P), 166, 370
Pleasant, Carroll, 174
plotters as output devices, 60
plug-ins, Web, 153
plunge approach, systems implementation, 330
Point-of-Sale (POS) devices as input devices, 57
point-of-sale TPS (fig.), 207
Point-to-Point Protocol (PPP), 150
policies
    corporate privacy, 372–373
    for preventing computer-related mistakes, 349–351
Popkin Software, 314
Porsche AG, 47
portable computers, 62
Porter, Michael, 21
portfolio trackers, 196
power blackout of Aug. 14, 2003, 8
PPP (Point-to-Point Protocol), 150
predicting. *See* forecasting
predictive analysis, 119
prescription drug Web sites, 192
Priceline.com, 198–199
primary keys in databases, 98
Prime Market Finance Corp., 120
printers, 60–62
privacy
    and the federal government, 367, 369
    information systems issues, 367–373
    and the Internet, 166
    spyware, 380–381
    and work, 369–370
Privacy Act of 1974, 346
probabilistic safety analysis (PSA), 297
problems
    highly structured, 244
    solving, and decision making, 229–232
Probst, Jack, 307
procedures described, 12
procedures for preventing computer-related mistakes,
    349–351
processes

best practices, 214
described, 6
value-added, 37
processing
    in information systems, 8
    and memory devices, 49–52
    in networks, strategies, systems, 142–144
    online. *See* online transaction processing
    online analytical processing (OLAP), 123–124
    tasks of OSs, 68–69
procurement, electronic (e-procurement), 15
product configuration software, 201
productivity
    described, 27–28
    paradox, 30
products
    creating new, improved, 26
    delivering e-commerce, 189
    food design at Kraft, 260–261
program trading systems, 9
programmed vs. nonprogrammed decisions, 230, 255
programmers, 308
programming languages
    choosing for task, 87
    data manipulation language (DML), 110
    summary of various, 84–85
    Web, 156
programs
    See also software
    described, 65
project management in systems development, 307
projecting data, 103
proprietary software vs. off-the-shelf, 75–76
protocols
    See also specific protocol
    communications, 146–147
    Internet, 170
prototyping in systems development, 312, 335
Public Broadcasting Service (PBS), 156
public key infrastructure (PKI), 362
purchase orders, e-commerce process for (fig.), 15
purchasing systems, 208–209
Purdue University's audio database, 125

Q

Qantum View Manage, 92
QBE (Query-By-Example), using for report generation, 110
Queen Mary 2, 41
Query-By-Example (QBE), 110
Quicken financial software, 77, 109
Quicken Loans, Inc., 54

R

RAD (rapid application development), 311–313, 335
Radianz, 334
radio on the Internet, 163
radio-frequency IDs (RFIDs), 26, 58, 291, 293
RAID (Redundant array of independent/inexpensive
    disks), 54, 86

random access memory. *See* RAM
rapid application development (RAD), 311–313, 335
Rational Software (IBM), 311
real estate, virtual reality applications, 290
RealEstate.com, 222
Recording Industry Association of America (RIAA), 163
records in databases, 126
Red Bull North America, Inc., 364
Red Hat Linux
    for IBM mainframes, 74
    AS network operating system, 73
    operating system, 71
redundancy, data, 99
Redundant array of independent/inexpensive disks
   (RAID), 54, 86
registration of software products, 85
regulations, cost of data-related, 120
relational database model, 101–106, 126
relationships
    of data, 5
    in databases, 102–103
    in technology-enabled relationship management, 196
replicated databases, 122
reports
    demand, 234–235
    developing effective, 236
    exception, 235
    generating database, 110–112
    key-indicator, 234
    scheduling, 233–234
    systems analysis, 322
    systems design, 326
    systems investigation, 317–318
    from transaction processing, 208
Request for Proposals (RFPs), 325, 336
requirements analysis in systems development, 320–321
Reservoir Dog hackers, 352
resource-based view, 24
rest estate market, online sales, 222
Restoration Hardware, 271
retail e-commerce, 193, 217
retrieving data from database, 109–110
return on investment (ROI), 27–31
Richter, Scott, 348
Ridge, Tom, 346
Riggs, Janice, 95
right to know issues, 371–372
Robertson, Hymans, 332
robotics, 270–271, 289
ROI (return on investment), 27–31
ROM (read-only memory), 51
Ross Systems, 213
routers, 147, 170
Royal Bank of Canada, 334

S

Sabre Airline Solutions, 312
SABRE airlines reservations system, 26
SABRE (Surveillance and Automated Business Reporting
   Engine), 237

sales tax and e-commerce, 13
Salomao, Miriam, 131
Samsung Electronics, 26
Sandia Laboratories, 274
Sandover, Alex, 296
Sanoma Magazines Belgium, 261–262
SANs (storage area networks), 54
SAP software company, 305
Sapiens Americas, 29
Sarbanes-Oxley Act of 2002, 120, 211, 351
SAS Corporation, 95, 309
satisfying model, 231
scalability
    of computers, 69
    and DBMS selection, 113
    of servers, 63
scams, computer-related, 359–360
Scan Eagle drone, 10
ScanSoft, Inc., 286
scenarios in object-oriented design, 323
scheduling
    feasibility analysis of systems development projects, 317
    reports, 233–234
    trade show appointments, 231
Scheduling Appointments at Trade Events (SATE), 231
schemas in databases, 107
SCM (supply chain management (SCM), 22
Scottish Intelligence Database (SID), 113
Scripps Institution of Oceanography, 9
script bunnies, 353
search engines, 155, 163
searching for jobs online, 161
secondary storage access methods, 52–56
security
    computer crime. *See* computer crime
    e-commerce concerns, 13
    with encryption, firewalls, 167–168
    responding to incidents (table), 355
    software vulnerabilities, 379–380
    wireless communications, 186
Security and Exchange Commission (SEC), 120
segmentation of markets, 195–196
Segway scooter, 290–291, 293, 341
selecting data, 103
semantic Webs, 117
sequence diagrams, 323–324
sequential access, 52
Serial Line Internet Protocol (SLIP), 150
servers
    described, 63
    HP e3000, 73
    Web, described, 11
services
    telecommunications, 139–140
    Web, 156
SETI@home screen saver, 52
shell, expert system, 277, 283–284
shopping on the Web, 162, 189
shopping carts, electronic, 201
Short Message Service, 159
Shuttle Project Engineering Office, Kennedy Space Center,
  248

Simplest-Shop, 117
site preparation, 329
Slammer virus, 356
SLIP (Serial Line Internet Protocol), 150
smart
    cards, 203
    dust, 291
    phones, 62
    tags, 58
sniffing, password, 358
social engineering, 353
Social Security Number theft, 167
societal issues
    open source software, 72
    phishing fraud, 14
software
    antivirus programs, 357–358
    bugs, 87
    bugs, copyright, 85
    communications, 146
    described, 10, 86
    for detecting phishing fraud, 14
    ERP vendors (table), 212
    filtering, 365
    fraud-detecting, 274
    GSS, 249–250
    make-or-buy decision, 328
    network management, 146
    online analytical processing (OLAP), 123–124
    overview of, 65–66
    piracy, 359
    proprietary vs. off-the-shelf (table), 76
    suites described, 81
    systems, types of, 65–74
    voice recognition, 271–272
    Web page construction cataloging, 200–201
Software Engineering Institute (SEI), 352
Software Publisher's Association (SPA), 362
Sony Corporation, 271
Sony Music Mobile Products Group, 185
Soquimich, 275
source code and programming languages, 84–85
source data automation in transaction processing, 207
Southwest Airlines, 25
spam, 232, 347, 348
Specialized Bicycles, 48
speech, and natural language processing, 271
speed
    clock speed, 50
    Internet issues, 166
    printer output, 60
    telecommunications, 137
sphere of influence
    described, 65
    enterprise, described, 66
    popular OSs and (table), 70
sponsored links, 155
SPOT (Smart Personal Objects Technology), 291
spreadsheet analysis software, 79, 109
Sprint, 185
Sprint Corporation, 341
SPS Commerce, 188

spyware, 166, 380–381
SQL, 110–111, 127
stakeholders in systems development, 307
start-up process, systems implementation, 330, 336
static analysis, 380
static Web pages, 200
stock, online trading of, 9, 196–198
Stockwell, Ashley, 11
storage
    of data, and transaction processing, 208
    data medium capacity and cost (table), 53
    in information systems, 8
    secondary devices, 53–56
storage area networks (SANs), 54
store, 108
storing data in databases, 109–110
Stouffer, Debra, 34
strategic alliances, 26
strategic planning, 253–254
strategies leading to competitive advantage, 27
streaming Web content, 162
Structured Query Language (SQL)
    and DBMSs, 127
    described, 110–111
subnotebook computers, 62
subschemas in databases, 107
Sullivan Street Bakery, 48
Sun Trust Banks, 351
supercomputers, 64
supply chain management (SCM), 37
    described, 22, 217
    and e-commerce, 187–188
    use of integrated software (fig.), 83
support
    component of IS departments, 33–34
    global for software, 85
    in problem solving phases, 244
support systems groups, 247–250
Surveillance and Automated Business Reporting Engine
    (SABRE), 237
sustaining changes, 23
Sutter Health, 16
switches, 169
synchronization, data, 122
syntax, programming language, 84
system resources, access to, 69
system units, 49
systems analysis, analysts, 20, 307, 322
systems development
    component of IS departments, 32–33
    described, 20, 37
    expert, 281–282
    IS planning and, 309–310
    life cycles (SDLC), 310–316
    overview of, 307–308
    systems analysis, 317–322
    systems design, 323–326
    systems implementation, 326–331
    systems investigation, 316–317
    systems operation, maintenance, review, 331–333
    team, 335
systems implementation, 20

systems review, 332–333
systems software, 10, 65–74
Systest, 35

T

tablet PCs, 62
tags, HTML, 154
talking, and natural language processing, 271
task management and operating systems, 68
tax preparation programs, 77
TaylorMade, 240
TCP (Transmission Control Protocol), 149, 170
technology
    acceptance model (TAM), 23, 38
    certifications, 36
    diffusion, infusion, 23
    e-commerce, 191–199
    infrastructure, 9
technology
technology-enabled relationship management, 196
telcos, 140
telecommunications
    comparison of line, service types (table), 142
    described, 10, 169
    devices, carriers, services, 139–142
    and networks, 137–147
    services and the Internet, 157–158
teleconferencing, decision support, 250–251
telephones, IP, 135
Tennant, Teresa, 116
TEPCO (Tokyo Electric Power Corporation), 297
terminal-to-host architecture, 143
terminals
    "dumb," 143
    as input devices, 57
terrorism, and passenger screening technology, 345–346
Terrorism Intelligence Database, 119
testing new systems, 330
Texas Instruments BA-35, 68
theft
    identity, 351, 353–354
    information, 358
thin clients, 62
Thorhauge, C.F., 309
time-driven reviews, 333
time-sharing capabilities, 69
titles in IS departments, 34–36
TopCoder, 33
Torvalds, Linus, 70
total cost of ownership (TCO), 30
touch-sensitive screens, 57
Toyota Financial Services, 20
tracking stock prices, 196–197
trail, audit, 211
training
    to avoid computer-related mistakes, 350
    virtual reality applications, 290
transaction processing cycle, 206
transaction processing systems (TPSs)
    audits, 211

described, 15–16, 37, 218
    integration of, 210
    in Japan, 223
    management and control of, 209–211
    online transaction processing (OLTP) systems, 117
    order processing, purchasing, accounting, 208–209
    overview of, 204–209
transactions described, 15
Transmission Control Protocol (TCP), 149, 170
transmission media, types of, 138
Transportation Security Administration (TSA), 345
Trend Micro, 232
Trojan horses, 355, 381
tunneling, 165
Turbo-Tax, 77
Turning Test, 267

U

Uioreanu, Calin, 117
UK Biobank, 368
Uniform Resource Locators (URLs), 147
United Parcel Service (UPS), 326–327
UNIX operating system, 71
unstructured problems, 244
UPC codes and data collection, 206–207
upgrading technology infrastructure, 216
UPS shipping services, 198
UPS (United Parcel Service), 92
UpToDate expert system, 285
URLs (Uniform Resource Locators), 147
U.S. Bureau of Customs and Border Protection, 205
U.S. Computer Emergency Readiness Team, 168
U.S. Department of Health and Human Services, 120
U.S. Department of Homeland Security, 346, 362
U.S. Steel, 241
US-Visit, 362
USA Patriot Act, 364
user acceptance documents, 331
user IDs on Internet, intranets, extranets, 165
user interface of OSs, 67
users in systems development, 307

V

value
    characteristics of valuable information (table), 7
    of information, 6, 36
value-added software vendor, 75
value-added processes, 37
value chains, 21, 22
Varsavsky, Martin, 151
Versant, 124
video on the Internet, 162, 163
videoconferencing services, 161–162
ViewSonic V1250, 62
Virgin Group, the, 11
virtual database systems, 125
Virtual Learning Environment (VLE), 179
virtual private networks (VPNs), 165
virtual reality

described, 293
described, types of devices, 19–20
systems, forms and applications of, 287–290
and winemaking, 296
virtual workgroups, 250–251
viruses
antivirus programs, using, 357–358
described, 355
detection, 287
most active (table), 356
Visa credit cards and e-commerce, 14, 203
visual database systems, 125
visual systems, 271
vocations in IT, 31–36
Voeller, John, 34
voice-over-IP (VoIP) technology, 161–162
voice recognition software, 19, 57, 271–272
VoIP technology, 161–162
VPL Research, 287
vulnerabilities, 379

W

Wal-Mart operational systems, 4
wallets, electronic, 202
WANs (wide area networks), 145
WAP (wireless application protocol), 190
We Blog: Publishing Online with Weblogs (Hourihan), 160
Web auctions, 163, 198–199
Web browsers, 155
Web logs. *See* blogs
Web pages
authoring programs, 200–201
and HTML, 154
Web programming languages, 156
Web servers, 11, 63, 200
Web services, 156, 174
Web shopping, 12
Web sites
development tools, 200
stock tracking, 196–198
useful for investors (table), 196–197
visitors to shopping (table), 162
Web, the
described, 11
developing content for, 156
shopping on, 162, 189
Webcasts, 162
WebCT product platform, 179–181
Webizens, 160
Webs, semantic, 117
Wells Fargo & Company, 136–137, 156

Werner, Al, 132
Western States Information Network (WSIN), 161–162
wholesale e-commerce, 193
Wi-Fi (IEEE 802.11b) protocols, 146
wide area decision network, 250–251
wide area networks (WANs), 145
Wiener, Scott, 121
Wildman, Harrold, Allen & Dixon, 54
Windows Embedded, 74
Windows Mobile, 74
Windows Server 2003, 71
Windows XP, 10, 69–70
wine industry, and virtual reality, 296
Winslow, Raimond, 125
WINTEL, 70
wireless application protocol. *See* WAP
wireless communications security issues, 186
wireless media, types of, 138–139
wireless networks
Barilla's implementation, 135–136
handheld entertainment, 185
and m-commerce, 190
wiring types, 138–139
Woodward Aircraft Engine Systems, 79
word processing programs, 77–78
wordlength, 50
workgroup application software, 82
workgroups
described, 66
virtual, 250–251
workstations, 63
World Economic Forum, 151
World Wide Web Consortium, 370
World Wide Web (WWW)
components, browsers, uses of, 153–157
described, 11
free access to, 151
and the Internet, 149
semantic Webs, 117
WorldWide Retail Exchange (WWRE), 193
worms, 355

Y

Yahoo!, 189
Yesawich, Pepperdine, Brown & Russell, 175

Z

z/OS operating system, 73
Zelos Group, Inc., 286
Zilvitis, Patrick, 341